Handbook of Research on Improving the Natural and Ecological Conditions of the Polesie Zone

Anatoliy Rokochinskiy
National University of Water and Environmental Engineering, Ukraine

Lyudmyla Kuzmych
Institute of Water Problems and Land Reclamation of the National Academy of Agrarian Sciences, Ukraine

Pavlo Volk
National University of Water and Environmental Engineering, Ukraine

A volume in the Advances in Environmental Engineering and Green Technologies (AEEGT) Book Series

IGI Global
PUBLISHER of TIMELY KNOWLEDGE

Published in the United States of America by
IGI Global
Engineering Science Reference (an imprint of IGI Global)
701 E. Chocolate Avenue
Hershey PA, USA 17033
Tel: 717-533-8845
Fax: 717-533-8661
E-mail: cust@igi-global.com
Web site: http://www.igi-global.com

Library of Congress Cataloging-in-Publication Data

Names: Rokochinskiy, Anatoliy, 1954- editor. | Kuzmych, Lyudmyla, 1976- editor. | Volk, Pavlo, 1985- editor.
Title: Handbook of Research on improving the natural and ecological conditions of the Polesie zone / edited by Anatoliy Rokochinskiy, Lyudmyla Kuzmych, Pavlo Volk.
Description: Hershey, PA : Engineering Science Reference, [2023] | Includes bibliographical references and index. | Summary: "The monograph has been prepared based on the generalization of the results of many years of research and production experience of leading scientists and specialists of the water management and melioration profile regarding natural, historical, social, economic, constructive, regime-technological, ecological, economic, and other aspects of land reclamation and development of the Polesie zone"-- Provided by publisher.
Identifiers: LCCN 2022056262 (print) | LCCN 2022056263 (ebook) | ISBN 9781668482483 (h/c) | ISBN 9781668482506 (eISBN)
Subjects: LCSH: Reclamation of land--Pripet River Watershed (Ukraine and Belarus) | Watershed management--Pripet River Watershed (Ukraine and Belarus) | Agricultural landscape management--Pripet River Watershed (Ukraine and Belarus) | Environmental management--Pripet River Watershed (Ukraine and Belarus) | Pripet Marshes (Belarus and Ukraine)--Environmental conditions.
Classification: LCC TC978.P77 I47 2023 (print) | LCC TC978.P77 (ebook) | DDC 333.91/6215309477--dc23/eng/20230106
LC record available at https://lccn.loc.gov/2022056262
LC ebook record available at https://lccn.loc.gov/2022056263

This book is published in the IGI Global book series Advances in Environmental Engineering and Green Technologies (AEEGT) (ISSN: 2326-9162; eISSN: 2326-9170)

British Cataloguing in Publication Data
A Cataloguing in Publication record for this book is available from the British Library.

For electronic access to this publication, please contact: eresources@igi-global.com.

Advances in Environmental Engineering and Green Technologies (AEEGT) Book Series

Sang-Bing Tsai
Zhongshan Institute, University of Electronic Science and Technology of China, China & Wuyi University, China
Ming-Lang Tseng
Lunghwa University of Science and Technology, Taiwan
Yuchi Wang
University of Electronic Science and Technology of China Zhongshan Institute, China

ISSN:2326-9162
EISSN:2326-9170

Mission

Growing awareness and an increased focus on environmental issues such as climate change, energy use, and loss of non-renewable resources have brought about a greater need for research that provides potential solutions to these problems. Research in environmental science and engineering continues to play a vital role in uncovering new opportunities for a "green" future.

The **Advances in Environmental Engineering and Green Technologies (AEEGT)** book series is a mouthpiece for research in all aspects of environmental science, earth science, and green initiatives. This series supports the ongoing research in this field through publishing books that discuss topics within environmental engineering or that deal with the interdisciplinary field of green technologies.

Coverage

- Policies Involving Green Technologies and Environmental Engineering
- Sustainable Communities
- Waste Management
- Industrial Waste Management and Minimization
- Alternative Power Sources
- Green Technology
- Biofilters and Biofiltration
- Renewable Energy
- Radioactive Waste Treatment
- Pollution Management

IGI Global is currently accepting manuscripts for publication within this series. To submit a proposal for a volume in this series, please contact our Acquisition Editors at Acquisitions@igi-global.com or visit: http://www.igi-global.com/publish/.

The Advances in Environmental Engineering and Green Technologies (AEEGT) Book Series (ISSN 2326-9162) is published by IGI Global, 701 E. Chocolate Avenue, Hershey, PA 17033-1240, USA, www.igi-global.com. This series is composed of titles available for purchase individually; each title is edited to be contextually exclusive from any other title within the series. For pricing and ordering information please visit http://www.igi-global.com/book-series/advances-environmental-engineering-green-technologies/73679. Postmaster: Send all address changes to above address.

Titles in this Series

For a list of additional titles in this series, please visit: www.igi-global.com/book-series

Multidisciplinary Approaches in AI, Creativity, Innovation, and Green Collaboration
Ziska Fields (University of Johannesburg, South Africa)
Engineering Science Reference • © 2023 • 405pp • H/C (ISBN: 9781668463666) • US $270.00

Opportunities and Challenges in Climate-Friendly Clean Water and Energy Technologies
J. Sadhik Basha (Department of Process Engineering, National University of Science & Technology, Oman) Jimoh Kayode Adewole (Department of Process Engineering, National University of Science & Technology, Oman) Anteneh Mesfin Yeneneh (Department of Process Engineering, National University of Science & Technology, Oman) and Luqman Kolawole Abidoye (Department of Process Engineering, National University of Science & Technology, Oman)
Engineering Science Reference • © 2023 • 344pp • H/C (ISBN: 9781668473030) • US $250.00

Transcending Humanitarian Engineering Strategies for Sustainable Futures
Yiannis Koumpouros (University of West Attica, Greece) Angelos Georgoulas (University of West Attica, Greece) and Georgia Kremmyda (University of Warwick, UK)
Engineering Science Reference • © 2023 • 338pp • H/C (ISBN: 9781668456194) • US $270.00

Food Sustainability, Environmental Awareness, and Adaptation and Mitigation Strategies for Developing Countries
Ahmad Ni'matullah Al-Baarri (Universitas Diponegoro, Indonesia) and Diana Nur Afifah (Universitas Diponegoro, Indonesia)
Engineering Science Reference • © 2023 • 293pp • H/C (ISBN: 9781668456293) • US $110.00

Perspectives on Ecological Degradation and Technological Progress
Veli Yilanci (Faculty of Political Sciences, Canakkale Onsekiz Mart University, Turkey)
Engineering Science Reference • © 2023 • 300pp • H/C (ISBN: 9781668467275) • US $225.00

Climate Change and the Economic Importance and Damages of Insects
Ahmed Karmaoui (Moulay Ismail University of Meknes, Morocco & Moroccan Center for Culture and Science, Morocco)
Engineering Science Reference • © 2023 • 279pp • H/C (ISBN: 9781668448243) • US $215.00

Implications of Nanoecotoxicology on Environmental Sustainability
Rafiq Lone (Central University of Kashmir, India) and Javid Ahmad Malik (Guru Ghasidas Vishwavidyalaya, India)
Engineering Science Reference • © 2023 • 360pp • H/C (ISBN: 9781668455333) • US $250.00

IGI Global
PUBLISHER of TIMELY KNOWLEDGE

701 East Chocolate Avenue, Hershey, PA 17033, USA
Tel: 717-533-8845 x100 • Fax: 717-533-8661
E-Mail: cust@igi-global.com • www.igi-global.com

List of Contributors

Table of Contents

Detailed Table of Contents

areas in the Lublin Province and in Poland. To preserve the region's natural resources, including the diversity of natural habitats and the sites of rare species of plants, animals, and fungi, various protection measures have been taken in this area for many years

This chapter presents the results of long-term (1932-2019) stationary studies of peat lands after drainage reclamation work and agricultural exploitation of river floodplains and wetlands of Polesie of Ukraine. Certain regularities in the transformation of organogenic soils and some ecological aspects were discovered. The rates and features of the processes of activation of the organogenic soil layer, changes in its agrochemical, water-physical, and biological characteristics were established. Thereupon, conclusions were drawn regarding the efficient and environmental use of drained organogenic soils in Ukraine.

This chapter considered qualitatively different approaches to the water regime model of drained lands implementation. The fundamental difference of these approaches is that the regime forecast calculations on a long-term basis are performed for schematized natural reclamation conditions by typing meteorological regimes in the calculation periods of vegetation, appropriate schematization, and parameterization of water regulation of drained lands. Secondly, it is proposed in the framework of the balance equation to perform based on certain patterns, to calculate the process of water regime based on the relationship and interdependence of its elements at the boundaries and in connection with the process of redistribution of moisture within the active layer of soil in relation to its moisture content and storage capacity. Considered methods and models of water regime allow determining both potentially possible due to meteorological and agrotechnical conditions and effective values of irrigation norms in the development of design modes of irrigation by irrigation and subsoil at drained lands.

In modern conditions of climate change and agricultural use of drained lands, there is a need for efficient production with the introduction of new or improved technologies for growing economically attractive crops that can ensure sustainable harvests, environmentally safe production, and minimize all costs, including water resources. By implementing a resource-saving technology of water regulation of drained

Based on the analysis and generalization of the data of domestic and foreign scientists and specialists, as well as the results of the authors' respective research, an assessment of the current state of changes in weather and climatic conditions in the Polesie zone and their impact on agricultural production, fertility, and moisture supply of soil; the conditions of functioning of water management and reclamation facilities, the natural and reclamation state of the drained lands was also given. The general recommendations for the development of adaptive organizational and economic, operational, agrotechnical, construction, and project activities that have their own clear goals and are closely interrelated were considered. On this basis, a complex of adaptive agromeliorative, agrotechnical and hydrotechnical, measures has been developed and proposed, aimed at a gradual transition to the cultivation of new varieties and types of agricultural crops, effective regulation of the water regime of drained lands and the functioning of drainage systems in variable climatic conditions.

Chapter 9
The Improvement of the Design Technology of Reclamation Projects for the Optimization of the Drained Lands Water Regulation Based on BIM Technologies 147
Anatoliy Rokochinskiy, National University of Water and Environmental Engineering, Ukraine

Improvement principles of development reclamation projects technologies considering drained lands water regulation optimization based on BIM were observed. Modern methodological basis was used, and, at the same time, a universal technical tool that allows the improvement of the practice of designing complex objects and systems that is successfully used and developed in almost all fields of science, technology, and industry is computer aided design building information modelling (BIM). In structural terms, BIM is an organizational and technical system consisting of a large number of interrelated and interacting components. The main function of BIM is to carry out object aided design and their components through the use of mathematical models and other automated design procedures and computer technology. So, transition to optimization design methods will help to improve the overall ecological and economic efficiency of drainage reclamation, especially in working out new construction, reconstruction, and modernization of irrigation and drainage systems.

Chapter 10
Scientific and Practical Principles of System Optimization of Various Parameters of Drainage Systems 158
Pavlo Volk, National University of Water and Environmental Engineering, Ukraine

Approaches to the systemic optimization of technological and design solutions for the creation and operation of drainage systems have been determined. It seems that the drainage system is a complex natural and technical ecological and economic system. Finding the general optimum in such a system based on the system optimization consists in substantiating local optima for all its main components of heterogeneous elements in the system effect - mode - technology - design in their interconnection. The practical implementation of a complex of predictive-simulation and optimization calculations in projects for new construction, reconstruction and modernization of drainage systems based on the developed scientific, methodological, information and software can be carried out using the appropriate toolkit, which is CAD and modern BIM technologies.

Chapter 11
Drainage Systems Project Development Taking Into Account Drained Lands Topography 167
Roman Koptyuk, National University of Water and Environmental Engineering, Ukraine

The results of assessing the nature and degree of influence of the terrain on drained lands are presented in making a design decision on water management technology, type and design of drainage system based on the indicator of land reclamation efficiency using BIM-technologies, forecasting and optimization calculations. The schematization of the relief conditions of drained lands according to the slopes and differences of the surface of the earth is considered. Schematization was performed and the classification of the drained lands was developed according to the index of reclamation efficiency of the relief, which depends on the degree of development of the relief by the slopes and differences of the surface of the land. The generalized results of the estimation of the land reclamation efficiency of the terrain of a separate section of the drainage system, the zonal conditions of the Rivne region and the Polesie of Ukraine within their administrative limits are presented.

The urgency of improving the methods of calculation of the groundwater flow regulation due to global warming as applied to the soil-climate conditions of the Polesie zone of Ukraine is discussed. The calculation schemes of drainage, which are typical for Polesie, are systematized. Recommendations on accounting for the hydrodynamic resistance in the calculations of imperfect drainage are given. A number of engineering methods are presented. They are intended, firstly, for prediction of the water table (WT) dynamics against the background of a single and systematic horizontal drain laid in the uniform ground or in the upper or lower layers of the two-layer ground, secondly, theoretical substantiation of the drainage design parameters (depth of laying drain, distance between regular drains). The dependencies for the calculation of regular drainage systems operating in the draining or subsurface wetting regimes are given. Proposals on additional consideration of influence of open conductive network and molehills system on the WT regime of temporarily waterlogged lands.

The necessity and scientific and methodological approaches to improving the calculation of a closed collector-drainage network of drainage systems are considered. Based on the application of the system methodology, the structure of the hierarchical and hydraulic connection between the water regime of the field and the operation mode of the closed collector-drainage network has been established. On the basis of the theoretical and experimental studies performed, relatively new scientific provisions have been proposed. This will allow, based on the application of the obtained universal equations, to construct a distribution profile of the total turbulent kinematic viscosity and the average flow velocity in the collector pipe. It becomes possible to evaluate the efficiency of the flow movement both in the constituent elements (collector pipes) and in the closed collector-drainage network as a whole. And it also becomes possible to improve the methods of designing and calculating their technological and design parameters.

This section analyses the technologies and means of traditional deep loosening as an agro-ameliorative measure, and their inconsistency with the requirements for improving the technology of water regulation of ameliorated lands in changing climatic conditions is determined. Therefore, it is advisable to switch from the traditional and most widespread technologies of slot and strip loosening to continuous loosening in profile and area, as well as differentiated loosening of deep soil horizons. It is proved that improved deep loosening can be the most effective means of ensuring the necessary filtration and simultaneous accumulation of excess soil moisture to ensure maximum productivity of ameliorated land throughout the entire growing season. Its implementation in the conditions of Polesie of Ukraine contributed to the improvement of water-physical properties, water permeability, and accumulating ability of the most common drained sod-podzolic soil, which provided an increase in the yield of cultivated crops by 20...40%

In changing climatic conditions, the deep continuous loosening of soils can be the most effective alternative measure for expensive reconstruction of existing reclamation systems, and, at the same time, an effective adaptive means for the accumulation of free-soil moisture. The accumulation efficiency depends on the structure of the loosened soil. The value of the proportionality coefficient of the Bond comminution hypothesis was experimentally established, which makes it possible to determine the energy saturation of technical means for loosening a certain soil mass. It has been established that the required capacity of technical means for deep loosening is approximately 2 times less for sandy soil than for loamy soil. So, with a total working width of 2 m, a working speed of 2 m / s for deep loosening (by 0.75 m) and 4 m / s for the basic tillage (by 0.25 m), it is necessary to have the power of 170- 240 kW of technical equipment for sandy soil and 390-440 for loamy soil

Drainage systems in the Chornobyl exclusion zone (CHEZ) are considered in the context of this zone's performance of barrier functions to contain the water transport of radionuclides beyond its borders. The principles of assessing the barrier resistance of catchment basins are described, taking into account landscape features and the dynamics of 90Sr removal; the role of drainage systems in the formation of removal of radionuclides at different stages of post-accident works and during periods of different water

The article is devoted to a topical scientific problem in modern conditions – valuation of land in Ukraine. A methodology for determining the market value of reclaimed land based on a differentiated assessment of its productivity through crop yields is proposed, taking into account natural and climatic zones and other conditions of a particular region. The basis of the methodology is the application of long-term forecast and a set of forecast and simulation models, in particular the model of area climatic conditions and the model of water regime and water regulation technologies on reclaimed land. The proposed approaches were tested by the method of large – scale machine experiment using a land plot in the zone of Western Polesie of Ukraine as the example. The obtained results indicate that there is a differentiation in land value, which is a proportional derivative of the yield of cultivated crops depending on the conditions of their cultivation.

Chapter 23
Mathematical Modeling of Economic Losses Caused by Forest Fire in Ukraine 372
Ivan Openko, National University of Life and Environmental Sciences of Ukraine, Ukraine
Ruslan Tykhenko, National University of Life and Environmental Sciences of Ukraine, Ukraine
Oleksandr Shevchenko, National University of Life and Environmental Sciences of Ukraine, Ukraine
Oleg Tsvyakh, National University of Life and Environmental Sciences of Ukraine, Ukraine
Yanina Stepchuk, National University of Life and Environmental Sciences of Ukraine, Ukraine

In this chapter, the authors analysed the current state of economic and environmental damage caused by forest fires in Ukraine. Uncontrolled spread of fire not only causes significant economic damage, but also has dire consequences for the ecology. The burning of large areas of forest leads to a radical change in the ecosystems of the affected region, which can eventually lead to unpredictable results. Thus, the negative indicators of forestry are: quantity of forest fires, units; area of forest lands covered by fires; losses caused by forest fires; burned and damaged the forest. It should be noted that the indicators of the consequences of forest fires do not mathematically correlate with the loss of wood cover in Ukraine, which are determined by the data remote sensing of the earth.

Chapter 24
Ecological Condition of River Basins of Western Polesie of Ukraine 384
Tetiana Basiuk, National University of Water and Environmental Engineering, Ukraine
Igor Gopchak, National University of Water and Environmental Engineering, Ukraine
Artem Yatsyk, National University of Water and Environmental Engineering, Ukraine

This section presents the results of the assessment of the ecological state of the river basins of the Western Polesie of Ukraine according to the level of anthropogenic load. The study was carried out based on systematic observations within the basins of small rivers, using the system model of the hierarchical structure «Basin of a small river». This model classifies the basins of small rivers according to the degree of anthropogenic loads exerted on them and makes it possible to monitor the state of the river basin according to various indicators within individual subsystems (radioactive contamination of the territory; land usage; river runoff usage; water quality). In order to improve and restore the already disturbed natural states of small river basins, a number of water protection measures have been proposed. The water management and ecological zoning of the basins of small rivers in the territory of the Western Polesie of Ukraine was carried out taking into account the quantitative and qualitative ecological and managerial indicators of the states of rivers and river basins.

This is an analysis of the state of natural waters of the Shatsk lake in the conditions of climate change. Analyzed are hydrological regime of water bodies and hydrogeological state of areas of formation and feeding. It is noted that there is a slight increase in the levels of ground and pressure aquifers, and, accordingly, surface aquifers, which is mainly due to atmospheric factors. It is noted that the chemical composition of the water of the lakes of the Shatsk group has remained stable enough for decades and the formation of its composition is determined not only by the peculiarity of the water supply of the lakes, but also by the level of anthropogenic load on the water area. It was established that in order to improve the condition and water supply of the Shatsk lake, it is necessary to develop and implement measures to prevent and minimize the possible impact of harmful effects of water, develop integrated approaches and measures to optimize water use, improve water quality, conserve and restore water resources, ensure sustainable use, and development of natural reserve fund objects.

The chapter underlines the necessity and suggests the pre-conditions of the introduction of the integrated management of water resources in Ukraine, and in Polesie region in particular. The authors state that the current state of the water resource management, water protection and water resource restoration of Ukraine are classified as a complicated system of branch, department and local functions and structures of the public administration which have a branch and administrative-territorial direction, an imperfect and unbalanced mechanism of water use, water protection and restoration of water resources. In view of this, prior trends in the implementation of the principles of stable water use were worked out, in particular, the introduction of the principles of the integrated management of water resources. A strategic aim was formulated, namely, to supply a basin balance and development of water management, water protection and restoration of water resources, to ensure stable water supply for the population and the branches of the national economy.

Foreword

Polesie is a unique natural complex. Land reclamation work has become an important factor in the socio-economic transformation of this region and plays an important role in the development of economic sectors. Reclaimed waterlogged lands in the humid zone of Ukraine and Poland are important, and in many regions, the only guarantor of stable farming and the creation of a reliable fodder base to meet the needs of animal husbandry, regardless of weather and climate conditions.

Therefore, the monograph presents an analysis of the results of long-term research from the past, or from the beginning of the first intensive research on land reclamation in the zone of excess moisture to the present, which is already associated with climate change, the deepening of the problem of soil degradation, the deterioration of the water balance in Ukraine and the world, ecology the environment in general and in the Polesie zone in particular. The authors highlighted the technological and technical aspects of water regulation on the drained lands of the Polesie zone: traditional and modern optimization methods for substantiating drainage parameters and other structural elements of drainage systems; elements of drainage technology and subsoil moistening, taking into account the influence of the structure and parameters of closed drainage; theoretical aspects of optimization of water regulation of drained lands on ecological and economic grounds, including taking into account the developed topography of the area; means and effectiveness of deep loosening of drained mineral soils; thermal land reclamation using discharged warm water from energy and industrial facilities; protection against pollution of territories and water bodies in the solid household waste storage area; means of mechanization for carrying out reclamation works; assessment of ecological and economic efficiency of investments in water management and land reclamation projects; development and implementation of a complex of adaptive organizational, agrotechnical, agromelioration and hydrotechnical measures.

You can find reflections and answers to these questions in the presented monograph, which was prepared on the basis of a generalization of the results of many years of research and production experience of leading scientists and specialists in the water management and reclamation profile. The monograph describes the main natural, historical, socio-economic, constructive, regime-technological, ecological, economic and other aspects of reclamation of the Polesie zone of Ukraine and Poland.

Vasyl Stashuk
The National Academy of Agrarian Sciences of Ukraine, Ukraine

Preface

Any civilization has its challenges. The priority challenges of the beginning of the 21st century are demography, lack of water, energy, and food resources, and climate changes occurring in the world.

Humanity faces the problem of not just providing water, but of guaranteeing the necessary water quality and quantity.

The problem of food for mankind has always existed. This is one of the oldest questions and problems relevant to the present time. Solving the issue of food security concerns the international category. It is related to a number of environmental and energy aspects, shortages of water resources, climate change, etc.

Among the main indicators affecting the formation of food resources, the following are distinguished such as natural conditions and population placement, land resources, total area, and structure of agricultural land, conditions of agricultural production, climate, the ratio between the quantity population and mass of food, etc.

Targeted improvement of properties is necessary to ensure guaranteed yields of natural and territorial complexes in order to optimally use the potential of soils, waters, climate, relief, and vegetation, and this can be implemented by implementing land reclamation. Land reclamation (Melioration) differs from conventional agricultural techniques with a long-term and more intense impact on land reclamation objects through drainage and irrigation lands, regulation of rivers and surface water flow, consolidation of sands and ravines, etc. Reclamation is a complex of hydraulic, cultural, chemical, agrotechnical, agroforestry, and other meliorative measures carried out with the aim of regulating water, heat, air, and the nutrient regime of soils, preservation, and improvement of their fertility and ecological formation balanced rational structure of the land.

Agricultural production, even with all other favorable factors, to a large extent depends on climatic conditions. Agricultural science and practice of many countries have proven that only for conditions for implementing the regulation of the water regime of soils based on the use of hydromeliorative systems, the dependence of agricultural production on the available natural resources significantly reduced moisture supply. A number of countries are using advanced methods and technologies of land reclamation implementations ensured the improvement of the conditions of agricultural production and fully solved the problem of own food security. Irrigation and drainage systems in Nethrlands, for example, 85% of agricultural land are covered, in the USA - 60%, and in Germany - 50%. In general, in the world, up to 20% of arable land is irrigated, which already amounts to more than 300 million hectares, on which up to 40% is grown agricultural products. At the same time, it should be noted that, despite economic efficiency and water reclamations, in recent decades the pace of their development in the world has decreased significantly, and in Ukraine, this industry has almost declined.

The challenges of modern times regarding energy, water, and food problems, which are intensifying as a result of climate change, as well as russia's military aggression, show how much our existing food potential affects the food security of Ukraine and other countries of the world. Such changes show the need to change approaches to the implementation of water engineering and water technologies in the Polesie region, as a possible guarantee of increasing the volume of marketable products of agricultural production in the regions and Ukraine as a whole.

Today's agrarian policy envisages intensive use of reclaimed land under certain conditions and mandatory provision of a favorable ecological state of the environment. There are issues of environmental protection extremely relevant in connection with the ever-increasing level of anthropogenic load, in that including agroecosystems and natural landscapes. A certain share of the anthropogenic load is caused by the consequences of large-scale hydraulic reclamation. Certainly, man-made systems influence the change of the water-air regime in both drained and irrigated areas as well as the territories adjacent to them. However, criticism of land reclamations regarding only negative impacts is not always justified and objective. And the reason for this is, as a rule, a lack of awareness among the public about the technical condition and the need to comply with regulatory operating conditions hydromeliorative systems.

Waterlogged lands of the humid zone of Ukraine occupy more than 5.5 million hectares of territory, of which drained 3.3 million hectares. More than 60% of the reclaimed fund and more than 70% of peat bog soils. The wetlands of the Polish Polesie zone in previous centuries had a negative impact on the economic development of the region and were an obstacle to social transformations. Even in ancient times, the human was aware of the need for drainage measures, as evidenced by archaeological excavations and traces of preserved reclamation work dating back a thousand years. Unfortunately, little has survived documents testifying to the scale of hydromelioration until the second half of the 18th century. Unfortunately, little has survived documents testifying to the scale of hydromelioration until the second half of the 18th century. In the 19th century, as a result of significant population growth and the need to meet its needs in food products, government bodies, and public organizations paid attention to waterlogged areas with the aim of turning them into cultural lands.

Forests yesterday... We remember the edge of lakes, swamps, and forests. The insignificant population of the region, unfavorable living conditions and conducting economic activities, small settlements, without amenities, schools only initial, dirt roads - all this formed not very attractive living conditions compared to more developed other regions. Peasant farms single-handedly "reconquered" the small ones from the swamp plots of land for growing agricultural products.

In 1873, the General Plan for the drainage of lands in Polesie was drow up for an area of more than 8 million hectares. In accordance with this, the Western Expedition was organized under led by Y.I. Zhilinsky. This work, begun in 1874, was the first great public work land drainage program.

In the period from 1874 to 1898, more than 4,000 km of canals were built in the Pripyat River basin with a volume of earthworks of about 16 million m^3. About 100,000 hectares of intensively drained land was used under arable land: locks, dams, and bridges for management (regulation) of the water regime. Experimental fields were created on well-drained lands for the cultivation of various agricultural crops, "... to show the peasant in reality ways and means of developing bogs for arable land and gardens..." - wrote Y.I. Zhilinsky. One of the disadvantages of this period was that, despite the large volumes of drainage works, at that time no not a single specialized scientific institution was created in this direction. Obviously, it was considered enough that outstanding scientists took part in the scientific support of the expedition: V.V. Dokuchaev, A.P. Karpinski, O.I. Voeikov, P.I. Panfilyev, E.V. Oppokov, et al. Evidence of deep scientific research decisions made by the Western Expedition is the fact that in 1878 the project received a gold medal at the World Exhibition in Paris, although Europe at that time already had a lot of experience in draining land.

The most extensive works on the expansion and drainage of agricultural land in the zone of Ukrainian Polesie took place during the Soviet period of the region's development, starting with the second half 60s of the last century.

In general, for almost a 20-year period, about 1.5 million hectares of drained land were brought into operation in the Ukrainian Polesie zone, including more than 700 thousand hectares with irrigation, which, respectively, accounted for about 50% and 60% of the total area of drained agricultural land in Ukraine.

This caused the rapid development of agricultural enterprises for the production and processing of plant and animal products. Villages have been revived, new regional settlements, educational and cultural institutions are growing, the population is growing, the socio-economic conditions of its living are significantly improving, and the difference between the city and the countryside is disappearing.

Reclamation of Polesie... Today, we repeatedly hear many opinions regarding the expediency, advantages, disadvantages, miscalculations, and harmfulness of hydrotechnical reclamation in general, in particular active regulation of the water and air regime of the soil. These are an agrotechnical background for the use of other types of reclamation - improvement, maintenance, and increase of the fertility of the grounds for receiving guaranteed crops. Someone even considers reclamation to be harmful to nature and humans, negatively assessing its impact on the hydro-ecological state of the territory and changing the natural conditions of the ecosystem. And they talk about this when world experience convincingly proves that only on reclaimed lands (under the conditions of applying a complex of agro-reclamation measures) it was possible to increase the yield level several times regardless of natural and climatic conditions and thereby ensure the reduction of arable land areas, their transformation into fodder, pasture, forest, recreational and hunting grounds.

Undoubtedly, land reclamation (all kinds of reclamation, including hydrotechnical ones) does not completely eliminate the influence of weather conditions on the growth and development of cultivated plants, but significantly reduces it, as evidenced by both domestic and foreign experience of agricultural production. In order to give a reliable assessment, it is necessary to take into account the multifaceted nature of this phenomenon.

Reclamation systems, like any technical systems, definitely affect the surrounding natural environment. At the same time, it must be remembered that when these systems were designed and built, mandatory compliance with operating conditions was assumed. Today, we are faced with the fact that land reclamation systems do not belong to one owner as a result of unsoldering land, and the conditions of their operation are significantly disturbed. So, what is the problem: drainage systems were built, or that they are left to their own devices today?

The emergence of new requests for reclamation science and production and the receipt by authors and researchers of scientific institutions of Ukraine of new experimental data in response to them make it necessary to generalize and highlight them in a separate publication.

You can find reflections and answers to these questions in the presented monograph. It was prepared on the basis of a generalization of the results of battery research and production experience of leading scientists and specialists in the water management and reclamation profile. This monograph is about the natural, historical, socio-economic, constructive, regime-technological, environmental, economic, and other aspects of the reclamation and development of the Polesie zone.

The monograph consists of ***26 chapters***. The presentation of the content of the sections is based on the author's principle. They are completely independent of each other, but the structural parts of the monograph and the sections included in them are subordinated to a single general idea - a systematic consideration of all the main aspects of improving the natural and ecological conditions of the Polesie zone on the basis of land reclamation and adjustment in their logical situation and relationship.

Chapters 1 and 2 present an analysis of the results of long-term research on the peculiarities of the geographical location and origin, the specifics of the formation and appearance, the characteristics of the hydrographic network, and the climatic and weather conditions of Polesie. An analysis of the results of buttery studies from the past, or from the beginning of the first intensive studies on land reclamation in the zone of excess moisture to the present, which is already related to climate change in the Polesie zone, is given. It is noted that drainage reclamation works were carried out in the free regions of modern Ukraine and Poland, but they achieved an extremely important role in the formation and improvement of agricultural efficiency in the Polesie a zone. ***Chapter 3*** presents the results of long-term (1932-2019) stationary studies of peat lands after drainage reclamation works and agricultural use of river floodplains and marshy massifs of Polesie of Ukraine. Certain regularities of the transformation of organogenic grounds were revealed and some ecological aspects were established. ***Chapter 4*** presents the results of studies on the assessment of the impact of modern and projected climate changes on the water regime of drained lands of the Polesie zone and related changes in the water use of cultivated crops, forecast and simulation modeling was planned and carried out, based on the results of which the water balances of drained lands were obtained land in relation to soil types, cultivated crops, and water management technologies under current and projected climate change conditions. The results of determining the total evaporation, water consumption, and technological efficiency of the use of drained lands in the changed natural-climatic and meliorative conditions of Polesie indicate an effective increase in the water consumption of cultivated crops by 1.5-2.0 times. In ***Chapter 5***, it is stated that in order to conduct the necessary effective agricultural production, the introduction of new technologies for growing economically attractive crops, which are able to ensure their sustainable harvests, the economical use of available material and water resources, the minimization of non-productive discharges of water beyond the boundaries of reclaimed territories, and the ecologically safe production of agricultural products. When implementing the technology of resource-saving water regulation on the drained lands of the humid zone of Ukraine, when growing perennial grasses, the economy of water resources on an experimental plot with an area of 245 hectares is 48.7 thousand m^3 per year, the consumed electricity is 45.7 kWh per 1 ha of drainage area, which is 5.7% per year. The saving of water resources when growing corn for silage amounted to 670 m^3/ha, while the operating costs for sluicing were reduced by 2-3 times. Adherence to the technological parameters of water regulation during the cultivation of paiza, amaranth, and fodder beans made it possible to increase the yield of cultivated crops by 15-20%, ensure the saving of water resources by up to 25%, and obtain an economic effect of up to 9-10 thousand UAH/ha. ***Chapter 6*** examines the principles of organization and implementation of environmental improvement monitoring of drained lands using the example of the Ukrainian Polesie zone. To control the changes that occur in natural complexes under the influence of drying, representative (typical) systems are used that characterize the complex natural and economic conditions of the studied object as fully as possible. ***Chapter 7*** provides an analysis of the current state of weather and climate conditions and an assessment of their predicted changes in the near and distant future in the drained territories of the Ukrainian Polesie. The main trends of changes in meteorological characteristics and their possible influence on the conditions of operation of water management and reclamation facilities and complexes, as well as on the natural and reclamation state of drained territories are determined. The research used the method of forecasting and simulation modeling with the use of models of predictive assessment of the normalized distribution of the main meteorological characteristics in perennial and annual vegetation sections. ***Chapter 8*** proposes a complex of adaptive organizational and economic, agrotechnical, agromelioration and hydrotechnical measures that make it possible to minimize the negative impact of climate change on justified processes and regimes, growing conditions, and productivity of agricultural crops grown on drained lands, in particular, through the introduction of resource-saving technologies taking into account modern economic and environmental requirements. ***Chapter 9*** presents the principles of improving the technologies of reclamation development projects, taking into account the optimization of land water management based on BIM technologies. The main function of BIM is performed in the execution of object design and its components through the use of mathematical models and other automated design procedures and computer technologies. Therefore, the transition to optimization design methods will help to improve the overall ecological and economic efficiency of drainage reclamation, especially when developing new construction, reconstruction, and modernization of irrigation and drainage systems. ***Chapter 10*** considers an approach to justifying the type and design of drainage systems based on the combination of a modern progressive optimization approach with traditional water balance, hydrodynamic and hydraulic methods of their calculation. Approaches to system optimization of technological and constructive solutions in the creation and functioning of drainage systems are defined. Finding the general (global) optimum in such a system on the basis of system optimization occurs in the substantiation of intermediate local optimums for all its main components of disparate elements (effect, mode, technology, design) in their relationship. The practical implementation of complex forecasting-simulation and optimization calculations in projects of new construction, reconstruction, and modernization of drainage systems based on the created scientific-methodical, information and software can be carried out with the use of appropriate tools, which are automated design and calculation system (ADCS) and modern BIM technologies. ***Chapter 11*** provides the results of assessing the nature and degree of influence of the developed relief of the area on drained lands when making a design decision regarding water regulation technology, the type and design of the drainage system based on the indicators of the relief effect of the relief using BIM technologies and forecasting and optimization calculations. The schematization of relief conditions of drained lands according to slopes and gradients of the land surface is considered. According to the research results, it was established that the nature of the influence of the topography of the area on the formation of the water regime and the yield of cultivated crops with different water regulation technologies has a pronounced optimum, which is differentially formed depending on the multiple natural and agromelioration conditions of the real object. Schematization was carried out and a classification of drained lands was developed according to the indicator of the relief efficiency of the relief, which depends on the degree of development of the relief according to the angles and gradients of the land surface. ***Chapter 12*** provides the improvement of methods for calculating the regulation of groundwater flow as a result of global heat supply in relation to the justified climatic conditions of the Ukrainian Polesie. Systematized water drainage design schemes are characteristic of Polesie. Recommendations are provided for taking hydrodynamic resistance into account when calculating non-ideal drainage. A number of engineering methods are presented. They are intended, first of all, for predicting the dynamics of groundwater (GW) against the background of single and systematic horizontal trees laid in a homogeneous foundation or in the upper or lower layers of a two-layer foundation, and secondly, the theoretical justification of the drainage design (the depth of laying the drain, the distance between regular drains). ***Chapter 13*** requires scientific and methodological approaches to improving the calculation of a closed collector-drainage network of drainage systems that operate in drainage and subsoil moistening modes, based on taking into account the effectiveness of its work on the effectiveness of regulating the water regime of drained lands. The existence and structure of the hierarchical and hydraulic connection between the water regime of the deposit and the mode of operation of the closed collector-drainage network based on the application of the system methodology was established. This water regime is determined by the flow regime in the drainage pipe as its main element. In ***Chapter 14***, we offer possible approaches to improving technologies and designs to obtain more effective means of deep loosening and the design of an improved tiered deep loosener. Thanks to the new principle of deep loosening, it became possible for the first time to control the loosening process of the foundation and obtain the desired degree of loosening in each horizon of the developed vertical profile of the foundation. Its implementation in the conditions of Polesie of Ukraine contributed to the improvement of water-physical properties, water permeability and storage capacity of the most expanded turf-podzolic soil drained there, which ensured an increase in the yield of cultivated plants by 20...40%, and also increased the duration of its aftereffect to 3-4 years with a payback of no more than 1 year. In ***Chapter 15***, it is presented that due to the large volume of work with continuous deep loosening and the need for significant changes in the structure, this process is justified as quite energy-consuming. We have determined the energy consumption for crushing the foundation and comparing it with all the energy predicted by the deep loosener for loosening the foundation array, we will add its efficiency coefficients, analyzing which, in time, will allow us to assess the effectiveness of possible ways to improve technological and economic indicators of work. So, with a total width of the working bodies of 2 m, a working speed of movement of 2 m/s for deep loosening (by 0.75 m) and 4 m/s for basic processing (by 0.25 m), it is necessary to have a power of technical means of 170- 240 kW for sandy foundations and 390-440 for loamy foundations. ***Chapter 16*** describes the principles and methods of assessing the barrier resistance of catchment basins, taking into account landscape characteristics and the dynamics of 90Sr removal (according to monitoring data); the role of drainage systems in the formation of removal of radionuclides at various stages of post-accident "liquidation" works and during periods of different water levels is analyzed, the effectiveness of water protection measures is evaluated. According to the results of the statistical analysis and comparison of the long-term dynamics of water removal of 90Sr, it follows that the performance of barrier functions mainly depends on the natural landscape-geochemical features of catchments, and not on water protection measures and structures. ***Chapter 17*** provides general data on an important scientific and applied problem, which is manifested in the development and justification of effective ways, technologies, and technical means for the use of waste heat water of industry and energy in agriculture, which is solved by the scientific justification of the construction of a hydrotechnical system of surface heating of the soil with water-filled plants-sleeves in the form of pipelines, located between the rows of plants and combined into separate block sections, which allows increasing the efficiency of thermal melioration of the grounds. Natural field research established that heating the base with hoses-pipelines with a diameter of 10 cm, in which warm water with a temperature of 20...30°C circulates, increases the temperature of the arable layer by 12...16°C in the base and 7...12 °C in the air, which provides sufficient conditions for the growth and development of plants in early spring. ***Chapter 18*** presents the results of monitoring studies of groundwater quality parameters, which are the source of water supply for settlements in the northwestern regions of Ukraine. The trend of their progression under the influence of the anthropogenic factor has been revealed according to the following parameters: the degree of aggressiveness, the content of iron compounds, manganese, ammonium nitrogen, and in a number of settlements - phenols. The analysis was carried out and the low efficiency of traditional technologies based on the "simplified aeration-filtering" method was noted. On the basis of the conducted research, a physical model and mechanisms of complex groundwater purification on the matrix structures of biominerals in the interpore space of the contact loading of the bioreactor were developed. ***Chapter 19*** considers the actual scientific and practical problem of taking into account the influence of meteorological and climatic factors in technical and economic calculations in the field of nature management. It has been proven that the implementation of scientifically based and effective methods of using meteorological and climatic information in economic calculations significantly reduces damages from weather conditions and improves the implementation of the optimal strategy for conducting agricultural production on reclaimed lands. Such calculations are based on economic and statistical modeling of various options that take into account normative hydrometeorological information when implementing design, management, and economic decisions. This increases the validity and reliability of calculations, as well as their compliance with the actual operating conditions of ecological and economic objects. Accordingly, it arouses the increased interest of both public and private investors. ***Chapter 20*** shows the specific features of the implementation of projects in the field of nature management, at the current stage the concept of environmental risk has become extremely popular. Based on the analysis of the content of various types of environmental risk, it is substantiated that the weather-climatic risk is decisive at the stage of planning project decisions since it is the meteorological conditions that determine the results of economic activity within the project. The economic content of such a risk reflects the absolute or relative deviation of the actual effect according to the design decision option from its potentially possible value as a result of the discrepancy between the real meteorological conditions and the optimal ones. The proposed approaches were tested during the planning of the project for the reconstruction of the reclamation system in the forest-steppe zone of Ukraine. ***Chapter 21*** presents the improvement of methodological and methodical approaches to the evaluation of project decisions in the water management and reclamation industry on ecological and economic grounds under conditions of uncertainty. On the basis of general recommendations and scientific-methodical principles used in global practice for evaluating the effectiveness of investments in various spheres of economic activity, the authors developed and implemented an improved methodology for such evaluation for water management and reclamation projects. It is based on methodical approaches, which include the following elements as mandatory: multivariate; taking into account changes in the value of money over time and specific features of the project implementation environment regarding the assessment of the impact of weather, climate, and environmental factors on the operation of the project; multi-level and phased evaluation of the project according to the relevant criteria and in accordance with the photosynthetically active radiation (PAR) of the project cycle. ***Chapter 22*** offers a methodology for determining the market value of reclaimed land based on a differentiated assessment of the level of their productivity due to the yield of agricultural crops, taking into account the natural and climatic zones and other conditions of a specific region. The basis of the methodology is the use of long-term forecasting and a complex of forecasting and simulation models, in particular a model of local climatic conditions and a model of the water regime and water regulation technologies on reclaimed land. At the same time, the model of the harvest of agricultural crops in the form of a complex model of the multiplicative type takes into account all the main factors influencing the formation of the crop: weather-climatic and soil conditions, agricultural cultivation techniques, the water regime of reclaimed land, etc. Approbation of the proposed approaches was carried out by the method of a large-scale machine experiment on the example of a land plot in the Western Polesie zone of Ukraine. ***Chapter 23*** presents the current state of economic and environmental damage from forest fires in Ukraine. The author's approach to determining damage from forest fires is proposed, which is based on statistical and mathematical methods (correlation analysis, mathematical modeling) of scientific research. The proposed mathematical model makes it possible to estimate the number of economic losses from forest fires in Ukraine. ***Chapter 24*** presents the results of the assessment of the ecological state of the river basins of the Western Polesie of Ukraine according to the level of anthropogenic load. The research was carried out based on systematic observations within the basins of small rivers, using the system model of the hierarchical structure "Basin of a small river". In order to improve and restore the disturbed natural state of small river basins, a number of water protection measures have been substantiated. The water management and ecological zoning of the basins of small rivers of the territory of the Western Polesie of Ukraine was carried out, taking into account the quantitative and qualitative water management and ecological indicators of the state of rivers and river basins, which will contribute to the concentration of efforts of all management and management objects for the development and implementation of basin-wide measures and will ensure the formation of priority areas of nature protection activities in the studied river basins. ***Chapter 25*** presents an analysis of the state of the natural waters of Shatsk Lake under climate change conditions. The hydrological regime of reservoirs and the hydrogeological state of the areas of formation and feeding were analyzed. It is noted that there is a slight increase in the levels of soil and pressure aquifers, as well as surface aquifers, which is mainly due to atmospheric factors. It was established that in order to improve the condition and water supply of Shatsk Lake, it is necessary to develop and implement measures to prevent and minimize the possible impact of harmful effects of water, to develop comprehensive approaches and measures to optimize water use, improve water quality, preserve and restore water resources, ensure sustainable use and development of objects of the nature reserve fund. ***Chapter 26*** formulates the priority directions for the implementation of the principles of sustainable water use, in particular - the implementation of the principles of Integrated Water Resources Management. Formulated by the most important conceptual provisions of the water management system, the defining principles of water supply and water use during the transition to the sustainable socio-economic development of the state, directions for the development of the economic mechanism of nature use and nature protection activities, the implementation of which must be ensured in the near future, the stages of the implementation of the basin management system. For a more complete implementation of the basin principle, the need for the creation of management structures that will carry out planning, coordination, and control functions within the catchment basins with the separation of powers in matters of water resource use between state authorities and local self-government bodies is substantiated.

Therefore, we hope that familiarization with the materials presented in this publication will contribute to the deepening of reclamation and ecological knowledge, and will be extremely useful for specialists in related fields of science and production in water management, nature protection, the agricultural sector, scientific workers, graduate students and students of the relevant specialties, land users of all forms of ownership, as well as increasing the efficiency of production, highly productive and nature-protecting use of reclaimed land and the agricultural resource potential of Polesie in general.

Anatoliy Rokochinskiy
National University of Water and Environmental Engineering, Ukraine

Lyudmyla Kuzmych
Institute of Water Problems and Land Reclamation of the National Academy of Agrarian Sciences, Ukraine

Pavlo Volk
National University of Water and Environmental Engineering, Ukraine

Chapter 1
History of the Polesie Development

Vasyl Stashuk
National University of Water and Environmental Engineering, Ukraine

Stepan Vozniuk
National University of Water and Environmental Engineering, Ukraine

Liubov Volk
https://orcid.org/0000-0003-1033-6715
National University of Water and Environmental Engineering, Ukraine

ABSTRACT

This chapter provides an analysis of the results of long-term research on the peculiarities of the geographical location and origin, the specifics of formation and existence, the characteristics of the hydrographic network, and the climatic and weather conditions of Polesie. The chapter presents the analysis of the results of many years of research from the beginning of the first intensive research on land reclamation in the zone of excess moisture to the present, which is already connected with the change in the climate of the Polesie zone. It was determined that drainage reclamation works were carried out in most regions of modern Ukraine, and they are very important in establishing and increasing the efficiency of agricultural production in the Polesie zone.

BACKGROUND

Geographical Position and Origin of Polesie

Polesie is a physical-geographical region located on the territory of the Polesie lowland. As for the location and distribution of the Polesie zone, different literature and authors do not agree. According to the latest research [19], "… Polesie (Polesie lowland, Desnianska lowland) in the basins of the Pripyat River, middle Dnieper, Desna has an area of 300 thousand km^2, where 1/3 is occupied by forests (pine

DOI: 10.4018/978-1-6684-8248-3.ch001

forests). The region is located in the southwestern part of the Eastern European plain with the main basins of the Western Bug (Baltic Basin), Pripyat, Dnieper, Desna (Black Sea Basin), Zhizdra (Caspian Basin). There are Lublin Polesie (Poland), Belarusian Polesie (Belarus), Ukrainian Polesie (Ukraine), Bryansk-Zhyzdryn Polesie, Orel-Kaluga Polesie (Russia). Within the Belarusian part - Brest Polesie, Pripyat Polesie, Zagorodye Polesie, Mozyr Polesie, and Gomel Polesie. In Ukraine - Volyn Polesie, Zhytomyr Polesie, Kyiv Polesie, Chernihiv Polesie, Novgorod-Siverskoye Polesie…" Figure 1.

Figure 1. Physical and geographical position of Polesie

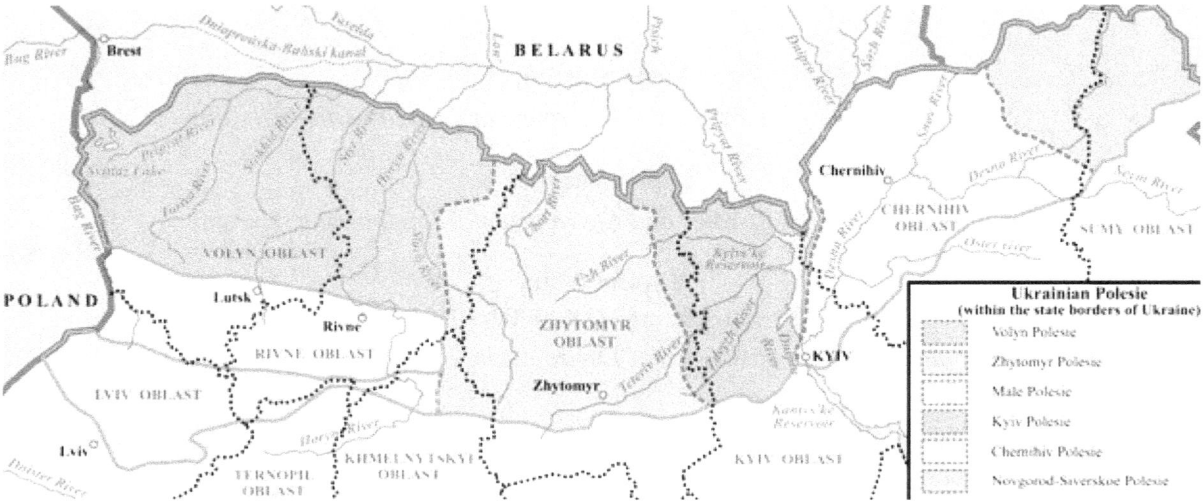

The Polesie lowland is a kind of natural province of the mixed forest zone of the Eastern European plain, divided into three large regions. They differ in geographical conditions: northern (left bank of the Pripyat River, located entirely in Belarus), southern (right bank of the Pripyat River and the Dnieper River, mostly located within modern Ukraine), and eastern (left bank of the Dnieper River, the northern part of which lies on the territory of Belarus and the Russian Federation, and the south - Ukraine) (Pledge, 1990).

There is no consensus among researchers about the origin of the toponym Polesie. Most of them are of the opinion that the term is based on - *lis*. Then Polesie is a forest area bordering the forest. There is an alternative point of view, according to which the toponym originates from the Baltic root *pol- / pal-*, which denotes a wetland. According to Polesie could be called an area where forest areas alternate with open wetlands (Gadzala, Stashuk, Rokochynskiy, et. al 2017).

Polesie (*Polish: Polesie*) is a kind of landscape region of the Eastern European physical-geographical country: in the north, it is bounded by the hilly plains of central Belarus, in the east - the Dnieper lowlands, in the south Prydniprovska and Volyn uplands, in the west Pribuzka river.

The origin of Polesie is due to the paleogeographical conditions of the anthropogenic period. It is a flat plain composed of the river, lake, and glacial sediments, with separate heights (Ovruch, up to 316 m), in some places ancient dunes, temperate continental climate, high water content, dense river network with wide valleys, the dominance of turf and podzolic coniferous and deciduous forests, which occupy up to 30% of the area, meadows, swamps, and wetlands. Natural complexes, without occupying large areas, often change each other, which is a characteristic feature of this region.

Hydrographic Characteristics of Polesie

All the rivers of Polesie are in the Dnipro basin. The main water artery is the Horyn river (386 km long within the Rivne region) with its tributaries: on the right - the Sluch river with the Stav, Bober, and Serehivka rivers, as well as the Zamchisko river; on the left - the river Melnytsia, the river Vyrka. From other large rivers flow Styr, Lva, and Stvyga, in the extreme northwest - Pripyat. Among the largest lakes of the territory are Nobel (499 hectares, Zarichnensky district), Bile (453 hectares, Volodymyretskyi district), Luka, Velike Pochaivske lake, Richytske lake, and Ozertse lake. The main direction of river flow is from south to north, which is due to the general lowering of the surface in this direction.

On the territory of Polesie, the rivers have small slopes (0.3 - 0.6 m/km), wide and swampy floodplains. There are many old houses and lakes in the river valleys. The average density of the river network in Polesie is 0.3 - 0.43 km/km^2. The main nourishment of rivers is melted snow water (55-60%). The rest of the runoff is underground and rainwater. The general flow of water levels in the rivers of the region is characterized by an intensive rise during the spring flood and a low level in the summer low tide. In autumn and winter, water levels in rivers are higher than in summer. Low-water discharge is often affected by floods: in the summer from downpours, and in the winter from thaws.

Hydrologically, Polesie is located within three artesian groundwater basins: the Volyn-Podilsky Basin, the Pripyat Basin, and the Ukrainian Fractured Water Basin. The underground water resources of the region are 345 thousand m^3/day, including: Volyn-Podilskyi artesian basin - 265 thousand m^3/day, Pripyat artesian basin - 12.0 thousand m^3/day, the Ukrainian Fractured Water Basin – 68.8 thousand m^3/day".

The Specifics of The Formation and Existence of Polesie

Polesie is a physical-geographical province of a mixed forest zone with peculiar features of nature: predominance of low relief, significant distribution of sand deposits, a large number of swamps, large number of swamps, a large number of rivers with low banks, presence of various rocks and fluvioglacial deposits.

The soil cover is mainly represented by sod-podzolic soils, which grow pine and mixed forests, meadows, swamps, and agricultural lands. A characteristic feature of this region is that natural complexes, without occupying large areas, often change each other.

Polesie is surrounded by hills on almost all sides. In general, it is a significant reduction in the relief of the Russian plain. According to physical and geographical conditions, Polesie province is currently divided into four subprovinces: Northern, Eastern, Southern, and Western (Konishchuk et, al., 2012).

The northern sub-province—Belarusian Polesie—is located on the left bank of the Pripyat River and the right bank of the Dnieper River. The northern sub-province is characterized by a much greater thickness of anthropogenic sediments (up to 100 m), the presence of large wetlands, large areas of forest, as well as climatic conditions.

The northern sub-province stretches from west to east along the Pripyat River for more than 500 km, and from north to south - about 200 km. In the west, Belorussian Polesie gradually turns into the Pribuzhskaya plain, in the east - in the Dnieper lowland, in the north - in the hilly-plain part of Belarus, in the south - in Ukrainian Polesie.

The main landforms in this area were formed as a result of the Dnieper Glacier, and especially the influence of its meltwater, which flowed from under the glacier.

An important role in the formation of the relief of the Belarusian Polesie was played by: the accumulative activity of the waters of the Pripyat River and its main tributaries, as well as the processes of waterlogging of low-lying areas of the relief.

Absolute heights of 130-150 m above sea level prevail within the Belarusian Polesie. Watersheds are flat and swampy. In the lowest part of the Belarusian Polesie (with almost flat relief) are the largest wetlands.

The main rivers of the Belarusian Polesie: the river Pripyat with its tributaries, the river Dnieper with the rivers Berezina and Sozh, the river Western Bug with the river Mukhavets. Small slopes of these rivers and their wide floodplains - create favorable conditions for the accumulation of river water during floods and summer downpours. Therefore, in this part of Polesie, along with large lakes, there are many small lakes.

The climate of Belarusian Polesie is warm, unstable, and humid, in the southeast it is transitional to forest-steppe.

The eastern sub-province—Eastern Polesie—occupies the left bank of the Dnieper River within Belarus, Russia, and Ukraine. Compared to the Northern and Southern sub-provinces, this area has a more continental climate and wetlands. It is also characterized by the presence of islands of loess.

The Dnieper River forms a clear natural western border of Eastern Polesie.

The western part of Eastern Polesie, known as the Chernihiv Polesie region, is located within the Dnieper-Donetsk basin.

The extreme eastern part of Eastern Polesie - Nerussa -Desnyansky's physical-geographical area - is located in the basin of the middle course of the Desna River and covers the area between the Desna River and its left tributaries - the rivers Nerussa and Navlya, as well as the left bank of the river Nerussa. In the east, the territory of the district is limited to parts of the Central Uplands.

Eastern Polesie has an extensive hydrological network. The Dnieper River receives the Sozh and Desna rivers here, as well as their small left-bank tributaries. Compared to other Polesie regions, the rivers of Eastern Polesie have a shorter duration of spring floods.

The southern sub-province—Southern Ukrainian Polesie—is located south of the Pripyat River and west of the Dnieper River.

Its nature is very different from the Northern and Eastern subprovinces. Southern Polesie has a general slope of its surface to the north and northeast. Here is the edge of the Ukrainian crystalline massif and the Slovenian-Ovruch ridge.

Most of the territory of Southern Polesie is characterized by absolute elevations above sea level of less than 200 m. The maximum altitudes are here on the Slovenian-Ovruch ridge, where they reach heights up to 320 m.

River valleys in most parts of the territory are wide, with low sloping banks. The exceptions are the river valleys of Zhytomyr Polesie, which cut into the Ukrainian crystalline massif. The valleys of the rivers Teteriv, Uzh, Zherev, Norin, Slovechna have steep banks and insignificant width.

In the southern part of Volyn Polesie, there are large hilly hills composed of Cretaceous rocks. This area was once described by P.A. Tutkovsky and to this day it is known as the "Tutkovsky ridge". Its origin is associated with the activities of glaciers.

In the north of Zhytomyr Polesie, the Slovechansko-Ovruch ridge stands out sharply.

The territory of Southern Polesie is dominated by a temperate-continental climate with warm humid summers and mild cloudy winters.

Western sub-province. Directly to Volyn Polesie, which is part of Southern Polesie and is located on the territory of Ukraine, adjoins Lubelskie Polesie (Lublin Polesie), located in Poland. It is allocated to the Western sub-province.

The border between Southern and Lubelskie Polesie runs along the Bug River. The natural center of this part of Polesie is Polesie National Park. It is located in the middle part of the Lenčinsko-Włodawa Lake District - part of Lubelskie Polesie.

Water plays a significant role in the natural conditions of this region. Lubelskie Polesie is an area with a very dense network of reservoirs. Most lakes are eutrophic, some are dystrophic. The lakes are characterized by rich vegetation and fauna.

Polesie is located on the territory of four states: Poland, Russia, Belarus, and Ukraine (Figure 1). The total area is about 130 thousand km^2.

In Poland, Polesie includes some regions of the Lublin Voivodeship, namely the valley of the Bug River in the area of Wola-Urguska and the Lenczyce-Włodawa Lake, known as Lublin or Western Polesie. The western border of Lublin Polesie runs along the Veps River. In the southern part of the Lublin Voivodeship, in the eastern part of Poland, there is the Polesie National Park - a small natural park, part of the biosphere reserve of the Polish region "Western Polesie". It was formed in 1990. The name of the park comes from the historical region of Polesie, in which it is located, and its area is 47.8 km^2.

Ukrainian Polesie occupies the southern part of the Polesie lowland and covers almost entirely the southern subprovinces (right bank of the Pripyat and Dnieper rivers) and the southern part of the eastern subprovinces (left bank of the Dnieper). It stretches from west to east for 750 km, from north to south - for 120-150 km. In the north-eastern direction, its territory gradually narrows to 50-80 km. The area is over 113 thousand km^2 (19% of the territory of Ukraine). Predominant heights are 150-200 m, maximum - 316 m (within the Slovenian-Ovruch ridge). The surface of the lowland has a slope to the rivers Pripyat and Dnieper. The relief is mostly flat and slightly undulating, there are some denuded moraine ridges and hills of the Dnieper age (Slovechansko-Ovrutsky ridge, Bilokorovytsky ridge, Ozeryansky ridge). A characteristic feature is the presence of Polesie dunes. The watershed is flat and swampy. Forests occupy about 33% of the territory, in some areas up to 60%.

Ukrainian Polesie consists of two parts separated by the Volyn Upland: the larger - Polesie, located in the north (within the Polesie lowland), and the smaller - Maly Polesie, which lies south in the basins of the Upper Bug and Styr, between Volyn and Podolsk Uplands.

Ukrainian Polesie includes the northern parts of Volyn, Rivne, Zhytomyr, Kyiv, Chernihiv, and Sumy regions, as well as part of the Lviv, Ternopil, and Khmelnytsky regions.

In the west, the border of Ukrainian Polesie coincides with the state border of Ukraine with Poland, in the north - with the state border with Belarus, in the northeast - with the state border with the Russian Federation. Its southern border with the Forest-Steppe passes through the following settlements (from west to east): *Ustyluh, Volodymyr-Volynskyi, Lutsk, Rivne, Shepetivka, Zhytomyr, Kyiv, Nizhyn, Baturyn, Krolevets*, the state border of Ukraine with Russia.

Depending on the location relative to the Dnieper, Polesie is divided into Right Bank and Left Bank (sometimes the toponyms Western and Eastern or Pripyat Polesie and Naddesnyanskoe Polesie are used).

According to the peculiarities of landscapes, Ukrainian Polesie is divided into parts - physical and geographical areas. To the west lies Volyn (Western) Polesie (Volyn and part of Rivne region); Maly Polesie (Rivne, Lviv, Ternopil and Khmelnytsky regions); further to the east - Zhytomyr Polesie (part of Rivne and Zhytomyr regions), Kyiv Polesie (Kyiv region), Chernihiv (Chernihiv region) and Novgorod-Siversk Polesie (Sumy and parts of Chernihiv region). The borders of Maly Polesie, which is located south of the Volyn Upland, can be traced in relief. In the north it is limited by the southern ledge of the Volyn Upland, in the south - by high (130-200 m) ledges of the Podolsk Upland (Gologory, Voronyaki, Kremenets Mountains). The western border of Maly Polesie is bounded by the north-eastern slopes of the Roztocze hills. In the east, the territory of Maly Polesie gradually narrows to a few kilometers or less; near Berdychiv a narrow strip of Maly Polesie connects with the main part of Polesie. The lowland plain of Lesser Polesie in the west separates the forest-steppe territory of the Volyn Upland from the rest of the Forest-Steppe. In general, Maly Polesie is less swampy than Volyn Polesie.

According to landscape features, Ukrainian Polesie is divided into five physical-geographical regions: Volyn, Zhytomyr, Kyiv, Chernihiv, and Novgorod-Siversky Polesie Fig. 2.

The geological structure of the territory of Polesie (in all its subprovinces) is unique. Tectonic conditions had a significant impact on the formation of the current relief of Polesie. The geomorphological boundaries of Polesie are mostly tectonic. The southern border of Polesie mainly corresponds to the Volodymyr-Volyn discharge of tectonic faults within Maly Polesie and the Zhytomyr tectonic zone. Each geostructural district of Polesie has distinctive features in the structure of the relief. The geological structure has a particularly great influence on the general plan of the river network and on the geomorphological structure of the river valleys of Polesie.

Volyn Polesie. This natural area is located between the Western Bug River in the west and the Sluch River in the east. Volyn Polesie includes most of the Volyn region and the north-western part of the Rivne region. The eastern border is located on the outcrops of crystalline rocks of the Ukrainian Crystal Shield near the cities of Korets, Sosnove, Klosov.

The main physical and geographical features of Volyn Polesie, which distinguish it from other Polesie regions, are: the presence of Cretaceous rocks under the Quaternary deposits, significant development of glacial landforms (Volyn moraine range), the presence of karst (lake basins, etc.), wide development of valley landscapes, warmer and wetter climate than in other Polesie physical-geographical areas, significant prevalence of swamps and wetlands, greater forest cover, which reaches 45% of the total area.

There is a regularity in the territorial distribution of landscapes. In the north of the region, in the upper reaches of the Pripyat River, and on the tributaries of this river, there are *young natural-territorial complexes - floodplain meadows and swamps*. The width of the Pripyat floodplain averages 3-5 km, and in some places up to 10 km. It has two levels: low level with heights of 0.5-2.0 m above the water level in the riverbed, and high level - 2-3 m, which is filled with water not every year. It is dominated by meadows with grasses and grasses. Large areas are occupied by swamps and wetlands, especially in low-level floodplains. Old houses, lakes, straits, and a dense network of meandering remains are common. In the riverbeds, there are common sands occupied by shrubs, and the central parts of the floodplains are covered with meadows, which are used for haymaking. High-level floodplains are used for vegetable and vegetable crops, especially after reclamation activities. The Upper Pripyat reclamation system operates here.

Figure 2. Physical and geographical zoning of Ukraine
1. Region of Volyn Polesie; 2. The region of Maly Polesie; 3. Region of Zhytomyr Polesie; 4. Region of Kyiv Polesie; 5. Region of Chernihiv Polesie; 6. Novgorod-Siversky Polesie region; 7-50 other physical-geographical regions of Ukraine

The next landscape level is formed by areas of floodplain terraces and ancient valleys. There are two floodplain terraces on the Pripyat River and its large tributaries. The first, pine, has an average height of 5-8 m, and in some places - 10-12 m. On the Pripyat River, it is 10-15 km wide. Here are important pine areas with sod-slightly podzolic soils, occupied by dry pine forests with sand embankments, ridges, and scattered sands. Even larger areas in this landscape-level are areas of poorly drained floodplain terraces with slightly podzolic sandy and sod-gley soils. They are occupied by arable land, meadows, alder and aspen forests, and swamps. There are many of them on the second floodplain terraces, which are 12 to 22 m high. Its width on the right bank of the Pripyat reaches 8-24 km. This landscape-level includes ancient valleys and, in particular, the great Styr-Slovechna valley, which stretches in the sublatitudinal direction from the Stokhod valley near the village of Holoby through the whole of Southern Polesie. Its length is over 400 km and a width of 20-25 km. Within its borders is the town of Kolky, the village of Stary Czartorysk, the town of Sarny, and other settlements. There are common landscapes that are similar to the landscapes of other floodplain terraces, but with a significant proportion of wetlands, including the marshes of Lutemetsky (between the rivers Stokhid and Styr), Halo and Lebed - between the rivers Gorin and Stviga.

Higher landscape level is associated with moraine-zander plains, mainly with sod-medium-podzolic soils, occupied by agricultural lands (cereals, potatoes, flax).

In the middle part of Volyn Polesie from the town of Lyubomlya to the town of Dubrovytsia, there are landscape areas connected with the Volyn moraine range. Here are moraine hills, shafts composed of sandy and loamy deposits with granite, quartzite, and silicon boulders, and inter-strip depressions and valleys.

Near the surface of the earth lie Cretaceous deposits, which improve the quality of sod-podzolic soils and their use in agriculture. Some of them are gravelly. The forests are dominated by pine and oak, spruce is found. Inter-strip depressions are occupied by wetlands.

In the south of Volyn Polesie, especially in the interfluves Western Bug - Turia - Stokhid - Styr, landscape areas of denudation plains with sod-carbonate soils on Cretaceous deposits with sod-medium-podzolic soils predominate. These soils are used for growing cereals and vegetables and even sugar beets. Oak-hornbeam forests and oak-pine forests have been preserved here. Chalk deposits lie near the surface. Their fracturing and a significant amount of precipitation contributed to the development of karst processes, including karst lakes, funnels, karst “window springs”.

Of the natural resources of Volyn Polesie, the most important are land, water, forest, mineral resources, and recreational resources. Significant reclamation measures have been taken here to improve the land. Rivers, lakes, and groundwater are used for water supply and recreational purposes. The total forest area of Volyn Polesie exceeds 2,300,000 hectares. Forests are about 23%. Mineral raw materials are used for building materials - marl, chalk, clay, sand, and, especially, basalt (Rivne region), and from combustible materials - peat.

In Volyn Polesie is a significant part of nature conservation facilities, which also confirms the richness and diversity of its natural environment. In the north-western part of the region is the Shatsk Natural National Park (established in 1983) in the Luboml administrative district, with a total area of 32.5 thousand hectares. The park preserves the natural complexes of the Shatsk Lakes region. Scientists are conducting research in the park, as well as maintaining the ecological balance.

There are 22 lakes in the park, mostly of karst and glacial origin. The largest of them are Lake Svityaz, Pulemetske, Lucimir, Crimne. Almost half of the park is covered with pine, pine-oak, hornbeam-oak, alder, and birch forests. Significant areas are occupied by meadows and swamps. The flora of the park includes 825 species of plants. Sod-podzolic soils predominate in the soil cover. There are also sod-carbonate, meadow-swamp, swamp, and peat. Diverse fauna: moose, roe deer, wild boar, squirrel, muskrat, weasel, wolf, otter, marten, many birds, and fish, among the latter - European eel. The territory of the park, in particular the shores of lakes, is used for recreation.

In Volyn Polesie there are also landscape reserves (Nechimne, Pochaiv), hydrological reserves (Ostrovsky, Dubrovsky), and a unique geological object - basalt pillars - in Rivne region (villages Berestovets, Zlazne, Ivanova Dolina) near Kostopol. Significant areas are occupied by protected wetlands: Perebrody (19600 ha) in Dubrovytsia district of Rivne region, Syra Pohonya (10000 ha) in Rakytniv district of Rivne region (Gadzala, Stashuk, Rokochynskiy, (2017).

Zhytomyr Polesie. This is an area of the mixed forest zone, which occupies most of Zhytomyr and the eastern part of Rivne administrative regions. The western border of Zhytomyr Polesie passes near the settlements of Klesiv, Sosnove, Korets, Shepetivka, and the eastern - near Narodychi, Malyn, Radomyshl, Kornyn. These boundaries are related to the access to the surface of the rocks of the Ukrainian Precambrian Shield, which mainly led to the separation of Zhytomyr Polesie.

The peculiarity of its natural environment is manifested in a higher hypsometric position (180-200 m), structure and depth of relief, hydrological and hydrogeological features, fewer wetlands and woods compared to other Polesie areas, and the originality of minerals associated with the crystal shield. Paleogeographical conditions of the territory in the anthropogenic have a noticeable influence on modern landscapes. The western, higher part of Zhytomyr Polesie was not covered by glaciers and moraines. To the east of the line, which passes near the settlements of Slovechno, Staryi Velednyky, Luhyny, Kryvotyn, Zhytomyr, Ivnytsia, there is a moraine part of the region, dominated by sandy sod-medium podzolic soils and pine-oak forests.

The river valleys of Zhytomyr Polesie are deeply incised into Precambrian rocks and in some areas have the character of breakthrough valleys. Thus, the Teteriv River near Denyshiv, Zhytomyr, and Korostyshiv has a narrow valley with steep rocky slopes up to 25-30 m high. Yankee near Zhytomyr and others. Erosion forms of relief have been developed on the forest "islands" in the area of the Slovechansko-Ovruch ridge, near Novohrad-Volynskyi, Horodnytsia, and Korostyshiv. On the Slovechansko-Ovruch ridge, the ravines are 20-30 m deep and 3-3.5 km long.

The climate of Zhytomyr Polesie occupies an intermediate position between the wetter and warmer climate of Volyn Polesie and the more continental climate of the eastern regions.

The rivers of Zhytomyr Polesie differ from other Polesie rivers by faster flow, and in some areas - gullies and rapids, higher water rise during floods. The vegetation here is less preserved, the plowed area varies from 15-30% in the north-western part to 50% in the eastern and central. Of all Polesie regions, Zhytomyr Polesie is the least swampy. The total area of wetlands is only 2.9% of the territory. Large wetlands are common only in the northwest and north of the region.

The landscape structure of Zhytomyr Polesie consists of the following landscape areas:

- plain-sandy on a crystalline basis with a predominance of sod-slightly podzolic soils under forests of pine and sub-forest types. Water-glacial sands have a thickness of 4-6 m and lie on granites, gneisses. In some places there are sand shafts and hills of Aeolian origin, and in the depressions - small swamps and wetlands. These areas are most common in the northwest in the area of Rokytne, Gorodnytsia, Yemilchyna;
- plain-zander and valley-zander wetlands with sod podzolic muddy and swampy soils on Precambrian crystalline, Cretaceous, and Paleogene sedimentary rocks under pine forests are common in the Olevsk, Perga, and Bolotnycia River. There are upland and transitional swamps with an area of over 1000 hectares, with a thickness of peat layer up to 4-5 m. The largest of the swamps are Ozeryanske and Gvozd. Sedge-hypnotic bogs are developed in the depressions and valleys. Reclamation works have been carried out in these areas;
- moraine-zander and moraine-hilly with sod-slightly podzolic and sod-medium-podzolic soils landscape areas are common in the eastern subregion of Zhytomyr Polesie near Potievka, Gorbulev, Modilev, Divochky, Torchin, etc.;
- denudation plains on crystalline rocks with sod-slightly podzolic gravelly soils of the area are widespread in interfluves with high Precambrian rocks. Water-glacial sands are low-power or absent at all, there are eluvial-diluvial gravelly deposits - products of weathering of crystalline rocks. Such landscapes are common in the area of Korosten, near the villages of Krayevshchina, Toporyshch, Pekershchina, and other areas.

In Zhytomyr Polesie, in contrast to Volyn Polesie, there are widespread landscape areas of forest islands with gray forest soils, deforested and occupied mainly by agricultural land. They have developed ravines and gullies, there are intense erosion processes. These areas have features of forest-steppe landscapes. They are found on the Slovechansko-Ovruch ridge, near the cities of Novograd-Volynsky, Zhytomyr, Korostyshiv, and others.

The landscapes of Zhytomyr Polesie have undergone significant changes due to anthropogenic activities - deforestation, land reclamation, plowing, and especially in connection with the extraction of minerals - granite, labradorite, pegmatite, quartzite, piezo quartz, placer deposits of ilmenite, and others.

Landscape reserve Plotnytsia (464 hectares) is located in Olevsk district with unique natural and territorial complexes. Didove Lake Hydrological Reserve with lake ecosystems (54 hectares) is located in Ovruch district. Poyaskovsky Forest Reserve (113 ha) in Olevsk district with oak and hornbeam forests was founded in 1926. Giant oaks aged 300-400 years have been preserved here (Gadzala, Stashuk, Rokochynskiy, (2017).

Kyiv Polesie. This natural area is located between Zhytomyr and Chernihiv Polesie. The western border runs along the outcrops of Precambrian crystalline rocks characteristic of Zhytomyr Polesie. It is held east of the settlements of Narodychi, Malyn, Radomyshl, Kholodkiv. The eastern border is located near the Dnieper River.

Geographically, Kyiv Polesie occupies the northeastern slope of the Ukrainian Crystal Shield to the Dnieper-Donetsk basin. The crystalline basement gradually decreases to the Dnieper river valley, where it is located at a depth of 300-400 m. It contains marine sediments of the Jurassic, Cretaceous, and Paleogene, as well as continental Neogene and Anthropogenic formations. Above the local base of erosion, there are Paleogene, Neogene, and Anthropogenic deposits, which are directly involved in landscapes. Of particular importance are anthropogenic sediments, which are mainly represented by the glacial, water-glacial, alluvial, lake, and aeolian formations. Their total thickness averages 15-20 m. The surface of Kyiv Polesie is an accumulative lowland with the dominant absolute heights of 120-170 m and the depth of dismemberment of wide river valleys - 25-50 m. The main rivers - Dnieper, Pripyat, Uzh, Teteriv, Irpen, Zdvyzh.

Climatic conditions of the region have typical Polesie features. In particular, most of the radiant heat (up to 60-70%) is spent on evaporation, clearly showing the seasonality of climatic processes and phenomena. Winter in Kyiv Polesie lasts from the end of November to March. January average temperatures range from -6.4 °C (Chernobyl) to -5.9 °C (Kyiv), and the minimum reaches -37 °C. In winter there are often thaws. Snow cover lasts up to 100 days, with an average height of 25-30 cm. Spring lasts 2.0-2.5 months. The transition of the average daily temperature through 0 ° C takes place in the second half of March, and through + 10 °C - at the end of April - beginning of May. Summer lasts from late May to early September. The average monthly temperature in July is +18 ... + 19.5 °C, and the maximum is +37 ... + 39 °C. In summer, about 40% of the annual rainfall is 530-570 mm. Autumn lasts from mid-September to the third decade of November. It is characterized by siege rains and fogs.

The inland waters of Kyiv Polesie consist of a dense hydrographic network of rivers, the Kyiv Reservoir, lakes, swamps, groundwater, and groundwater.

The main types of soils are different types of sod-podzolic, which are dominant, peat-swamp, and gray forest on forest islands. Among the natural vegetation, the main role is played by pine, hornbeam-oak-pine, alder forests, shrubs, meadow, and swamp vegetation. Forests and shrubs occupy about 50%, meadows - 2.2%, swamps, and wetlands - 3.2% of the territory. Plowing reaches 36.8%.

As a result of the Chornobyl catastrophe of 1986, significant areas of agricultural lands, forests, water bodies of Kyiv Polesie were withdrawn from economic use due to contamination with radioactive elements.

In *the landscape structure* of Kyiv Polesie, the largest areas are occupied by moraine-zander low-undulating lowland plains with sod-medium-podzolic soils covered with pine and sub-forest forests; moraine-hilly plains with sod-medium-podzolic soils with sub and subsoil forests; sandy lowland plains with sod-slightly podzolic soils and pine forests; floodplain-terrace lowlands with sod-slightly podzolic soils and forests of mostly pine type; floodplain meadow-swamp lowlands with slightly podzolic and swampy soils and forest eroded "islands" with gray forest soils.

Water-glacial sands, which are mostly found on moraines, are very common in the moraine-zander lowlands. On the surface, there are depressions with gleyed soils. In the sandy areas, forests consist mainly of pine, and in moraine, oak and hornbeam grow. Moraine-hilly plains are characterized by the presence of glacial and water-glacial forms in the form of moraine hills and belts such as rocks and lakes with a relative height of 15-20 m. Declines between hills and ridges are occupied by meadows, sometimes lowland swamps. These landscapes are common near the villages of Chistogolivka, Lelova, Korogoda, Andrievka near Chernobyl. Zander plains occupy large areas in Kyiv Polesie, especially in the northern part. On the surface of the sandy plain, there are sand shafts and ridges, the upper part of which is processed by the wind. Most Zander plains are forested with pine, birch, alder, or aspen. Some areas are occupied by meadows. Rye and potatoes predominate among crops. In many places, the Zander plains turn into floodplain sand terraces of the rivers Pripyat, Uzh, Teteriv, Irpin, and others. Floodplain-meadow-swamp landscapes occupy floodplains that reach a width of 1.5-2.0 to 7-8 km. Reclamation measures (Irpin, Zdvizh) were carried out in many valleys, which made it possible to use these lands for growing vegetables. Forest "islands" occur in small areas near the settlements: Chernobyl, Ivankov, Pryborsk, Borodyanka, Dymer, Vyshgorod, and others.

Among the nature protection objects of Kyiv Polesie, the largest is the Dnieper-Teterivka State Forestry (area over 30 thousand hectares) and Ilyinsky Hydrological Reserve (over 2 thousand hectares).

Kyiv Polesie suffered significant radioactive contamination as a result of the Chornobyl disaster in 1986. The main role in the pollution of the region belongs to long-lived radioactive elements with a long period of decay - cesium-137, strontium-90, plutonium-239. The precipitation and distribution of radionuclides on the earth's surface is associated with the movement of air masses and in some way with the morphology and properties of landscapes, especially components such as relief. For example, an increase in pollution is registered within Kyiv Polesie on the Chornobyl-Chistoholiv moraine ridge, swamps, floodplains, and in some forests. Landscape conditions have a significant impact on the migration of radionuclides. Landscape pollution with cesium-137 is from $18.5 \cdot 10^{10}$ to $148 \cdot 10^{10}$ Bq/km^2, strontium-90 - from $3.7 \cdot 10^{10}$ to $11 \cdot 10^{10}$ Bq/km^2. Therefore, scientists in all fields of knowledge are actively involved in research aimed at developing measures to eliminate the consequences of the Chernobyl disaster.

Chernihiv Polesie. This is a physical and geographical area of the mixed forest zone, which stretches from the Dnieper in the west to its eastern border, passing near the mouth of the river Revna (Snov river basin), west of Holmy and Ponornytsia, Obolonnya, east of Korop and to the west of Krolevets. The region is located within the Dnieper-Donetsk basin, which largely determined its physical and geographical features.

The crystalline basement is at a depth of several thousand meters and is covered by Paleozoic, Mesozoic and Cenozoic deposits. Paleogene, Neogene and anthropogenic deposits lie above local erosion bases. The latter are most often represented by glacial (moraine), water-glacial, alluvial, lake, aeolian and forest-like deposits. By the nature of the relief Chernihiv Polesie is an accumulative lowland with significant areas of modern and ancient river valleys.

The climate of Chernihiv Polesie can be characterized as moderately continental, with significant humidity during the year, high relative humidity, and light winds. Compared to the western oblasts, Chernihiv Polesie is characterized by a larger amplitude of annual temperatures, lower winter temperatures, and a longer period of snowfall. The average annual rainfall is 500-610 mm.

The river network of the region is dense. Rivers are characterized by slight falls, calm river flow, and meandering. There are many floodplain lakes in the valleys of the Dnieper, Desna, Snova, and Seimas rivers.

The wetlands of Chernihiv Polesie are quite large. The area of peat bogs is more than 4.5% of the total area. Almost all swamps belong to the lowland type. Transitional and upland bogs are very rare. The largest bogs are Zamglai (8334 ha), Osterske (10 558 ha), Snovske (9400 ha), Smolyanka (4288 ha), Doch-Gali (3600 ha), Otter (2458 ha), Pariste (2340 ha).

Most of the swamps have been reclaimed and turned into agricultural lands. The soil cover is dominated by sod-podzolic soils, large areas are occupied by swampy and gray forest soils. A feature of the natural conditions of Chernihiv Polesie is the common saline soils. These soils are confined to forest "islands" on terraces with near-surface groundwater levels. In the valleys of the rivers Desna and Ostra meadow saline and sweetened soils have developed.

The forest cover of Chernihiv Polesie is the lowest in comparison with other Polesie oblasts and is 15-18%. The main areas are occupied by pine and oak-pine forests. Less common are linden, maple, elm. The eastern limit of continuous hornbeam distribution passes in the region. The largest forests have been preserved in the interfluve of the Dnieper-Desna, on the left-bank terrace of the Zamglai, on the interfluve of the Snova-Desna.

Large areas are occupied by floodplain and dry meadows, which are a significant forage base. Wetlands are covered with such moisture-loving as various sedges, reeds, cattails, sedges, etc.

The lands of the region are distributed as follows: arable lands occupy more than 42%, orchards, and berries - 0.7%, hayfields - 15.5%, pastures, and pastures - 6.5%, forests, and shrubs - 21%, swamps - 4.5%.

The peculiarity of *the landscape structure* of the region is that Polesie areas occupy 63% of the entire territory, and more than 18% of the area has forest-steppe characteristics.

Moraine-zander lowlands with various sod-podzolic soils in combination with muddy and swampy soils are significantly developed in the north and northeast. In many places, they are deforested and occupied by agricultural land.

Zander lowlands are characterized by a predominance of sandy and loamy sod-medium-podzolic soils in combination with wetlands.

In Chernihiv Polesie, floodplain-terraced areas with sod-medium-podzolic soils are developed. They stretch in broad bands along the rivers of the Dnieper, Desna, Snova, Seimas, and others. They are used for agricultural land, and forests are preserved in small areas. More forested are the first floodplain terraces, where pine forests are developed.

Sandy-swamp valleys are characterized by peculiar landscapes. The largest of them is the ancient passage valley of the Dnieper - Zamglai. It stretches in the submeridional direction from the Dnieper River to the Desna River for 60 km, 8-10 km is wide. The bottom of the valley is occupied by lowland swamps with peatlands. The thickness of peat reaches on average 1.5-2 m, and the maximum - is up to 5-6 m. Peat deposits are being developed. Part of the Zamglai swamps is reclaimed and used for agricultural land. Sandy meadows and swamps of passable valleys are also common in the north-eastern part of Chernihiv Polesie on the border with Novgorod-Siversky Polesie, in the interfluve of Snov-Desna.

Meadow-wetlands of modern floodplains occupy large areas in the valleys of the Dnieper, Desna, Snova, Ubeda, Mena, Ostra, and their tributaries. Their height is 1.5-2 m above the water level. On the surface of the floodplain are many lakes, old rivers, streams .

Novgorod-Siverskoye Polesie. This is a natural area of Polesie physical-geographical province. It is located in the east of the Dnieper lowland and on the low slope of the Middle Russian Upland, within the Chernihiv and Sumy regions.

Geographically it is connected with the south-western slope of the Voronezh massif. The landscape features of Novgorod-Siversky Polesie are due to the peculiarities of geological and geomorphological conditions. There are common Cretaceous deposits, which in many places come to the surface with a low thickness of anthropogenic sediments, up to 100 m deep, the density of erosive surface dismemberment (density of ravine-beam network exceeds 1 km / km^2), low depth of groundwater and more continental climate.

Among mixed-forest landscapes (over 80% of the area) are dominated by moraine-water-glacial, characterized by afforestation (over 30%) and wetlands (up to 20%), the development of passable valleys and karst depressions. The background tracts here are flat and slightly undulating interfluves, composed of low-density sands with layers of clay sands with sod-slightly podzolic soils and wide swampy valleys occupied by lowland peatlands.

Forest-steppe landscapes (occupying more than 15%) are represented by forest plains, located in small areas, mainly on the right bank of the Desna River. Typical here are complex tracts of fresh and deep beams with steep slopes with hornbeam-oak and maple-linden-oak forests.

Floodplains (about 5%) are well developed in the valley of the Desna River and its left-bank tributaries. This area is mostly used for hayfields and pastures. Widespread high undulating, narrow floodplains with turf low-gley soils under grassy meadows and low, flat, extended floodplains, composed of lowland peatlands under wet-grass-sedge groups, with swampy terraced complexes under forests and sedge-swamp vegetation.

It is dominated by agricultural and mining nature management. Within the region, there is a reserve of national importance - Velykyi Bir.

In general, the soil cover of Polesie is characterized by mosaic. This is explained on the one hand - by the high diversity of soil-forming rocks (valleys are wide, interfluves are not covered with homogeneous forests, as in the southern regions of Ukraine). On the other hand, there are significant differences in moisture, which are associated with a high average level of groundwater and mostly light particle size distribution of soils, which for this reason are poorly water-retaining and easily dry. Increasing the soil surface by only a few tens of centimeters in Polesie causes major changes in soil-forming processes and vegetation composition. Polesie occupies the southwestern part of the podzolic soil zone with the development of sod-podzolic soils (6.9% of the total area), swamp, and peat-swamp soils (13%), sod-meadow (10%), and others (7-8%). The granulometric composition of sod-podzolic soils of Polesie is sandy. They are represented by sod-slightly podzolic (river terraces, sandstone valleys of the middle part of Polesie), sod-medium-podzolic (watersheds of water-glacial and moraine deposits of almost all regions of Polesie), sod-strongly podzolic (Zhytomyr Polesie).

These soils are infertile and require high agricultural techniques. Fertile soils of Polesie are sod-humus-carbonate (southern and south-western part of Volyn Polesie), gray forest, and podzolic chernozems (Turov plain, separate islands in Chernihiv and Zhytomyr Polesie). The main soils are characterized by a number of common adverse properties: acid reaction, low humus content, excessive permeability, and being significantly susceptible to erosion.

According to (Kozlovsky, 1995; Kovalenko, 2007), Ukrainian Polesie, most of which is in the lower reaches of the Pripyat river basin, is swamps and forests. Until the 19th century, the deep basin of the Pripyat basin was a series of swamps and marshy reservoirs and was not used for agriculture in vast areas. The population of villages and hamlets, which are far apart from each other, in the spring and autumn was isolated from active communication with neighboring villages, towns, and cities.

Reclamation as a Necessary Condition for the Development of the Polesie Region

Polesie lowland is sloping to the Pripyat River and the Dnieper, heavily swampy, forested. It has many freshwater lakes, swamps and peatlands. The reclamation fund of lands is 3.7 million hectares, of which peat-swamp soils occupy almost 0.95 million hectares, mineral waterlogged soils - 2.75 million hectares. This is 68.5% of swamps, wetlands and wet mineral lands of Ukraine, so the Polesie zone of Ukraine is the main area for drainage reclamation.

The problem of land use in the Polesie lowland began to be studied in the 16th century. The history of drainage in Polesie is closely connected with the history of drainage works in Europe. The history of land reclamation in Europe dates back to ancient Rome. There is information that Emperor Caesar himself was interested in the problem of draining the Pontic swamps near Rome. These swamps have been draining for centuries, and even until recently. The Padano-Venetian plain in northern Italy is also a major object of centuries-old land reclamation.

Drainage of swamps and wetlands was initiated in European countries by Roman soldiers. In modern England, Roman soldiers were ordered: "If the ground is wet, it must be drained through dug canals four feet (1.22 m) deep." After the Romans, no new drainage works were carried out until the 11th century. In the XI century after the conquest of the country by the Normans, drainage work was continued. In 1252, the royal charter of King Henry III passed the first law on the drainage of agricultural land, which became the basis for the development of land reclamation in the following centuries. England's example of the development of drainage is unique, it is from here that the experience and technique of reclamation spread to other countries.

The growth of the country's population prompted the state and the church to turn to the experience of land drainage. In the twelfth and thirteenth centuries, due to the entrepreneurship of peasants and landowners, drainage work continued. In the middle of the 17th century, drainage works were carried out in the swamps, despite the opposition of "swamp people" who lived among the swamps and wanted to preserve them.

In 1652, W. Blade's book on drainage was published in England, in which, along with the construction of canals, the construction of closed drainage was recommended. In 1724 the first "drainage plow" was constructed for the construction of mole drainage. The first closed drainage in its modern form was laid in Suffolk County in 1727. In 1764, farmer J. Elkington discovered the pressure of groundwater in the swamps and proposed to lay selective drainage with unloading wells, which were immersed in the pressure aquifer.

A special role in the development of this Polesie region was played by Mateusz Butrymowicz (1745-1814), Pinsk city judge and deputy mayor, ambassador to the four-year Sejm in Warsaw, holder of the Order of St. Stanislaus. He was the first landowner of that era to start extensive reclamation of wetlands, financed the construction of new roads, and provided all possible assistance in the construction of the Ogino and Royal (now Dnieper-Bug) canals. In 1773, estates began to be built on reclaimed lands.

A century later, another attempt was made to drain the swamps and wetlands of the Polesie lowland. In May 1872, a commission was set up "to study the state of agriculture," which concluded that "drainage of swamps is extremely necessary for many provinces, including Volyn." In 1873, the Western (Polesie) expedition was organized, which was to cover an area of about 9 million hectares, under the leadership of a surveyor, Lieutenant General Joseph Ipolitovich Zhilinsky (1834-1916). J.I. Zhilinsky proposed to drain large areas of Polesie and turn them into agricultural lands.

Based on previous research in 1873-1874, J.I. Zhilinsky developed the "General Plan for the Drainage of Polesie". J.I. Zhilinsky intended, "by rejecting the causes of wetlands, to achieve a more even distribution of moisture and thus free the wetland from excess water." It was proposed to carry out drainage by draining water from wetlands through the construction of reclamation canals, diversion of existing watercourses, which was to ensure a reduction in groundwater levels and the possibility of forest rafting. During the years of the expedition, 450 thousand desyatinas were drained, including 221 thousand desyatinas of state dachas and 229 thousand desyatinas of landowner's lands.

As a result of the work was formed (Gadzala, Stashuk, Rokochynskiy, (2017):

a) new meadows in previously inaccessible swamps – 355,000 desyatinas;
b) new arable, garden and manor lands – 115,000 desyatinas;
c) drained swampy forest areas – 530,000 desyatinas;
d) the area of valuable forests growing on dry islands, for the rafting of which roads are arranged - 640000 desyatinas;
e) the area of land that did not require direct drainage, but for them the best operating conditions - 1215,000 desyatinas.

3806,000 rubles were spent on drainage in 1897, which was 1 ruble 33 kopecks per 1 desyatina (1.0925 ha) of the whole region. The cost of draining 1 des swamp was about 4 rubles. The cost of land in Polesie has risen sharply: before the sewer, the average selling price of tithing was 4 rubles, and after - 28 rubles. In 1878-79 the area of hayfields was 10,238 desyatinas, and the treasury received from them an income of 7,552 rubles; in 1888-89 they were already 51,376 desyatinas, and the income was 10,743 rubles; in 1896-97 the area of hayfields increased to 59,203 desyatinas, and the revenue of the treasury reached 149685 rubles. These figures show that most of the drained areas belonging to the treasury have been converted into meadows. The same can be said about drained lands belonging to private landowners and peasants, although there are no accurate statistics on these lands.

During the harvest of 1891-92 in the middle chernozem provinces and in the Volga region, it was widely believed that the drainage of Polesie was the cause of this crop failure. This cannot be accepted both because the places affected by drought and crop failure are very far from Polesie (from 500 to 1000 versts and more) and because drainage would affect only surface waters. Soils of meadows and forests of Polesie are moist; the groundwater level is kept high, and canals and ditches make it possible to raise it even more when needed. Therefore, the lush vegetation of meadow grasses grows until autumn and these grasses evaporate a lot of water. Numerous studies have shown that with the full development of cereals and meadow grasses from this area evaporates no less water than at the same time and other things being equal from the area covered with water.

At the same time, research was conducted on the cultivation of winter and spring crops, fodder grasses, and vegetables on reclaimed lands, "in order to show the Polesie peasant methods and techniques of developing wetlands and gardens, a good example to make him want to learn from experience." In particular, experiments were performed using imported mineral soil.

As a result of the work of the Western Expedition, the climate improved, and epidemic diseases common in the area lost their epidemic character.

This project had no analogs in the world. For the drainage works in Polesie J.I. Zhilinsky, even before their completion, was awarded the Gold Medal at the World's Fair in Paris in 1878.

In 1903, General J.I. Zhilinsky wrote: "The sewage work done so far in Polesie can be considered sufficient under the current economic conditions of the country, but as these conditions change, as the population and culture of the country increase, it will be necessary to expand the network of canals. It is impossible to consider the work completely completed - only the beginning of this kind of work in the future ".

The fate of Polesie swamps has constantly caused sharp controversy - the need for drainage was justified by many arguments. Transformation of Polesie, through land reclamation, scientists said, will increase the area of land suitable for agricultural work, and increase their value. However, opponents of this plan were immediately found. Waterway engineers say drainage is harmful because it will lower water levels in rivers and complicate navigation on the Pripyat and Dnieper rivers. Before the First World War, the journal Bolotoznavstvo wrote: "In recent years, the issue of over-drying of swamps has become fashionable. There are complaints everywhere about the sad consequences of the drain: the swamps, which provided poor, rough fodder, began to produce only a meager amount of hay after the drain. Many scientists believe that in fact the swamps stabilize the hydrological regime, accumulate moisture, preserve the water of rivers, lakes, wells, reduce droughts in neighboring areas, are a good home for many birds, animals, a real paradise for vegetation, a huge storehouse of biological resources... So even then the question arose - should all the swamps be drained? ... ".

Two world wars and other historical troubles suspended land reclamation in the first half of the twentieth century. Large-scale work was resumed only in the 1960s. During this period, 1,130 drainage systems were built and operated, including 527 inter-farm systems and 603 on-farm systems. Only between 1980 and 1990, the total area of land with a drainage network increased by 37%, the composition of agricultural land and the structure of sown areas improved. The share of arable land increased from 65.7% in 1980 to 70.7% in 1990. Of the total area of land with a drainage network, 1,976.5 thousand hectares were used in active agricultural production, which is 89.6%, and under the forest, bushes, peat, canals, roads were 222 thousand hectares or 10.1%. The structure of sown areas on drained lands in 1990 was dominated by grain and industrial crops (43.7%).

According to there are three stages of development of design and construction of drainage systems in the postwar period in the western regions of Ukraine, taking into account changes in design features:

1) 1946-1960 - design and construction of drainage systems with an open network of canals;
2) 1961-1975 - design and construction of systems with material horizontal drainage;
3) 1976-1990 - design and construction of modern systems with two-way regulation of water-air regime, water return, and drainage-humidification systems with automated water regulation.

Drainage projects included regulation of water regime and, in some cases, construction of dams along riverbeds, within drained floodplains, interception of the slope, diluvial waters by upland canals, acceleration of surface runoff, the outflow of infiltration waters, and lowering of groundwater levels by both open and canal networks. pottery drainage. By the nature of the action on the water regime of land reclamation systems of Ukraine, mainly unilateral action. The drainage of drainage systems is not regulated, and the use of locks is limited. Of the total drainage area (2.2 million hectares in 1990), only 450.000 hectares were moistened, of which 174 thousand hectares were covered by mole drains, 264 thousand hectares by infiltration from open canals, and 12.000 hectares by sprinkling. An open network of canals, built at different times, drains the floodplains of all major rivers and their numerous tributaries.

Pottery drainage is the main way to drain wetlands and wetlands of Ukraine. The area of lands drained by pottery drainage is 1.4 million hectares.

The use of drainage has been known in reclamation practice for a long time. For example, the ancient Romans used small canals filled with stone or bushes to drain the land. This method they borrowed from the older peoples of the East.

Drainage systems for reclamation of wetlands and to combat salinization of irrigated soils were known in ancient Egypt, India, Assyria, Babylon. They were discovered by excavations in the Parthian state with its capital Ness near the modern city of Ashgabat.

On the territory of ancient Kievan Rus the first drainage works were carried out in the XI-XII centuries in the Novgorod, Vladimir and Moscow principalities.

Under Peter the Great, the drainage of lands adjacent to the Gulf of Finland began in Russia.

Beginning in 1650, drainage systems made of wood, fascines and stone (gravel) began to appear in England.

Later in Scotland and Germany began to use drains from shaped tiles. The invention of the press for the manufacture of pottery drainage pipes in 1840 in England contributed to the rapid use of drainage throughout Europe (Gadzala, Stashuk, Rokochynskiy, (2017).

In 1853 the first results of drainage works were shown in Riga.

During 1879-1897 in the western regions of Ukraine (Lviv, Ivano-Frankivsk, Chernivtsi and Zakarpattia regions) 7575 hectares of land were drained. A total of 20,000 hectares in Western Ukraine were drained 50 years ago. Drainage of land was carried out only by closed horizontal drainage in Chernivtsi, Odessa, Donetsk regions.

The largest areas of drainage, which was performed using bilateral regulation of water regime in Luhansk region (91.9% of the drainage area), Chernihiv region (80.3% of the drainage area), Poltava region (70.4% of the drainage area), Sumy region (68.4% of the drainage area). The largest areas with guaranteed regulation of the water regime of agricultural lands are located in the Polesie region Table 1.

CONCLUSION

Therefore, the technology of development and agricultural use of drained soils has been developed to a certain extent. However, in recent years, as a result of the crisis in agriculture, attention to the use of reclaimed land has decreased, and certain elements of the technology of their productive use are often ignored. One of the reasons for this situation is the actual state of the intra-economic network. In many cases, it does not meet the requirements of reliable management of the water regime, environmental safety during intensive agriculture.

Table 1. Methods of land drainage in Ukraine (Gadzala, Stashuk, Rokochynskiy, (2017)

N°	Administrative entities (regions)	Total area of drained lands					
		Total		With closed drainage		With bilateral water regime regulation	
		thousand hectares	% of the area	thousand hectares	% of the area	thousand hectares	% of the area
1	Vinnytsia region	57.3	2.2	51.0	89.0	3.3	5.8
2	Volyn region	416.6	20.7	236.6	56.8	157.2	37.7
3	Donetsk region	4.7	0.2	0.1	2.1	-	-
4	Zhytomyr region	425.4	14.3	296.7	63.4	188.7	44.3
5	Transcarpathian region	183.8	14.4	148.5	80.8	1.9	1.0
6	Ivano-Frankivsk region	195.5	14.0	195.2	99.8	2.9	1.5
7	Kyiv region	188.8	6.7	91.4	48.4	68.0	36.0
8	Luhansk region	11.1	0.4	0.4	3.6	10.2	91.9
9	Lviv region	513.2	23.5	390.1	76.0	36.4	7.1
10	Odesa region	4.4	0.1	0.1	2.3	-	-
11	Poltava region	37.2	1.3	8.4	22.6	26.2	70.4
12	Rivne region	390.4	19.5	275.0	70.4	234.3	60.0
13	Sumy region	106.6	4.5	55.7	52.3	72.9	68.4
14	Ternopil region	165.6	12.0	151.4	91.4	10.2	6.2
15	Kharkiv region	11.8	0.4	4.2	35.6	5.6	47.4
16	Khmelnytsky region	117.5	5.7	97.8	83.2	24.9	21.2
17	Cherkasy region	55.7	2.7	4.6	8.3	31.5	56.6
18	Chernivtsi region	121.8	15.0	121.4	96.7	-	-
19	Chernihiv region	300.0	9.4	167.8	55.9	241.0	80.3
	Total:	3307.4	5.5	2296.4	69.4	1115.2	33.7

In modern conditions, meliorative agriculture should be based exclusively on a biological, ecological and economic basis, namely by implementing:

- resource-saving soil tillage systems, which includes agromelioration measures;
- crop rotation systems (with obligatory saturation of them with legumes) in order to limit the use of nitrogen fertilizers.

Such a complex should ensure the arrival and accumulation of organic matter in the soil to a level that ensures a deficit-free or positive balance of humus.

The ideology of the reconstruction of reclamation systems should be based on the following principles:

- the creation of the latest constructions of reclamation systems of the block-modular type, adapted to specific natural conditions and land use requirements;

- the creation of resource-saving technologies of water regulation, construction of reclamation systems with two-way regulation of the water regime, and technical means of managing technological processes, which will allow obtaining an economic effect in the first years after their implementation;
- ensuring the rational use of water and land resources, restoring and increasing soil fertility;
- ensuring ecological balance in reclamation areas, in the basins of rivers and watercourses in the zones of influence of reclamation systems, prevention of negative consequences, especially such as wind and water erosion of soils, and pollution of water resources;
- reduction of the area of use of overmoistened lands due to their partial renaturalization, temporary conservation of unproductive lands, renewal of the structure of lands with priority expansion of fodder rotations, hayfields, and pastures in connection with new forms of management;
- implementation of scientifically based technologies for restoration and agricultural use of radiation-contaminated territories;
- accounting for changes in land relations, forms of management, features of temporary land management in the area of land reclamation systems, technological conditions of their operation, and the balance of interests of land users and water management organizations;
- limitation of terms of carrying out works regarding reconstruction, minimization of capital costs, and maximum use of effective elements of the existing reclamation system.

In order to legally support the productive use of reclaimed land in the zone of excessive moisture, it is necessary to develop the following normative documents, namely:

- the procedure for using drained lands and reclamation funds of intra-economic systems;
- the procedure for financing expenses for maintenance of intra-household systems;
- rules of on-site inspections, certification, assessment of the technical condition of land reclamation systems and the ecological condition of reclaimed agricultural landscapes, determination of renaturalization volumes of reclaimed lands.

The implementation of such approaches will make it possible to ensure the further development of such a unique in all respects region, which is Ukrainian Polesie, taking into account modern socio-economic and ecological requirements.

REFERENCES

Savchyn, O. (2013). A little about Polesie and Volyn. *UA Chronicle*. http://hronosua.blogspot.com/2013/02/blog-post_7831.html

Gadzala, S. (2017). Reclamation and development of Ukrainian Polesie: a collective monograph. NIH.

Konishchuk, V.V., Andrienko, T.L., Bondar, O.I., Grummo, D.G., Davidyuk, V.F., Yevstigneev, O.I., Konovalchuk, V.K., Landin, V.P., Lukash, O. .V., Serebryakov, V.V., Solovei, T.V., Tytar, V.M., Furdychko, O.I., Chobotko, H.M., & Shershun M.H. (2012). *Polesskaya ecological convention – an epistemological paradigm of formation, 9*, pp. 289-299.

Kotlyakova V. M. (2006). *Dictionary of modern geographical names*. Getty.

Kovalenko P.I. (2007). Conceptual approaches to land reclamation in the Ukrainian Polesie. *Amelioration of waterlogged lands, 2*.

Kozlovsky, B. I. (1995). *Scientific basis of monitoring drained lands*. Missioner.

Pledge, F. D. (1990). *Geography of Ukraine*. World.

Geografica. (2023). Poliski landscape region. *Geographica.* http://geografica.net.ua/publ/galuzi_geografiji/fizichna_ geografija_ukrajini/poliskij_landshaftnij_kraj/39-1-0-544

Chapter 2
Characteristic Features of the Polish Part of Polesie

Jerzy Jeznach
Warsaw University of Life Sciences, Poland

Danuta Urban
Institute of Soil Sciences and Environmental Engineering, Poland

Zdzisław Michalczyk
Maria Curie-Skłodowska University, Poland

Mariusz Kulik
University of Life Sciences, Lublin, Poland

Piotr Sugier
https://orcid.org/0000-0002-1448-1517
Maria Curie-Skłodowska University, Poland

Grzegorz Grzywaczewski
University of Life Sciences, Lublin, Poland

ABSTRACT

The chapter presents the most important data of the Polish Lublin Polesie. Polesie Lubelskie is a unique natural area. It is the peripheral western part of Polesie proper – a vast geographical and historical region that spans the territory of southern Belarus and northern Ukraine, and has western outposts in Poland and eastern ones in Russia. Polish Polesie is home to a great variety of species of vascular plants and bryophytes. This is influenced by the diversity of biological communities which include water, peatland, meadow, poor grassland, moor, forest, shrub, and synanthropic ecosystems. Polish Polesie has 1,305 species of vascular plants and about 200 species of bryophytes. The fauna of Polesie Lubelskie is also species-rich. Polesie Lubelskie has for many years been considered one of the most valuable natural areas in the Lublin Province and in Poland. To preserve the region's natural resources, including the diversity of natural habitats and the sites of rare species of plants, animals, and fungi, various protection measures have been taken in this area for many years

DOI: 10.4018/978-1-6684-8248-3.ch002

BACKGROUND

Polesie Lubelskie is a unique natural area. It is the peripheral western part of Polesie proper – a vast geographical and historical region that spans the territory of southern Belarus and northern Ukraine, and has western outposts in Poland and eastern ones in Russia. Polesie Lubelskie is distinguished by exceptional bio- and geodiversity, with a wealth of unique plant and animal species and morphogenetically diverse (denudational, karst, glacial, lacustrine, organogenic) landforms found in the apparent monotony of the landscape. The area is dominated by wetlands, which are considered to be the most endangered components of this environment, and at the same time ones that are extremely sensitive to its changes. Wetlands determine the stability and ecological resistance of entire ecological landscapes. Among them stands out a compact group of lakes with a complex genesis (which bears the features of a lakeland) and accompanying functional systems with valuable and unique natural characteristics: extensive peat bog complexes, small, wooded peat bogs, and mires, and more or less strongly transformed river valleys. Due to the role they play in the environment, wetlands are counted among areas of extremely high natural value.

In the last few decades, the area of Polesie Lubelskie has been subject to increasing human interference in the naturally formed biotic and abiotic structures, which has disrupted the functioning and the biocenotic balance of the area's ecosystems. In the mid-twentieth century, most of the peat bogs located here were drained, including the largest peat bog complex of Krowie Bagno. In the western part of Polesie Lubelskie, the Wieprz-Krzna Canal was built in the zone of the surface watershed of the rivers Vistula and Bug to regulate and stabilize the area's water balance. Several lakes were incorporated into the Canal's system. In the south-western part of Polesie Lubelskie, in turn, the hard coal mine "Bogdanka" has been operating since mid-1970s, in which coal is excavated using the caving method. Excavation leads to a far-reaching transformation of the hydrological regime of this part of the area (uncontrolled development of subsidence depressions and the resulting changes in the course of local watersheds). Unfortunately, more mines are planned to be constructed in the vicinity of the border of the Polesie National Park. Despite considerable anthropogenic pressure, Polesie Lubelskie remains an area of high natural and cultural value, which leads to the development of tourism in this area.

Location of Polesie Lubelskie

Polesie Lubelskie is the western part of the large region of Polesie, which extends over the territories of southern Belarus and northern Ukraine. It lies within Western Polesie (Polesie Zachodnie), which covers part of the Vistula and Bug inter-river basin in Poland, with an area of 5559 km^2. It is situated between the northern edge of the Lublin Upland (Wyżyna Lubelska) and the valley of the river Krzna, which flows into the Bug. Polesie Lubelskie includes the catchment area of the upper course of the river Tyśmienica, which flows into the river Wieprz, and in the east it reaches the valley of the Bug. On the southern side, it adjoins the Lublin Upland (the Central European Highlands zone), and in the west and north it borders on the area of Middle-Polish Lowlands.

The bottoms of river valleys and the large depressions outside those valleys are filled with Holocene sands, peat and peaty silts. Among the organogenic Holocene sediments, fen peats occupy the biggest area. They form large, compact peat bogs, especially in the western part of the Łęczna-Włodawa Lakeland (Pojezierze Łęczyńsko-Włodawskie) (Harasimiuk, Dobrowolski 2010). The largest of them are Krowie Bagno, Durne Bagno, Bagno Bubnów and Bagno Staw. The thickness of the peat bogs is very diverse and ranges from several dozen centimetres to over 10 meters.

The contemporary relief of Polesie Lubelskie is the product of complex morphogenetic processes, mainly erosion and accumulation, which have been shaping this area from the recessional phase of the Odranian Glaciation to the Holocene. Morphogenetically, the region is clearly divided into two parts. The northern part consists of strongly denuded, flat ground-moraine plains cut by a complex system of valleys with very poorly developed watershed zones, with numerous bifurcation zones between the basins of the rivers Wieprz and Bug (Dobrowolski, Harasimiuk 2002). The differences in elevation are small and do not exceed 25 m.

Peat Bogs in Polesie Lubelskie

The distribution and the ecological type of the peat bogs found in Polesie Lubelskie are derivatives of the morphology of the area and its geological structure (Harasimiuk, Dobrowolski 2010). The largest concentration of peatland geo-ecosystems is found in the central and southern part of the region, i.e. the Łęczna-Włodawa Lakeland, the Dubienka Depression (Obniżenie Dubienki) and the Dorohuck Depression (Obniżenie Dorohuckie). These territories boast almost 800 peat bogs with a total area of approximately 45 thousand ha. The peat bogs here clearly differ from one another in their surface area, shape, type of deposit, development stage, plant cover, hydrological regime, and the level of anthropogenic transformation.

Polesie Lubelskie is the largest expanse of hydrogenic landscapes to the south of the zone of the maximum extent of the Vistulian Glaciation. The largest number of lake and peat bog complexes are found in the central and southern part of this area, especially in the Łęczna-Włodawa Lakeland. These lake and peatland ecosystems have different sizes, depths, and biolithofacial succession, and in the case of lakes also a different degree of area decline caused by gradual shallowing and overgrowth. It is agreed that contemporary organogenic accumulation reservoirs in this area have been developing since the Late Glacial (Dobrowolski 2006). At that time, there occurred significant changes in the climatic, hydrological and hydrogeological conditions, which shaped the mechanisms responsible for the formation of peatlands and/or lakes and their subsequent functioning.

Plants

Polish Polesie is home to a great variety of species of vascular plants and bryophytes. This is influenced by the diversity of biological communities which include water, peatland, meadow, poor grassland, moor, forest, shrub, and synanthropic ecosystems. Polish Polesie has 1,305 species of vascular plants and about 200 species of bryophytes (Fijałkowski, Izdebski 2002), a floristic richness that, according to Fiałkowski (2007), is also associated with the presence of numerous species representing different geographical elements (boreal, Pontic, Atlantic, Central European, mountainous, Mediterranean and South Siberian species). A large group of vascular plant species in Polish Polesie are rare for the region or for Poland in general. Am9ong them, 103 are protected species. Some of the legally protected or rare species, such as Nymphaea alba, Helichrysum arenarium, Lycopodium annotinum and L. clavatum, occur in abundance or even grow in masses. Others, such as Linnaea borealis, Adenophora liliifolia, Polemonium coeruleum, Liparis loeselii, Hammarbya paludosa, Dianthus armeria, Diphasistrum complanatum, Salvinia natans and Empetrum nigrum are found in only one or a few sites. In the last few decades, no records have been made of the following species: Chamedaphne calyculata, Pulsatilla patens, Carex heleonastes, Saxifraga hirculus, Litorella uniflora Rosa gallica, Petasites albus, Juncus atratus, Eleocharis quinqueflora and Isöetes lacustris (Urban et al. 2020).

Polish Polesie has the following Habitats Directive Annex II species (Council Directive 92/43/EEC): *Aldrovanda vesiculosa, Adenophora liliifora, Caldesia parnassifolia, Cypripedium calceolus, Ostericum palustre* and *Liparis loeselii*. Other species growing in this region that deserve special attention are those listed in the Polish Red Book of Plants (Kaźmierczakowa et al. 2014): *Botrychium multifidum, Salix myrtilloides* and *S. lapponum, Betula humilis, Nymphaea candida, Aldrovanda vesiculosa, Caldesia parnassifolia, Trapa natans, Adenophora liliifolia, Ostericum palustre, Pinguicula vulgaris* subsp. *bicolor, Succisella inflexa, Carex buxbaumii* and *C. chordorrhiza, Cypripedium calceolus, Dactylorhiza incarnata* subsp. *ochroleuca, Cephalanthera rubra, Liparis loeselii, Hammarbya paludosa, Orchis morio, Pedicularis sceptrum-carolinum, Alisma lanceolatum* and *Nymphaea candida*.

Compared to the last century, this century has seen a considerable spread of invasive species such as *Acer negundo, Padus serotina, Quercus rubra, Robinia pseudoacacia, Echinocystis lobata, Impatiens glandulifera, Rumex confertus, Solidago spp., Helianthus tuberosus, Echinops sphaerocephalus* and *Aster spp.* Other invasive species, such as *Asclepias syriaca, Althaea officinalis* and *Heracleum sosnowskyi* are relatively rare. These plants lead to the degradation of indigenous communities by eliminating native species from their age-old habitats.

Changes in the flora of Polesie Lubelskie can be counteracted by taking appropriate measures aimed at restoring some species to their natural sites (restitution) and strengthening some plant populations (active protection, shaping of habitats). Other threats to the plant communities of Polesie Lubelskie include the shading and overgrowth of the stands of numerous species, e.g. species associated with meadow ecosystems.

Fens and transition mires are threatened by the encroachment of shrubs and trees (natural succession towards forest communities). Most species of transition mire and raised bog ecosystems are light-loving species.

In the case of species with a stable current status, this status should be maintained by ensuring a proper water regime, maintaining habitats in an unwooded state free of shrubs, and by periodic mowing for some species.

The water ecosystems of Polesie Lubelskie are diversified both typologically and limnologically, as they include lakes, rivers, ditches, ponds, flooded peat workings, old river beds, and sinkholes. Their waters exhibit different trophic levels and different physicochemical properties. The plant communities occurring in this area represent four phytosociological classes: *Lemnetea, Charetea, Potametea* and *Phragmitetea*. The most widespread *Charetea* complexes include those of *Charetum intermediae, Ch. hispidae, Ch. delicatulae* and *Ch. fragilis*, which are associated with mesotrophic and eutrophic lakes as well as flooded peat workings. A relatively large group are communities of macrophytes from the class *Potametea*. Among elodeids, the most frequently encountered phytocoenoses are those of *Ceratophylletum demersi* and *Myriophylletum spicati*. Less widespread are communities of *Elodeetum canadensis, Hydrocharitetum morsus-ranae, Potametum lucentis* and *Ranunculetum circinati*, and the least common are communities of *Potametum pectinati* and *Myriophylletum verticillati*. Among water communities of nympheids, the most frequently found are *Nupharo-Nymphaeetum albae* and *Potametum natantis*, and the rarest – *Nymphaeetum candidae*. Lakes, of course, are not the only hydrogenic areas of Polesie Lubelskie that have valuable floristic and phytocoenotic elements and characteristics. Other water bodies, such as rivers, ponds, artificial lakes and peat flooded workings also provide habitats for interesting and unique plant species. Worthy of attention are communities of *Hippuridetum submersae* and *Lychnothamnetum barbati*, which are relatively rare in Poland as well as *Najadetum intermediae* and *Trapetum natantis*, which occur at only one site – in the Polesie National Park (Lorens, Sugier 2020).

The processes of organogenic accumulation lead to the gradual filling of lake basins, which has recently resulted in a drastic reduction in the surface area of many lakes. However, changes in the properties of hydrogenic habitats in Polesie Lubelskie in the last few decades have also been caused by human activity, in particular the construction and use of the Wieprz-Krzna Canal and the related network of canals and ditches. The hydrological conditions of many lakes, and thus the ecological conditions of macrophyte vegetation, were changed when they were connected to this Canal. Also, hydrotechnical measures taken to reclaim fens adjacent to lakes and transform them into grasslands very often resulted in periodic dehydration of parts of lake catchments, and in some cases a long-term decrease of the water level in lakes leading to rapid changes in aquatic and rush vegetation. Many lakes suffer the effects of strong recreational pressure manifested by a reduction in the diversity of submerged macrophytes, an increase in the share of helophytes, and, in extreme cases, algal bloom, which effectively eliminates underwater vegetation. Despite anthropogenic pressure, the vegetation of the lakes of Polesie Lubelskie, including the Polesie National Park, is relatively well preserved due to the fact that both the reservoirs themselves and the adjacent peatlands were placed under protection as natural reserves very early on (Lorens, Sugier 2020).

The region of Polesie Lubelskie, and especially the Łęczna-Włodawa Plain, is exceptionally rich in peat bogs. It has about 800 peat bogs of different sizes (Borowiec 1990). Most of the peat bogs in Polesie Lubelskie have great natural value. The prevailing type of peat bog in the region are fens. There are also transition mires and raised bogs (mainly in the Łęczna-Włodawa Plain). In phytosociological terms, most of the region's peatland communities fall into one of the two classes: *Scheuchzerio-Carictea nigrae* and *Oxycocco-Sphagnetea*. Many rush communities from the class *Phragmitetea* and some forest communities from the classes *Alnetea glutinosae* and *Vaccinio-Piceetea* also have peat-forming properties.

Particularly valuable plant communities include *Caricetum limosae*, *C. chordorrhizae*, *Rhynchosporetum albae*, and *Sphagno-Caricetum rostratae* found in transition mires, *Ledo-Sphagnetum*, *Sphagnetum magellanici* and *Sphagnum fallax-Eriophorum vaginatum* growing on raised bogs, as well as *Cladietum marisci*, *Caricetum buxbaumii*, *C. davallianae* and *Schoenetum ferruginei*, which occur in calcareous fens (Urban et al. 2020).

Many protected species of vascular plants have been recorded in fen, transition mire, and raised bog communities of Polesie Lubelskie. The protected fen species include *Ostericum palustre*, *Ligilaria sibirica*, *Polemonium caeruleum*, *Trollius europaeus*, *Iris sibirica*, *Dianthus superbus*, *Veratum lobelianum*, *Carex davalliana*, *Schoenus ferrugineus*, *Liparis loeselii*, *Epipactis palustris*, *Pedicularis sceptrum-carolinum*, *P. palustris*, and *Gentiana pneumonanthe*. Transition mires are home to species such as *Salix lapponum*, *S. myrtilloides*, *Betula humilis*, *Liparis loeselii*, *Hammabrya paludosa*, *Epipactis palustris*, *Pedicularis sceptrum-carolinum*, *Scheuchzeria palustris*, *Drosera rotundifolia*, *D. anglica*, *D. intermedia*, *D.×obovata*, *Andromeda polifolia*, *Empetrum nigrum* and *Carex limosa*. The protected species growing on raised bogs are *Drosera rotundifolia* and *Ledum palustre* .

The most important threats to raised bogs and transition mires include changes in the water regime (soil drainage), changes in the species composition (succession) and the related reduction or loss of certain features of the habitat (a great threat is the encroachment of shrubs and trees, as well as common reed onto these wetlands), alien invasive species, and environmental pollution related to tourism and cranberry picking. The communities of alkaline fens are most acutely threatened by changes in the water regime (soil drainage) (Urban et al. 2020).

Active protection measures for peat bog ecosystems are taken in Polesie Lubelskie, which include the removal of trees and shrubs. As a result of such treatment, the groundwater level has been successfully raised in some of the peatlands. In some transition mires (e.g. in the Polesie National Park), reed is mowed along with the removal of biomass. The well-preserved patches of alkaline fens do not require active protection. However, when the water regime of a peat bog is disturbed, measures should be taken to restore the proper level of groundwater and inhibit its accelerated outflow by backfilling and blocking the existing drainage ditches. The changes taking place in some peat bogs of Polesie Lubelskie (including the Polesie National Park) have been much less prominent in the recent years as active protection measures for water retention, preservation of natural conditions in the peat bogs and elimination of invading tree and shrub species have been taken (Sugier, Różycki 2010).

The forests of Polesie Lubelskie are highly diversified both in terms of their distribution and size, which depend on the local geological structure, soil diversity and the development of settlements. In terms of phytosociological affiliation, the forests of Polesie Lubelskie belong to four classes. The largest group are forest communities from the classes *Vaccinio-Piceetea* and *Alnetea glutinosae,* followed by a smaller group of *Querco-Fagetea* forest communities and a small group of forests of the *Salicetea purpureae* type. Coniferous communities of the *Vaccinio-Piceetea* class are the dominant forest communities in Polesie Lubelskie. They have different trophic levels and show large hydrological and floristic differences (Urban et al. 2020).

Meadows are found in Polesie Lubelskie mainly in river valleys, but also in several large fen complexes and in peat bogs surrounding some lakes. The largest area is occupied by semi-natural meadow and pasture communities of the *Molinio-Arrhenatheretea* class. In humid and wet habitats, one can find communities from the order *Molinietalia caeruleae*, and in fresh habitats – communities from the order *Arrhenatheretalia*, which are usually used for mowing or grazing. Meadows which are not used for those purposes are overgrown with shrub communities that include willows.

The most valuable meadow communities in Poland, and also in Polesie Lubelskie, are meadows from the *Molinion* association. They are anthropogenic communities of one-cut unfertilized meadows with variable groundwater levels. They are characterized by a wide ecological amplitude and occur in both acidic mesotrophic habitats and in fertile alkaline habitats containing calcium carbonate. *Selino-Molinietum* meadows in particular are of great landscape, aesthetic and floristic value. Some molinia meadows in Polesie Lubelskie are characterized by high floristic diversity; for instance, in the meadows of the Polesie National Park, 173 plant species have been recorded, which include: *Betonica officinalis, Dianthus superbus, Galium boreale, Gentiana pneumonanthe, Gladiolus imbricatus Inula salicina, Iris sibirica, Ostericum palustre, Trollius europaeus, Carex davalliana, Veratrum lobelianum Laserpitium prutenicum, Molinia caerulea, Selinum carvifolia* and *Succisa pratensis.* The most interesting phytocoenoses of Polesie Lubelskie, plant- and landscape-wise, are the communities of previously fertilized moist and wet meadows from the *Calthion palustris* association found in drained complexes. Rarely recorded in Polesie Lubelskie is the *Trollio-Cirsietum* community. Communities of *Deschampsia caespitosa* develop on grazed, rarely mown and poorly fertilized meadows. The only Cnidion meadows of Polesie Lubelskie grow in the valley of the river Bug, where patches dominated by *Allium angulosum* or *Cnidium dubium* are found. Foxtail meadows of the *Alopecurion pratensis* association are most often found in river valleys on peat-muck and silt-muck soils. Fresh meadows of the order *Arrhenatheretalia* are characterized by a fairly large diversity and are represented by several communities occupying habitats of different soil moisture and fertility (Kulik, Urban 2020).

Polesie Lubelskie is home to the following habitats listed in the Habitats Directive Annex II: 3160 Natural dystrophic lakes and ponds, 4030 Dry heaths (*Calluno-Genistion*, *Pohlio-Callunion*, *Calluno-Arctostaphylion*), 6120 Xeric sand calcareous grasslands (*Koelerion glaucae*), 6230 *Nardus* grasslands in mountain and lowland areas (*Nardion* - floristically-rich patches), 6410 Molinia meadows with varying humidity (*Molinion*), 6510 Extensively used lowland and mountain fresh meadows (*Arrhenatherion elatioris*), 7110* Active raised bogs, 7140 Transition mires and quaking bogs (mostly vegetation of the *Scheuchzerio-Caricetea*), 7150 Depressions on peat substrates of the *Rhynchosporion*, 7210 Calcareous fens (*Cladietum marisci*, *Caricetum buxbaumii*, *Schoenetum nigricantis*), 7230 Mountain and lowland alkaline fens, 9170 Central European and subcontinental *Galio-Carpinetum* oak-hornbeam forests, 91D0* Bog woodland, 91E0 * Alluvial forests with willows, poplars, alders and ashes (*Salicetum albae*, *Populetum albae*, *Alnenion glutinoso-incnae*), 91I0 * Euro-Siberian steppic woods with *Quercus* spp. (*Quercetalia pubescenti-petraeae*), 91T0 Central European lichen Scots pine forests.

Animals

The fauna of Polesie Lubelskie is also species-rich. So far 62 species of mammals, including introduced and migrating taxa, have been found here. Of these, the largest group, both in terms of the number of species and the frequency of occurrence, are small mammals (*Micromammalia*, 21 species). In the Lublin Province, Polesie Lubelskie is the main habitat of one of the region's most important rodents, which has a large impact on the human economy – the beaver *Castor fiber*. A noteworthy example of insect-eating mammals (*Insectivora*) is *Neomys fodiens*, Poland's only poisonous mammal. Predatory mammals are represented by 14 species. The most valuable predators of Polesie are *Canis lapus* and *Lynx lynx*. Among the ungulates, the most abundant species are *Capreolus capreolus Cervus elaphus* and *Sus scrofa*. Next to the Kampinoski National Park and the Biebrza National Park, Polesie is one of the three most important sites for *Alces alces* in Poland (up to approx. 22 specimens/1000 ha of forest-swamp areas (Piskorski, Bielek-Bielecki 2020).

Polesie Lubelskie is one of the species-richest areas when it comes to avifauna. The great diversity of natural habitats provides living space for a huge variety of bird species, including birds associated with wetland habitats. The key area for birds in Western Polesie is the Poleski National Park and its buffer zone (Grzywaczewski, Bednarz 2020). The Poleski National Park, its buffer zone and the Special Area of Conservation Krowie Bagno are home to a total of 244 bird species, i.e. 53% of all bird species observed in Poland (463 species in Poland, as of 01/10/2020, http://komisjafaunistyczna.pl). The quite commonly occurring bird species of this area are *Ardea alba, Egretta garzetta*, *Botaurus stellaris*, and *Circus pygargus*. The rare species nesting in this area include *Crex crex*, *Gallinago media, Aythya nyroca* and *Haliaeetus albicilla* (Poland's largest bird of prey). The inconspicuous *Acrocephalus paludicola*, which also lives in this region, is threatened with global extinction.

There are 13 species of amphibians in Polesie Lubelskie (12 of them are breeding species). As for typical mountain species and those whose ranges do not include Central-Eastern Poland, all potentially encountered domestic species can be observed here. Interesting rare species include *Bufo calamita*, *Pelobates fuscus* and *Triturus cristatus* (Gosik 2020) . Reptiles are represented by eight species. *Emys orbicularis*, which is a relict species in Poland and the country's endangered reptile deserves special attention. Snakes are represented by *Viper berus* and *Natrix natrix*, as well as by the recently encountered *Coronella austriaca* (Różycki 2020).

The rivers, lakes, ponds and flooded peat workings of Polesie are inhabited by 39 fish species. Among them, six are protected species, four are listed in Annex II of the Habitats Directive (DSII) and five feature in Annex V of the Habitats Directive (DSV). Special attention should be paid to *Rhynchocypris percnurus*, one of the rarest fish species in Poland, which lives in small water reservoirs on peat substrates (Rechulicz, Kolejko 2020).

System of Protected Areas

Polesie Lubelskie has for many years been considered one of the most valuable natural areas in the Lublin Province and in Poland. To preserve the region's natural resources, including the diversity of natural habitats and the sites of rare species of plants, animals and fungi, various protection measures have been taken in this area for many years. The system of protected areas of Polesie Lubelskie consists of: the Polesie National Park, seven landscape parks, five protected landscape areas, twenty four nature reserves and numerous forms of individual protection, such as: ecological land, natural monuments, nature-and-landscape complexes, and documentation stands.

Since Polesie Lubelskie has very diverse and valuable natural habitats and many species of plants and animals, as documented in numerous scientific studies, over twenty Natura 2000 areas have been established in this region. The most valuable and the most diversified part of Polesie is the Łęczna-Włodawa Lakeland. In its central part, there are two special bird protection areas: Bagno Bubnów PLB060001 and Polesie PLB060019. Both of these areas provide habitat for the aquatic warbler and other species associated with open areas such as the crane or the ruff. An important component of the system of protected areas in Polesie is the Cross-border Biosphere Reserve "West Polesie" which spans parts of Poland, Ukraine and Belarus and includes one of Europe's most valuable hydrogenic landscape complexes (Urban et al. 2020).

REFERENCES

Borowiec, J. (1990). *Torfowiska Regionu Lubelskiego. Prace Wydz. Nauk o Ziemi i Nauk Górniczych.* Wydaw. PWN.

Dobrowolski, R. (2006). *Glacjalna i peryglacjalna transformacja rzeźby krasowej północnego przedpola wyżyn lubelsko-wołyńskich (Polska SE, Ukraina NW).* Wyd. UMCS.

Dobrowolski, R., & Harasimiuk, M. (2002). Geologiczne uwarunkowania rozwoju rzeźby Polesia. [In:] S. Radwan et al. (eds.), Środowisko przyrodnicze Polesia – stan aktualny i zmiany.

Fijałkowski D. (1994). Flora roślin naczyniowych Lubelszczyzny. *Środowisko przyrodnicze Lubelszczyzny.* LTN.

Fijałkowski, D., & Izdebski, K. (2002). Flora Poleskiego Parku Narodowego, rośliny naczyniowe. In S. Radwan (ed.), Poleski Park Narodowy. Monografia przyrodnicza.

Harasimiuk, M., & Dobrowolski, R. (2010). Geneza krajobrazów hydrogenicznych Polesia Zachodniego. [In:] T. Chmielewski, D. Piasecki (eds.), The Future of hydrogenic Landscapes In European Biosphere Reserves. Poleski PN, PAN Oddz. Lublin.

Yury, M. Rokochinskiy A., Aleksandr, A., Aleh, M., & Jerzy, J. (2020). POLESIE – Środowisko Melioracje. Monograph.

Urban, D., Grzywaczewski, G., Jeznach, J., & Mażajski, J. (2020). Ochrona przyrody polesia lubelskiego, 854-854. .

Sugier, P., & Różycki, A. (2010). Wybrane obszary hydrogeniczne Pojezierza Łęczyńsko-Włodawskiego. In A. Obidziński (Ed.), *Mazowsza na Polesie i Wileńszczyznę*. Zarząd Główny.

Chapter 3
Transformation and Ecological Aspects of Organogenic Soils Under Drainage and Agricultural Exploitation

Ivan Slyusar
National Science Center "Institute of Agriculture of NAAS", Ukraine

Olena Solyanik
National Science Center "Institute of Agriculture of NAAS", Ukraine

Viktor Serbenyuk
National Science Center "Institute of Agriculture of NAAS", Ukraine

Hanna Serbeniuk
National Science Center "Institute of Agriculture of NAAS", Ukraine

ABSTRACT

This chapter presents the results of long-term (1932-2019) stationary studies of peat lands after drainage reclamation work and agricultural exploitation of river floodplains and wetlands of Polesie of Ukraine. Certain regularities in the transformation of organogenic soils and some ecological aspects were discovered. The rates and features of the processes of activation of the organogenic soil layer, changes in its agrochemical, water-physical, and biological characteristics were established. Thereupon, conclusions were drawn regarding the efficient and environmental use of drained organogenic soils in Ukraine.

DOI: 10.4018/978-1-6684-8248-3.ch003

BACKGROUND

The issue of swamp reclamation requires a comprehensive study since they are an important component of local natural resources and environmental ecology. The main condition for the agricultural development of swamps is the drainage process. As a result, significant changes in the basic properties of the soil occur. In the interstices of the upper layers, under the influence of draining, aeration significantly increases, aerobes begin to function, the plant's nutritive regime changes, and at the same time the vegetation composition changes as well.

According to many researchers (Himbarzhevsky, 1971; Tsiupa, 1990; Yatsyk, 2004), under the conditions of meadow-field exploitation of peatlands, an average of 7-9 tons per 1 ha of peat is irretrievably mineralized annually. As a result, in recent years there has been a decrease in peat resources in Ukraine by 120-150 million tons. In addition, up to 20 million tons of peat are extracted annually in Ukraine for various needs and, especially, for fertilization. There is excursive exploitation of valuable peat resources, that in the foreseeable future can lead to their complete depletion. This requires urgent measures for the transition to an environmentally balanced system of rational and economical peatlands exploitation.

Many researchers have established a close correlation between the settlement of peat mass and the lowering of groundwater. Significant peat losses are also observed from wind erosion. On average, under inter-tilled crops, they are 2-3 t/ha, and under cereals – 1 t per 1 ha. Wind erosion is not observed under perennial grasses (Tsiupa, Bystrytskyi, Slyusar, 1990).

As for the introduction of crop rotations with a long meadow period, the degree of peat decomposition also depends on the groundwater depth level. According to Tsiupa (1990), the annual mineralization of peat under inter-tilled crops (potatoes, mangel-wurzel) is 14-16 t/ha, and under perennial grasses – within 3-4 t/ha per year.

Consequently, peat mineralization takes place under the grasses as well, but compared to other crops, grasses compensate for almost half of the destroyed peat with stubble and root residues. The unproductive losses of mineral nitrogen in the process of peat mineralization due to its removal by drainage waters and in the form of gaseous losses into the air are of concern as well.

Thus, the study's purpose was to examine the effect of draining and agricultural development in different time cycles on the transformation of peat, and changes in the fertility and productivity of peat soils.

The most characteristic peat soil locations for the Polesie and Forest-Steppe zones were chosen for the research. These are the reclamation systems "Zurno" and "Olshanka" in the Western Polesie of the Rivne region; the Zamhlai swamp in the Left-bank lowland Polesie of the Chernihiv region; in the forest-steppe zone on drained peatlands of the floodplain of the Romen river in the Sumy region and the floodplain of the Supii river in the Kyiv region.

The reclamation systems "Zurno" and "Olshanka" are low-level bogs, which were covered with bushes of different densities, hummocks, and rotted stumps before the start of development. The herbage consisted mainly of forb-sedge vegetation, in the grass canopy of which about 20% of marsh horsetail, marsh bedstraw, puddle grass, and creeping buttercup were found. The grain varieties consisted of bent grass, red fescue, and swamp meadow grass. There was up to 40-50% of carex in the grass canopy. A significant part of the site was covered with bog peat. Data on the properties of peat are presented in Table 1.

The construction of the "Zurno" system on a site of 4870 hectares was completed in 1957. Drainage was carried out by a grid of open channels at a distance of 300 m between them. The Sluch River serves as the system's water intake. Peat sedge with a transition at a depth of 0.5 m to sedge-reed, slightly decomposed. The degree of decomposition of peat is 20-25%, the ash content is 5.2-8.1%. The significantly lower density of the solid phase compared to mineral soils and very low bulk density. The reaction of the soil solution is slightly acidic (pH 5.3-5.7%). The depth of the peat layer is 3-4 m, and directly under the experimental sites is 3.1 m. The construction of the "Olshanka" reclamation system was completed in 1959 (drainage was carried out by open channels with a distance of 300 m between them). The total area of drained swamps is 940 ha. The Sluch River also serves as a water receiver.

The reasons for their swamping, as for most swamps of this zone, were:

- shallow occurrence of water-proof rocks and groundwater;
- very low surface runoff due to the specific terrain unfavorable for natural runoff;

Table 1. The main agrochemical properties of peat bog soils of the "Zurno" and "Olshanka" reclamation systems before the start of development, 1962

Soil layer, cm	pH KCI	Ash content,%	Degree of peat decomposition, %	Organic matter	N	P_2O_5	K_2O	CaO	MgO
				% for absolutely dry matter					
reclamation system "Zurno"									
0-20	5.5	6.47	20	98.5	3.51	0.37	0.09	3.27	0.11
20-40	5.4	8.12	25	91.9	3.24	0.85	0.10	3.13	0.07
40-60	5.6	5.25	25	94.8	3.07	1.24	0.04	2.86	0.09
reclamation system "Olshanka"									
0-20	5.3	13.81	20	86.2	3.42	0.54	0.11	2.20	0.14
20-40	5.6	9.05	25	90.9	3.12	1.14	0.03	2.42	0.13
40-60	5.7	6.56	25-30	93.4	2.97	0.93	0.06	2.14	0.19

- excess of precipitation over total evaporation.

The climate of the studied area is moderately warm and humid. The average annual air temperature is about +7°C. The winter period lasts almost 3.5 months and is relatively mild with an average monthly temperature of - 2.5 to - 5°C. The average annual precipitation is 634 mm. Most of them fall during the summer months. Spring and autumn frosts are observed, which does not allow the cultivation of warm-season crops.

The total area of the swamp is 17718 ha, it is of low-leveled type, peat of carex origin. The average thickness of peat is 2.5-3.0 g, and in some places 9 m. The reaction of the soil solution is close to neutral, the ash content is about 20%. The swamp was drained by open channels in 1928–1930. In 1949, the deepening of the main channel by 0.6 m was carried out. The swamp was often flooded with spring waters, mainly in April for 15-20 days. As an object of research in the Zamhlai swamp, we chose the Buriv experimental field, which has existed since the autumn of 1926, first as a branch of the Rudnia-Radovelska experimental bog station, and since 1932 as an independent point, transformed in 1935 into an experimental field.

In the Forest-Steppe zone, the object of research is the Romen swamp in the floodplain of the Romen River in the Sumy region, where the Sula research field of the Institute of Water Problems and Land Reclamation of the National Academy of Agrarian Sciences of Ukraine is located. The Romen swamp has an area of 12000 hectares. The banks of the river in the area of its location are steep, with deep beams and ravines. The bottom of the swamp consists of clayey Post-Tertiary layers. The average thickness of peat is 2.5-3.0 m, the maximum is 7.1 m. The ash content is high, more than 20%. Contains lots of shell impurities. The peat is weakly decomposed. The swamp is low-lying, grass-carex. The vegetation is dominated by carex, in some places by thickets of reeds and green mosses. The average amount of precipitation is 510 mm. In addition, the swamp is fed by the huge amount of spring waters, which is a characteristic feature of almost all swamps of the Forest-Steppe.

Panfyl experimental field (later experimental station) was organized in 1936 in the floodplain of the Supii river in the forest-steppe zone of the Kyiv region. Carbonate peat of varying thickness is underlain by light loam rich in nutrients. A periodically reconstructed drainage-humidifying reclamation system operates here. The swamp is not flooded and is fed exclusively by atmospheric and spring waters.

Changes in the Water-Physical Properties of Organogenic Soils Under the Influence of Anthropogenic Factors

As already noted, the "Zurno" and "Olshanka" reclamation systems were built in 1957–1959 by a system of open channels with a distance of 300 m between them. In 1978–1979, the main canals were deepened and the side network was cleaned, which until 1992 was carried out regularly according to the maintenance and repair plan of the operation department.

At the research sites, founded in 1962, the level of groundwater was constantly monitored. The researchers studied ways of their agricultural use as well. Until 1986, at the swamp where the experimental sites are located, fodder crop rotation was observed with the following set of crops: 1-4 – perennial grasses, 5 – root crops (mangel-wurzel, potatoes), 6 – winter ruttishness, 7 – corn for silage, 8 – barley for summer-autumn sowing of perennial grasses. In subsequent years, the drained soils on the "Olshanka" system were used haphazardly, and on the "Zurno" system – mainly for perennial grasses and pastures.

Groundwater levels in the systems depended mainly on meteorological conditions and varied within a fairly wide range that is from 38 to 78 cm in spring and up to 110-142 cm in summer in the drained Zurno bog and, respectively, 34-69 and up to 116 cm in the Olshanka bog.

The analysis of studies showed (table 2) that the filtration coefficient (C_f) of drained peatlands in the 0-30 cm layer on the "Zurno" system was within 2.1 m/day before the development, and significantly decreased under the influence of drainage and agricultural use. Already 8 years after development, it was 1.6, and after 27 years equaled 1.3 m/day. The same is observed in the "Olshanka" system, where the filtration coefficient decreased from 1.9 m/day in 1962 to 0.8 m/day in 2002. In the soil layer of 40-60 cm, the filtration coefficient changed quite little during this time.

Before the start of development, it was 1.9 m/day on the "Zurno" system, and 1.8 m/day on the "Olshanka" system. 40 years after the development of peatlands, it is 1.4 m/day in the "Zurno" system and 1.1 m/day in the "Olshanka" system.

Attention is drawn to a noticeable increase in density and a decrease in the maximum water capacity of peatlands under the influence of drainage and agricultural exploitation. The density of peat soil on the "Zurno" system before development in a layer of 0-30 cm was in the range of 0.138-0.166 g/cm^3, and the maximum water capacity was 656-572%.

Table 2. Changes in the water-physical properties of bog-peat soil in the process of development and agricultural exploitation

Soil layer, cm	Reclamation system									
	"Zurno"					"Olshanka"				
	density, g/cm^3		moisture-holding capacity		C_f, m/day	density, g/cm^3		moisture-holding capacity		C_f, m/day
	solid body	soil	Max WC, %	Min WC, % від Max WC		solid body	soil	Max WC, %	Min WC, % від Max WC	
before the development, 1962										
0-10	1.44	0.166	572	71.8	2.2	1.47	0.170	530	75.6	2.1
10-20	1.40	0.149	605	73.8	2.1	1.43	0.165	545	78.4	1.9
20-30	1.40	0.188	656	85.0	1.9	1.41	0.146	618	86.0	1.7
40-60	1.38	0.120	755	93.3	1.9	1.40	0.138	656	91.1	1.8
after 8 years after development, 1970										
0-10	1.51	0.198	455	62.6	1.6	1.53	0.212	416	69.7	1.4
10-20	1.46	0.189	478	66.4	1.5	1.50	0.204	438	71.3	1.2
20-30	1.44	0.151	596	79.8	1.7	1.47	0.172	524	81.3	1.5
40-60	1.40	0.129	706	67.2	1.9	1.43	0.144	627	89.1	1.7
after 27 years after development, 1989										
0-10	1.55	0.229	373	54.7	1.2	1.56	0.246	349	58.2	1.1
10-20	1.50	0.221	412	58.9	1.1	1.52	0.238	358	68.4	1.0
20-30	1.47	0.182	495	74.5	1.4	1.48	0.191	478	74.5	1.3
40-60	1.42	0.131	695	81.4	1.6	1.44	0.149	606	86.6	1.5
after 40 years after development, 2002										
0-10	1.61	0.247	348	52.1	0.9	1.60	0.273	312	54.9	0.8
10-20	1.55	0.234	364	54.2	0.8	1.56	0.265	328	57.2	0.7
20-30	1.48	0.194	465	64.9	1.1	1.53	0.231	369	73.4	0.9
40-60	1.45	0.137	662	95.3	1.4	1.47	0.152	592	80.1	1.7

Over the first 8 years, the peat density in the 0–30 cm layer increased by 0.013-0.032 g/cm^3 in the "Zurno" system and by 0.042–0.026 g/cm^3 on the "Olshanka" system, while the maximum water capacity decreased by 117–60 and 126–94% respectively. In the 40–60 cm peat layer, such changes were less noticeable. Peat density increased from 0.120 to 0.129 g/cm^3 on the "Zurno" system, and from 0.138 to 0.144 g/cm^3 on the "Olshanka" system.

Therefore, taking into consideration the changes in density and maximum water capacity in different time cycles, it can be seen that over the first 8 years, the peat density on the "Zurno" system in the 0-10 cm layer increased by 18.5%, 10-20 cm – by 26.8% and in the layer of 20-30 cm – by 9.4%. On the "Olshanka" system, respectively, for the same layers of peat by 24.7; 23.6 and 17.8%. In the peat layer of 40-60 cm, its density increased by 7.5% in the Zurno bog and by 4.3% in the Olshanka bog. In 27 years after development, the density of peat, in comparison with the previous one in the "Zurno" system, was the highest by 15.6-16.9%, and in the 0-30 cm layer only by 2%. After 40 years of agricultural use, peat compaction on the "Zurno" system reached 7.8-6.6% in the 0-30 cm layer and 4.5% in the 40-60 cm layer.

A similar dependence was observed on the "Olshanka" system. Soil density 27 years after development increased in the 0-20 cm layer by 31-47%, and after 40 years - by 58.6-60.6%, in the 40-60 cm layer - by 89%, respectively, after 27 years and by 10.1% after 40 years, compared to the density before draining.

The regulation of the water regime of peat bogs is a very important and complex process. The main goal is to create optimal water-air regime conditions for growing crops. As already noted, the Zamhlai swamp was drained in 1926-1930 by a system of open channels at a distance of 60 m between each of them. In 1952, scientific report gives data on some water-physical properties of the peat soil of the Buriv experimental field (Table 3). The depth of the peat layer is 2.3 m, the soil density in the 0-20 cm layer is 0.240 g/cm^3, the total moisture capacity is 355%, and in the 20-40 cm layer it is 0.178 g/cm^3 and 506%, respectively.

Table 3. Changes in the basic water-physical properties of peat soils of the Buriv experimental field, August 2002

Peat layer, cm	In 1952			In 1982			In 2002			
	peat thickness, m	density, g/cm^3	Max WC,%	density, g/cm^3		Max WC,%	peat thickness, m	density, g/cm^3		Max WC,%
				d	D			d	D	
0-20	2.3	0.240	355	0.268	1.85	318	1.74	0.321	1.88	261
20-40	-	0.178	506	0.216	1.46	408	-	0.284	1.71	301
40-60	-	-	-	-	-	-	-	0.233	1.35	366
60-80	-	-	-	-	_	-	-	0.192	1.31	470

Note. MaxWC – maximum water capacity; D – solid body density; d – soil sample density

After 30 years (in 1982) at the Buriv experimental field, the reconstruction of the drainage system and the soil survey were carried out. It was found that in the 0-20 cm peat layer the density was 0.268 g/cm^3, the maximum water capacity was 318%, and the solid body density – 1.85 g/cm^3, in a layer of 20-40 cm respectively 0.216 g/cm^3; 408% and 1.46 g/cm^3. According to the latest data, the thickness of the peat layer over 49 years has decreased by 56 cm, or by 24.3% of the original thickness of 2.3 m. If we extrapolate the peat triggering process according to these figures into the future, then the loss of peat per year reaches 1.14 cm.

An analysis of the studies carried out shows that the highest soil density is in the upper (in cultivated soils – in the arable) layer. With depth, its value decreases and in the lowest, the density is the smallest; a fairly significant increase is observed in the site that has been cultivated since 1932. The water capacity of the soil in the developed sites in the arable layer is much lower compared to the areas drained, but not developed. In deeper layers, this difference is somewhat levelled.

Thus, the term and intensity of the development of peat soils affect the change in their water-physical properties: with an increase in the terms of use, the degree of decomposition, ash content, and density increase, and the maximum water capacity decreases. It should be noted that even in an undeveloped but drained area, noticeable changes in water-physical properties occur over time.

In 1956 the depth of peat in this area varied between 181-197 cm, averaging 189 cm. A survey conducted in 2002 showed that the average thickness of the arable layer is 141 cm. This indicates that even in an undeveloped swamp, range of peat for 45 years has decreased by 48 cm, which is primarily due to the compaction of peat, as well as its mineralisation. Soil density increased markedly both in the arable and undersoil layers. A significant decrease in the maximum water capacity is observed (Table 4).

Table 4. The effect of the duration and intensity of development on the density and maximum water capacity of peat soil, Sula experimental field 2002

Peat layer, cm	Drained, not developed		Drained and developed					
			In 1932-1934		In 1949		In 1954	
	D, g/cm³	Max WC, %	D, g/cm³	Max WC, %	D, g/cm³	Max WC, %	D, g/cm³	Max WC, %
0-10	0.248	347	0.351	237	0.342	243	0.233	250
10-20	0.221	397	0.316	264	0.324	258	0.301	279
20-30	0.202	444	0.291	294	0.293	291	0.273	312
30-40	0.196	460	0.280	305	0.269	217	0.241	354
40-50	0.127	717	0.208	426	0.198	455	0.185	488
50-60	0.119	765	0.149	605	0.166	542	0.176	512
60-70	-	-	0.126	722	0.137	662	0.149	605

In 1946, 1956, 1958 conducted a survey of the swamp in the Romen River floodplain (Sula experimental field). A survey at the same points and an analysis of the indicators obtained by us indicate (Table 5) that over 68 years of use of peat soils, significant changes have occurred in their water-physical properties. The density of peat has almost doubled, respectively, the maximum water capacity has decreased. On the site, which was drained and began to be developed since 1949, stationary field experiments were carried out to study the effect of methods of agricultural development of high-ash peat soils on changes in their fertility and productivity of crop rotations and overall loss of organogenic mass.

Table 5. The effect of agricultural exploitation of peat soils on changes in its properties, Sula experimental field (The Romen River floodplain)

Peat layer, cm	At the beginning of development,1933		1956		2002	
	density, g/cm³	Max WC, %	density, g/cm³	Max WC, %	density, g/cm³	Max WC, %
0-25	0.188	497	0.305	0.278	0.348	238
25-50	-	-	0.156	0.850	0.319	261
50-75	-	-	0.145	0.623	0.292	292
75-100	-	-	-	0.284	0.301	-

Before laying the experiments in each crop rotation, a detailed measurement of the depth of peat was made by sounding. As can be seen from the studies (Table 6), on the drained floodplain peat soils of the Forest-Steppe, with the systematic use of tilled crops, the peat thickness annually decreased by 1.4 cm, while the loss of organogenic mass reaches 13 t/ha. In fodder crop rotations with 4-6 fields of perennial grasses, the peat thickness decreased by 1.03 and 0.8 cm per year, respectively. Losses of organogenic matter amounted to 9.5 and 4.7 t/ha per year, respectively.

On the site with the permanent cultivation of perennial grasses, the annual decrease in the depth of peat was 0.65 and the loss of organogenic mass was 3.8 t/ha per year. Attention should be paid to the fact that in terms of capacity, field crop rotations do not have a significant advantage over fodder ones.

An analysis of the observations of changes in the physical and agrochemical properties of peat soil in the site drained in 1949 demonstrated (Table 7) that over the first 19 years after the laying of drains and in the process of agricultural use, the density of peat in the arable layer increased from 0.213 to 0 .2 g/cm^3, and over the next 14 years up to 0.352 g/cm^3. The subsequent process of soil compaction slowed down, and over 10 years (1963-1992) the density of peat increased only to 0.372 g/cm^3 and, finally (1993-2002) it reached 0.384 g/cm^3. There are no data on the density of peat in the gift layer for 1950, but over the period from 1969 to 1983 it increased in the 30-50 cm layer from 0.215 to 0.234 in 1983, to 0.241 in 1993 and to 0.24 in 2002.

The site, drained and developed in 1954, is represented by low- and medium-thick peat. The average peat depth was 86 cm in 1954, 55 cm in 1973, and 51 cm in 2002. Consequently, over the first 19 years, the peat depth decreased by 31 cm (1.63 cm per year), over the next 28 years by only 4 cm (0.14 cm per year).

Data on the change in peat thickness show that in the field crop rotation saturated with tilled crops, the peat depth decreased by 45 cm over 32 years. In fodder crop rotations with 4 and 6 fields and permanent cultivation of perennial grasses, by 33, 26 and 26, respectively.

Table 6. The effect of the exploitation duration of peatlands and the type of crop rotation on the peat erosion (the site was drained in 1949), Sula experimental field (The Romen River floodplain)

Crop rotation	Peat thickness, cm			In the period of years; peat power reduction, cm			Annual loss of organogenic mass, t/ha	Crop rotation capacity, t/ha
	1969	1984	2002	15	18	the whole period (33 p.)		
Field	107-111	81-85	61-67	26	19	45	13,0	7,6
	109	83	64					
Fodder with four fields of perennial grasses	102-106	86-84	66-76	19	14	33	9,5	7,1
	104	85	71					
Fodder with six fields of perenni-al grasses	123-127	111-113	96-102	13	13	26	4,7	7,2
	125	112	99					
Permanent cultivation of perennial grasses	124-126	113-117	101-107	10	11	21	3,8	7,0
	125	115	104					

Note: top cell are the fluctuation limits, bottom cell is the average value.

Table 7. The effect of drainage and agricultural exploitation on the physical and agrochemical properties of peat, Sula experimental field

Peat layer, cm	Soil density, г/см³	Ash content, %	Labile forms, mg/100 g of soil		
			N	P_20_5	K_20
In 1950					
0-25	0.210	36.2	32.7	12.0	31.0
In 1969 p.					
0-30	0.293	39.7	30.2	22.4	37.9
30-50	0.215	27.1	17.4	18.3	24.2
In 1983					
0-30	0.352	50.9	32.1	25.4	39.1
30-50	0.234	30.4	17.4	17.2	19.4
In 1993					
0-20	0.372	42.2	28.2	20.4	11.0
20-40	0.241	32.1	17.3	8.2	11.0
In 2002					
0-20	0.379	46.4	24.6	21.7	12.3
20-40	0.249	135.5	18.4	10.3	9.4

Changes in the Agrochemical Characteristics of Peat Soils in the Process of the Development and Agricultural Exploitation

Our analysis of the agrochemical indicators of drained peatlands over many years shows (Table 8) that before development in the peat soils of the Zurno and Olshanka bogs in the 0–20 cm layer, the total nitrogen content was 3.42–3.51%, phosphorus - 0.37-0.54, potassium – 0,09-0,10%. The comparatively high total content of phosphorus in the soil attracts attention. On the "Zurno" system, an increase in the phosphorus content down the profile is quite noticeable (0.37, 0.85 and 1.24% for the indicated layers). Approximately the same soil characteristics were indicated on the "Olshanka" system (0.54-1.14-0.93% for the same sources). Evidently, the soils of the studied reclamation systems of the forest-steppe zone are provided with phosphorus that is more than average for the peat soils of Polesie. A similar characteristic was found in carbonate peatlands located in the Central Forest-Steppe of the floodplain of the Supii River (Panfyl Experimental Station) (Slyusar et al., 2020).

Studies on the research of fertilisation in the first years of development showed that the application of phosphorus fertilisers for various crops, depending on the doses of application and methods of soil preparation, increased the yield by an average of 27-51%, and potassic fertilisers provided a yield increase of 2.1-2, 4 times. Was found that the agricultural use of peatlands during the first 8 years caused an increase in the ash content of peat from 5.47 to 8.6%, in the 0-20 cm layer and from 8.12 to 9.3% in the 0-20 cm layer.

Table 8. Agrochemical indicators of the Zurno and Olshanka bogs peatlands and their change in the process of development, Sula experimental field

Soil layer, cm	pH	Degree of decomposition, %	Ash content, %	Organogenic matter, %	% from the mass of absolutely drained soil				
					N	P_2O_5	K_2O	CaO	MgO
1	2	3	4	5	6	7	8	9	10
The Zurno bog, 1970, 8 years after development									
0-20	5.4		8.6	79.9	3.84	0.39	0.14	2.81	1.32
20-40	5.3	26	9.3	82.7	3.49	0.87	0.18	3.44	2.44
40-60	5.6	25	6.1	90.6	3.19	1.12	0.09	3.08	2.16
The Olshanka bog, 1970									
0-20	5.2	25	14.8	79.4	3.97	0.84	0.17	2.04	1.72
20-40	5.4	25	12.4	83.0	3.52	0.93	0.06	2.55	1.87
40-60	5.7	32	6.6	89.7	3.12	12.16	0.09	2.47	3.16
The Zurno bog, 1989									
0-20	5.3	32	13.3	74.6	3.81	0.91	0.21	2.75	1.21
20-40	5.3	30	11.7	78.5	3.76	0.96	0.1-0	3.56	2.56
40-60	5.5	26	8.9	88.1	3.57	1.22	0.08	3.27	2.44
The Olshanka bog, 1989									
0-20	5,1	33	18.8	72.1	3.86	1.11	0.19	1.87	1.57
20-40	5.2	27	16.2	75.6	3.80	0.95	0.08	2.64	1.92
40-60	5.6	32	10.7	86.4	3.49	1.31	0.11	2.43	2.11
The Zurno bog, 2002									
0-20	5.2	34	21.4	73.2	3.92	0.96	0.27	2.31	0.41
20-40	5.2	32	18.5	75.8	3.24	1.11	0.223	4.15	0.54
40-60	5.6	28	12.3	86.3	3.76	1.23	0.09	2.10	0.73
The Olshanka bog, 2002									
0-20	5.2	37	27.1	71.3	3.97	1.13	0.28	1.87	0.32
20-40	5.0	34	23.2	74.5	3.92	0.94	0.20	1.66	0.26
40-60	5.6	34	16.1	86.1	3.87	1.26	0.11	2.16	0.37

More noticeable changes in peat soil occurred 27 years after the start of agricultural exploitation. The acidity of the soil slightly increased, remaining within the weak range (pH decreased from 5.4 to 5.1 in the 0-20 cm layer and from 5.3 to 5.2 in the 20-40 cm layer). Ash content doubled compared to 1970 and tripled compared to the beginning of development (1962). The degree of peat decomposition and the organogenic matter content increased as well. The mineralisation of peat organogenic matter and the application of mineral fertilisers significantly contributed to an increase in the content of total phosphorus.

Agrochemical indicators of peat soils change markedly over the years. The reaction of the soil solution in 1932 was neutrally mildly alkaline (pH water 7.2), and in 2001 it was mildly acidic – 6.2. As a result of the mineralisation of organogenic matter and the annual application of mineral fertilisers, the content of labile forms of phosphorus and potassium increased significantly. In 1949, the 0–20 cm layer contained 8 mg of labile PO_5 and 6.4 mg/100 g of K_2O; in 1982 their content increased in accordance with 15.5 and 11.1, and in 2001 up to 33.6 and 19.4 mg/100 g of soil.

Valuable results of soil survey of peatlands in the Buriv experimental field were obtained by various studies. Our evaluation of these observations over many years shows that the agrochemical indicators both in terms of the total content of nutrients and labile forms have changed significantly regardless of the method of development of peatlands.

The soils of the Buriv experimental field (Table 9) in 1949 had the following indicators: the depth of the peat layer was 3.2 m, the pH of the water extract was 7.2; the top layer of soil (0-27 cm) is yellow-brown, dusty and well decomposed. In the raw state, it was very smeared, and when compressed, a brown liquid was released from it. The remains of peat-forming plants were almost invisible to the naked eye. The peat layer of 27-32 cm retained its natural state, undisturbed by cultivation; the remnants of carex, hypnum moss and reeds are more noticeable in it.

Table 9. Agrochemical characteristics of the arable layer (0-20 cm) of peat soil of the Buriv experimental field, 1949

Total content, %							**Labile forms, mg/100 g of soil**		
N	P_2O_5	K_2O	CaO	MgO	AI_2O_3	Na_20	N	P_2O_5	K_2O
3.84	0.37	0.21	15.1	1.19	0.13	0.25	9.8	8.3	6.8

For 1952 some agrochemical properties of the soil of the Buriv research field are given, and after 30 years a new soil survey was conducted (Table 10).

Table 10. Agrochemical characteristics of the peat soil of the Buriv experimental field, labile forms of nutrients, mg/100 g of drained soil

Peat layer, cm	**1952**				**1982**				
	ash content, %	**N**	P_20_5	K_2O	**ash content, %**	**N**	P_2O_5	K_2O	Fe_2O_3
0-20	18.7	22.4	8.7	6.4	31.7	26.8	15.5	11.1	1.39
20-40	12.4	18.7	8.2	4.5	16.4	21.1	13.2	8.2	1.29

From the data obtained. it can be seen that it is difficult to establish a certain dependence on gross reserves. But compared to 1949, when in the soil layer 0-20 cm nitrogen content was 3.84 mg / 100 g, and in 2001 (Table 11) - 2.43 mg / 100 g of soil, indicating a decrease in nitrogen reserves, the content of movable forms of nitrogen, on the contrary, increased (19.8; 22.4; 26.8; 56.8 mg / 100 g, respectively, in 1949, 1952, 1982 and 2001). The same was observed in the layer of 20-40 cm (1952 - 12.7; 1982 - 21.1; 2001 - 39.2 mg / 100 g). Compared to 1949, in the soil layer 0-20 cm of phosphorus and potassium content in 2001 was found in three and twice as much.

Table 11. Agrochemical characteristics of peat soils of the Buriv experimental field, 2001

Peat layer, cm	Total content, %			Labile forms, mg/100 g			pH_{KCl}	Ash content, %	Degree of decomposition,%
	N	P_2O_5	K_2O	N	P_2O_5	K_2O			
0-20	2.43	0.31	0.24	56.8	26.7	14.8	6.2	37.5	27.2
20-40	2.27	0.29	0.21	39.2	20.4	24.5	6.1	39.4	25.8
40-60	1.86	0.25	0.18	22.4	10.1	11.1	6.2	22.7	17.9
60-80	1.81	0.17	0.09	13.2	17.4	8.4	6.4	20.1	-

The term and intensity of use of peat soils affect the change in both agrochemical and water-physical properties. With an increase in the period of use of drained peat soils, the content of nutrients, the degree of decomposition, ash content, density increase in them, and the maximum water capacity decreases.

A survey of peat soils in 1956 and 1993 showed that the first site has a deep peat bog of carex composition, underlain by alluvial gley loams. An analysis of long-term observations (Table 12) shows that more total forms of nitrogen, phosphorus, and potassium are currently found in the soil, but labile forms, on the contrary, even decreased compared to the content before the development of floodplains. The second site, was drained in 1932-1934. Before development it was covered mainly with carex and reeds. The site has a peat depth of 2.3-1.4 m, an ash content of 31.3-49.0%, and a degree of peat decomposition of 25-35%. Agrochemical characteristics of the site in 1958, shown in table 13.

Table 12. The effect of peat soil drainage on its agrochemical properties, drained and undeveloped site, Sula experimental field

Peat layer, cm	Ash content, %	pH	Total content, %			Labile forms, mg/100 g of soil		
			N	P	K	NO_3	P_2O_5	K_2O
				In 1956				
0-25	27.9	7.3	2.26	0.26	0.14	18.0	6.1	5.3
25-50	16.1	7.2	2.17	0.34	0.11	10.2	6.9	4.7
50-75	17.7	7.2	2.93	0.31	0.06	4.8	5.3	4.3
				In 1993				
0-25	29.6	6.8	2.43	0.29	0.23	4.21	5.6	3.1
25-50	19.7	6.5	2.28	0.20	0.16	2.07	2.0	6.5
50-75	18.2	6.5	2.56	0.12	0.09	1.51	2.7	4.0
				In 2001				
0-25	30.4	6.7	2.54	0.29	0.27	5.16	5.4	3.6
25-50	21.2	6.5	2.37	0.27	0.14	3.02	3.7	5.7
50-75	18.6	6.5	2.55	0.19	0.11	0.98	1.8	5.0
				In 2019				
0-25	32.9	6.8	2.55	0.30	0.28	5.09	5.5	4.7
25-50	21.9	6.6	2.38	0.27	0.15	2.99	3.6	6.1
50-75	18.8	6.5	2.59	0.21	0.14	0.98	2.1	5.2

Table 13. Agrochemical characteristics of the peat, Sula experimental field, drained in 1932-1934

Soil type	Peat layer, cm	pH of aqueous solution	Soil composition, %			Total forms, %			Labile forms, mg/100 g of soil		
			Humus	Plant residues	Mineral particles	N	P	K	NO_3	P_2O_5	K_2O
Peat (deep and high-ash)	0-15	7.6	50.5	17.5	32.6	2.19	0.37	0.16	35.0	7.9	5.1
	15-30	7.6	51.0	18.0	32.0	2.15	0.43	0.14	24.0	8.2	4.1
Idle field	30-40	7.6	59.5	18.9	21.6 ‘	2.20	0.40	0.02	18.3	8.6	5.2
	40-60	7.6	77.5	17.0	5.5	3.15	0.33	0.08	6.8	5.7	6.2
Peat (deep and high-ash, acquired for 13 years)	0-10	7.5	54.0	8.5	37.5	2.56	0.26	0.20	69.3	8.2	8.7
	10-20	7.5	59.0	8.5	32.0	2.10	0.31	0.15	53.6	8.5	2.5
	20-30	7.6	58.0	7.0	35.5	2.23	0.30	0.15	50.3	7.9’’	3.1
	31-61	7.6	77.0	19.0	4.0	3.09	0.05	0.05	38.4	10.0	-
	61-100	7.6	77.5	19.0	11.5	3.23	0.03	0.03	8.6	6.6	3.4

Studies have established (Table 14) that over 60 and 68 years of use of peat soils there have been significant changes in their agrochemical properties. The ash content of peat in the upper layer of 0-20 cm increased by almost 1.5 times, the degree of decomposition by 12.7%. The reserves of total forms of nitrogen, phosphorus and potassium increased as a result of the application of mineral fertilisers and the mineralisation of peat organogenic matter. The reaction of the soil solution from alkaline changed to mildly acidic. The content of trace elements in these peats, which was as a percentage for a completely dry sample: $CaC0_3$ – 4.5-18.7; SO_4 – no more than 0.065; Cl – 0.009-0.150; SiO_2 – 12.8-85.5% of ash.

On the site that was used for stationary field experiments, it was found that the ash content of peat in the 0-20 cm layer increased from 36.2% in 1950 to 46.4% in 2001, and in the 20-40 cm soil layer, respectively 27.1% in 1969 to 35.5% in 2001. With regard to labile potassium compounds, there is no clear pattern over the years of development, but the amount of phosphorus and potassium increased markedly from 1950 to 1983. The availability of total nitrogen in 2001 significantly decreased compared to 1950. A similar dependence was also observed in site 4, where the field was drained in 1954 (Table 15).

As for the main elements of plant nutrition, their content has changed markedly. If in 1954 the total nitrogen in the peat layer 0-20 cm was 2.86%, and in the layer 25-50 cm 2.71%, then after 39 years of use, in 1993, the content was respectively 2.48 and 2.30, and in 2001 – 2.41 and 2.34%. The content of total phosphorus and potassium almost doubled, but the content of labile forms of nitrogen, including its nitrate form, NO_3, decreased significantly. If in 1954 its content was 18.4 and 9.2 mg, then in 1993 and 2001, respectively, 1.13-2.81 mg and 2.11-2.76 mg/100 g of soil. In general, peatlands are poor in potassium.

Observations show that even in old-arable areas, despite the annual application of high doses of potash fertilizers, the effectiveness of their use remains quite high, which has a different explanation (Tsiupa, Bystrytskyi, Slyusar, 1990). Labile forms of potassium on peat soils make up 40-80% of the total reserve and generally characterise the state of the potassium regime. Observations showed (Table 16) that the potassium content in the studied samples is quite low and ranges from 25.6-56.5 mg/100 g of soil. More of it is in the soil, which has been acquired for more than 40 years. This is due to high doses of potash fertilisers, as well as a more intense degree of mineralisation of peat organogenic matter.

Table 14. Changes in the main agrochemical properties of peat soils in the process of development, site No. 2

Peat layer, cm	Ash content, %	Decomposition degree, %	pH	Total content, %			Labile forms, mg/100 g of drained soil		
				N	$P_2 0_5$	K_2O	$N0_3$	$P_2 0_5$	K_2O
In 1933									
0-25	31.3	25.0	8.2	2.58	0.21	0.10	21.0	6.3	3.2
In 1956									
0-25	32.6	31.1	7.5	2.19	0.37	0.16	35.0	7.9	5.1
25-50	20.9	20.9	7.2	2.15	0.43	0.14	24.0	8.2	4.1
50-75	15.5	14.3	6.5	2.20	0.40	0.02	18.3	8.0	6.2
In 1993									
0-20	44.6	37.2	6.5	2.67	0.39	0.47	4.23	11.0	8.0
20-40	37.9		6.5	2.37	0.37	0.35	4.08	9.1	8.0
40-60			6.4	2.76	0.20	0.15	1.77	5.9	6.7
60-80			6.4	3.58	0.10	0.10	2.63	8.8	6.7
In 2001									
0-20	46.3	37.7	6.5	2.64	0.38	0.32	5.06	10.6	6.6
20-40	38.5	29.4	6.4	2.47	0.44	0.39	3.4	9.7	6.2
40-60	21.1	-	6.6		0.27	0.19	2.1	5.8	5.1

Table 15. The effect of durability of drained peatland on its agrochemical characteristics, experimental site of the bog Nº4, Sula experimental field

Peat layer, cm	Ash content, %	pH of aqueous solution	Total content, %, dry sample			Labile forms, mg/100 g of drained soil		
			N	P	K	NO_3	$P_2 0_5$	K_2O
In 1954								
0-25	20.7	7.3	2.86	0.18	0.16	18.4	9.1	6.0
25-50	16.2	7.2	2.71	0.11	0.07	9.2	7.8	4.7
In 1993								
0-20	32.3	6.6	2.48	0.30	0.35	2.81	6.9	5.0
20-40	27.0	6.8	2.30	0.22	0.20	1.13	4.9	12.5
In 2001								
0-20	33.0	6.5	2.41	0.32	0.30	2.76	7.4	6.1
20-40	29.7	6.8	2.34	0.20	0.22	2.11	5.7	9.8

Table 16. The effect of the agricultural exploitation durability on the potassium group composition in peat soil, system "Zurno"

Durability	Soil layer, cm	Potassium content, mg/100 g soil				% available potassium from total	Potassium reserves, kg/ha
		water-soluble	exchangeable	non-exchangeable	in total		
1961p (beginning)	0-30	1.8	6.2	17.6	25.6	25	36
	40-60	4.1	15.9	12.5	32.7	18	42
1970 (cultivated for 9 years)	0-30	4.6	11.2	20.6	36.4	32	112
	40-60	3.0	8.7	28.1	39.8	24	43
2001 (cultivated for 40 years)	0-30	9.1	39.4	8.0	56.5	69	389
	40-60	8.7	32.3	3.5	44.5	72	137
Drained. not developed	0-30	2.1	7.7	18.2	28.0	26	69
	40-60	4.4	19.6	12.0	36.0	44	86

The same pattern is observed for the exchange form of potassium. At the same time, the non-exchangeable form of potassium is the smallest in the site of long-term agricultural exploitation, and the largest in the undeveloped and used in culture for nine years. If the share of non-exchangeable potassium in the site of long-term use accounts for 9%, then at the beginning of development – 54% and in the undeveloped area – 56%. This evidences that the mineralisation of peat (its earthing) does not contribute to the fixation of potassium by the colloidal complex into a non-exchangeable (inaccessible) form for plants, but, on the contrary, promotes its release from peat-forming plants and transfer to more labile forms that become easily accessible for cultivated crops.

The percentage of the available potassium in the soil in the site of long-term use reaches 71%, and in sites in the first years of development and under grasses with weak peat mineralisation, labile forms of potassium barely reach 25-26%.

Thus, the intensity of agricultural production does not worsen the potassium regime of peat soils, and with scientifically substantiated doses of mineral fertilisers, high and stable yields can be obtained here. Naturally, the natural soil reserves of potassium cannot ensure high yields, therefore, enhanced annual potash fertiliser is necessary. Against the background of $K_{120\text{-}150}$ application, more than 10 t/ha of hay or 3-4 t/ha of potatoes can be obtained.

On poorly cultivated areas under natural hayfields and pastures, potassium is 2-3 times less than under tilled crops. In contrast to the labile forms of phosphorus, the seasonal dynamics of potassium is quite noticeable and ranges from 7.3 to 67.4 mg/100 g of soil. Its decrease was noted during the growing season from spring to autumn.

The reaction changed more in the alkaline peatlands of the Forest-Steppe. By the beginning of the development of the peat bog of the Sula experimental field in the floodplain of the Romenka River in 1933, the soil reaction in the 0-25 cm layer was 8.2, after 23 years, in 1956, it decreased to 7.5, after 60 years, in 1993 to 6.5 and after 70 years, in 2003 to 6.5. In the 25-50 cm layer, the peat soil reaction changed from 7.2 in 1956 to 6.5 in 1993, to 6.5 in 2003 and to 6.8 in 2019. Even deeper (50-75 cm, 60-80 cm) peat reaction almost did not change, remaining alkaline, which is due to the carbonate content and composition of groundwater.

As is known, the intensity of mineralization and physical dispersion of peat depends on the degree of drying, the method of cultivation and the crops grown. The highest mineralisation of peat occurs under tilled crops during deep annual plowing and intensive soil draining. Studies have established that under perennial grasses, mineralisation gradually slows down and already at 4–5 years the amount of labile nitrogen compounds in the soil is sharply reduced (Table 17).

Accordingly, the productivity of drained soils significantly depends on the method of their use. The highest productivity was achieved when using organogenic soils in a grass-field crop rotation with two annual crops for the application of mineral fertilisers – about 7 tons of dry mass per 1 ha, which is almost 1.6 times higher than the productivity of natural hayfields and pastures. At the same time, the most productive (more than 7.5 tons of dry matter per 1 ha) turned out to be a crop rotation with full fertiliser with a set of such crops: perennial grass mixtures 1-6 field + in the last year after the second slope, oil radish or green manure mustard; field 7 – annual culture; Field 8 – grain spring crop + summer sowing of perennial grass mixtures.

When applying phosphorus-potassium fertiliser, the highest productivity of crop rotations (8.69-11.5 tons of drained weight per 1 ha) was for the cultivation of tilled crops, but in such a crop rotation there was an intensive mineralisation of peat, which was almost two times higher compared to grass-field crop rotations.

Based on the results of the above studies, for peat soils with a well-regulated water regime in Polisie, environmental conditions develop according to the structure of sown sites: perennial grasses – 65-70%, cereals – up to 10%, potatoes and vegetables – up to 5%, silage and annual grasses – 15%, and in 7-9-field crop rotations, perennial grasses should occupy 5-7 fields, in addition, it is advisable to sow intermediate crops in order not to leave the soil without vegetation during the warm growing season.

Table 17. Change in the characteristics of the peat soil layer 0-30 cm, depending on the method of its use, floodplain of the Supii River, Panfyl Experimental Station

Method of use	pH aqueous solution, 2015	Ash content, %, 2015	Content of nutrients total forms, %						Biotic soil activity, %, 2015
			In 1982			In 2015			
			total nitrogen	P_2O_5	K_2O	total nitrogen	P_2O_5	K_2O	
Grassland crop rotation	7.6	61	2.5	0.5	0.1	1.6	0.7	0.1	73.5
Permanent perennial grasses cultivation	7.5	55	2.5	-	0.1	1.8	0.6	0.1	55.7
Birch wood	7.7	27	-	-	-	2.0	0.5	0.2	-
Haymaking with periodic reloading	7.6	47	2.3	0.4	0.1	1.7	0.7	0.1	47.9
Herbage	7.7	45	-	-	-	2.0	0.6	0.1	-
Natural waterlogging	7.8	34	-	-	-	2.1	0.4	0.1	-

Biological Activity of Drained Peat Soil Depending on the Intensity of its Exploitation

The study of biological processes occurring in peat soils is one of the main methods for diagnosing their agro ecological state. It is important for assessing the effectiveness of the farming system and individual agro technical measures in the cultivation of crops. The total biological soil activity was determined by the method of applications from flax fabric, laid at a certain depth, with an exposure of 30 and 60 days. The essence of this method lies in the quantitative determination of the intensity of decomposition of cellulose in the soil for a certain time.

There are various alternatives of the described method, but the field method can be considered preferable, which makes it possible to study the decomposition of cellulose along the soil profile without digging a hole. As follows from the data given in Table 18, tillage has a great influence on the intensity of biological processes. The rate of cellulose decomposition during plowing is higher compared to the disk. If by plowing under winter ruttishness the percentage of decomposition of linen fabric a month after laying was 21.6%, then by disking it was 19.1%, and after two months it was 28.3 and 23.5%, respectively. The same phenomenon is observed 60 days after laying the flax fabric.

Table 18. Destruction of cellulose under different agricultural crops in the 0-30 cm arable layer of peat soil, Sarny experimental station, 2003

Agricultural crops	Loss of tissue mass, %			
	in a month		in 2 months	
	plowing	disk plowing	plowing	disk plowing
Winter ruttishness	21.6	19.1	28.3	23.5
Barley	23.2	18.4	39.0	25.4
Corn for silage	59.3	45.5	64.8	60.1
Perennial grasses of the 4th year of cultivation	18.9	14.7	21.6	18.7
Permanent cultivation of perennial grasses	13.7	13.7	15.9	15.9

Quite noticeable is the difference in the intensity of cellulose decomposition during various treatments of peat soil under corn. Here, after a month, the degree of tissue decomposition for plowing reached 59.3%, and after two months – 64.8%; with surface treatment, respectively, 45.5% after a month and 60.1% after two months.

If we analyse the results of microbiological processes under different crops, then their slowest activity was on the site with permanent cultivation of perennial grasses (13.7% per month and 15.9% in two months). Under perennial grasses of the fourth year of use, the corresponding figures were 14.7% and 21.6%, they also depended on the methods of soil preparation for perennial grasses. This indicates that the introduction of perennial grasses into crop rotation reduces the rate of cellulose destruction and the rate of mineralisation of peat organogenic matter.

The ratio of the main components of humic substances changes towards an increase in more condensed forms such as humic and ulmic acids. The largest number of them was noted under herbs of a long period of use (unchanging cultivation of herbs); in crop rotations with field crops, their content is reduced by 3-8%.

A quantitative measure of the type of humus is the ratio of the carbon content of humic acids to the carbon content of fulvic acids (C_{ha}:C_{fa}). The data indicates that all types of humus in cultivated peatlands are classified as fulvate-humate type (C_{ha}:C_{fa} within 0.8-1.2). Interestingly, in the arable layer of 0-30 cm, this indicator somewhat decreases compared to the gift layer, which is explained by increased aeration of the former.

Therefore, for peat soils used only under perennial grasses, the quantitative expression of the humus type is lower than for crop rotations where row crops and cereals are grown with herbs.

Nutrients Migration Depending on Geochemical Conditions, Drainage, and Exploitation

According to the results of studies (Tsiupa, 1990; Slyusar and ets., 2020) regarding the background hydro-chemical situation within the humid zone of Ukraine, both surface and fresh groundwater have a mineralisation below 1 g/l. Moreover, the contained salts are predominantly bicarbonate-calcium. However, depending on the geostructural and geomorphological features of the territory, there may be certain differences in the hydrochemical composition. For example, in Chernihiv Polesie, the second anion in the composition of water is mainly chlorine, and in Volyn Polisie, both chlorine and sulphate ion are equally divided. Among the cations in the Chernihiv Polesie are sodium and calcium equally. In the Volyn region there are calcium and magnesium predominate. In Zhytomyr Polesie, calcium and bicarbonate ions predominate in the composition of water, while sulphate ions and chloride ions, as well as sodium and magnesium, are in small quantities.

At the Sula experimental field, drainage water was sampled on high- and medium-ash peat soils under tilled crops and perennial grasses from the main canal, the Romen River, in spring, summer and autumn. Studies have shown that in peat soils under tilled crops, drainage water contains from 135 to 207 mg/L of dissolved substances. Noticeably less is taken out from the soil under perennial grasses 99-114 mg/L. It should be noted that we observed a similar pattern in the floodplain of the Supii River (Forest-steppe), but with significantly higher rates (Tsiupa, 1990; Kurhak et al., 2020) and in the water of the main drain, the bicarbonate ion, calcium, chlorine and sulphate ions predominate, to a lesser extent magnesium, sodium, potassium, nitrogen and phosphates. It should be noted that the amount of phosphates in both the surface and drainage waters of the main drain is very small. This is due to their low solubility, on the one hand, and significant absorption of phosphorus by mineral colloids, on the other.

The observations made indicated (Table 19) that bicarbonate, calcium, phosphates, chlorides, and nitrogen are most washed out by water. However, interestingly, the drainage itself does not greatly affect the increase in salts in the water and their migration. The total amount of salts in the drainage water of a drained and undeveloped bog increases by 23.1-30.2% compared to a non-drained bog, depending on the time of year. At the same time, an increase is observed for all determined components. However, agricultural development has the greatest effect on increasing the salinity of drainage waters, it includes plowing, fertilising, and enhanced decomposition of peat.

Table 19. Chemical composition of groundwater and peat soils at different intensity of their development, Sula experimental field, 2003

Sampling interval	Content, mg/L of water										
	Ca	M_8	K	Na	MH_4	$N0_3$	HCO_3	$S0_4$	CI	PO_4	Сума
Non-drained peat bog											
Spring	18.1	1.3	0.2	2.0	1.2	0.3	59.9	7.7	3.1	0.11	93.8
Summer	16.2	1.9	0.3	1.5	1.x	0.6	37.4	11.2	2.6	0.07	73.8
Autumn	21.6	1.6	0.5	1.8	1.5	0.4	40.1	9.2	3.5	0.06	79.3
Drained, undeveloped peat bog											
Spring	23.8	2.2	0.5	2.4	2.7	0.4	67.9	13.8	4.7	0.08	115.5
Summer	21.7	1.9	0.5	2.5	1.7	1.6	35.2	22.7	8.2	0.07	96.1
Autumn	21.8	1.8	0.5	2.0	0.5	0.2	40.4	9.8	3.5	0.08	-
Drained, developed peat bog											
Spring	31.2	4.5	7.4	4.4	4.5	2.7	88.5	24.1	14.9	0.16	182.4
Summer	41.4	3.9	3.5	4.9	3.2	0.5	67.8	27.7	22.6	0.10	174.7
Autumn	72.3	3.5	2.8		1.1	1.6	74.3	28.8	12.9	0.09	173.3

As noted earlier, in peat bogs occupied by perennial grasses, the content of chemical elements in drainage water is noticeably lower than in the area occupied by mangel-wurzel. The obtained data correlates with the data of other researchers (Slyusar et al, 2021). They found that the amount of salts in the water of an undrained swamp decreases in spring (up to 65 mg/L), drained – up to 115 mg/L, drained developed – up to 177 mg/L, and in autumn their content was respectively 78. 91 and 165 mg/L.

Soil Conservation Measures

As a result of drying and subsequent agricultural exploitation, organogenic soils significantly change their properties and become vulnerable to the development of degradation processes, especially where the recommended structure of sown areas, crop rotations, optimal cultivation systems and water regimes are not adhered to. The first place in the series of these destructive processes in terms of the scale of manifestation and consequences is occupied by wind erosion, which on drained peat soils manifests itself in the form of dust storms mainly in the spring, when the soil is not yet protected by vegetation, and the upper dried layer is loosened by processing Less often, dust storms occur in summer and autumn, when soil is cultivated for sowing perennial grasses and winter crops (Kurhak et al, 2020; Bohovin et al, 2005; Slyusar, 2013; Konishchuk, 2015; Furdychko, O., 2010).

The main measures are the use of shallow peatlands exclusively for crops of perennial grasses with replanting every 5-7 years. In crop rotations on medium and deep peatlands under perennial grasses, 50-70% of the area should be occupied with a short field period (one-two years) and with minimal plowing.

In all cases, the recommended crop rotations, systems of differentiated tillage and optimal water regime should be observed. Fields with perennial and annuals grasses, including tilled crops, should be placed in strips across the direction of the prevailing winds. Along the permanent boundaries of the fields, as well as across the backwater, it is necessary to arrange plantings of windbreak trees in the form of two-row strips with a distance between the strips of 600-1000 m. (Slyusar, Solyanik, 2005; Truskavetsyi, 2010).

To prevent water erosion on river floodplains, it is better to cultivate the soil in spring across the direction of flood waters. Fires can pose a great danger to peatlands, especially overdrained ones. To protect against fire, fire-fighting forest belts 20-100 m wide are planted, as well as fire-fighting reservoirs with a volume of water sufficient to continuously extinguish the fire for three days. Reservoirs should be placed so that they separate peatlands from settlements, railways, electrical substations and other important facilities.

CONCLUSION

Therefore, in this chapter, based on the research, observations, and analysis of literary sources and archival materials, the authors draw the following conclusions:

As a result of drainage and agricultural exploitation of peatlands, the activated soil-forming process, accompanied by a change in their qualitative and quantitative properties: the degree of decomposition, ash content, density, total reserves of KPK, and the water-reducing and water-holding capacity decreases.

The nature and intensity of changes in the properties of peat soils depends on the length of cultivation and how they are used.

The most intensive compaction of peat (33% of the initial) and the wear of organogenic matter (13.1 t/ha per year) occur under tilled crops and less under perennial grasses respectively 12% and 3.9 t/ha per year. An intermediate place among these indicators is occupied by peat, which was used in grass-field crop rotations with one or two tilled fields, one field of annual and 4-5 fields of perennial grasses (20% and 4.9 t/ha per year).

The regulation of soil processes by ameliorative and agrotechnological methods in order to preserve peat soils and increase fertility should become the basis for their rational use. Changes in the agrochemical properties of organogenic soils under the influence of agricultural use are more intense in the Forest-Steppe zone than in Polisie.

Further research. Long-term studies of the transformation of waterlogged floodplain lands after drainage work require subsequent long-term monitoring observations. Especially environmental protection, taking into account global warming, would prevent pollution of ground and river waters, help reduce greenhouse gas emissions and would provide for the conservation and conservation of peatlands, the development of biodiversity and recreational areas.

REFERENCES

Bohovin, A., Slyusar, I., & Tsarenko, M. (2005). Herbaceous biogeocenoses, their improvement and rational use. K.: Agrarian science.

Furdychko, O., Lavrov, O., & Konishchuk, V. (2010). Agroecological aspects of environmental protection on the basis of sustainable development. Agroecological journal, pp. 5-10.

Himbarzhevsky, V. (1971). Cultural works on drained lands. *Visn. c.-g. science, 8*, 28-32.

Konishchuk, V. (2015). Concept and strategy of balanced development of wetland landscapes and peat ecosystems of Ukraine. K.

Kurhak, V., Panasyuk, S., Asanishvili, N., Slyusar, I., & Slyusar, S. (2020). Influence of perennial legumes on the productivity of meadow phytocenoses. *Ukrainian Journal of Ecology, 10*(6), 310–315. https://www.webofscience.com/wos/woscc/full-record/WOS:000608622800047. doi:10.15421/2020_298

Shmatok, V. (1994). Qualitative changes in organic matter of dried peat soils under the influence of agricultural use. *Reclamation and water management, 80*, 39-40.

Slyusar, I. (2013). Conceptual foundations of environmental use of land and water resources of the gumid zone of Polesie. In V. Kaminsky (ed.) Adaptive farming systems and modern technologies are the basis of rational land use of soil conservation and reproduction. VP Edelweiss.

Slyusar, I., Solyanik, O., Serbenyuk, V., Viryovka, V., & Tarasenko, O. (2020). Effect of the water regime, crop rotation and fertilizers in biogenic matters leaching into ground water and surface water. *Ukrainian Journal of Ecology, 10*(3), 197–200. https://www.webofscience.com/wos/woscc/full-record/WOS:000576691500013. doi:10.15421/2020_154

Slyusar, I., Solyanik, O., Serbenyuk, V., Zadubinna, E., & Perets, S. (2021). Energy crops safe cultivation on drained organogenic soils. *Ukrainian Journal of Ecology, 11*(1), 415–418. https://www.webofscience.com/wos/woscc/full-record/WOS:000631814800042. doi:0.15421/2021_60

Slyusar, I., & Solyanyk, O. (2005). *Features of the farming system on the drained lands of the gumid zone of Ukraine: problems, solutions. Zb. Ecology: problems of adaptive landscape farming*. State Ecological University.

Truskavetskyi, R. (2010). *Peat soils and peatlands of Ukraine*. City Printing.

Tsyupa, M., Bystrytsky, V., & Slusar, I. (1990). Agriculture on drained lands. Harvest..

Yangol, A. (1970). *Two-sided regulation of the authorities for drainage. M.* Colossus.

Yatsyk, A. (2004). Water management ecology. Vol. 4. K. *Genesis (New York, N.Y.)*, 325–339.

Chapter 4
Formation and Predictive Assessment of Drained Lands Water Regime of Ukraine Polesie Zone

Nataliia Prykhodko
https://orcid.org/0000-0003-1424-2628
National University of Water and Environmental Engineering, Ukraine

Lyudmyla Kuzmych
https://orcid.org/0000-0003-0727-0508
Institute of Water Problems and Land Reclamation NAAS, Ukraine

Roman Koptyuk
https://orcid.org/0000-0002-7086-3608
National University of Water and Environmental Engineering, Ukraine

Anna Kuzmych
National University of Water and Environmental Engineering, Ukraine

ABSTRACT

This chapter considered qualitatively different approaches to the water regime model of drained lands implementation. The fundamental difference of these approaches is that the regime forecast calculations on a long-term basis are performed for schematized natural reclamation conditions by typing meteorological regimes in the calculation periods of vegetation, appropriate schematization, and parameterization of water regulation of drained lands. Secondly, it is proposed in the framework of the balance equation to perform based on certain patterns, to calculate the process of water regime based on the relationship and interdependence of its elements at the boundaries and in connection with the process of redistribution of moisture within the active layer of soil in relation to its moisture content and storage capacity. Considered methods and models of water regime allow determining both potentially possible due to meteorological and agrotechnical conditions and effective values of irrigation norms in the development of design modes of irrigation by irrigation and subsoil at drained lands.

DOI: 10.4018/978-1-6684-8248-3.ch004

BACKGROUND

In general, at the territory of Ukraine and the Polesie zone, in the twentieth century there was a general tendency to increase air temperature and increase the amount of precipitation: the annual temperature increased by 0.3…0.7° C, and precipitation – by 50…100 mm, significantly (1,5 times) increased the frequency of significant anomalies of both indicators. Their intra-season variability has also increased.

According to all climate change scenarios, the amount of precipitation will also increase. In some seasons, this increase may exceed the existing level by 20% (Romashchenko, Rokochynsky, Halik et al., 2007). In the warm period of the year, as a result of a sharp rise in air temperature, much of the precipitation will be spent on evaporation. The main feature of the warming period is the constant uneven precipitation in the middle of the year and in some years.

At the same time, due to global warming, the zone of Polesie feels and will feel the greatest impact during the winter and spring seasons (Shawcroft, Lewon, Allen et al., 1974).

For drained areas with close groundwater, climatic conditions are directly involved in the formation of soil water regimes, both in the natural state and in certain technological periods of cultivation.

The projected increase in aridity in climate change (rising air temperatures, uneven distribution of rainfall in the warm season, inefficient accumulation of moisture in the soil, etc.), which are observed today, will inevitably lead to changes in the formation of water regime of drained lands agrolandscapes of the Polesie zone, in particular, due to the reduction of the natural moisture supply of the territory and the increase of potential evaporation and, accordingly, the total water consumption in the cultivation of crops on drained lands.

The shortage of natural moisture will require additional moisture for cultivated crops through the implementation of appropriate water regulation technologies on drained lands.

The direction and degree of impact on the natural water regime in the reclamation of excessively moist lands should meet the solution of the main task – to create through the implementation of hydromelioration measures a favorable regime of moisture of the active root layer of soil in compliance with environmental requirements.

When regulating the water regime of drained lands, there are mainly two often separate or periodically changing processes – drainage and wetting, which use interrelated, but at the same time independent principles and techniques of designing reclamation systems according to relevant technical solutions.

Drainage is intended to ensure the removal of excess water from wetlands during spring floods or rain floods. For this purpose, as a rule, hydrodynamic methods are used to calculate the movement of groundwater to drains (water intakes) and hydraulic calculations of parameters of drainage lines, canals, rivers, etc., which determine the position of groundwater level (GWL) in critical periods of wetting.

When reclamation of wetlands (overwet lands and periodically overwet lands), such periods are the most dangerous in the development of these lands for use in agricultural production, and hydrodynamic methods of calculating the movement of groundwater detail the regime of maximum allowable wetting of agricultural reclamation fields in these although very responsible, but relatively short periods. and floods.

Since the outlined issues form the basis of drainage reclamation, the appropriate methods of hydrodynamic and hydraulic calculations to determine the parameters of the regulatory and conductive network of drainage systems are currently developed for almost all major soil reclamation conditions at the appropriate scientific level, tested in practice. in the relevant special and regulatory and technical literature and we are not considered.

According to another, a much longer period of vegetation of cultivated crops on drained lands, soil moisture is determined by a much more complex set of factors that include meteorological, agrotechnical, soil reclamation, and other local conditions of water regime formation.

In this case, taking into account local hydrological characteristics and processes of formation of profitable and expendable elements of moisture exchange, which characterize the direction and intensity of both biological and geological cycles and, accordingly, the conditions of formation of water and general natural reclamation regimes of drained lands. due to the fact that they are mainly determined by the hydrometeorological regime of current conditions or typical (calculated in terms of humidity) growing seasons and can not be considered arbitrarily, regardless of natural conditions.

Due to significant stochastic variability of hydrometeorological elements in intra-vegetation and perennial cross-section on drained lands, periods of waterlogging alternate with periods of lack of soil moisture, so to establish the regime of humidity as a basis by forecasting) water regime during the growing season, ie obtaining characteristics of changes in the time of GWL, soil moisture, moisture exchange, etc.

This regime should be determined and designed on the basis of detailed calculations on the set of relevant regime characteristics of profitable and expendable elements of the water balance due to their change during the growing season. The regularities of their change, in turn, are due to both reclamation and, most of all, zonal and azonal factors of general physical and geographical processes of the region where the object is located (Shebeko, Mozheem, Kiselyova, 1980).

Such requirements are sufficiently met by theoretically substantiated and tested in practice methods of water balance modeling, which are used to solve various problems that arise in the design and operation of reclamation systems. In recent decades, modeling the balance of soil moisture has received considerable attention both in our country and abroad (Afanasik, Armonik, Skvortsov, 1976; Afanasik, Zhukov, 1983; Ostapchik, Kostromin, Koval et al., 1989; Lazarchuk, 1992; Maslova, 1985; Aidarov, Golovanov, Nikolsky, 1990; Skripnik, Soroka, Kubyshkin, 1992; Hanks, Ashcroft, 1985; Shebeko, 1983; Shebeko, Mozheem, Kiselyova, 1980; Yangol, 1970; Yangol, 1990).

The use of water balance models provides an opportunity not only to assess, but also to predict (predict) the impact of one or more factors on soil moisture conditions within a particular reclaimed field and the system as a whole.

In particular, in Ukraine O. Skrypnyk with colleagues were actively involved in the development, implementation and development of the water balance method for forecast assessment of water regime, parameters of technologies and modes of water regulation of drained lands, etc., starting with A. Yangol (Yangol, 1970; Yangol, 1990), A. Yatsyk (Yatsyk, 2001; Yatsyk, 1997), M. Lazarchuk with colleagues (Lazarchuk, 1997; Lazarchuk, 1992) and many other researchers.

They have made a significant contribution to the application of the balance method in the general theory and practice of drainage reclamation in the area of sufficient and unstable moisture.

Despite the fact that many models of water balance that exist today are generally developed at a sufficient level and are widely used in practice, they are constantly being improved. This process is due to the initial imperfection of our ideas about reclamation systems as complex natural and technical systems and the need to find a permanent compromise between the practicality of the model, the available source information and the level of tasks.

Currently, there are mainly two approaches to calculating the water regime of soils according to the relevant models of soil moisture balance (Lihatsevich, 1991).

The first one is to find numerical solutions of differential equations of moisture conductivity – the differential form of water balance (DFWB). The second one – in the discrete use of the equation of water balance of the active (root-containing) layer of soil (or aeration zone as a whole) – an integrated form of water balance (IFWB).

The first method of priming on the principle of the greatest full consideration of the mechanism of movement of soil moisture along the soil profile and over time, so from a theoretical point of view it is the most reasonable.

Starting with E. Childs and S. Averyanov, a significant contribution to the development of DFWB of agricultural reclaimed fields was made by G. Afanasyk, O. Budagovsky, O. Golovanov, O. Oliynyk, V. Polyakov and others (Budagovsky, 1981; Oleynik, 1985; Oliynyk, 1993; Oleynik, Polyakov, 1987).

But this method has not been widely used in practice due to the significant difficulties in developing an algorithm for its implementation.

Given the additional assessment of the effectiveness of DFWB and IFWB, performed by A. Likhatsevich (Lihatsevich, 1991), a necessary condition, performing as effective as possible use of DFWB, is the necessary accuracy of the components of the water balance in accordance with their size, taking into account their dynamic properties.

This condition is that in addition to the initial values of the elements of the water balance and their value at the end of time (which is required for traditional IFWB), there must be a message with the necessary accuracy of their dynamics over time. And this, in this case, you need to know the analytical relationship of water balance elements with the coordinate time.

This is probably the main difficulty, which can not be used in principle due to the stochastic nature of natural processes for detecting elements of water balance. Indeed, such a defining component of the water balance as precipitation is discrete in time for its nature, ie either they are or they are not. has the same argument can be attributed to watering when building land.

Therefore, based on the fact that the transition from integrated to differential indicators, as required by the system of differential levels, is always accompanied by significant errors in calculations, A. Likhatsevich (Lihatsevich, 1991) believes that traditional IFWB in calculating the water regime is more practical.

Really, water balance models that calculate the water regime of soils, which provide solutions to numerical problems performed in the design and operation of systems, and are very common in practice.

According to V. Shebeko (Shebek, 1980), the method of water balance is a universal tool for assessing and forecasting the water regime of reclaimed lands, which is noted by:

- validity of decisions based on the law of conservation of mass;
- reliability of results that reflect the conditions of operation of real objects;
- maximum approximation of the decisions made to specific natural conditions and means of reclamation;
- wide application and estimation of simplicity of its realization on condition of schematization of water balance calculations.

At the same time, the calculations of water balance in the set of profitable and expendable elements of moisture in their dynamics and relationship are sufficient components that determine the components of balances act as random processes. the connection is not expressed unambiguously and explicitly.

Therefore, by analogy with the model of V. Shebeko (Shebeko V.F., Mozheem I.I., Kiselyova, 1980), the water regime of drained lands due to its main characteristics – the regime of GWL $Hg=f(\tau)$ and soil moisture $W=f(\tau)$ – with fixed water-physical properties of the environment and finite periods of time $\{\tau\}$, $\tau = \overline{\underline{\tau}, \overline{\tau}}$, within the estimated (p to moisture) vegetation period in general can be schematically represented as:

$$F = \left(H_g, W, \tau\right) = N_0 + \sum_{\tau=\underline{\tau}}^{\tau=\overline{\tau}} f_\tau \left(P_\tau, E_\tau, V_\tau, m_\tau\right) \tag{1}$$

where: N_0– initial conditions; P_τ– precipitation; E_τ–total evaporation; $V_\tau=\phi_\tau(H_{g,W})$– function of moisture exchange of aeration zone with GWL; $m\tau$– ir_{ri}gation rate for moistening drained lands; $\tau = \overline{\underline{\tau}, \overline{\tau}}$ – change of the process of formation of the water regime in time from the estimated interval of the beginning $\underline{\tau}$ to its end $\overline{\tau}$.

The process of formation of the water regime differs on the basis of the relationship and interdependence of its defining components at the boundaries of the accepted calculation layer of soil (CLS) and in connection with the process of redistribution of moisture throughout it.

In the forecast of soil moisture and when checking the optimal conditions of the call under the influence of the designed different measures, the water balance approach can be used when

$$Wh_\tau = Wh_{\tau-1} + \sum_{\tau-1}^{\tau} Q_m \leq Wh^0 \tag{2}$$

where Wh_τ, $Wh_{\tau\text{-}1}$– moisture reserves of CLS h in accordance with the beginning (τ–1) and the end τ of the estimated time period $\Delta\tau$; Q_m– profitable and expendable elements of moisture at the boundaries of the CLS, which change the moisture exchange and moisture in this layer; Wh^0– maximum moisture content for water holding capacity of CLS.

Similar to the general model (1), for the GWL:

$$Hg_\tau = Hg_{\tau-1} + \sum_{\tau-1}^{\tau} \Delta Hg_n \tag{3}$$

where $Hg_{\tau\text{-}1}$, Hg_τ– the appropriate GWL at the beginning (τ–1)and the end τ of the estimated time period $\Delta\tau$; ΔHg_n–change of GWL within the estimated time period $\Delta\tau$ under the influence of external n–factors.

Thus, recommended by V. Shebeko, the structure of the relationship in the models of water balance calculations developed by it, in principle, allows to solve the problem of designing the water regime of drained lands for calculations of climatic conditions of vegetation.

Calculations for models (1), (2), (3) are implemented in the set of income and expenditure elements of moisture with their changes over time and in the process of interaction of discrete intervals that have changed from any initial moment with these initial conditions.

This approach to the conditions of moisture exchange for the current estimated time interval τ takes into account the features of the previous period (τ–1) and the features of the intra-seasonal course of individual components of the water balance (Afanasik, 1976; Afanasik, 1983; Anderson, Bertha, 1988; Shebeko, 1980; Shebeko, 1983), which, in principle, generally corresponds to our as a basis for the structure of construction and implementation of the general model of the reclaimed field at drained lands.

However, as shown by the analysis and practice of the balance method, and based on its forecast estimates of the water regime of drained lands on a long-term basis, they are as follows, still not finally resolved fundamentally very important issues:

1) the difficulty of implementing the model while considering its main components – moisture reserves CLS and GWL – for the uncertainty of the GWL regime on ways to regulate it and the impossibility of differentiated assessment of the latter under different natural reclamation conditions;
2) impossibility of adequate assessment of conditions of water regime formation and efficiency of water regulation due to application in forecast calculations, as a rule, only potential (climatically provided) values of water balance level components without taking into account changes of real moisture content of CLS.

In addition, according to (Afanasik, Omelko, Zhukov, 1989; Lazarchuk, 1992; Shebeko, Zakrzhevsky, Bragilevskaya, 1980), in the design of the optimal humidification regime is not enough to focus only on the drying rate (normalized depth of GWL). The connection between GWL and soil moisture reserves is insufficient, and the regulation of the latter due to the creation of the necessary GWL by locking for many systems does not provide water-physical properties, soil water conductivity and topography.

For the warm (most appropriate) period of agricultural development, the water regime of drained lands (regimes of GWL and moisture reserves of the active root-containing soil layer) largely depends on the combination of profitable and cost elements of the water balance of the territory. In this case, the dynamics of GWL during the growing season can not be estimated by calculations due to sufficiently reliable levels.

Needed soils with moisture and vegetation period GWL, usually moisture exchange on the surface and in the aeration zone, ie infiltration of precipitation to replenish water reserves and upward flow of moisture in the upper soil layers and in the zone of evaporation from GWL. These processes are also about the action of network channels and drains, but it is, in cases, the internal growing season, reduced to a minimum.

All this gives grounds to consider the possibility of abandoning the mandatory model of changes GWL in dynamics during the growing season for long-term distribution of drying with atmospheric-soil type of water supply when performing predictive modes of calculation on an urgent basis, if in the form of the normalized value of the average for the growing season its value depending on the method of water regulation and moisture supply of the estimated growing season.

Then it will significantly simplify the model of water regime of drained lands, and the impact of GWL on the formation of moisture (or moisture reserves) of the active root-containing soil layer can be determined through the components of its moisture exchange in the implementation of the model.

But the application of this approach requires a comprehensive analysis of completely diverse in nature regime, technological and technical aspects of water regulation of drained lands in their relationship, schematization and parameterization of natural reclamation conditions of their implementation due to the predominant stochastic nature of formation.

METHODS AND MODELS

The model of water regime and water regulation technologies is a necessary component of the complex forecasting and optimization models and plays an extremely important role because it combines the parameters of modes and technology parameters, and therefore the necessary link between effect parameters and design parameters. ecological and economic principles.

Based on the previously considered general provisions and prerequisites for the development of this model, such requirements are sufficiently met by the equation of water balance of the recurrent type, which is presented in an integral form similar to models (1) and (2) (Rokochinsky, Zubik, 1999).

The equation of water regime and water regulation technologies of drained lands

$$\vec{W}_\tau = W_\tau\left(W_{\tau-1}, q_\tau, p_\tau, a_\tau^g, \tau\right),\ \tau = \overline{1, n_\tau} \tag{4}$$

(a component of the integral model of the reclaimed field) can be used for modeling the water regime of the reclaimed field at drained lands in the general two-dimensional case, when its implementation is considered in relation to the following variable coordinates: time τ and space – changes in the depth of the CLS h_τ over time in the vertical plane. However, this, in turn, is associated with certain practical difficulties.

As for the first variable, coordinate, model (4) can be implemented for any time intervals (day, pentad, week, ten days, month, vegetation period, etc.). The required accuracy of such a solution is achieved at the cost of significant time. But since modern PCs have significantly accelerated the calculations that used to be done manually, the time required to perform them has ceased to be a serious problem.

Regarding the second variable coordinate, this issue needs to be considered in several aspects. On the one hand, the task of assessing and forecasting the water regime of soils is considered by us in relation to the need for moisture supply of crops. The possibility of this one is determined by the development (change over time) of the depth of the root system of the crop and soil moisture. And this necessitates the adoption of a variable value of CLS $h(\tau)$ for the balance equation, which in the general case should be determined by the depth of penetration of the root system of the crop into the soil.

On the other hand, the desire to take into account the uneven water consumption of crops from different soil layers causes a significant complication in the model by increasing the number of its components, as evapotranspiration is an n-dimensional function of soil moisture (n-number of layers to break the root layer). n-dimensional model of moisture dynamics in relation to the processes of redistribution of the root system in the soil during the growing season.

There are already examples of the development of such models for drained soils (Afanasik, Zhukov, 1983; Lazarchuk, 1992; Oliynyk, 1993; Cherenkov, Cherenkova, Lazarchuk, 1993), but as the practice of their application has shown, the number of necessary parameters to be introduced for the construction and implementation of these models is growing. to control production conditions. In addition, the question of the accuracy of calculations in the implementation of the layered balance model for forecasting the water regime of drained lands on a long-term basis remains open. All this indicates the inexpediency of using such a model in production.

Therefore, if we want to build a model of the water regime of drained lands, based on specific requirements for building a general model of reclaimed fields in relation to the appropriate level of forecasting and optimization problems, we must abandon the desire to take into account water consumption and moisture exchange from different soil layers. It's a simple single-layer version.

In addition, numerous studies, including ours, have shown that in conditions of the shallow location of GWL is 0,8…1,4 m, the greatest change and variability over time are moisture reserves in the soil layer up to 0,5 m depth (up to GWL) they are almost unchanged or change very little due to capillary nutrition.

Thus, based on the above and accepted as the basis of the calculation scheme (Fig. 1), the model of forecasting on a long-term basis of the water regime of reclaimed fields at drained lands is based on the "simple" equation of water balance of the most active root layer h = 0,5 m (h_τ=*const*), which is quite common in the practice of creating such models of production nature both in our country and abroad (M. Andersen, Y. Dirse, A. Konstantinov, M. Lazarchuk, A. Likhatsevich, V. Ostapchik, O. Skripnik, R. Hanks, S. Kharchenko, V. Shebeko, A. Yangol, A. Jacyk, and others) (Afanasik, Armonik, Skvortsov, 1976; Afanasik, Zhukov, 1983; Anderson, Bertha, 1988; Ostapchik, Kostromin, Koval et al., 1989; Lazarchuk, 1997; Lazarchuk, 1992; Lihatsevich, 1991; Ostapchik, Kostromin, 1976; Skripnik, 1992; Kharchenko, 1987; Hanks, Ashcroft, 1985; Shebeko V.F., Mozheem I.I., Kiselyova, 1980; Rokochinskiy, Zubyk, 2000; Shebeko, Zakrzhevsky, Bragilevskaya, 1980; Yatsyk, 2001; Yatsyk, 1997).

Figure 1. Calculation scheme of the water regime model of drained lands

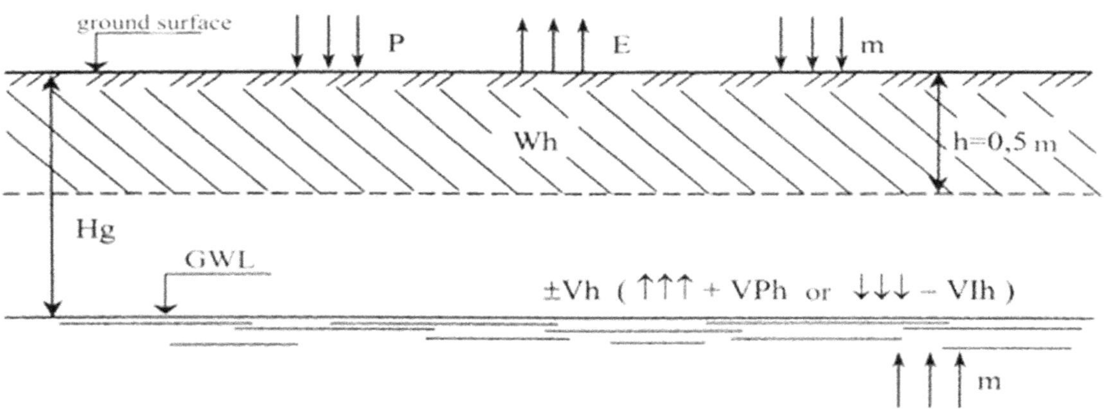

For the accepted structure of calculations and the set sampling step τ, $\tau = \overline{1, n_\tau}$ (pentad, week, decade – according to the implementation of the model of meteorological regimes), the specified model has the following general form (Rokochinskiy, Zubyk, 2000):

$$Wh_\tau = Wh_{\tau-1} + P_\tau - E_\tau \pm Vh_\tau + m_\tau, \quad \tau = \overline{1, n_\tau} \tag{5}$$

where Wh_τ, $Wh_{\tau-1}$– the corresponding reserves of productive moisture CLS at the end of the estimated current τ and previous (τ–1) time intervals at a given initial value Wh^0; P_τ– the amount of precipitation over time τ; E_τ – the corresponding value of the total evaporation; Vh_τ–the amount of moisture exchange CLS h with the lower layers and GWL in the form of power (+) VVh_τ or infiltration (-) VIh_τ; m_τ – irrigation rate with the appropriate method of humidification.

The water regime model in the form of equation (5) describes a quasi-stationary process, when all changes in the studied complex phenomenon occur instantaneously at the end of the estimated time τ. It, in principle, allows to perform water balance calculations when modeling the dynamics of changes in water regime and its determining components in the reclaimed field within the reclamation system for possible variable multiple conditions in the aggregates: meteorological stations and posts $\{\omega\}$, $\omega = \overline{1, n_\omega}$; estimated on the moisture content of the growing seasons $\{p\}$, $p = \overline{1, n_p}$; types of common drained soils $\{g\}$, $g = \overline{1, n_g}$; project types of cultivated crops $\{k\}$, $k = \overline{1, n_k}$; possible ways of water regulation of drained lands $\{s\}$, $s = \overline{1, n_s}$, etc. At the same time, natural reclamation conditions should be considered as averaging due to the typification of meteorological regimes in relation to the calculated vegetation periods, schematization and parameterization of possible methods of water regulation of drained lands.

As for the methods of water regulation of drained lands as a technological aspect of land reclamation, it means that the implementation of the method is provided by the necessary parameters of a drainage system and its constituent elements (regulatory and conductive networks, necessary hydraulic structures, etc.) (Oliynyk, 1993; NTD 33.63-074-87, 1987).

The difficulties of implementing such a "simple" model begin with the technical feasibility of implementing a very large amount of forecast calculations due to the need to consider numerous variables that affect the conditions of formation of water regime of drained lands.

Secondly, the difficultie is need for sufficiently reliable and effective methods for determining the components of the water balance equation and the necessary information support.

As for the first difficulties, as we have already noted, with the advent of modern high-speed PCs, the problem of time to perform calculations is virtually eliminated. As for the second difficultie, the calculation of the water balance on the set of profitable and expendable elements of moisture in their dynamics and relationship, even with a relatively "simple" form of equation (5), is quite complex.

All this necessitated the application in practice in the area of both irrigation and drainage reclamation of simplified approaches to the implementation of balance equations of the form (5) when performing forecast calculations on a long-term basis.

This used, as a rule, potentially possible under the current climatic conditions, the maximum allowable values of such components of the water balance equation as total evaporation, moisture exchange (by the way, this value is often and not always justified due to the complexity of its definition) and water consumption (irrigation rates) for moistening reclaimed lands.

In this case, the solution of equation (5) was mainly to determine the relative (not related to the method of water regulation) the value of irrigation rate m_τ or wetting rate M_p for soil moisture during dry growing seasons due to water balance deficit (or deficit of total water consumption), when the indicator of productive moisture reserves of the active layer of the soil at the end of the estimated time interval τ took negative values, so $Wh_\tau<0$. This approach cannot be considered correct in relation to the essence of the studied phenomenon, but it was generally accepted as quite acceptable in practical water balance calculations.

Today, the method of determining the design regime of land humidification in the area of irrigation and drainage reclamation is practically based on the application of such an approach. Therefore, this structure of the relationship between the components of equation (5) was used by us as a basis in the previously proposed methods of water balance calculations for the development of design regimes for moistening crops on drained lands, discussed in detail in (Lazarchuk, Rokochinskiy, Cherenkov, 1990; Lazarchuk, Rokochinskiy, Gus, 1991).

But in essence, the applied principle is quite conditional, because in the implementation of such an approach it is impossible to adequately assess the actual or actual conditions of formation of the water regime of reclaimed lands because it does not take into account the transformation or change of potential values of almost all components of equation (5) the ratio between profitable and expendable components of the water balance in relation to available moisture or soil storage capacity. This certainly leads to a significant distortion of the results of forecast calculations and, accordingly, the insufficient level of validity of technical and technological engineering solutions in the design and operation of hydromeliorative systems.

Due to the relatively slow in the natural conditions of the formation of the water regime of drained lands and the associated inertia in the relationship of individual phenomena and elements of these processes (precipitation, absorption and inflow of water into the soil, infiltration to GWL, evaporation and nutrition of the active layer of soil with GWL, etc.) it is possible to develop and expediently apply relatively simple, but quite effective methods that allow to take into account the transformation of the components of the water regime within the accumulating capacity of the soil., but quite effective methods that allow to take into account the transformation of the components of the water regime within the accumulating capacity of the soil.

Based on the defined and generally accepted principles of interconnection and interdependence between almost all components of equation (5), we considered a qualitatively different approach to the implementation of the model (Rokochinsky, Zubik, 1999; Rokochinskiy, Zubyk, 2000).

The fundamental difference of this approach is, firstly, that the regime forecast calculations on a long-term basis are performed for schematized natural reclamation conditions by typing meteorological regimes in the calculation periods of vegetation, appropriate schematization and parameterization of water regulation of drained lands.

Secondly, it is proposed in the framework of the balance equation (5) to perform not a simple summation of rigidly set, potentially possible (climatically provided) values of its components, as has been done so far, but, based on certain patterns, to calculate the process of water regime based on the relationship and interdependence of its elements at the boundaries and in connection with the process of redistribution of moisture within the active layer of soil in relation to its moisture content and storage capacity.

This approach allows to determine the really possible (effective) values of total evaporation EP_τ (including its transpiration component ET_τ), moisture exchange of the active layer of soil in the form of food or infiltration VIh_τ, water consumption m_τ when necessary to moisten drained lands in an appropriate way and, finally, soil moisture within the estimated time interval $\tau - \overline{WPh_\tau}$.

Thus, the problem of forecasting the water regime of drained lands on a long-term basis is formulated as the need to simulate during the growing season dynamic change in discrete of all components included in the adopted model (5) calculation of water balance for variables. reclamation conditions of the studied object.

Therefore, based on the adopted general structure of forecasting and optimization calculations, the practical implementation of the water regime model should be based on the use of previously obtained results on relevant models of forecasting meteorological conditions (typical distribution of meteorological factors) and development of crops (phases of their development, ontogenesis), data on the performed schematization and parameterization of methods of water regulation of drained lands under different natural reclamation conditions, and the model itself is considered and implemented in several stages in this sequence (Fig. 2).

At the first stage, for fixed values of multiple variables of the object with $\{\omega\}$, $\omega=\overline{1,n_\omega}$; $\{g\}$, $g=\overline{1,n_g}$; $\{k\}$, $k=\overline{1,n_k}$; $\{s\}$, $s=\overline{1,n_s}$; $\{p\}$, $p=\overline{1,n_p}$ – potentially possible (climate-justified) values of model components are determined (Fig. 2, block 3), such initial model of water regime is considered and implemented (Fig. 2, block 4)

$$Wh_\tau = WPh_{\tau-1} + P_\tau - EV_\tau + VVh_\tau \qquad (6)$$

under the initial condition: $WPh_0 = WPh_g^0$ (where WPh_g^0 is the initial reserve of productive moisture, which corresponds to the maximum field moisture content of the CLS h of the corresponding soil type g within the reclaimed field); $Vh_\tau = VVhs_\tau$; $ms_\tau = 0$.

This is quite true, because the water balance calculations begin in early April of any estimated growing season, when the moisture reserves of the drained soil are within optimal limits.

In this case, according to the results of calculations according to the original model (6), the value Wh_τ can take any value ($0 \le Wh_\tau \le WPh_g^0$; $Wh_\tau<0$; $Wh_\tau > WPh_g^0$), but then it does not always (in two extreme cases) characterize the effective values of moisture reserves due to violation of the physical content of the phenomenon.

Therefore, in the second stage (Fig. 2, block 5) the determination of the effective values of the components of equation (6) is carried out through the corresponding correcting coefficients by the method of sequential iteration (the corresponding sequence of approximation):

- $VVh_\tau \rightarrow VIh_\tau$, when at the first stage we receive $Wh_\tau > WPh_g^0$; $VVh\tau \rightarrow VPh_\tau$ or values $m_{s\tau}$ are determined by the appropriate method of moistening drained lands when $Wh_\tau < WPh_g^0$, and, moreover, when $Wh_\tau \le 0$;
- $Wh_\tau \rightarrow WPh_\tau$ in relation to the average moisture reserves $\overline{WPh_\tau}$ and storage capacity $\left(WPh_g^0 - \overline{WPh_\tau}\right) / WPh_g^0$ of the CLS within the estimated time period $\Delta\tau$.

The application of such an approach is in fact not new in the practice of performing water balance calculations on reclaimed lands. It is substantiated and considered in the papers of M. Lazarchuk (Lazarchuk, 1992), V. Ostapchik (Ostapchik, Kostromin, Koval, 1989), V. Platonov (Platonov, 1982), V. Shabanov (Shabanov, 1973), V. Shebeko (Shebeko, 1983) and others. However, as a rule, it was used to adjust the potential values of individual components of equation (6): either total evaporation EP_τ or fertilization of CLS from the lower layers and GWL VVh_τ – and was practically not related to determining the value of effective soil moisture WPh_τ.

Figure 2. Structural block diagram of water regime and technologies of water regulation of drained lands

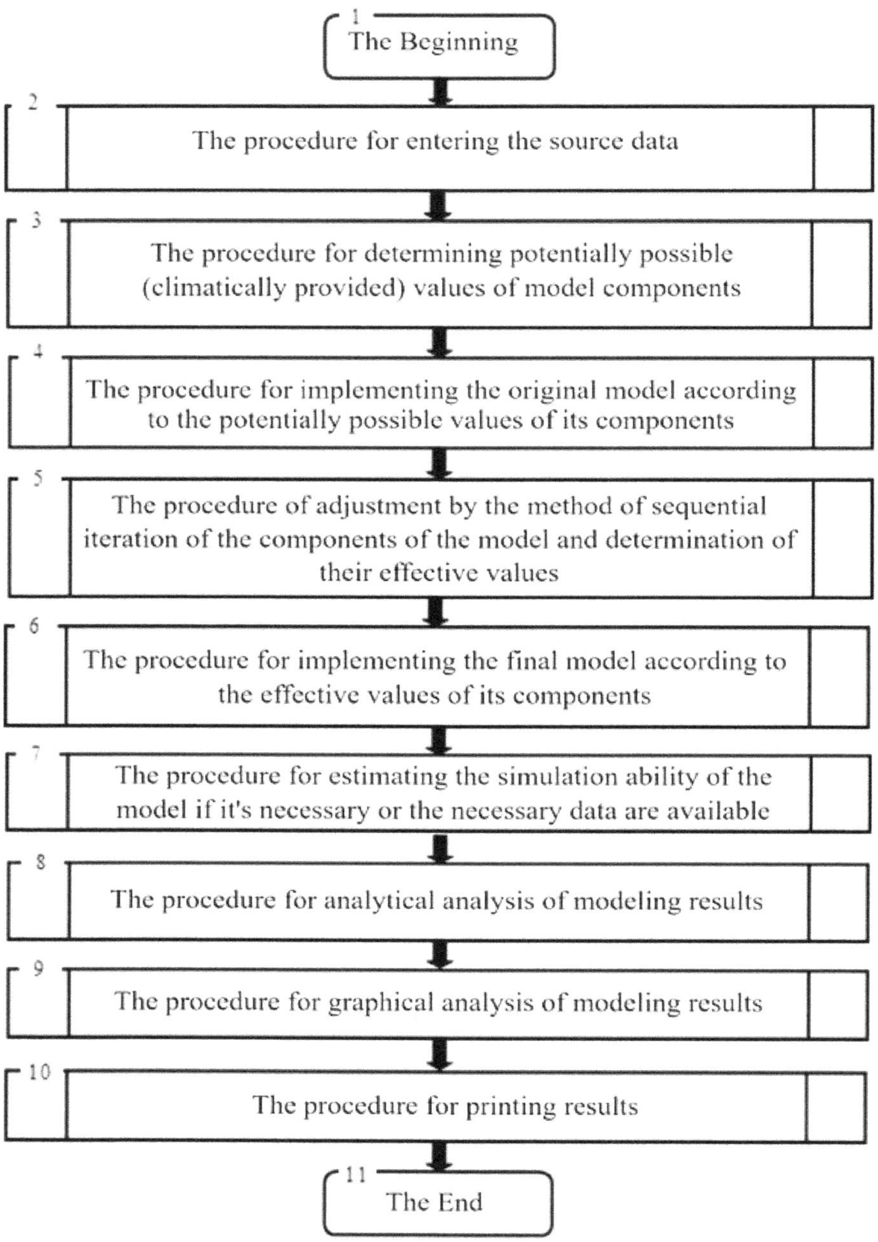

We have developed this approach and apply it to determine the really possible (effective) values of all components of the water balance equation and, first of all, to determine the effective value of productive moisture reserves of CLS, which allows to predict and adequately assess the water regime of drained lands under different natural reclamation conditions.

Therefore, at the third, final stage (Fig. 2, block 6) the following final model of water regime is considered and implemented:

$$WPh_\tau = WPh_{\tau-1} + P_\tau - EP_\tau + VPh_\tau + m_\tau, \ \tau = \overline{1, n_\tau} \quad (7)$$

The subject to the requirement that under any current natural reclamation conditions will be obtained:

$$0 \le WPh_\tau \le WPh_g^0 \quad (8)$$

This achieves a more accurate solution of the balance equation, which is the basis of the model, which does not violate the physical content of a complex phenomenon, which is the water regime of drained lands and natural reclamation factors that determine it.

Other blocks in the structure of the model implementation (Fig. 2, blocks from 7 to 11) include a set of standard procedures for model identification, analytical and graphical analysis of modeling results, and so on.

In addition, we have made the necessary changes and additions to the methods of determining potentially possible and effective values of the main components of the model by regime and technological parameters of water regulation of drained lands, which are discussed below.

Initial climatic components. It is generally accepted that climatic or meteorological (weather) conditions have a decisive influence on the formation of water and general natural reclamation regimes of drained lands, technological and technical solutions for their water regulation.

Therefore, the climatic components of the water balance equation underlying the construction and implementation of the water regime model, such as precipitation directly included in water balance calculations, as well as temperature and relative humidity (or humidity deficit), which determine the potential values of total evaporation and the derived value of the active soil layer are set and determined in accordance with the level of implementation of our models of meteorological regimes on a long-term basis in the form of a typical distribution of major meteorological factors over estimated time periods $\{\tau\}, \ \tau = \overline{1, n_\tau}$ (decade, week, pentad), depending on the presence or absence of the consumer's database of long-term observations.

Therefore, the climatic components of the water balance equation that underlie the construction and implementation of the water regime model, such as precipitation directly included in water balance calculations, as well as temperature and relative humidity (or humidity deficit), which determine the potential values of total evaporation and the derived value of the nutrition of the active layer of soil. These parameters are set and determined in accordance with the level of implementation of our models of meteorological regimes on a long-term basis in the form of a typical distribution of major meteorological factors over estimated time periods (decade, week, pentad), watersupply of vegetative periods $\{p\}, \ p = \overline{1, n_p}$, estimated data on moisture consumer database of long-term observations.

Total evaporation. Data on water consumption of crops in different periods of their development, depending on current weather conditions are the basis for developing design and formation of operational modes of water regulation by selecting and justifying the necessary methods of water regulation, types, structures and modes of hydromeliorative systems, calculation of their parameters.

Therefore, the task of studying the patterns of water consumption of crops and determining its value in reclaimed lands is one of the most important in hydraulic sciences.

Due to the complex dependence of evaporation with many factors that determine it, there are currently many different models of the relationship between the intensity of evaporation and the indicators that affect it. According to (Ostapchik, Kostromin, Koval, 1989; Lazarchuk, 1992; Ostapchik, 1981; Yatsyk, 2001) such models were developed by I. Sharov, G. Lgov, S. Kharchenko, A. Konstantinov, M. Budiko, M. Danilchenko, D. Shtoiko, H. Penman, L. Turk and others – for the irrigation zone; in the practice of drainage reclamation use the formulas of A. Kostyakov, A. Ivytsky, A. Sharov, V. Shebeko, A. Yangol and others.

However, all available empirical dependencies, despite the possible, at first view, their universality, have, as the practice of their use has shown, a pronounced zonal nature and need to be clarified in each specific conditions of the real object.

In addition, the realization of the potential (climatic) value of total evaporation is limited by the available actual soil moisture, and this issue can be solved only by considering the relationship and interdependence between all components of the water balance equation by matching their values in time and space.

According to the generalizing conclusions of M. Lazarchuk (Lazarchuk, 1992) and the results of our experimental studies, the most convenient and practical, in terms of its use on drained lands, is the model for determining total evaporation, proposed by V. Ostapchik and colleagues (Ostapchik, 1989; Ostapchik, 1981).

Thus, when implementing the initial model of the water regime (6), the potential value of total evaporation $EV_{\omega kp\tau}$ can be determined by the following general formula, according to (Ostapchik, 1989):

$$EV_{\omega kp\tau} = f_{k\tau} \cdot ETV_{\omega kp\tau} + (1 - f_{k\tau}) \cdot EGV_{\omega p\tau},\ f_{k\tau} \le 1,\ m^3 \bullet ha^{-1} \quad (9)$$

where $f_{k\tau}$ – characteristics of the degree of projective coverage of the reclaimed field by the cultivated k crop, which is determined from the ratio $f_\tau = a_{?\tau} / a_?^{,}$ ($a_{n\tau}$ – observed in the estimated day τ the degree of coverage of the field by plants; $a_?^{,} = 0,6$ – its critical value, from which you can neglect the impact on the total evaporation $EV_{\omega kp\tau}$ of its constituent particle – evaporation from the soil surface (evaporation) $EGV_{\omega p\tau}$; $ETV_{\omega kp\tau}$ – the potential value of the transpiration component of the total evaporation.

The interpretation of the model indices (9) and their dependence on multiple natural reclamation conditions is given in the previous sections.

According to V. Ostapchyk (Ostapchik, 1989) transpiration component:

$$ETV_{\omega kp\tau} = 10 \cdot k_{\delta(E^0)_{k\tau}} \cdot E^0_{\omega p\tau}\ \text{m3•h}^{a}\text{-1} \quad (^{1}0)$$

where $k_{\delta(E^0)_{k\tau}}$ – the biological coefficient of total evaporation as a function of evaporation E*0*· which changes over time according to the phases of development of the *k* crop for the term τ; $E^0_{\omega p\tau}$ – evaporation for the estimated period τ, determined by the well-known formula N. Ivanov (Rokochinskiy, Frolenkova, Voloshchuk et al., 2008).

Accordingly, the magnitude of the evaporation:

$$EGV_{\omega p\tau} = 10 \cdot E^{0}_{\omega p\tau} \tag{11}$$

Then finally, taking into account (10) and (11):

$$EV_{\omega kp\tau} = f_{k\tau} \cdot 10 \cdot k_{\delta\left(E^{0}\right)_{k\tau}} \cdot E^{0}_{\omega p\tau} + \left(1 - f_{k\tau}\right) \cdot 10E^{0}_{\omega p\tau} = 10E^{0}_{\omega p\tau}\left[1 - f_{k\tau}\left(1 + k_{\delta\left(E^{0}\right)_{k\tau}}\right)\right], \text{ m}^{3}\text{•ha}^{-1} \tag{12}$$

Since the realization of the potential (climatically provided) value is due almost exclusively to the available soil moisture content (Ostapchik, 1989; Ostapchik, 1981), its effective value can be defined as

$$EP_{\omega gksp\tau} = EV_{\omega kp\tau} \cdot \lambda^{W}_{\omega gksp\tau}, \text{ m}^{3}\text{•ha}^{-1} \tag{13}$$

where $\lambda^{W}_{\omega gksp\tau}$ – a coefficient similar in content to the reduction coefficient (according to V. Ostapchyk), which determines the efficiency of the potential evaporation relative to the available moisture content of the active layer of soil.

Value λ_{W} can be calculated according to our correction function. The effective value of total evaporation under any natural agro-ameliorative conditions must vary within:

$$0 < EP_{\omega gksp\tau} \leq EV_{\omega kp\tau} \tag{14}$$

Thus, the proposed method allows to determine the potential (climatically provided) total evaporation *EV* and effective (relative to the available soil moisture) its value *EP* in the implementation of the model of water regime of drained lands under variable natural reclamation conditions.

Mo*isture exchange.* Determining the potentially possible and effective value of moisture exchange of the active soil layer with lower layers and GWL is the most difficult and perhaps the key issue of the balance model of water regime of drained lands, as this value significantly affects almost all other components of the model and their effective implementation.

Difficulties in solving this problem are due to the fact that the process of moisture exchange can occur depending on the ratio between precipitation and total evaporation in relation to the storage capacity of the soil (ie its available moisture content) in almost opposite directions and thus form the appropriate implementation of the overall model.

Currently, there are quite a number of methods and models for determining this characteristic, which have both positive and negative aspects of their application, due to the goals, objectives and conditions for which they were obtained by various authors (Kharchenko, 1987; Tsipris, Seleznev, Plavnik et al., 1986; Cherenkov, Cherenkova, Lazarchuk, 1993; Shebeko, Zakrzhevsky, Bragilevskaya, 1980).

Therefore, based on the accepted structure of forecast calculations for models of water regime (6) and (7), we tested and proposed to use such a dependence to determine the potential value of CLS supply from the lower layers and GWL taking into account variable natural and agro-ameliorative conditions relevant indicators of indices (as in the consideration of total evaporation), based on the generally accepted model of S. Kharchenko (Kharchenko, 1987)

$$VVh_{\omega gksp\tau} = EV_{\omega gksp\tau} \cdot e^{-m_{hg}\bar{H}g_{gsp}}\text{, mm} \tag{15}$$

where $EV_{\omega gksp\tau}$– potential (climatically provided) total water consumption, *mm*; *e*– the basis of the natural logarithm; mh_g– coefficient that takes into account the influence of certain soil conditions *g* and variable in the ontogenesis of the k crop of the root-containing soil layer h; $\bar{H}g_{gsp}$ – normalized average vegetation depth of GWL depending on the type of soil, method of water regulation and moisture supply of the estimated vegetation period, m.

Based on the analysis of numerous data of other authors and normative-reference sources (Afanasik, Armonik, Skvortsov, 1976; Afanasik, Omelko, Zhukov, 1989; Tsipris, Seleznev, Plavnik et al., 1986; Maslova, 1985; NTD 33.63-074-87, 1987; Skripnik, Soroka, Kubyshkin, 1992; Shebeko, 1983; Shebeko, Zakrzhevsky, Bragilevskaya, 1980; Yangol, 1970; Yatsyk, 2001), as well as the performed schematization and parameterization of existing methods of water regulation of drained lands for different natural-ameliorative conditions, the average normalized values $\bar{H}g_{gsp}$ and the corresponding average values m_{hg} of the characteristics for CLS *h*=0.5 m in relation to the type of soil *g* by particle size distribution were obtained: for clay m_{hg}=1; loam m_{hg}=1,4; list m_{hg}=1,8; sand m_{hg}=2,2; peat m_{hg}=1,5.

These characteristics, if necessary, can be easily refined in the process of model identification (15) for the conditions of the real object under study.

In turn, this made it possible in the forecast calculations of the water regime according to models (6) and (7) to consider the GWL regime during the growing season of the estimated moisture supply as an appropriate background, which greatly simplified the structure of calculations without changing their content. precision.

Thus, the model can be represented as:

$$VVh_{\omega gksp\tau} = a^V_{gsp} \cdot EV_{\omega kp\tau}\text{, m}^3\text{•ha}^{-1} \tag{16}$$

where

$$a^V_{gsp} = e^{-m_{hg}\bar{H}g_{gsp}} \tag{17}$$

and $EV_{\omega kp\tau}$ is determined by model (12).

The values of the indicator a^V_{gsp} can be easily determined according to the recommended normalized values m_{hg} and $\bar{H}g_{gsp}$ for different types of drained soils *g* under different natural reclamation conditions (water regulation methods *s* and estimated vegetation periods *p*) when performing long-term water balance calculations.

The potential value $VVh_{\omega gksp\tau,}$ determined by (16) should be adjusted to its effective (really possible) value $VPh_{\omega gksp\tau}$ taking into account the real ratio between the profit and expenditure components of the water balance equation relative to the storage capacity of CLS (Fig. 1) according to the following basic schemes (taking account in the further designations only a variable τ for the purpose of simplification of statement of necessary considerations).

For all possible methods of water regulation of drained lands in the case when the implementation of the original model (6) we get the value $Wh_\tau > WPh_g^0$, we accept $Wh_\tau = WPh_g^0$.

In this case, the effective value of moisture exchange will be determined as:

$$VPh_\tau = WPh_g^0 - WPh_{\tau-1} - P_\tau + EV_\tau \text{, m}^3\text{•ha}^{-1} \tag{18}$$

Then, if in the formula (18) we obtain $VPh_\tau>0$, then in the balance model (6) there is a transformation $VVh_\tau \rightarrow VPh_\tau$. And if $VPh_\tau<0$, so, $VVh_\tau \rightarrow VIh_\tau$, what happens, as a rule, when $P_\tau > (Wh_\tau + EV_\tau)$ and there is an infiltration of precipitation through the CLS.

In all other cases, when $0 \le Wh_\tau \le WPh_g^0$ or $Wh_\tau<0$, the effective value of moisture exchange in the form of feeding CLS can be determined by the formula:

$$VPh_\tau = k_\tau^V \cdot VVh_\tau \text{, m}^3 \text{ / ha} \tag{19}$$

where k_τ^V – the adjustment factor determined by the following balance ratio:

$$k_\tau^V = \frac{WPh_g^0 - Wh_\tau}{WPh_g^0} = \frac{WPh_g^0 - \left(WPh_{\tau-1} + P_\tau - EV_\tau + VVh_\tau + m_\tau\right)}{WPh_g^0} \tag{20}$$

where m_τ can be only when applying irrigation by sprinkling at drained lands.

In this case, under any circumstances, the requirement is:

$$VPh_\tau \le EP_\tau, \tag{21}$$

where EP_τ – the effective value of total water consumption.

Thus, the considered methods and models allow a fairly simple and effective way to determine the potentially possible and effective values of moisture exchange of the active root layer of soil with lower layers and GWL in the implementation of the model of water regime and water regulation technologies.

Moisturizing (irrigation) norms. Irrigation norms for the implementation of humidification measures at reclaimed lands in general, both in the irrigation zone and in the drainage zone also act as a complex character that is determined and depends on a significant number of factors: natural and climatic; agro-technical; soil and reclamation conditions of the area; humidification techniques, etc.

It should be noted that the issue of determining irrigation norms should be considered today in several such aspects.

Firstly, it is the determination of the values of irrigation norms according to the need due to natural, climatic and agronomic factors. They are considered biologically optimal irrigation and, accordingly, total moisturizing (irrigation) norms.

Secondly, it is the definition of irrigation norms according to the possibilities of their implementation, due to the available resources of water for humidification, land reclamation, relief and economic conditions of the object, and etc. They are considered technically possible, respectively, irrigation and humidification standards.

And, finally, the definition of rational (economically and environmentally sound) irrigation and humidification norms and modes of humidification of reclaimed lands.

All three aspects are closely interrelated and should be considered in the general case in the hierarchical sequence we have outlined, so from the necessary biologically optimal to technically possible, and then economically and environmentally optimal or rational.

However, since the last of these three aspects is today a separate independent, broad, and extremely complex problem in reclamation production, which is still insufficiently developed, even in theoretical terms, we do not consider it in this section, and we will only discuss the first two types of irrigation norms in the design regime of moistening of drained lands on a long-term basis by appropriate methods – with subsoil moistening and irrigation by sprinkling.

In the general case, the calculation of the design mode of humidification is recommended to perform on the basis of consideration of the water balance of the active layer of soil, which allows for determining the values of the irrigation norm as needed.

Therefore, when applying irrigation by sprinkling at drained lands, as practice shows (Lazarchuk, 1992; Maslova, 1985; Oliynyk, 1993; NTD 33.63-074-87, 1987), the value of the net irrigation norm depends on the value of CLS, available meteorological conditions, particle size distribution, type and phase of cultivation, the type used sprinkler equipment and $m_\tau^{7@}$=200…400 m^3•ha^{-1}.

Provided there are no restrictions on the quantity and quality of irrigation water, irrigation is always assigned when $Wh_\tau \leq 0$. And it is also necessary to provide for the possibility of their appointment when $0 < Wh_\tau < WPh_g^0$, provided that $m_\tau^{7@} \leq \left(WPh_g^0 - Wh_\tau\right)$.

It is recommended to determine the value of the net irrigation norm, in contrast to the previously used method (Oliynyk, 1993; Maslov, 1970; Lazarchuk, Rokochinskiy, Cherenkov, 1990; Lazarchuk, Rokochinskiy, Gus, 1991), when $m_\tau^{7@}$ it was considered and accepted as a constant value during the estimated growing season ($m_\tau^{7@}$ = const), more flexible calculation from the general balance equation (6):

$$m_\tau^{7@} = WPh_g^0 - Wh_\tau = WPh_g^0 - \left(WPh_{\tau-1} + P_\tau - EV_\tau + VVh_\tau\right), \text{ m}^3\text{•ha}^{-1} \tag{22}$$

It must meet the requirement

$$m_{min}^{7@7@7@ max}, \tag{23}$$

where, $m_{min}^{7@}, m_{max}^{7@}$ – the minimum and maximum possible values of irrigation norm for a given sprinkler technique, taking into account soil and relief conditions of the reclaimed field; $\tilde{m}_\tau^{7@}$ – gross irrigation rate for sprinkler irrigation:

$$\tilde{m}_\tau^{7@} = m_\tau^{7@} \cdot \eta^{7@}, \tag{24}$$

where η^{3p}– coefficient of performance of irrigation at drained lands.

When $\tilde{m}_\tau^{7@} \leq m_{min}^{7@}$, then in further calculations we accept $m_\tau^{7@} = 0$.

If $\tilde{m}_{\tau}^{7@} > m_{max}^{7@}$, so, then several waterings n_{τ}^{m} are assigned during the estimated term of the condition:

$$n_{\tau}^{m} = \frac{2\tilde{m}_{\tau}^{7@}}{m_{min}^{7@max}} \tag{25}$$

where a certain number of waterings is rounded to whole values.

Determined by formulas (22), (23), (24), (25), the necessary and technically possible final value of the irrigation norm for irrigation by sprinkling can be considered an effective value, as there are no other restrictions for its implementation.

As for the determination of irrigation norm for subsoil moisture by a sluice m_{τ}^{72}, the generally accepted approach of expression (18) can not be considered effective, because the deficit of moisture reserves of CLS is covered by its supply from the lower layers and GWL, the value of which, as practice shows, limited to water and physical properties of soil and GWL.

It includes several components:

$$m_{\tau}^{72} = m(\Delta Hg)_{\tau} + m(\Delta Wh)_{\tau}, \text{ m}^3\text{•ha}^{-1} \tag{26}$$

where $m(\Delta Hg)_{\tau}$ – the amount of water required to raise the GWL in the implementation of subsoil moisture from a depth of $Hg_{\tau-1}$ up to $Hg\tau$, m³•ha⁻¹; $m(\Delta Wh)_{\tau}$ – the amount of water entering the CLS by raising and maintaining a higher corresponding GWL, m3•ha⁻¹.

Based on the adopted structure of the forecast regime calculations on a long-term basis and performed for this scheme of methods of water regulation of drained lands under different natural reclamation conditions, the effective amount of water required m_{τ}^{72} for groundwater recharge by raising and maintaining appropriate GWL for the estimated time, can be defined as the difference between the effective power supply value of the CLS VPh_{τ}^{72} according to this control method and the possible power supply value obtained for the operation of the control network at the reclaimed field in precautionary lock or drainage mode by slues $VVh_{\tau}^{?H}$ or draining regime $VVh_{\tau}^{>A}$ (depending on which of these methods was used in the previous τ–1 decade), so it will be:

$$m_{\tau}^{72} = VPh_{\tau}^{72} - VVh_{\tau}^{?H}\left(01 > VVh_{\tau}^{>A}\right), \text{ m}^3\text{•ha}^{-1} \tag{27}$$

where the values VPh_{τ}^{72}, $VVh_{\tau}^{?H}$, $VVh_{\tau}^{>A}$ are determined by methods and models discussed previously.

The total amount of irrigation water at subsoil moisture (gross irrigation norm) is equal to:

$$\tilde{m}_{\tau}^{72} = m_{\tau}^{72} \cdot \eta^{72}, \text{ m}^3\text{•ha}^{-1} \tag{28}$$

where η^{72} – coefficient of performance of subsoil moisture, which depends on soil and relief conditions of the reclaimed field.

It must also consider the requirement:

$$0 < \tilde{m}_{\tau}^{72} \leq m_{max}^{72} \qquad (29)$$

where m_{max}^{72} – the maximum value of the irrigation norm at subsoil moisture, due to the terrain, soil conditions, design, and capacity of the regulatory network of a dranage system and its water supply.

In this case, if the value is obtained $\tilde{m}_{\tau}^{72} > m_{max}^{72}$, then in further calculations we accept $\tilde{m}_{\tau}^{72} = m_{max}^{72}$.

FUTURE RESEARCH DIRECTIONS

In order to effectively meet the current challenges associated with existing and projected climate change, it is necessary to develop and justify a set of adaptive agro-ameliorative and hydraulic measures aimed at effective regulation of water regime, regulation, and accumulation of moisture into the soil within the system, the transition from traditional periodic to the implementation and provision of regular wetting of drained lands, improvement of water management technologies, types and designs of hydromeliorative systems, methods of their design, and calculation that meets modern economic and environmental requirements.

CONCLUSION

1. The direction and degree of impact on the natural water regime in the reclamation of excessively moist lands should meet the solution of the main task – to create through the implementation of hydromelioration measures a favorable regime of moisture of the active root layer of soil in compliance with environmental requirements.
2. When regulating the water regime of drained lands, there are mainly two often separate or periodically changing processes – drainage and wetting, which use interrelated, but at the same time independent principles and techniques of designing reclamation systems according to relevant technical solutions.
3. The use of water balance models provides an opportunity not only to assess, but also to predict (predict) the impact of one or more factors on soil moisture conditions within a particular reclaimed field and the system as a whole.
4. The model of water regime is a necessary component of the complex forecasting and optimization models and plays an extremely important role because it combines the parameters of modes and technology parameters, and therefore the necessary link between effect parameters and design parameters. ecological and economic principles.
5. The model of forecasting on a long-term basis of the water regime of reclaimed fields at drained lands is based on the "simple" equation of water balance of the most active root layer h = 0,5 m ($h_{\tau=}$const), which is quite common in the practice of creating such models of production nature both in our country and abroad.
6. We considered a qualitatively different approach to the implementation of the model of the water regime of drained lands. The fundamental difference of this approach is, firstly, that the regime forecast calculations on a long-term basis are performed for schematized natural reclamation conditions by typing meteorological regimes in the calculation periods of vegetation, appropriate

schematization and parameterization of water regulation of drained lands. Secondly, it is proposed in the framework of the balance equation to perform not a simple summation of rigidly set, potentially possible (climatically provided) values of its components, as has been done so far, but, based on certain patterns, to calculate the process of water regime based on the relationship and interdependence of its elements at the boundaries and in connection with the process of redistribution of moisture within the active layer of soil in relation to its moisture content and storage capacity.

7. This approach allows to determine the really possible (effective) values of total evaporation moisture exchange of the active layer of soil in the form of food or infiltration, water consumption when necessary to moisten drained lands in an appropriate way and, finally, soil moisture within the estimated time interval.
8. Thus, the problem of forecasting the water regime of drained lands on a long-term basis is formulated as the need to simulate during the growing season dynamic change in discrete of all components included in the adopted model calculation of water balance for variables reclamation conditions of the studied object.
9. Therefore, based on the adopted general structure of forecasting and optimization calculations, the practical implementation of the water regime model should be based on the use of previously obtained results on relevant models of forecasting meteorological conditions (typical distribution of meteorological factors) and development of crops (phases of their development, ontogenesis), data on the performed schematization and parameterization of methods of water regulation of drained lands under different natural reclamation conditions, and the model itself is considered and implemented in several stages.
10. The considered in the chapter methods and models of water regime allow determining both potentially possible due to meteorological and agrotechnical conditions, and effective (according to soil reclamation and technical conditions) values of irrigation norms in the development of design modes of irrigation by irrigation and subsoil at drained lands.

REFERENCES

Afanasik, G., Armonik, O. & Skvortsov, N. (1976). Calculation of the water regime of reclaimed soils. *Constructions and calculations of drainage and humidification systems, 2*, 75–86.

Afanasik, G., Omelko, A., Zhukov, L. (1989). Water regime management of large reclamation systems. *Reclamation and environmental protection*, 17–26

Afanasik, G. & Zhukov, L. (1983). Layered balance scheme for calculating the water regime of the soil for the purposes of its optimal management. *Issues of operation of drainage and humidification systems*, 15–22.

Aidarov, I., Golovanov, A., & Nikolsky, Yu. (1990). *Optimization of reclamation regimes of irrigated and drained agricultural lands (recommendations).* Agrometeoizdat.

Anderson, M., Bertha, T. (1988). *Hydrogeological forecasting: Per. with English.* Mir.

Budagovsky, A. (1981). Scientific bases of an estimation of need for irrigation and calculation of irrigation systems. *Modern problems of hydrology of irrigated lands, 1*, 16–32.

Cherenkov, A., Cherenkova, L. & Lazarchuk, M. (1993). Layered-balance scheme of calculation of water regime in the conditions of close occurrence of groundwater level. *Hydromelioration and hydraulic engineering construction, 20*, 9–12.

Hanks, R., & Ashcroft, J. (1985). *Applied soil physics. Soil moisture and temperature: Per. with English.* Gidrometeoizdat.

Kharchenko, S. (1987). *Water regime management on reclaimed lands in the Non-Chernozem zone (Hydrological aspects)*. Gidrometeoizdat.

Lazarchuk, M. (1997). *Land drainage: A textbook.* IZMN.

Lazarchuk, N. (1992). Optimization of the calculation of drainage systems and their management. Water, 10(1).

Lazarchuk, N., Rokochinskiy, A., & Cherenkov, A. (1990). *Technical instructions for designing on a computer of calculated modes of moistening of agricultural crops on systems of bilateral action of the Ukrainian SSR: NTD 33.63-089-89*. Ukrgidrovodhoz.

Lazarchuk, N., Rokochinskiy, A., & Gus, V. (1991). *Technical instructions for the development of computer-aided system plans for water use for drainage and humidification systems of Ukraine: NTD 33-04-01-91*. Ukrgidrovodhoz.

Lihatsevich, A. (1991). On the assessment of the accuracy of the equations of the water balance of the irrigated field. *Melioration and water management, 9*, 25–26.

Maslov, B. (1970). *Groundwater regime of wetlands and its regulation*. Kolos.

Maslova, B. (1985). *Reclamation and water management. 3. Drainage: Handbook*. Agropromizdat.

NTD 33.63-074-87 (1987). *Guide to the design of drainage systems in the Ukrainian SSR*. Ukraine, Kiev: Ukrgidrovodhoz.

Oleynik, A. (1985). Regulation of water-air regime of soils on the background of drainage on the basis of mathematical modeling. *Melioration and water management, 54*, 72–75.

Oleynik, A., & Polyakov, V. (1987). *Drainage of wetlands*. Naukova dumka.

Oliynyk, A. (1993). Methods of calculation and design of drainage and humidification systems of Ukraine: Reclamation systems and facilities. Magazine of Civil Engineering (96)4.

Ostapchik, V. (1981). *Planning of irrigation regimes based on the biological method of calculating water consumption of agricultural crops*. Central Research Institute of the Ministry of Water Resources of the USSR.

Ostapchik, V. & Kostromin, V. (1976). Substantiation of the method of calculating the most probable seasonal dynamics of the water balance deficit of a given security. *Melioration and water management, 36*, 2–12.

Ostapchik, V., Kostromin, V., & Koval, A. (1989). *Information-advisory irrigation management system*. Harvest.

Platonov, V. (1982). *Optimization of conditions of moisture supply of agricultural crops*. Gidrometeoizdat.

Rokochinskiy, A., Frolenkova, N., & Voloshchuk, V. (2008). Meteorological support of engineering and reclamation calculations in projects of construction and reconstruction of drainage systems. Manual to DBN B.2.4-1-99. Reclamation systems and structures, Rivne.

Rokochinskiy, A. & Zubyk, L. (2000). Model for forecasting the effectiveness of various technologies of water regulation of drained lands in projects of construction, reconstruction and operation of reclamation systems. *Bulletin of Rivne State Technical University: Coll. Science, 2*, 63–71.

Rokochinsky, A. & Zubik, L. (1999). Principles of creation and implementation of models of optimization of water regulation of drained lands on a long-term basis. *Problems of reclamation and water management at the present stage: Sb. scientific of works Bel. s.-x. academy, 1*, 158–163.

Romashchenko, M., Rokochynsky, A., Halik, O., Savchuk, T. & Kolodych O. (2007). Modern climate change and their manifestations from global to regional manifestations. *Hydromelioration and hydraulic engineering construction: Coll. Science. work, 32*, 65–79.

Shabanov, V. (1973). Bioclimatic substantiation by reclamation. Gidrometeoizdat.

Shawcroft, R., Lewon, E., Allen, L., Stewart, D., & Jensen, S. (1974). The soil-plant-atmosphere model and some of it's predictions. *Agricultural Meteorology, 14*(1/2), 287–307. doi:10.1016/0002-1571(74)90025-9

Shebeko, V. (1983). *Influence of drainage reclamation on the water regime of the territory*. Harvest.

Shebeko, V. (2000). *Water management calculations for reclamation of wetlands*. NIH.

Shebeko, V., Mozheem, I., & Kiselyova, A. (1980). *Methodical instructions and computer programs for designing the water regime of drained lands on the basis of water balance calculations*. NIH.

Shebeko, V., Zakrzhevsky, P., & Bragilevskaya, E. (1980). *Hydrological calculations in the design of drainage and dehumidification systems*. Gidrometeoizdat.

Skripnik, O., Soroka, I., & Kubyshkin, V. (1992). *Technology of regulation of the water regime of drained lands*. Urozhay.

Stashuk, V. (1993). *Assessment and forecast of conditions for the formation of the water regime of drained lands with different ways of its regulation: Dis. Cand. tech. sciences in the form of sciences. add.: 06.01.02*. UIIVG.

Tsipris, D., Seleznev, V., Plavnik, G., & Shtakovsky, A. (1986). Calculation of norms of water consumption for the set crop. *GiM, 4*, 33–36.

Yangol, A. (1970). *Bilateral humidity control during drying*. Kolos.

Yangol, A. (1990). Degree of drainage of peat soils and its calculated elements. *Tr. UkrNIIGM, 70*, 18–26.

Yatsyk, A. (1997). *Ecological bases of rational water use (on the example of rivers of Ukraine)*. TS-NIIKIVR.

Yatsyk, A. (2001). *Environmental security in Ukraine*. Geneza.

Chapter 5
Environmentally Safe and Resource-Saving Water Regulation Technologies on Drained Lands

Lyudmyla Kuzmych
https://orcid.org/0000-0003-0727-0508
Institute of Water Problems and Land Reclamation NAAS, Ukraine

Halyna Voropai
Institute of Water Problems and Land Reclamation NAAS, Ukraine

ABSTRACT

In modern conditions of climate change and agricultural use of drained lands, there is a need for efficient production with the introduction of new or improved technologies for growing economically attractive crops that can ensure sustainable harvests, environmentally safe production, and minimize all costs, including water resources. By implementing a resource-saving technology of water regulation of drained lands during the cultivation of perennial grasses, the saving of water resources is 48,700 m3/year, the consumed electricity is 5.7%/ year. The saving of water resources during the cultivation of corn for silage was 670 m3/ha, and the costs of operating the sluices decreased by 2-3 times. Compliance with the technological parameters of water regulation during the cultivation of paiza, amaranth and fodder beans made it possible to increase the yield of cultivated crops by 15-20%, ensure the saving of water resources by up to 25% and obtain the economic effect.

BACKGROUND

Reclamation agriculture in the zone of drainage reclamation is a source of sustainable agricultural production, and the economic, ecological, and social stability of the region depends on the efficiency of the use of drainage systems (Land reclamation, 2015; Romashchenko M.I., 2019). Due to the higher level

DOI: 10.4018/978-1-6684-8248-3.ch005

of productivity and the possibility of ensuring the stability of agricultural production, reclaimed land is considered an insurance fund for the food security of Ukraine. Drainage systems, the majority of which are located in the Polesie Ecozone of Ukraine, have a total area of 3.2 million hectares and include 1,671 drainage reclamation systems (Voropai H.V. et al, 2019).

At the same time, climate changes that have been observed in recent decades are accompanied by the instability of water supply to reclaimed areas and create new conditions for growing agricultural crops. According to the calculations made by the scientists of the Institute of Scientific Research of the National Academy of Sciences of Ukraine, the territory of Ukraine with negative and sufficient atmospheric moisture has decreased by 10% over the past 25 years and occupies only 22.5% or 7.6 million hectares of arable land (Land reclamation, 2015; Romashchenko M.I. et al, 2019, 2020).

Taking into account the presence of a clear trend towards the further increase in the aridity of the climate in Ukraine at the present stage and, accordingly, a decrease in the conditions of natural moisture supply throughout the greater part of its territory, as well as the formation in the Polissia zone not only in conditions of over wetting of soils but also in their moisture deficit, especially in the second half of the vegetation period, the need and role of meliorative systems in the sustainable management of agriculture on drained lands are increasing (Irrigation and drainage strategy in Ukraine for the period up to 2030, 2019; Voropai H.V. et al, 2019).

At the same time, the value and importance of agricultural land are increasing precisely due to changes in climatic conditions in the humid zone (Datsko L.V., 2016; Voropai H.V. et al, 2021). Along with these climatic changes are accompanied by a change in the conditions for growing agricultural crops and, accordingly, transform the role of drainage systems. And if in the 60s and 80s of the 20th century, drainage systems mainly performed the function of removing the excess water in the spring period, now effective farming on drained lands requires expanding their capabilities with the ability to improve the moisture supply of cultivated crops throughout the growing season.

One of the important means of minimizing the impact of modern climate changes on agricultural production in humid zone is the effective use of the existing potential of drainage systems. At the same time, the restoration of their effective functioning becomes a defining component not only of modern intensive technologies for growing agricultural crops but also of creating favorable living conditions and protecting the rural population and rural areas from the harmful effects of water (Yatsyk M.V et al, 2015).

In modern conditions of climate change and the structure of agricultural use of drained lands, for effective agricultural production on drained lands, it is necessary to introduce new and improved technologies for growing agricultural crops, which are capable of ensuring a high yield and quality of products, under the condition of economical use of available material and water resources, minimization of non-productive discharge of water beyond the boundaries of reclamation areas, rational use of the natural potential of agro-landscapes and ensuring sustainable environmentally safe production of agricultural products (Voropai H.V. et al, 2021).

The water-air regime of the soil, which is favorable for agricultural crops, is maintained thanks to water regulation technologies, which should provide for the dynamics of the groundwater level (GWL), in which the necessary moisture in the root layer of the soil is maintained. At the same time, when managing the water regime of drained soils, it is necessary to ensure not only the requirements of agricultural crops for the water-air regime throughout the growing season but also ecological requirements for the preservation of the natural environment (Yatsyk M.V et al, 2016).

The introduction of advanced, cost-effective, ecologically safe regimes and technologies of water regulation on drained lands in the complex conditions of functioning of drainage systems at the current stage will allow providing conditions for the transition from extensive management of reclamation agriculture to intensive based on the use of the latest scientific achievements and best practices (Yatsyk M.V et al, 2009).

The need to combine agricultural intensification with environmental protection in view of the large-scale reclamation areas of the humid zone of Ukraine is the basis for rethinking the essence of reclamations and changing approaches to their justification on new scientific grounds.

Modern environmental problems and issues of resource-saving are closely related, first of all, to the problem of optimizing the water regime on drained lands. At the same time, the processes of regulating the regimes of nutrition and heat in accordance with the requirements of agricultural crops in the root layer of the soil depend significantly on the dynamics of the water regime (Voropai H.V. et al, 2021).

Optimizing the water regime of the soil and the moisture supply of plants is understood as a system of remedial measures that creates conditions for obtaining rational and economically sound crop yields. This problem was and remains the main task of water reclamation of soils.

The purpose of research is the development of environmentally safe and resource-saving water regulation technologies on drained lands, which take into account the peculiarities of the use of moisture by cultivated crops during the growing season, make it possible to ensure the necessary parameters of the water-air regime of the soil, the saving of water resources and ecological requirements for the preservation of the natural environment.

In order to achieve this goal, the scientists of the National Academy of Sciences of Ukraine developed a technology for regulating the water regime of the soil when growing agricultural crops on drained lands, which takes into account the peculiarities of the use of moisture by cultivated crops, makes it possible to substantiate rational water regulation regimes and simultaneously accumulate additional moisture reserves in the aeration zone. The method of calculating reclamation regimes is based on a comprehensive approach to the optimization of water and nutrient regimes, their impact on fertility and harvest, ecological stability, and minimization of non-productive discharges of water beyond the borders of reclamation territories. The basis of the methodology is the algorithm for determining resource-saving water regulation parameters and the calculation of optimal ranges of GWL for each calculation period, for which the necessary moisture reserves in the root layer of the soil must be provided during the growing season (Yatsyk M.V et al, 2016).

REGULATION OF THE SOIL WATER REGIME ON DRAINED LANDS

The technology of regulating the water regime of the soil when growing agricultural crops on drained lands includes separate functional elements of preventive, continuous, and cyclic sluicing, (Voropai H.V. et al, 2019; Yatsyk M.V et al, 2015). The block diagram of the algorithm for regulating the water regime of the soil is presented in Fig. 1.

In the spring, before sowing, from the moment the GWL drops to the norm h_n (the upper limit of the optimal GWL), which ensures the passage of agricultural machinery and tillage, during the preliminary sluicing. After sowing, the moisture reserves of the root layer of the soil are mainly used for evaporation from the soil surface (evaporation). The reclamation system works in the mode of passive reduction of RHV. From the moment of moisture reduction in the upper layers of the soil to the value of the pre-irrigation threshold at the end of moistening, the indicated moisture corresponds to the maximum permissible reduction of GWL h_k (the lower limit of the optimal GWL).

The reclamation system is transferred to the wetting mode, the duration of which depends on the dynamic characteristics of the "drain-soil" link. From the moment the soil moisture reaches the upper limit of the optimal regime (according to the value of soil water pressure Ψ_m in the Table 1), the RHV also reaches the upper limit h_n of water supply for moistening and stops (Voropai H.V. et al, 2021). The ameliorating system is transferred to the mode of passive reduction of RHV under the action of evaporation and transpiration. After completion of the process of formation of the root system of the cultivated agricultural crop, control of moisture reserves is created in the zone of its maximum absorption.

Table 1. The value of soil water pressure (Ψ_m), which has a minimum moisture capacity for different types of soils (Voropai H.V. et al, 2021)

Soil type	The value of the suction pressure (Ψ_m), that corresponds to the upper limit of the optimal moisture range, kPA
Sandy, peaty soils	0.5 ÷ 1.0
Light loam	1.0÷1.5
Medium loam	1.5÷2
Heavy loam	2÷2.5
Clay	2.5÷3

The optimal ranges of soil moisture and GWL are determined for each subsequent cycle, the frequency and duration of which depend on the water-physical characteristics of the soil, the "tension" of meteorological conditions, the type of cultivated crop, and its development phase. Cultivated crops require special attention during critical periods of certain phases of their development, when humidity deviations from optimal limits are not allowed, as it is shown in the Table 2 (Kovalenko P.I. et al, 1996; Kuzmych L. et al, 2021a).

The schematization of natural conditions and construction of calculated filtration schemes, which reflect the most important factors in the formation of water-physical processes in real conditions, is an important stage preceding the calculation of optimal reclamation regimes. In the process of schematization, the shape and structure of the groundwater flow area are established; the presence and intensity of pressurized power supply; nature of water exchange between saturated and unsaturated zones; significance, location, and nature of the action of the main elements of the reclamation system; initial and final values of flow characteristics, their relationship with other elements of the scheme.

The construction of calculated filtration schemes is based on the initial data of engineering and geological studies at the design stage, data from hydrological expeditions, and the results of research on the object according to the methodology outlined in works (Oliynyk A. Ya., 1981, 1985; Oliynyk A. Ya. et al, 1987).

Taking into account the characteristic natural conditions of reclamation facilities in the humid zone, we assume that the upper layer of the aeration zone m^* has a small capacity, therefore the lower limit of the root layer m_r is located in the lower layer (Fig. 2). In addition, the physical evaporation of E_u depends on the humidity of the soil θ^* near the surface and the degree of its shading (Kovalenko P.I. et al, 1996).

Figure 1. Block diagram of the algorithm for regulating the water regime of the soil

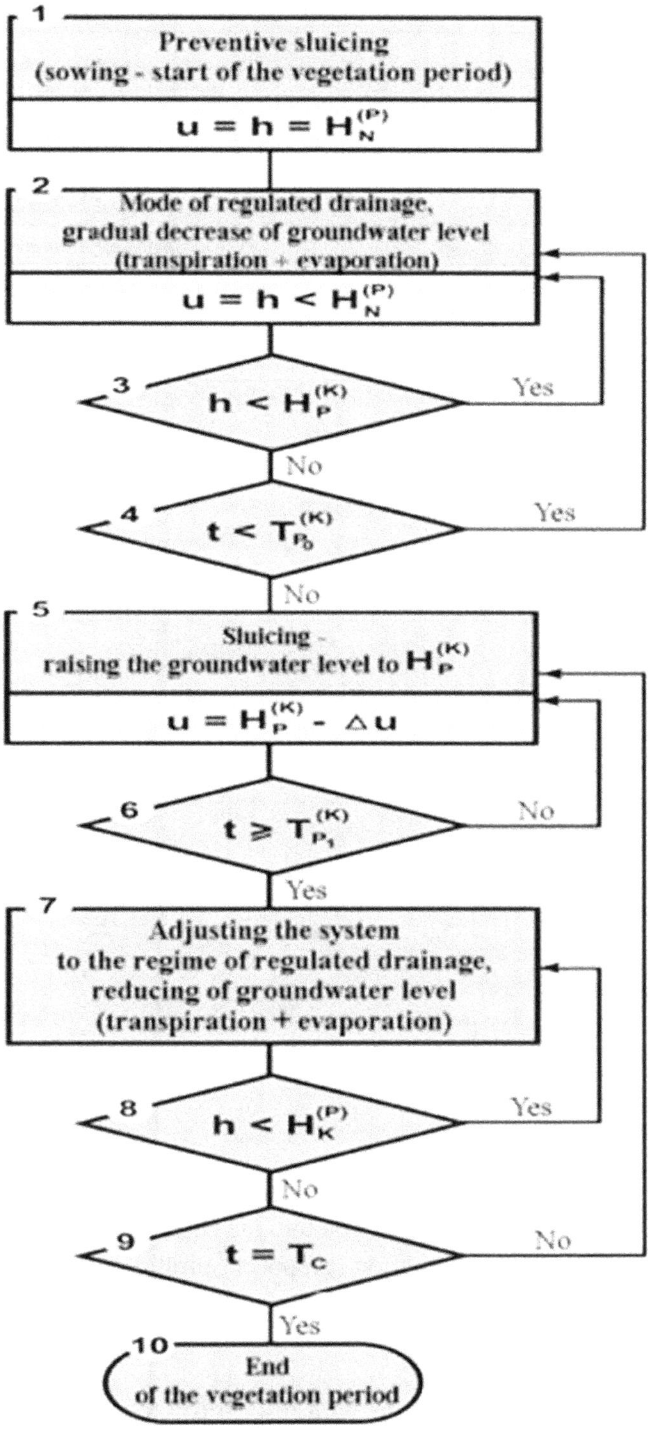

Table 2. Periods of high biological activity of agricultural crops during the growing season

Agricultural crops	Periods of mandatory observance of optimal humidity in the root layer
Perennial herbs: cereals legumes	Regrowth - exit into the tube Regrowth - stemming
Eared grains	Exit into the tube - pouring grain
Corn	The critical period lasts 30 days, 10 before and 20 after throwing out the panicle (water consumption is 40-45% of the total amount)
Potatoes	The beginning of flowering
Fodder beets	From sowing to intensive leaf growth
Flax	From sowing to flowering, the first two months
Oats	From tillering to dropping the panicle

Figure 2. Scheme to determine meliorative regimes during the growing season

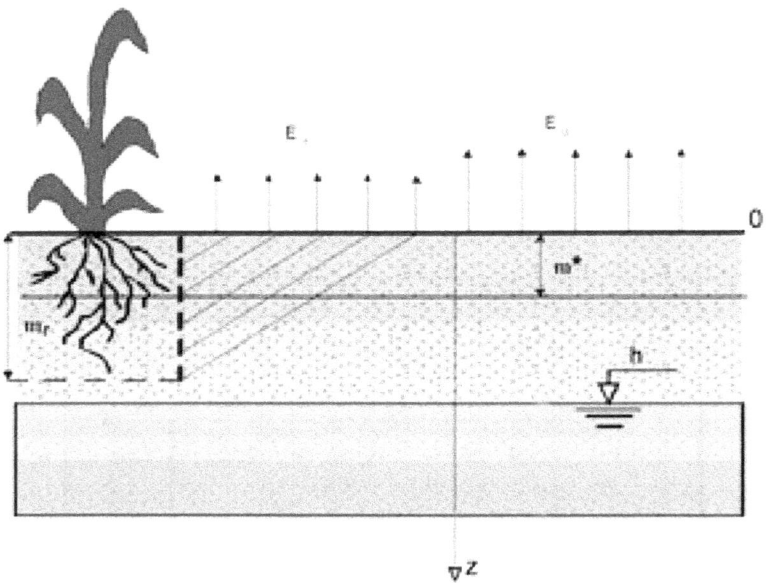

In general, the absorption of water by the root system of plants can be represented by the function $T(z,\psi)$.. Then the intensity of moisture transpiration into the atmosphere E_T is determined by integration T within the root zone, and the total evaporation (evapotranspiration) as the sum:

$$E = E_T + E_u \tag{1}$$

The initial system of steady-state equations with respect to groundwater pressure in the upper soil layer ψ_1, the lower part of the root layer ψ_2, and the lower part of the aeration zone ψ_3 has the form:

$$\frac{\partial}{\partial z}\left[k_{2_1}(\psi_1)\left(\frac{\partial\psi_1}{\partial z}+1\right)\right]-T(z,\psi_1)=0 \text{ at } -m^*\le z\le 0;$$

$$\frac{\partial}{\partial z}\left[k_{2_2}(\psi_2)\left(\frac{\partial\psi_2}{\partial z}+1\right)\right]-T(z,\psi_2)=0 \text{ at } -m^*\le z\le m_r^*; \quad (2)$$

$$\frac{\partial}{\partial z}\left[k_{2_3}(\psi_3)\left(\frac{\partial\psi_3}{\partial z}+1\right)\right]=0 \text{ at } -h\le z\le -m_r,$$

where k_{2_V} - coefficients of moisture conductivity in the *i*-layer ($0\langle k_{2_V}\le k_V$); for the definition k_e, in this case, is accepted $k_{2_V}=k_Ve^{\alpha\psi}$; to determine the volume of moisture, the expression $\theta(\psi)=\theta_m e^{\beta\psi}$ is used, where θ_m– the volume moisture at full saturation, α,β– the parameters that are determined for the soil experimentally.

Boundary conditions have the form:

$$z=0, k_{2_1}(\psi_1)\left(\frac{\partial\psi_1}{\partial z}+1\right)=\gamma E_u^{(m)};$$

$$z=-m^*, k_{2_1}(\psi_1)\left(\frac{\partial\psi_1}{\partial z}+1\right)=k_{2_2}(\psi_2)\left(\frac{\partial\psi_2}{\partial z}+1\right); \psi_1=\psi_2; \quad (3)$$

$$z=-m_r, k_{2_2}(\psi_2)\left(\frac{\partial\psi_2}{\partial z}+1\right)=k_{2_3}(\psi_3)\left(\frac{\partial\psi_3}{\partial z}+1\right);$$

$\psi_2=\psi_3$, z=-*H*, $\psi_3=0$.

Using the approximation for the coefficient of moisture conductivity $k_2=k_ie^{a_i\psi_i}$ the pressure distribution in the aeration zone is described by the dependencies:

$$\psi_1=\frac{1}{a_1}\ln\left[\left(\frac{U_1}{k_{2_1}}\right)^{\frac{a_1}{a_2}}e^{-a_1(z+m^*)}-\frac{a_1}{k_{2_1}}e^{-a_1z}F_1(z)\right];$$

$$\psi_2=\frac{1}{a_2}\ln\left[\left(\frac{U_2}{k_{2_2}}\right)^{\frac{a_2}{a_3}}e^{-a_2(z+z_*)}-\frac{a_2}{k_{2_2}}e^{-a_2z}F_2(z)\right]; \quad (4)$$

$$\psi_3 = \frac{1}{a_3}\ln\frac{-E+\left(k_{2_3}+E\right)e^{-a_3(z+H)}}{k_{2_3}},$$

where in the following dependencies:

$$U_1 = k_{2_2}\left(\frac{U_2}{k_{2_3}}\right)^{\frac{a_2}{a_3}} e^{-a_2\left(m_r+m^*\right)} - a_2 e^{a_2 m^*} F_2\left(-m^*\right),\ U_2 = -E+\left(k_{2_3}+E\right)e^{-a_3\left(m_r-H\right)};$$

$$F_1 = \int_{-m^*}^{z} f_1(\xi)e^{a_1\xi}d\xi,\ F_2 = \int_{-z_k}^{z} f_2(\xi)e^{a_2\xi}d\xi;$$

$f_1(z) = \gamma E_u^{(m)} + \int_z^0 Tdz$ at $-m^* \le z \le 0$, $f_2(z) = \gamma E_u^{(m)} + \int_z^0 Tdz$ at $-m_r \le z \le -m^*$.

The fixed moisture profile is determined by dependences (4) using the main hydrophysical characteristics (MHC) for each layer $\theta_j=\theta_j(\psi_j)$, j=1,2.

In the absence of vegetation:

$$F_1 = \frac{\gamma E_u^{(m)}}{a_1}\left(e^{a_1 z} - e^{-a_1 m^*}\right),\ F_2 = \frac{\gamma E_u^{(m)}}{a_2}\left(e^{a_2 z} - e^{-a_2 m^*}\right);$$

$$U_1 = k_{2_2}\left(\frac{U_2}{k_{2_3}}\right)^{\frac{a_2}{a_3}} e^{-a_2\left(m_r+m^*\right)} - \gamma E_u^m\left[1-e^{-a_2\left(m_r-m^*\right)}\right].$$

The obtained expressions are the basis for determining meliorative regimes during the growing season.

In practice, when regulating the water regime, it is important to establish the necessary range of GWL changes for each calculation period, which ensures the necessary moisture reserves in the root layer of the soil. Information support was developed to calculate the technological parameters of the water regulation algorithm (Table 3).

According to functional characteristics, information support is divided into six blocks. In the first three blocks (meteorological, soil, biological) regulatory and reference information is formed. Filtration and hydrophysical characteristics, including total and minimum moisture capacity included in the soil block, are determined based on existing reference data and, if necessary, supplemented with field determinations. The necessary data for calculations related to the growth characteristics of specific agricultural crops and their water consumption is contained in the biological block. In particular, in this block, averaged data from the leaf index, the power of the root system of cultivated crops according to the phases of their development, and the function of the distribution of moisture absorption by the depth of the aeration zone is formed.

Operational information (the following three blocks) about the actual parameters of water regime regulation (actual GWL for the previous decade and soil moisture), plant growth parameters, and current meteorological parameters are formed based on the results of monitoring the production process on the reclaimed field.

With the given humidity θm in the zone of maximum absorption $z_0=z_m^*$ in the i-th tendays of the growing season as a result of solving the system of equations (2) and with the given boundary conditions and the given function for T, the following dependencies were obtained for determining the limiting GWL (Voropai H.V. et al, 2021):

$$\text{at } z_m^* \rangle m^* \quad h_r^{(i)} = m_r - \frac{1}{\alpha_3}\ln\left\{E + k_{2_3} e^{\alpha_3(m_r - z_m)}\left[e^{\alpha_2\psi_m} + \frac{\alpha_2}{k_{2_2}} e^{\alpha_2 z_m} F_2(z_m)\right]\right\};$$

$$\text{at } z_m^* \langle m^* \quad h_n^i = m_r - \frac{1}{\alpha_3}\ln\frac{E + k_{2_3}\left[e^{\alpha_2(m_r - z_m)} e^{\alpha\psi_m} + \frac{\alpha_1}{k_{2_1}} e^{\alpha_1 z_m} F_1(-z_m)\right] + \frac{\gamma E_u^{(m)}}{k_{2_2}}\left(e^{-\alpha_2(m_r - m^*)} - 1\right)}{k_{2_3} + E}, \tag{5}$$

where m^* – the thickness of the upper layer of the aeration zone, m; m_r – the power of the root system, m; z_m^*– zone of maximum moisture absorption by the root system, m; E – the – total evaporation, m/day; E_u – the physical evaporation (evaporation), m/day when justifying melioration regimes, it is enough to accept a linear dependence of physical evaporation, which decreases to the maximum hygroscopicity $E_u = \gamma E_u^{(m)}$, where $\gamma = \frac{\theta_* - \theta_K}{\theta_r - \theta_K}$, θ_r is some threshold value of humidity, depending on which physical evaporation E_u changes as follows: if soil moisture near the surface θ_* is greater θ_r ($\theta_* \rangle \theta_r$), then E_u retains its maximum value $E_u^{(m)}$, which is determined by the energy resources of the atmosphere; when $\theta_* \langle \theta_r$ the intensity of evaporation decreases according to the aforementioned linear law.

ILLUSTRATIVE EXAMPLE

The implementation of the technology of resource-saving management of the moisture supply of agricultural crops was carried out at the research and production site of the Drainage System (DS) "Ikva" (Rivne region, Ukraine) during the cultivation of perennial grasses and corn for silage.

The main technological parameters necessary for the implementation of the combined control algorithm are the time of switching the control network to one or another mode; the duration of the transition process and its spatio-temporal characteristics. According to the results of the typification and schematization of the natural conditions of the humid zone of Ukraine, the most common filtration scheme is a non-homogeneous layered structure of an aquifer with an arable layer of small thickness (0.2–0.3 m), the water-physical characteristics of which differ significantly from the underlying horizons, and operational fluctuations GWL take place within the upper layer (1.2 m of a thick).

Table 3. Information support for calculations of optimal reclamation regimes

Nº	Parameter	Designation	Units of measurement
REGULATORY AND REFERENCE INFORMATION			
Meteorological block			
1	Atmospheric precipitation (multi-year average for the i-th tendays)	P^i	m/day
2	Air temperature (multi-year average for the i-th tendays)	t^i	°C
3.	Deficit of air humidity (multiannual average for the i-th tendays)	d^i	mb
Soil block			
1.	The full moisture content (FMC)	θn	% of the volume
2.	The lowest moisture content	θm	% of the volume
3.	Coefficient of filtration	k_ϕ	m/day
4.	Coefficient of moisture conductivity	k_e	m/day
5.	Basic hydrophysical characteristic	$\theta(\psi)$	-
Biological block			
1.	Type of cultivated crop, phase of development	-	-
2.	Average power of the root system (per tendays)	m_r	m
5.	Leaf index	ωL	-
6.	Absorption distribution function	$T(z)$	-
OPERATIONAL INFORMATION			
Meteorological block (current parameters)			
1.	Atmospheric precipitation	P^i	m/day
2.	Air temperature	t^i	°C
3.	Lack of air humidity	d^i	mb
Biological block			
1.	Type of cultivated crop, phase of development	-	-
2.	Average power of the root system	m_r	m
3.	Leaf index Листковий індекс	ωL	-
4.	Absorption distribution function	$T(z)$	-
Technological block			
	Current parameters of water regulation (GWL and moisture in the root soil layer)	$h^{(i)}$ $\theta(^{i)}$	m % of the volume

Complex studies related to the processes of regulating the soil water regime include long-term observations of weather conditions, the dynamics of crop growth, total evaporation, water exchange in the aeration zone (including infiltration, and capillary feeding from GWL), and soil moisture.

The research plot, where corn was grown for silage, has an area of 28 hectares. The domestic regulation network is represented by a closed pottery drain with a depth of 0.9 - 1.2 m. The calculation scheme is represented by a two-layer structure of the aeration zone with an upper layer of low-capacity peat (0.3 m), well decomposed with a volumetric weight of 0.60 – 0.65 t/m^3, which turns into moderately decomposed peat with a volumetric weight of 0.175 t/m^3 (Table 4).

Table 4. Characteristics of the research plot

N°	Name	Characteristics
1	The method of regulating the water-air regime	closed pottery drainage
2	Cultivated crop	Perennial grass corn for silage
3	Distance between drains, m	16.0
4	Drain laying depth, m	0.9-1.2
5	The number of layers to the water stop	2
6	Soil characteristics of the aeration zone: a) upper layer - soil type - filtration coefficient, m/day - power, m b) lower layer - soil type - filtration coefficient, m/day - soil capacity, m	well-decomposed peat mineralized 0.1-0.3 0.30 peat 0.45-0.6 5-5.5
7	Distance to the water stop, m	5.5

The technological parameters that characterize the process of subsoil moistening (dehydration) and are related to the regulation of the water regime are defined according to the developed methodology for calculating the wetting regimes.

Research on the dynamics of GWL and soil moisture during the cultivation of perennial grasses is divided into two stages: the cycle of passive reduction of GWL (duration is 53 days), when due to a small amount of precipitation, as well as evaporation and transpiration of plants, GWL decreases to the maximum possible - 1.26 m. According to traditional technology, GW were maintained at a depth of 0.7-0.9 m. This period continued until August 12. The end of the cycle of passive reduction of GWL and the beginning of moistening was determined by soil moisture in the zone of its maximum absorption by the root system of perennial grasses at a depth of 0.15 - 0.25 m (monitored by measurements of groundwater pressure). When the moisture in this soil layer decreased to the lower limit of its optimal value, the first cycle ended (passive reduction of GWL) and the drainage system was transferred to the moistening mode.

As a result of the regulation of GWL in different modes (passive reduction and humidification) taking into account natural conditions, in particular the amount of precipitation, the necessary parameters were obtained that characterize the water regime in the aeration zone. The analysis of the obtained results shows that the moisture regimes calculated according to the developed method made it possible to sufficiently provide moisture to the root layer. Accumulated precipitation in the active soil layer (0 - 0.6 m) in the mode of passive reduction of GWL to 1.26 m, when an accumulative capacity for moisture retention was formed in the upper layers of the soil, was used as efficiently as possible.

It was established that during the growing season in the cyclic humidification mode, the amount of infiltration to the GWL in the aeration zone was 9.65 mm (96.5 m^3), and when implementing resource-saving modes, it was 1.42 mm (14.2 m^3). In general, during the cycle of passive reduction of GWL, the amount of total evaporation was 19.58 mm (195.8 m^3), and in the mode of cyclic humidification - 141.33 mm (1413.3 m^3). It was determined that evaporation from GWL significantly decreases, and transpiration costs are carried out at the expense of moisture reserves in the aeration zone. At the same time, the non-productive consumption of water for evaporation is significantly reduced.

To evaluate the effectiveness of regulating the water regime from the point of view of minimizing the washing regime for years of research when implementing the developed regimes and with traditional cyclic moistening, calculations were made that confirm a significant decrease in the amount of infiltration on the GWL. On average, its value is 4 times smaller than with the cyclic sluicing scheme. The results of studies of water exchange in the aeration zone showed a decrease in the amount of evaporation from GWL on average from 932 to 328 m^3/ha, which is 2.8 times less than the amount of evaporation in the cyclic sluicing scheme.

The technical and economic indicators of the implementation of resource-saving modes of water regulation at the experimental site of the DS "Ikva" are given in the Table. 5.

Table 5. Technical and economic indicators of the implementation of resource-saving modes of water regulation (DS "Ikva")

N°	Name of indicators	Quantity
1	drainage area including polder drainage	245 ha 245 ha
2	irrigation area	140 ha
3	The total moisture deficit in a dry year is 75% of precipitation	366.32 thousand m^3
4	area covered due to moisture sluicing	242.5 thousand m^3
5	The source of irrigation	Ikva River
6	The capacity of the pumping station	0.77 m^3/s
7	The volume of pumped water per year	861.3 thousand m^3
8	Electricity consumption per year	198,000 kW/h
9	Water savings from the implementation of resource-saving water regulation regimes per year	48.7 thousand m^3
10	Savings in electricity consumption per year	11.2 thousand kW/h

The period of approbation when growing corn for silage falls on the period from the 2nd decade of June to the end of the 3rd decade of July. For the cultivated crop, corn for silage, the critical period for moisture supply is from June 23 to July 23 (the panicle shedding phase: 10 days before and 20 days after panicle shedding). During this period, it is necessary to ensure optimal moisture in the root layer of the soil.

The permissible range of soil moisture in other growing seasons (outside the critical period) is considered wider: the upper limit for mineralized peat in the 0-0.3 m layer is 0.7 FMC (51.8%), and the lower limit corresponds to the smallest permissible GWL in the summer period (moisture withering) – 1.0 m.

For the critical period in terms of moisture supply, the optimal GWL is determined by the method of calculating water regulation modes and we accept the required range of humidity within (60 - 70) % of FMC, i.e., 44.4 - 51.8%.

The results of calculations of the total evaporation of corn for silage are shown in the Table. 6.

Table 6. Total evaporation of corn for silage

Month	Tendays (decada)	Air temperature, °C	Deficit of air humidity, mb	corn for silage	
				bioclimatic coefficients, k_1	Total evaporation, *E*, mm/day
June	1	19.2	5.5	0.47	1.25
	2	20.2	6.7	0.52	1.79
	3	20.3	6.4	0.62	2.34
July	1	19.8	7.1	0.74	2.56
	2	18.0	4.9	0.90	2.46
	3	19.6	6.1	0.85	2.72

For the critical moisture supply period, we determine the amount of total evaporation (E_p = 2.39 mm/day).

The dynamics of the leaf index during the testing period are presented in Figure 3.

Figure 3. Dynamics of the leaf index on the experimental site (DS "Ikva", cultivated crop - corn for silage)

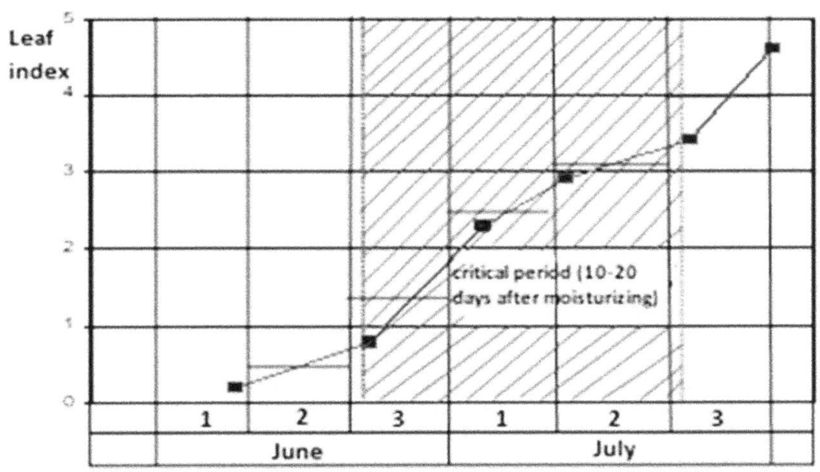

At the same time, we take into account that the actual date of the beginning of corn sowing is 24.04, and the end of the growing season is 30.08 (corn harvest). Thus, the duration of the growing season is 128 days. More than 70% of the root system is already formed by the end of June (0.35-0.40 m). At the end of July, the actual capacity of the corn root system was 0.45 m.

To determine the hydrophysical characteristics of the soil of the research and production area, in particular the main hydrophysical characteristics, we use the graph of the dependence of tensiometry pressure on soil moisture for peat soils.

The coefficient of moisture conductivity of the soil is calculated according to the dependence:

$$\kappa_{в} = \kappa \bullet \psi^{L} \tag{6}$$

where ψ – tensiometry pressure; coefficients k, L for the upper layer with a thickness of 0.3 m are k* = 0.003, L* = - 0.957; for the lower layer – k_1 = k* = 0.003, L_1 = - 1.121.

We formed the base of initial data of indicators of the total, physical evaporation, and transpiration in the decadal distribution. At the same time, this database is supplemented with actual data to determine the need to adjust the current GWL for each previous decade. The initial data for calculating the optimal water regulation regimes are given in the Table. 7.

Table 7. Output data for calculating optimal water regulation regimes for growing corn for silage

<table>
<tr><th rowspan="2">Month</th><th rowspan="2">Tendays (decada)</th><th colspan="2">E, mm/day</th><th rowspan="2">Precipitation P/P^c, mm/day</th><th rowspan="2">m_r, m</th><th rowspan="2">ω_L</th></tr>
<tr><th>E_T/, mm/day</th><th>E_u/ E_u^c, mm/day</th></tr>
<tr><td rowspan="6">June</td><td rowspan="2">1</td><td colspan="2">1.25/1.4</td><td rowspan="2">5.1/1.6</td><td rowspan="2">0.21</td><td rowspan="2">0.13</td></tr>
<tr><td>0.1/0</td><td>1.15/1.4</td></tr>
<tr><td rowspan="2">2</td><td colspan="2">1.79/1.6</td><td rowspan="2">9.9/1.7</td><td rowspan="2">0.28</td><td rowspan="2">0.46</td></tr>
<tr><td>0.4/0.1</td><td>1.39/1.5</td></tr>
<tr><td rowspan="2">3</td><td colspan="2">2.34/1.9</td><td rowspan="2">0.0/2.6</td><td rowspan="2">0.37</td><td rowspan="2">1.27</td></tr>
<tr><td>0.9/0.2</td><td>1.44/1.7</td></tr>
<tr><td rowspan="5">July</td><td rowspan="2">1</td><td colspan="2">2.56/2.2</td><td rowspan="2">0.1/2.4</td><td rowspan="2">0.41</td><td rowspan="2">2.47</td></tr>
<tr><td>1.9/0.5</td><td>0.66/1.7</td></tr>
<tr><td rowspan="2">2</td><td colspan="2">2.46/2.9</td><td rowspan="2">1.9/2.6</td><td rowspan="2">0.43</td><td rowspan="2">3.1</td></tr>
<tr><td>2.0/1.0</td><td>0.46/1.9</td></tr>
<tr><td>3</td><td colspan="2">2.72/2.6</td><td>1.1/2.3</td><td>0.45</td><td>3.9</td></tr>
</table>

The optimal humidity range is maintained in the zone of maximum moisture absorption at the depth of 0.2 m.

Based on the fact that the thickness of the upper soil layer (m* = 0.3 m) exceeds the established depth of the zone of maximum moisture absorption (z_0=0.2 m), the optimal GWL that corresponds to the specified humidity ranges [$\theta_{m,}\theta_{n}$], we determine based on the solution of the stationary moisture transfer problem (Voropai H.V. et al, 2021; Kuzmych L. et al, 2021b, Kuzmych L. et al, 2021c). At the same time, to establish the appropriate range of tensiometry pressure [ψ_m,ψ_k], we use the dependencies $\psi=\psi(\theta)$ for peat soils. We use the dependence of soil moisture on tensiometry pressure to determine soil moisture based on tensiometer readings.

The results of model calculations of water consumption by the root layer depth are shown in Figure 4.

Figure 4. Average daily water consumption of corn at the depth of the root layer of the soil (the 3rd decade of July)

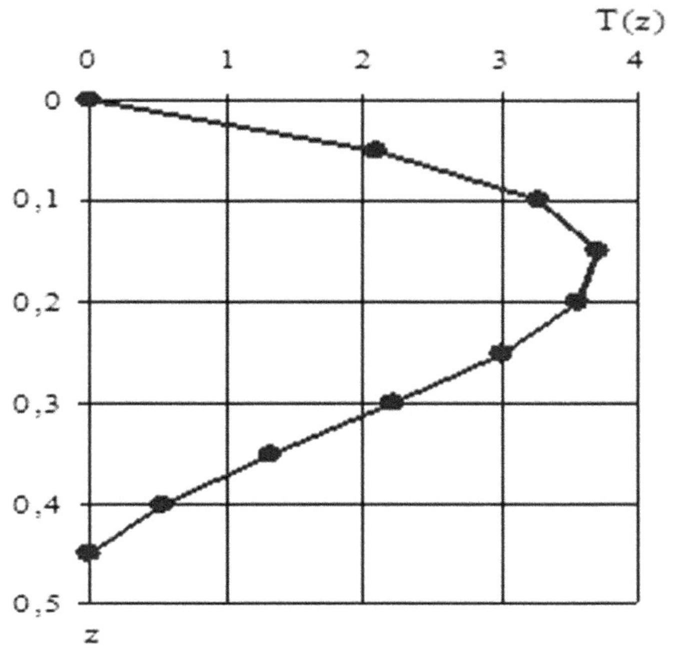

The determination of the upper and lower limits of optimal water regulation regimes (in the decadal distribution) is carried out taking into account the period critical for moisture supply, and the upper and lower limits of optimal humidity are determined in the zone of maximum moisture absorption. The results of the calculations are given in the Table 8.

Table 8. Recommended and actual parameters of water regulation when growing corn for silage

Meliorative regime	**June**			**July**		
	1	**2**	**3**	**1**	**2**	**3**
Groundwater level, m from the surface						
Recommended upper limit	0.75	0.75	0.79	0.80	0.80	0.80
Recommended lower limit	0.95	1.01	0.88	0.85	0.87	1.02
Actual GWL	0.79	0.6	0.79	0.85	0.80	0.80
Humidity in the zone of maximum moisture absorption (0.15-0.25 m), % of volume						
Recommended upper limit	51.8	51.8	51.8	51.8	51.8	51.8
Recommended lower limit	40.7	40.7	44.4	44.4	44.4	40.7
Actual humidity	49.5	57	50.3	49.6	53.2	45.0

During the testing period, weather conditions were especially "tense" at the end of the first decade and the first half of the second decade of June, when the amount of precipitation exceeded 120 mm.

In general, for the entire period of approbation of the resource-saving technology, with the purposeful creation of storage tanks in the upper horizons of the root layer of the soil, the moisture supply was optimal and there was no need for sluicing during the growing season (Figure 5).

Figure 5. Dynamics of humidity in the root layer of the soil (a) and GWL (b) for implementing resource-saving management technology moisture supply during corn cultivation (DS "Ikva", Rivne region)

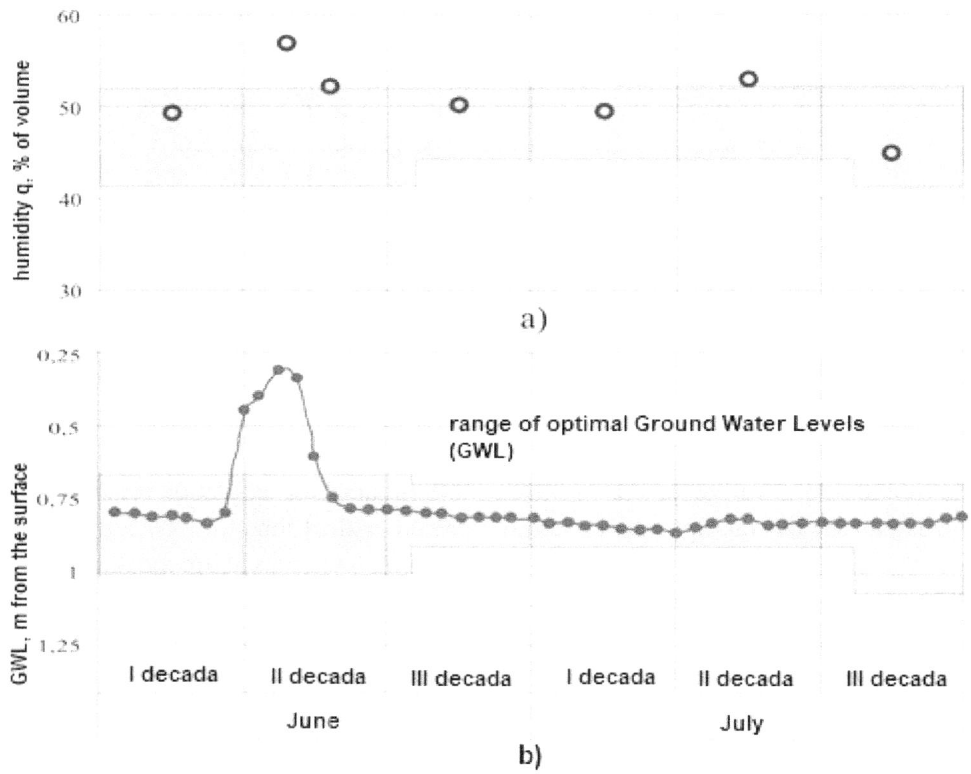

At the control plots with the traditional technology of regulating the water regime, in the same period, the reduction of GWL to the upper permissible limit occurred much later (by 9 days), which led to unacceptable flooding of the root system of corn; moisture in the root layer of the soil exceeded the permissible upper limit for a long time (until June 27). At the same time, the infiltration of atmospheric precipitation to GWL amounted to 67 mm, which on average exceeded this value by 35% in comparison with the resource-saving technology of moisture management.

In the first case, the drainage volumes are 17% higher than the same volumes determined by the traditional technology, which does not take into account the volumes of water that are additionally released due to the redistribution of moisture in the aeration zone (unsaturated zone) when reducing the GWL.

Research on the development of technological parameters of water regulation during the cultivation of promising fodder crops was carried out during 2016–2019 at the drainage systems of the Sarny Research Station (SRS) of the Institute of Water Problems and Reclamation of the National Academy of Agrarian Sciences of Ukraine (IWPLR NAAS), in particular peat swamp massif "Chemerne", Rivne region, and the drainage system "Romen" (Sumy region, Ukraine). These objects, taking into account natural and climatic conditions and structural and technological features of melioration systems, are representative of the territory of the humid ecozone (Polyssia and Forest Steppe of Ukraine).

It is considered that a promising direction in fodder production is the introduction of highly productive and profitable varieties of agricultural crops that could complement traditional fodder crops and that have significant adaptive potential when grown in the complex agro-climatic conditions of the humid zone (Polyssia and Forest-Steppe) of Ukraine. This will make it possible to increase fodder production and reduce the cost of plant and animal products. Since climate changes lead to the aggravation of droughts, which are also characteristic of the Polissia and Forest-Steppe zones, the rational use of water resources is also relevant. Therefore, when forming the structure of the fodder group of agricultural crops based on the assortment of their varieties and hybrids, it is important to introduce such crops that are characterized not only by high productivity but also by drought resistance (Korobiichuk, I. et al, 2017, 2020; Rokochinskiy A. et al, 2022).

Paiza, amaranth, and fodder beans occupy an important place among rare, but highly productive, fodder crops for cultivation on drained lands. The existing national and international experience shows that the introduction of these crops into production allows not only provides livestock with high-quality feed but also satisfies the needs of the food and pharmaceutical industry with valuable raw materials (Koval S.I., Zosimchuk O.A., 2014; Kuzmych L. et al, 2021a)

The method of calculating the technological parameters of managing the moisture supply of agricultural crops is based on the use of physically based models of filtration and moisture transfer in saturated-unsaturated zones under the conditions of drainage, taking into account the peculiarities of the formation of soil processes, their inertia, and the structural features of the reclamation network (Onyango C.M. et al, 2009; Rokochinskiy A. et al, 2020)

Taking into account the research experience of previous years on reclaimed lands of the SRS and the DS "Romen" regarding the cultivation of high-yielding crops that have high profitability and are promising for growing in the humid zone, experimental studies were conducted on the development of technological parameters of water regulation in the cultivation of high-yielding fodder crops.

The basis for the development of technological parameters of water regulation in the cultivation of paiza, amaranth, and forage beans are the results of field studies on the determination of meteorological parameters (precipitation, air temperature), dynamics of the groundwater level (GWL) and humidity in the root layer of the soil (Stetsyuk M. A., 2015).

According to the research results, resource-saving technological parameters of water regulation (standards of groundwater level and moisture in the root layer of the soil) during the growing season during the growing of high-yielding fodder crops (peas, amaranth, and fodder beans) on peat and mineral soils of Polissia and Forest Steppe of Ukraine were determined. Permissible terms have been established in which the drainage system must ensure the removal of excess water and ensure a timely reduction of the GWL to the recommended indicators (Table 9).

Table 9. Recommended GWL for growing paiza, amaranth, and fodder beans on drained lands

Vegetation phase	**Groundwater level, m (in the numerator - optimal; in the denominator - the smallest permissible)**					
	pre-sowing period	**seedling-tubbing**	**exit into the tube**	**shedding of panicles**	**blossom**	**seed ripening**
Paisa						
Peat soils	0.50-0.55	0.60-0.65 0.65-0.70	0.60-0.75 0.75-0.85	0.60-0.75 0.75-0.85	0.75-0.85 0.90-0.95	0.75-0.90 0.9-1.0
Mineral soils	0.50-0.60	0.60-0.70 0.70-0.75	0.65-0.75 0.80-0.85	0.65-0.75 0.80-0.85	0.80-0.90 0.90-1.0	0.85-0.90 1.0-1.1
Amaranth						
Vegetation phase	pre-sowing period	Sprouting- the appearance of a leaf	shedding of panicles		blossom	seed ripening
Peat soils	0.60-0.65	0.60-0.65 0.65-0.75	0.60-0.70 0.70-0.80		0.65-0.80 0.80-0.90	0.80-1.0 1.0-1.1
Mineral soils	0.50-0.60	0.60-0.70 0.70-0.75	0.65-0.75 0.75-0.80		0.70-0.85 0.90-1.0	0.85-1.0 1.0-1.1
Fodder beans						
Vegetation phase	pre-sowing period	Sprouting- the appearance of a leaf	budding	blossom	formation of beans	seed ripening
Peat soils	0.55-0.65	0.55-0.65 0.70-0.75	0.60-0.65 0.70-0.75	0.65-0.80 0.80-0.90	0.75-0.90 0.90-1.0	0.90-1.0 1.1-1.2
Mineral soils	0.55-0.65	0.60-0.70 0.70-0.75	0.60-0.70 0.70-0.75	0.65-0.75 0.80-0.90	0.80-0.90 0.90-1.0	1.0-1.1 1.1-1.2

Approbation of the technological parameters of water regulation during the cultivation of paiza, amaranth, and fodder beans (Table 10) was carried out at the Breeding and Demonstration Range of Highly Productive Fodder Crops of the Sarny Research Station (SRS, Rivne Region) and the State Enterprise “Nadia” of the National Academy of Agrarian Sciences of Ukraine (DS “Romen”, Sumy Region).

During the testing period, research was conducted on the dynamics of GWL and soil moisture during the growing season. Optimum reclamation regimes on the lands of the reclamation systems of the Drainage Systems “Sarny” and “Romen” were ensured by regulating the GWL and the humidity of the root layer of the soil according to the phases of the development of paiza, amaranth, and fodder beans and taking into account the critical periods of their optimal moisture supply.

According to the results of research, it was established that on the reclaimed lands of the Sarny District, due to the minimal amount of precipitation during the autumn-winter period of 2019–2020 and the spring period of 2020 for the sowing period (the second half of April), the GWL decreased to a depth of 114 cm -regulators on the meliorative system at the beginning of May (for the period of sowing paiza, amaranth and forage beans) it was possible to raise the GWL to 84–87 cm. In the future, ensuring optimal humidity and, accordingly, the moisture reserves of the soil according to the development phases of the studied crops was achieved by regulating the GWL and taking into account additional moisture due to the accumulation of precipitation in the root layer.

Optimum humidity and, accordingly, soil moisture reserve according to the development phases of the studied crops on the DS "Romen" were ensured during the critical periods of growing paiza, amaranth, and fodder beans and were achieved mainly due to the accumulation of precipitation in the root layer. Due to the insufficient filling with water resources of the existing water sources (accumulative capacity, Karabutiv reservoir), which are located in the zone of reclaimed lands of the DS "Romen", water supply for moistening the cultivated crops was not carried out (Kuzmych L. et al, 2021a).

Table 10. Approbation of technological parameters of water regulation during the cultivation of paiza, amaranth, and fodder beans

The name of the development	**Optimum reclamation regimes and technological parameters of water regime management when growing paiza, amaranth, and fodder beans on drained lands**
Developer:	Institute of Water Problems and Reclamation NAAS
The approval period	2020
The name of the agricultural enterprise, where the approval was carried out	State Enterprise "Nadia" of the National Academy of Agrarian Sciences (the Drainage System "Romen", Sumy Region). The area on which the approval was carried out is 15 hectares
	the Breeding and Demonstration Range of Highly Productive Fodder Crops of the Sarny Research Station (SRS) of the Institute of Water Problems and Reclamation of the National Academy of Agrarian Sciences of Ukraine (Rivne Region) The area on which the approval was carried out is 0.3 ha
A brief description of the development	Optimal meliorative regimes is provided by adjusting the GWL and soil moisture. For drained soils (both mineral and organic), the average moisture in the 0-30 cm layer at the beginning of sowing is 75-80% of the full moisture capacity (FMC). The recommended norms of soil moisture during the period of active vegetation are: - for peat soils, the optimum is 65–75%, and the lowest permissible in summer is 55–60%; - for minerals, the optimum is 65–80%, and the lowest permissible in the summer period is 55–60% of the FMC.

The results of the approbation of optimal reclamation regimes and technological parameters of water regime regulation in the cultivation of paiza, amaranthand fodder beans on drained lands show that average and higher-than-average yield indicators of these crops were obtained. The yield of fodder units on the reclaimed land of the DS SRS when growing fodder beans is 67.5–86.2 t/ha, amaranth – 34.9–48.8 t/ha, and paiza – 39.6–54.0 t/ha; on the DS "Romen" - when growing fodder beans is 56.8–49.3 t/ha, amaranth – 45.7–57.5 t/ha, and paiza – 65.8–77.5 t/ha.

THE DIRECTION OF FUTURE RESEARCH

Given that the main feature of the impact of modern climate changes on the water supply of reclaimed territories of the humid zone of Ukraine is the formation of conditions that cause a shortage of local drainage and surface runoff, and in conditions where a decrease in usable water resources due to climate change is predicted, for modern agricultural production it is an important development and implementation of scientifically based resource-saving and ecologically safe water regulation technologies when growing economically attractive crops on drained lands.

CONCLUSION

1. In order to solve the modern problems of soil water regime management, differentiated and reliable moisture supply of cultivated crops, scientific development, and implementation of resource-saving and ecologically safe water regulation technologies are necessary, which will ensure the requirements of cultivated crops to regulate the soil water regime during the growing season with minimal consumption of water resources, taking into account ecological today's requirements.
2. The implementation of the algorithm of resource-saving regulation on drained lands in experimental and production conditions during the cultivation of perennial grasses shows a significant reduction in water consumption for evaporation and an increase in the efficiency of the use of moisture reserves of the root layer of the soil.

Taking into account the spatial characteristics of the processes of moisture absorption by the roots of plants makes it possible to adequately predict the moisture regime of the root layer of the soil and, on this basis, to carry out drying and moistening measures and to use the available water resources more reasonably and economically, to minimize the possible negative impact of infiltration processes in the aeration zone on groundwater pollution.

3. Purposeful creation of accumulative tanks and the possibility of moisture accumulation due to precipitation in the upper horizons of the aeration zone, which is ensured by a justified reduction of the GWL, contributed to an increase in the moisture availability of perennial grasses by 15-20%, significantly reduced water consumption for infiltration beyond the root layer of the soil. On average, the amount of infiltration was 4 times smaller compared to the traditional regulation scheme (cyclic sluicing).

The saving of water resources from the implementation of resource-saving technologies of water regulation on the example of a module of an experimental plot with an area of 245 hectares is 48.7 thousand m^3 per year, consumed electricity is 45.7 kWh per 1 ha of drainage area, which is 5.7% per year.

4. Approbation of the resource-saving technology of water regulation when growing corn for silage on drained lands in the growing season "stressed" due to weather conditions showed the possibility of maintaining the level of groundwater within the optimal range and required moisture and, accordingly, moisture reserves in the root layer of the soil.

In the conditions of the average rainfall of the growing season and their uneven distribution and simultaneous accumulation of the majority of these precipitations in the upper layers of the root layer of the soil, there was no need for sluicing. Water savings during the June-July testing period amounted to 670 m^3/ha, while operating costs for sluices were reduced by 2-3 times.

5. Adherence to scientifically based technological parameters for regulating the water regime of the soil when growing high-yielding fodder crops (peas, amaranth, and fodder beans) allows you to increase the yield of cultivated crops by 15-20%, ensure the saving of water resources - up to 25%, and obtain an economic effect of up to 9000 -10000 UAH /ha.

REFERENCES

Datsko, L. V. (2016). Modern agricultural use of lands of the humid zone of Ukraine. *Reclamation and water management*. 2016. *Vip., 103*, 41–47.

Irrigation and drainage strategy in Ukraine for the period up to 2030. (2019). FAO.

Korobiichuk, I., Drevetsky, V., Kuzmych, L., & Kovela, I. (2020). The method of multy-criteria parametric optimization. *Advances in Intelligent Systems and Computing, 1140*. https://link.springer.com/chapter/10.1007/978-3-030-40971-5_9 doi:10.1007/978-3-030-40971-5

Korobiichuk, I., Kuzmych, L., Kvasnikov, V., & Nowak, P. (2017). The use of remote ground sensing data for assessment of environmental and crop conditions of the reclaimed land. Advances in Intelligent Systems and Computing (AISC), 550, 418–424. doi:10.1007/978-3-319-54042-9_39

Koval S.I., & Zosimchuk O.A. (2014). Productivity of fodder crop rotations from rare fodder crops on drained peat soils of Western Polissia. *Bulletin of the National University of Water Management and Nature Management*, *65*(1), 64 - 72.

Kovalenko P.I., Yatsyk M.V., Polyakov V.L. (1996). Management of moisture supply of agricultural crops on reclaimed land taking into account the dynamics of external environmental factors. *Land reclamation and water management, 82*.

Kuzmych, L., Furmanets, O., Usatyi, S., Kozytskyi, O., Mozol, N., Kuzmych, A., Polishchuk, V., & Voropai, H. (2022). Water Supply of the Ukrainian Polesie Ecoregion Drained Areas in Modern Anthropogenic Climate Changes. *Archives of Hydro-Engineering and Environmental Mechanics*, *69*(1), 79–96. doi:10.2478/heem-2022-0006

Kuzmych, L., Voropay, G., Moleshcha, N., & Babitska, O. (2021a). Improving water supply capacity of drainage systems at humid areas in the changing climate. *Archives of Hydro-Engineering and Environmental Mechanics.*, *68*(1), 29–40. doi:10.1515/heem-2021-0003

Kuzmych, L. V., Voropay, G. V., Shevchuk, S. A., Usatiy, S. V., Kozitsky, O. M., & Mozol, N. V. (2021c). Influence of climate change on the condition of the reclamation fund of the Ukrainian Polissya on the example of the Maryanivka drainage system. *Engineering: theory and practice: materials of the I International Correspondence Scientific and Practical Conference.* Polissya State University.

Kuzmych, L. V., Voropay, G. V., Shevchuk, S. A., Usatyi, S. V., Kozytsky, O. M., & Mozol, N. V. (2021b). The influence of climate change on the state of the reclaimed lands of the Ukrainian Polissya on the example of the drainage system "MARYANIVKA". *Proceedings of the IV Intern. scientific-practical. conf.* State Technical University.

Baluk, S. A. (2015). Land reclamation. *Collective monograph.* Kherson: Grin D.S.

Oliynyk, A. Ya. (1981). *Hydrodynamics of drainage*. Naukova dumka.

Oliynyk A. Ya. (1985). Regulation of the water-air regime of soils on the basis of mathematical modeling. *Improvement of the efficiency of drying-moistening systems*. K.: UkrNIIGyM.

Oliynyk, A. Ya., & Polyakov, V. P. (1987). *Drainage of waterlogged lands*. Naukova dumka.

Onyango, C. M., Imungi, J. K., Mose, L. O., & Harbinson, J. Olaf Van Koteen. (2009). Feasibility of commercial production of amaranth leaf vegetable by small scale farmers in Kenya. *African Crop Science Conference Proceedings,* 9, 767-772.

Rokochinskiy, A., Korobiichuk, I., Kuzmych, L., Volk, P., & Kuzmych, A. (2020). The system optimization of technical, technological and construction parameters of polder systems. *Advances in Intelligent Systems and Computing*. 1140. . https://link.springer.com/chapter/10.1007/978-3-030-40971-5_8 doi:10.1007/978-3-030-40971-5

Rokochynskiy, A., Volk, P., & Kuzmych, L. (2022). *Formation of Water Consumption at Drained Lands in Climate Changes. Monograph*. LAMBERT Academic Publishing.

Romashchenko, M. I. (2019). *Development of scenarios of climate change in Ukraine in the medium- and long-term using data from global and regional models*. UHMI. http: // uhmi.org.ua/project/rvndr/climate.pdf

Romashchenko M.I., Husyev Yu. V., Shatkovskyi A. P., Saidak R. V., Yatsyuk M. V., Shevchenko A. M., Matiash T. V. (2020). The impact of modern climate change on water resources and agricultural production. *Land reclamation and water management*, *1,* 5–22.

Romashchenko, M. I., Saidak, R. V., Matiash, T. V., & Knysh, V. V. (2019). Influence of climatic changes on moisture supply of the territory of Ukraine and production of agricultural products. *“Water for All”: dedicated to World Water Day: International. scientific-practical Conference: thesis add*. Kyiv, 2019. P. 179–180.

Stetsyuk M. A. (2015). *Promising payza* . The Ukrainian Farmer.

Voropai H.V. (2021). Modern aspects of scientific support for the restoration of drainage systems of the humid zone of Ukraine. *Reclamation and water management, 1*, 39-51.

Voropai H.V., Yatsyk M.V., Mozol N.V. (2019). Current state and prospects of development of drainage reclamation in the conditions of climate change. *Reclamation and water management*.

Yatsyk, M. V., Skrypnyk, O. V., Voroshnova, L. M., Voropay, G. V., & Mozol, N. V. (2009). Resource-saving technology of water regulation on reclaimed lands in the zone of excess moisture. Land reclamation and water management, 97, 68–76.

Yatsyk M.V., Voropai H.V., Moleshcha N.B. (2016). Improving water supply of reclamation systems of the humid zone. *Reclamation and water management, 103*, 63–68.

Yatsyk M.V., Voropai H.V., Topolnuk T.I., Shushkevich Yu.A. (2015). Methodological bases of increase of water supply of reclamation systems by creation of accumulating capacities. *Land reclamation and water management, 102,* 54–58.

Chapter 6
Ecological and Reclamation Monitoring of Drained Lands of Ukrainian Polesie

Viktor Moshynskyi
National University of Water and Environmental Engineering, Ukraine

ABSTRACT

The principles of organization and implementation of ecological and reclamation monitoring of drained lands are considered in the example of the Ukrainian Polesie zone. To control the changes occurring in natural complexes under the influence of drainage, representative (typical) systems are used, which fully characterize the complex natural and economic conditions of the studied object. On the basis of the mathematical model of forecast assessment of the condition of drained lands (FACDL) and the concept of ecological niche, the corresponding mathematical model and the methodology of the assessment of environmental sustainability of drained lands and natural and technogenic systems (ASDL) are developed to objectively interpret the data of ecological and reclamation monitoring, rational management of drained lands of Ukraine under the conditions of preservation of the environmental sustainability

BACKGROUND

Establishment and Development of Drained Lands Monitoring in the Humid Zone of Ukraine

Modern reclamation science has its own paradigm, methodology, principles, schools, etc. approaches that consider land reclamation as a complex system of environmental optimization, taking into account the creation of highly efficient agricultural production, which can be considered classic for drainage reclamation. Of great scientific and practical importance for land reclamation in the humid zone of Ukraine are the works of S. Perekhrest, A. Yangol, V. Kubyshkin, D. Tiutiunnyk, S. Sabochkina, V. Aleksieievskyi, H. Pototskyi, P. Kovalenko, A. Bulavko, P. Zarzhevskyi, F. Murashko, S. Skoropanova,

DOI: 10.4018/978-1-6684-8248-3.ch006

V. Shebeko, etc. (Hadzalo, Stashuk, Rokochynskyi, 2017), who covered the issues of drainage norms, hydraulic and filtration calculations of systematic drainage, deep loosening of soils of heavy granulometric composition, design of hydro-reclamation systems, features of water regime of the swamp- and wetlands in connection with their drainage, water properties of peat deposits, hydrological calculations of drainage systems.

The issues of ecological consequences, preservation, and protection of reclaimed soils have been studied much less. However, large-scale drainage of swamps- and wetlands has led to significant changes in natural complexes (Truskavetskyi, 1992) of the humid zone of Ukraine. The nature of the changes observed on lands with regulated water regimes is determined by the peculiarities of both engineering and physical and geographical factors. The changes recorded in the process of observations include: groundwater level decline, the transformation of natural landscapes, deterioration of surface runoff, reduced soil fertility and changes in water and physical properties of soils, reducing the number of species of flora and fauna, development of harmful geological processes, crop failure, etc. The intensity of changes depends on the strength and duration of reclamation measures, as well as on the sustainability of natural complexes and agricultural agrosystems (Hadzalo, Stashuk, Rokochynskyi, 2017).

In 1972-1974, the Scientific Committee of Problems of the Environment of the International Council of Scientific Unions (SCOPE) developed and proposed the idea of global monitoring. It was theoretically substantiated by the Canadian scientist R. Munn (Klymenko, Pryshchepa, Voznyuk 2006), who considered monitoring as a system of environmental control, which includes monitoring its condition, identifying its possible changes and developing possible measures for environmental management. The essence of the concept of global monitoring is the need for repeated observations of the elements of the environment in space and time with a specific purpose for specific programs. Based on this concept, various subsystems of environmental monitoring have emerged: monitoring of the surface and upper layers of the atmosphere; precipitation monitoring; hydrosphere monitoring; lithosphere monitoring; climate monitoring; ozone layer monitoring; ocean monitoring; geophysical monitoring; physical monitoring; biogeochemical monitoring (Duque, Zapico, Bugosh, Tejedor, Delgado, et al., 2021; Isaeva, Shevchenko, 2019; Ukhurebor, Aigbe, Onyancha, Ndunagu, 2022; Untenecker, Tiemeyer, Freibauer, Laggner, Luterbacher, et al., 2017).

The purpose of drained land management is to create such a condition that would meet the requirements of meeting the needs of agricultural products while maintaining a sustainable and favorable development of natural subsystems. Such a goal would be in line with the concept of "Sustainable Development" declared by the United Nations in the late 1980s. This goal is achieved by one of the leading subsystems of state monitoring – the departmental network of ecological and reclamation monitoring (ERM) of reclaimed lands of Ukraine, which has been operating on the drained lands of Ukraine (Figure 1).

Monitoring research is currently the main source of information for the development of agricultural and environmental management measures. The main unresolved problem of drained lands and water monitoring is the lack of scientific methods of data interpretation for the needs of their practical application in agricultural and environmental management practice (Novokhatin, Osipova, 2020; Schilling, Dinsmore, 2018; Richling, Solon 1998).

Figure 1. Scheme of drained lands management using feedback
S – adjustable subsystem (drained lands); R – control subsystem; X – input signal; Y – exit signal; B – ERM measurement unit; P – ERM central control unit; MB – block of management decisions implementation; X_c –information signal; X_0 -- optimal signal value; X_2 – regulating signal.

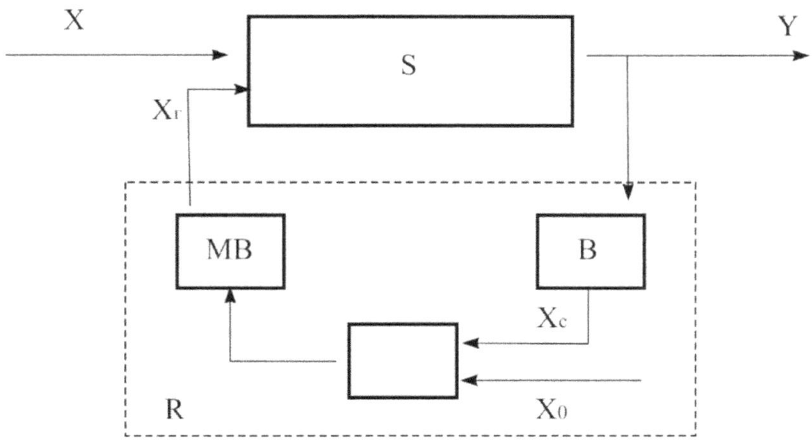

Ecological and Reclamation Monitoring in the Modern Structure of Environmental Monitoring

According to modern principles of monitoring, adequate assessment of the current state of the environment is the basis for forecasting changes in this state in the future, as well as the basis for developing effective environmental quality management programs. The implementation of these principles is the main task of monitoring (Moshynskyi, Sasyuk, 2004; 2010).

Monitoring is a way of introducing a negative feedback loop in ecological and economic systems (Fig. 2) as one that ensures the stability of these systems and guarantees the possibility of operational management (Matsniev, Protsenko, Sabliy, 2000).

According to the works of A. Isachenko, V. Krapivin, V. Medvedev, N. Moisyeyev, G. Motuzova, F. Ramada, N. Reimers, P. Trojan, etc., the diversity and complexity of natural systems provide a number of classification aspects. The main result of obtaining information is planning and conducting a full-scale or more often numerical optimization experiment (Hadzalo, Stashuk, Rokochynskyi, 2017).

For an objective assessment of the impact of drainage reclamation on natural complexes of Polesie during the operation of reclamation systems of different types, it is necessary to have a clear idea of the peculiarities of natural conditions of the drained area. This is crucial in determining the layout of the regime network, in analyzing the results of observations, determining and making decisions on the direction of drained lands use and environmental protection, and preventing the development of negative processes and phenomena (Moshynskyi, Sasyuk, 1995; 2004; 2010).

The ultimate goal of management on drained lands is to create such a condition that would meet the requirements for meeting the needs of agricultural products while maintaining a sustainable and favorable development of natural subsystems. Such a goal would be in line with the concept of "Sustainable Development" declared by the UN in the late 80's of the last century (Richling, Solon, 1998).

Figure 2. Negative feedback loop based on monitoring in ecological and economic systems

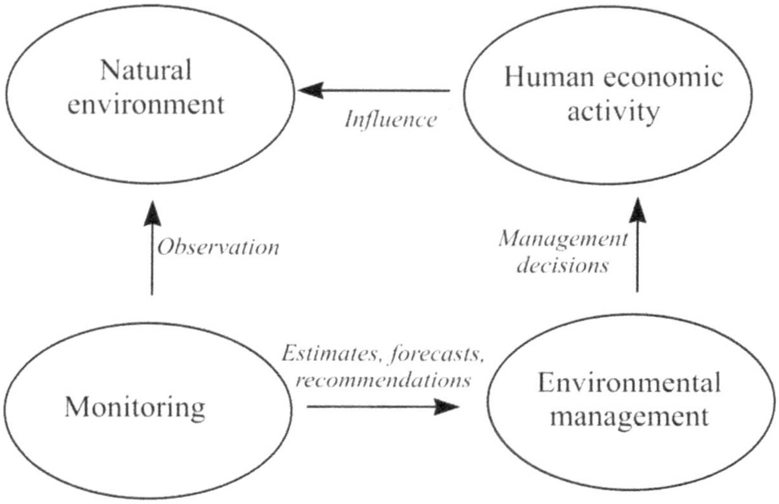

Ecological and reclamation monitoring (ERM) can be defined as a system of repeated observations in space and time and the interaction of natural and man-made factors on drained and adjacent lands, carried out to identify negative consequences, aims to assess and forecast the condition of drained lands for their management (Moshynskyi, Tsvietova, 2004; Moshynskyi, 2005).

The existence of an effective ERM system of drained lands, which was developed and operates in Ukraine through the efforts of many researchers and practitioners (V. Aleksyeyevskyi, S. Baliuk, I. Bilous, A. Bondar, P. Kovalenko, P. Kovalchuk, B. Kozlovskyi, V. Medvyediev, B. Maslov, I. Minaev, E. Nasiedkin, M. Romashchenko, etc.), is the guarantor of an objective assessment of reliable forecasting and rational management of drained lands.

The urgency of ecological and reclamation monitoring is determined by the fact that Ukraine has committed itself to create conditions for biodiversity and wetlands conservation, including in the Prypyat River basin, by signing a number of international treaties (Ramsar Convention, 1971), UN Development Program, EcoNetworks project, 1999, etc.). So, it is impossible to assess the impact of the anthropogenic intervention and make regulatory decisions without monitoring.

Methodology and Prospects of Ecological and Reclamation Monitoring of Polesie

Drainage and further use of reclaimed lands lead to predictable and unpredictable changes in natural processes that have different directions. First of all, these are changes in water exchange and soil formation, which affect the quantitative and qualitative indicators of these natural components. There are also changes in flora and fauna, water runoff, hydrographic network parameters, microclimatic indicators, etc. (Aleksyeyevskyi, Skrypnyk, Ryabtseva, 1992).

The basis of the ERM methodology is a system of indicators, the selection of which (from more than a hundred known soil indicators and dozens of water indicators) is determined by the need for the most accurate and complete description of the main properties and functions of soils, soil-forming and degradation processes, the ability of the soil to satisfy the needs of plants, the features of level, the hydrological, hydrochemical and radiological regime of natural waters (Moshynskyi, 2005; 2010).

Drainage and further use of reclaimed lands lead to predictable and unpredictable changes in natural processes that have different directions. First of all, these are changes in water exchange and soil formation, which affect the quantitative and qualitative indicators of these natural components. There are also changes in flora and fauna, water runoff, hydrographic network parameters, microclimatic indicators, etc. (Aleksyeyevskyi, Skrypnyk, Ryabtseva, 1992).

The basis of the ERM methodology is a system of indicators, the selection of which (from more than a hundred known soil indicators and dozens of water indicators) is determined by the need for the most accurate and complete description of the main properties and functions of soils, soil-forming, and degradation processes, the ability of the soil to satisfy the needs of plants, the features of level, hydrological the hydrochemical and radiological regime of natural waters (Moshynskyi, 2005; 2010).

In order to select the objects of observation for the system of ecological and reclamation monitoring, the current state of the existing observation network, and the nature of corresponding observations carried out by the hydrogeological and reclamation service to compile the cadastre of drained lands were analyzed. Taking into account the natural and economic conditions of the reclaimed territories, the typification of all drainage systems in the Ukrainian Polesie was carried out and 99 typical systems were selected. The conducted typification allowed unifying the hydrogeological and reclamation observations on other drainage systems of the Ukrainian Polesie as well (Moshynskyi, 1995; 2005; 2010).

The placement and number of plots on the system depend on the soil and reclamation conditions and the number of crop rotations (Moshynskyi, Tsvietova, 2004).

Long-term studies have shown that modern ERM and monitoring of a natural environment is no longer included in its classical definitions, and requires expansion and addition. This situation has developed as a result of the rapid development of computer technology, information technologies, and information systems in recent decades. In our opinion, the modern EM model has the structure of a complex cybernetic system, which can be depicted in the form of a block diagram (Fig. 3).

The described structure of ERM meets the existing requirements of information support of the management process, the current state, and the level of scientific support of ERM (Moshynskyi, 1995). Such a structure, equipped with the necessary computer equipment and modern means of telecommunications, is able to ensure reliable management of the land reclamation and environmental protection situation on drained lands, to bring nature monitoring in Ukraine to the forefront of world science and practice.

It allows monitoring the condition of drained lands, provides an opportunity to assess the impact of reclamation measures on the natural environment and its individual components, establishes the causes that disrupt the normal functioning of reclamation systems in specific natural conditions, allows forecasting and rational management of the ecological and reclamation condition and ecological stability of drained lands on the basis of modern information, computing and modeling support.

Indicators and Criteria for Assessing the Ecological and Reclamation Condition of Drained Lands

Various indicators are used to quantitatively and spatially express the ecological and reclamation condition of drained lands (Zakrzhevskyi P., 1984; Kantsiber Yu., Tilk A., Yemelyanova I., 1984; Tilk A., Yemelyanova I., Malysheva G., 1984; Alekseevskyi V., Nasedkin I., Korsunska I., 1987; Hadzalo, Stashuk, Rokochynskyi, 2017).

Figure 3. Conceptual block-diagram of a modern ERM system

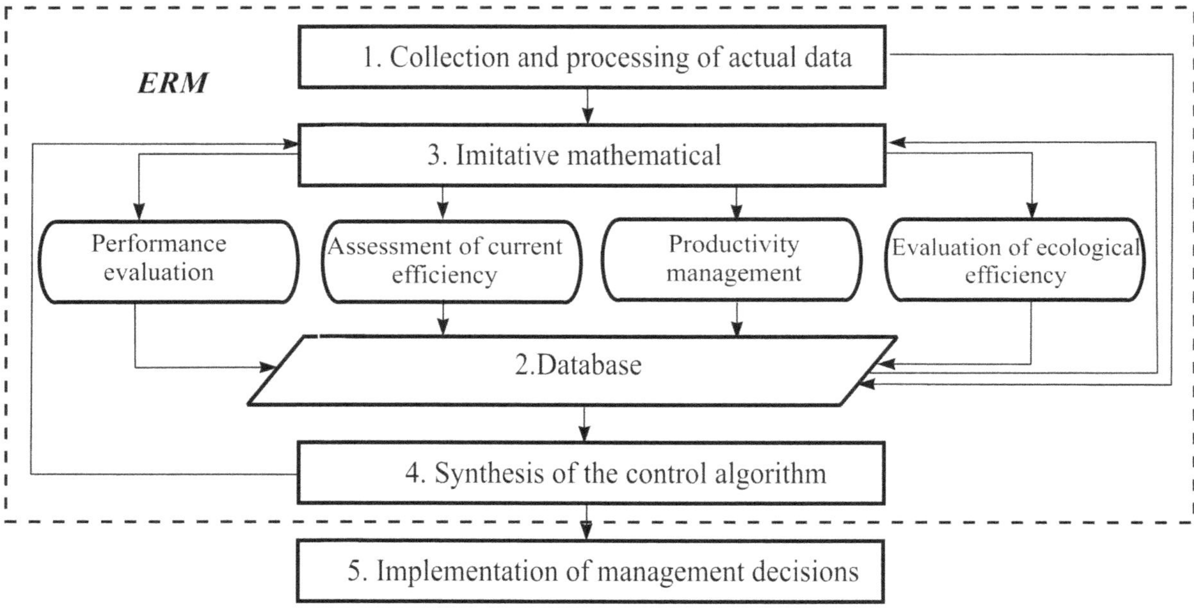

The criteria for assessing the ecological and reclamation condition according to the groundwater level (GWL) for drained lands are given in Tables 1–2.

The optimum depths of groundwater are differentiated, depending on climatic zones, soil types, crops, and seasons. The level of groundwater (upper waters) can also be used as a criterion for assessing the ecological situation, i.e. drying or overwetting of lands (Klymenko, Pryshchepa, Vozniuk, 2006).

Table 1. Assessing criteria of the ecological and reclamation condition of drained lands according to GWL

Ecological and reclamation condition	**Pre-sowing period**	**The middle of the growing season**
	depth to GWL, m	
Satisfactory	0.5-0.75 for all soil types	0.75-1.0 for peat and sandy soils
		0.75-1.5 for loamy and clay soils
Unsatisfactory	0.75-1.0 for peat and sandy soils 0.75-1.5 for loamy and clay soils	0.5-0.75 for all soil types
		1.0-1.85 for peat and sandy soils
		1.85-2.0 for loamy and clay soils
Threatening	0.0-0.5 for all soils	0.0-0.5 for all soils
	below 1.0 for peat and sandy soils	below 1.25 for peat and sandy soils
	below 1.5 for loamy and clay soils	below 2.0 for loamy and clay soils

Table 2. The optimum values of GWL depths for the Polesie zone

Provision of years with precipitation	GWL optimal for main agricultural crops, cm					
	Grains, legumes		Vegetables, fodder root crops		Perennial herbs	
	Development periods					
	sowing, planting	active vegetation	sowing, planting	active vegetation	sowing, planting	active vegetation
Polesie (mainly peat soils of diffent strength)						
Of medium humidity (50%)	60-80	90-100	65-85	100-120	50-55	80-100
Excessively humid (25%)	70-95	110-120	80-100	120-140	60-65	100-120
Arid (75%)	50-60	70-80	50-70	80-95	40-45	65-80
Small Polesie (heavy mineral soils)						
Of medium humidity (50%)	65	100	75-100	150	50-70	75
Excessively humid (25%)	75-80	120	90-120	160-180	60-85	90-95
Arid (75%)	50-55	80	60-80	100-120	40-55	60-70

Assessment of the Ecological and Reclamation Condition of Drained Lands

The evaluation of the ecological and reclamation condition of drained lands according to a set of indicators is carried out according to the actual data obtained on typical systems, comparing them with the accepted permissible and critical values.

The evaluation of the ecological and reclamation condition of drained lands is performed on the data of the ecological and reclamation monitoring, the results of which are given in the reclamation cadastre, according to the water and air regime and the terms of surface water removal (Table 3).

Table 3. Assessment of the ecological and reclamation condition of drained lands according to the terms of excess water removal during the growing season, days (DBC 33-5.5-01-97)

Agricultural use	Ecological and reclamation condition			
	favourable	satisfactory	unsatisfactory	very unsatisfactory
Vegetable and vegetable-forage crop rotations	$\frac{<0.5}{<2.0}$	$\frac{0.5-1}{1-2}$	$\frac{2-3}{3-5}$	$\frac{>3}{>5}$
Field, forage crop rotations, pastures	$\frac{<1}{<2}$	$\frac{1-2}{2-3}$	$\frac{3-5}{4-7}$	$\frac{>5}{>7}$
Haymakers	$\frac{<1}{<3}$	$\frac{1-2}{3-5}$	$\frac{3-5}{6-10}$	$\frac{>5}{>10}$

The gradations of the assessment of the ecological and reclamation condition by the level of groundwater are given in Table 4.

Table 4. Assessment of the ecological and reclamation condition of drainage systems depending on the average groundwater level depth (per day) (DBC 33-5.5-01-97)

Agricultural use	Period	Ecological and reclamation condition of soils according to GWL values, m			
		favourable	satisfactory	unsatisfactory	Very unsatisfactory
Vegetable and vegetable-forage crop rotations	Pre-sowing Sowing	0.5- 0.6	0.4- 0.5	0.3-0.4	below 0.3
Field. forage crop rotations, pastures	Vegetative and harvesting	1.0- 1.3	0.8-1.1	0.6-0.7	below 0.6
	Pre-sowing – sowing	0.5- 0.6	0.4-0.5	0.3-0.4	below 0.3
Haymakers	Vegetative and harvesting	0.8- 1.0	0.7-0.9	0.5- 0.6	below 0.6
	Pre-sowing – sowing	0.4- 0.5	0.3-0.4	0.1-0.2	below 0.1
	Vegetative and harvesting	0.6- 0.8	0.4-0.6	0.3-0.4	below 0.3

A deviation of no more than 30% (reaching the limit of permissible influence) from the accepted as a comparison basis must be evaluated as a satisfactory condition, more than 30% – unsatisfactory, up to 10% – favorable. The evaluation of the impact of drainage on adjacent lands is performed as a result of comparing the normative values of GWL decline and the width of the zone of active influence with the actual ones.

The technical condition of the drainage system *(Tc)*, as an indicator, determines the ability of the system to provide its most important function: to maintain the optimal humidity of the root layer of the soil in the design mode.

Forecast of the Ecological and Reclamation Condition of Drained Lands

The task of the ecological and reclamation forecast is to assess the prerequisites, opportunities and determine the initial data for creating a favourable water and air regime of drained soils.

The ecological reliability of reclamation systems is assessed by the ecological and economic feasibility and efficiency of their functioning. The set of evaluation parameters includes:

- technical condition of the reclamation system (T_c);
- state of environmental protection elements of the system (*PE*);
- ecological and reclamation condition of drained lands (*EMC*);
- yield of agricultural crops (A_y);
- radioecological condition of the drained area (*PC*).

The integral indicator of the environmental reliability of the drainage system is defined as the average of individual assessments in percentages and is calculated according to the formula:

$$En = \frac{Tc + MC + PE + Ay + PC}{5}, \ \% \tag{1}$$

The assessment of environmental reliability is carried out according to a specially developed methodology (Moshynskyi, 2005) with the following categories:

- reliable – over 75%;
- conditionally reliable – 75-55%;
- unreliable – less than 55%.

It is possible to determine the state of moisture of the arable layer of the soil of a certain composition by the values of the groundwater depth. To ensure optimal values of soil moisture in years with different water content, appropriate forecast deviations were made (Table 5).

Table 5. Predicted groundwater level depths in various soils to ensure optimal moisture in the arable soil layer

Soil moisture, %	GWL (cm) in the soils of different composition with different precipitation								
	Mostly peat soils of different strength			Mostly shallow peatlands, mineral sandy and light loamy soils			Mostly heavy mineral soils		
	25%	50%	75%	25%	50%	75%	25%	50%	75%
40	>180	>180	>180	>200	>200	>200	>200	>200	>200
50	140	130	130	150	140	140	170	160	160
60	100	95	90	110	105	100	130	125	120
70	75	70	65	80	75	70	95	90	85
80	55	50	45	60	55	50	70	65	60
90	45	40	35	45	40	35	50	45	40

On humid lands, the GWL depth H (m) can be determined at a distance x (m) from the drainage system by the formula:

$$H = H_0 \bullet f(z), \tag{2}$$

where H_0 is the value of GWL decline in the enclosure channel, i.e. at the boundary of the drainage system, m; $f(z)$ is a special function, the value of which is found depending on indicator z, which is determined by the formula:

$$z = \frac{x}{\sqrt[2]{\frac{K_f \cdot h \cdot t}{\mu}}}, \tag{3}$$

where K_f is a soil filtration coefficient, m/day; h is aquifer capacity, m; μ is a soil water yield coefficient; t is the duration of GWL decline from the spring maximum to the summer-autumn minimum.

The magnitude of the decrease in the groundwater level (Fig. 4) is determined at different distances from the system and a dispersion curve is constructed to establish the zone of influence of the drainage system. The boundary of the influence zone of the drainage system is determined where GWL decline does not exceed 0.2 m.

Figure 4. GWL decrease in the zone of influence of drainage systems

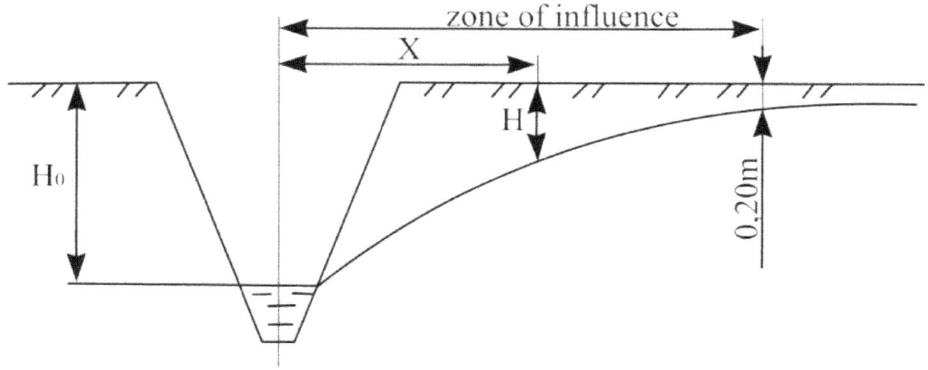

Formulation of the Model of Productivity and Assessment of Drained Lands Condition in Ukraine

The definition based on the system approach of a set of components and the corresponding system-forming connections makes it possible to formalize the modelling object. The amount of yield on DLs functionally depends on four arguments. This dependence in a general implicit form can be represented in the form of a function:

$$y = f(g, l, t, b), \tag{4}$$

where y is the yield (phytoproductivity) of the evaluated system; g is potential soil fertility and soil conditions; l is insolation level; t is thermal regime of the atmosphere and soil; b is botanical, morphological and biological features of a plant (culture, variety, hybrid) or plant association.

The object of modelling is S_{03}, a polycentric, polystructural system consisting of five complex elements of different nature, which interact with each other and with the environment, i.d. a person (Figure 5). So, we have an abstract natural and man-made system, which is a territorially integrated set of natural geographical components within agrogeosystems of a certain territorial level, which are under the constant influence of water regulation engineering structures and perform the social function of agricultural production.

Taking into account the properties of the modelled system as the basic operator (functional $y = f(x)$, measured and bounded on some limited set Ω^j of points x with measure μ) of the FACDL model, the measure functional can be applied in the form of a Lebesgue integral sum $y = \sum_{j=1}^{m} y_j \mu\left[\Omega^j\right]$ or in a dynamic form after the introduction of time variable *τ:*

$$y_\tau = \sum_{j=1}^{m} y_\tau^j \mu_{\Omega_\tau^j} \tag{5}$$

Figure 5. Graph of the studied system
g, l, t, b – elements of the system; P – a green plant; L – external system;

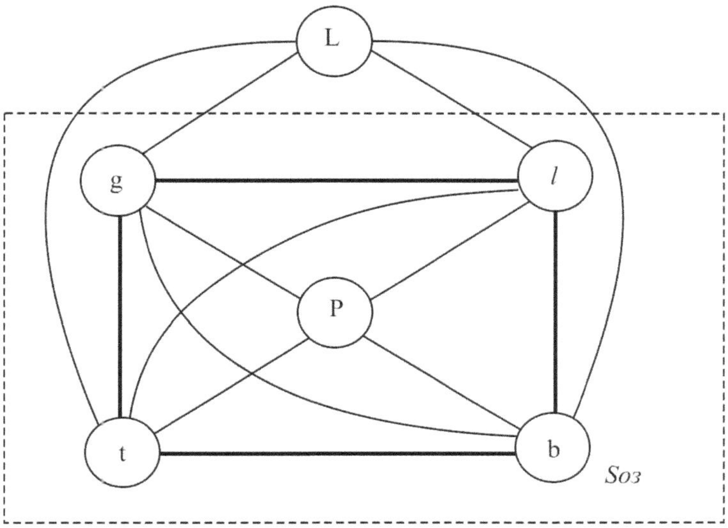

The basic operators of the model are formulated in the form of Lebesgue measure functionals for the discrete objective function

$$y_e = K\left(\varsigma_a y_a + \varsigma_s y_s\right),\ y_a = \sum_{f=1}^{f} \alpha_\tau^f y_\tau^f,\ y_e = \overline{0, y_b}, \tag{6}$$

where y_e is ecologically provided (created in the conditions of a specific combination of environmental factors) useful phytomass (harvest), t/ha; y_b is biological productivity (determined by potential productive possibilities of agrophytocenosis or natural phytocenosis), t/ha; ς_a and ς_s are weighting factors that determine the degree of participation of the combined effect of factors in the formation of an ecologically secure harvest, moreover *K* is a coefficient of relative productivity of cut grass (only for multi-cut grasses) according to the empirical formula of the type (13):

$$K = \frac{100 y_{0u} \exp\left(-k_u \left(n_u - \lambda_u\right)^2\right)}{1 + \left(100 - y_{0u}\right) \exp\left(-\mu_u \left(n_u - \lambda_u\right)\right)},\ K = \overline{0,1}, \tag{7}$$

where n_u *is* serial number of the cutting, $n_u = 1.5$; μ_u, k_u, λ_u, y_{0u} are parameters depending on the groundwater level on DLs; y_a is performance in terms of measure; y_s is productivity from the standpoint of the combined effect of factors; $y_\tau^f \supset \{y_l, y_g, y_t, y_b\}$ are partial yield factors; $\alpha_\tau^f \supset \{\alpha_l, \alpha_g, \alpha_t, \alpha_b\}$ are weighting coefficients of productivity factors.

The yield effect from the combined effect of the factors is calculated according to the empirical dependence of the type:

$$y_s = \frac{y_b 0.08 \exp\left(-1.66(q-0.24)^2\right)}{0.01 + 0.92\exp\left(-4.992(q-0.24)\right)},\ y_s = \overline{0, y_b}, \tag{8}$$

Element g is a complex subsystem of the system S_{O3}. Therefore, phytoproductivity (yield) is ensured by soil fertility and soil conditions in accordance with (5):

$$y_\tau^g = \sum_{p=1}^{n_p} \gamma_p y_\tau^p, \tag{9}$$

where y_τ^p is productivity according to the p-th soil indicator, t/ha; $p = \overline{1, n_p}$; γ_p are weighting coefficients of soil indicators (variables).

Since the partial values of phytoproductivity are values that depend not only on the instantaneous values of the variables l, t, and p, but also on the previous dynamics of these variables within the studied system of S_{O3} during the current growing season, these values are determined taking into account the background history of the growth and development of the phytocenosis according to the dependence:

$$y_\tau^{ltp} = \sum_{i=1}^{n} v_\tau^i y_\tau^{ij}, \tag{10}$$

where y_τ^{ij} is phytoproductivity (yield), calculated for the j-th variable at the i-th step of model discretisation; v_τ^i are weighting coefficients of the i-th step of model discretisation, which is determined for each crop or type of vegetation depending on the phase of development; n is the number of model discretization steps.

$$y_\tau^{ij} = \frac{n_f n_l n_s n_d k_{wt} k_l k_s k_d k_i k_h k_w y_b y_{0ij} \exp\left(-k_{ij}\left(x_{ij} - \lambda'_{ij}\right)^2\right)}{1 + \left(100 - y_{0ij}\right)\exp\left(-\mu_{ij}\left(x_{ij} - \lambda'_{ij}\right)\right)} \tag{11}$$

where x_{ij} is the value of j-th variable at the i-th step of model discretisation; μ_{ij}, k_{ij}, y_{0ij} are parameters of the tolerance function for the j-th variable (for energy variables determined at the end of the growing season); λ'_{ij} is the parameter of the phase position of the optimum of the j-th energy variable at the i-th step of discretisation by dependence:

$$\lambda_{ij}^{'} = \lambda_j - \left(x_j^* - \hat{x}_{ij}^*\right) \tag{12}$$

where λ_j is the parameter of the phase position of the optimum of the tolerance function of the j-th variable (for energy variables, it is determined at the end of the growing season); x_j^* is the optimal value of the j-th energy variable at the end of the growing season; $\hat{E}_{ij}^*$ is the optimal value of the j-th energy variable at the i-th discretisation step, calculated according to the submodel of the course of the optimal values of energy variables; k_f, k_{wt}, k_p, k_s, k_d, k_i, k_h, k_w are correction coefficients for the degree of crops damage by stressful conditions, respectively: frosts, adverse conditions of the summer and autumn phase of vegetation and overwintering (for winter crops), flattening (for cereals), hail, droughts, as well as for the degree of adverse effects of diseases, pests and weeds; n_f, n_p, n_s, n_d are the number of stress damages during the growing season due to frosts, flattening (for cereals), hail, droughts.

On the basis of Shelford's fundamental law of tolerance (minimum, optimum and maximum) and the experience of mathematical modelling of agricultural systems and phytocenoses, a new type of sigmoidal function was developed, suitable for describing the tolerance of plants and plant associations to the action of internal and external factors of the species:

$$y = \frac{y_0 y_b e^{-k(x-\lambda)^2}}{1+\left(100-y_0\right)e^{-\mu(x-\lambda)}} \tag{13}$$

where μ is the parameter of the growth rate of the function, $\mu = \overline{0.2}$, k is the parameter of the decay rate of the function on the interval $x^* < x < \infty$ or damage parameter, $k = \overline{0,1}$; λ is the variable optimum position parameter x, $\lambda = \overline{-xmax_{max}}$; y_0 is the parameter of plant adaptation to the action of the variable x, $y_0 = \overline{0, y_b}$. The advantages of the model: 1) it makes it possible to reliably approximate experimental data, 2) all parameters are physically interpreted, 3) the negative impact of exceeding the optimum is taken into account, 4) the graph of the function does not pass through the origin, 5) the simulated values of y cannot exceed the biological maximum of a given species, variety, phytocenosis, agricultural crop, 6) simplicity, which makes it possible to use standard software for modelling.

Further studies showed that the dependence (13) can be successfully applied to describe not only the production process, but also many other stochastic natural processes and regularities, and in particular: 1) the dependence between the volume of dry matter in the biomass of multi-cut grasses and the ordinal number of the cutting, 2) the course of optimal values of the sum of active air temperatures (higher than 10°C) in ontogenesis, 3) the average long-term course of the sum of active air temperatures (higher than 10°C) during the growing season, 4) the dependence of the cumulative effect coefficient on the relative index of the interaction of productivity factors.

A serious problem of modeling production processes in natural and man-made systems is the presence of a cumulative effect from the combined action of plant life factors. According to references and archive sources, we have developed a sub-model that allows simulating the cumulative effect of the action of an unlimited number of variables that determine the DLs condition with a level of adequacy sufficient for ERM practice and the DLs use. In order to achieve comparability of the values of physically and mathematically disparate variables of the model, the values of the variables and the goal criterion are expressed in a relative form in relation to the optimal values of the variables and the maximum values of productivity (y_b). To describe the cumulative effect, the relative indicator of the interaction of productivity factors, the geometric mean of the relative deviations of the values of the variables from the local optima, is used:

$$q = \sqrt[k]{\prod_{j=1}^{k} q_j}, \quad \begin{cases} q_j = 1 - \dfrac{\left|x_{ij}^* - x_{ij}\right|}{x_{ij}^*} \; at \; x_{ij} \prec 2x_{ij}^*; \\ q_j = 0 \; at \; E_{ij} \geq 2x_{ij}^*, \end{cases} \tag{14}$$

where x_{ij}^* is the value of the local optimum of the *j*-variable at the *i*-th step of discretisation of the general model, $j = \overline{1,k}$, where *k* is is the number of variables taken into account at the *i*-th discretisation step; x_{ij} is the actual value of the *j*-variable at the *i*-th discretization step. The geometric mean of relative deviations allows taking into account the following important regularity of production processes: *productivity approaches zero if the value of at least one of the factors approaches zero*. Finally, the productive effect of the combined action of factors is calculated according to dependence (7).

The result of the modelling can be the value of the "stored" or ecologically ensured harvest of agricultural crops (depending on the modelling needs). The "stored" harvest is determined by the dependence $y=k_m k_t y_e$, where k_m is the coefficient of losses during harvesting (calculated taking into account the type and brand of harvesting equipment, harvesting dates, weather conditions during the harvesting period, etc.); k_t is the coefficient of losses during product transportation; y_e is ecologically ensured productivity; τ_i is the time (days) elapsed from April 20 (the date of the average multi-year transition of average daily air temperatures through 0°C) to the *i*-th date.

Considering the above, based on the developed structure of the mathematical model of FACDL (see dependencies (1)-(11)), we developed an algorithm for its operation in the form of a block diagram, which is the basis of a computer program for implementing the FACDL model.

Application of the Model of Productivity and Assessment of the Condition of Drained Lands and its Functional Units

The application of the FACDL model for the practical needs of drained lands control and management based on the data of ecological and reclamation monitoring involves the development of algorithms and methods that would enable the user in real production conditions to evaluate and forecast the ecological and reclamation condition, the productivity of DLs, the efficiency of their use, the environmental sustainability of DLs, to develop rational management measures and recommendations for increasing the productivity and environmental sustainability of DLs. The basis of the development of these methods is such general principles of their construction as a single time scale and spatial uniformity of controlled natural and man-made processes on DLs. Since the characteristic time of drained agrogeosystems in connection with phytocoenotic and successional changes is 5-10 years, and the study of evolutionary transformations at this stage of ERM is impossible, the minimum representative duration of the period of monitoring observations for management needs is $t = t_{np} = \overline{5,10}$ years, and the more responsible and complex the researched object is, it should be the greater. For small DLs and their parts, t ³ 5 years.

The application of the mathematical model of FACDL on the example of five typical drainage systems of the western part of the humid zone of Ukraine showed that it can be effectively used for the interpretation of monitoring data by assessing the ecological and reclamation state of drained lands, as well as for evaluating the overall efficiency of the use of drained agricultural land.

Considering the available material and technical means, the developed simulation mathematical model of FACDL makes it possible to choose a rational (including minimum) combination of inputs (values of ERM indicators, data of other services, and possible measures at a given level of their implementation) in order to obtain the maximum output in compliance with current environmental restrictions. The development of management measures can be performed at any territorial level of the organization of drained agrogeosystems (after the appropriate adaptation of the model): at the level of a field or a farm with the same type of soil conditions, at the level of a part of DLs, or DLs as a whole, which usually coincides with geo-reclamation sub-districts, as well as at the local level (at the level of geo-reclamation microdistricts).

For the selected territorial system or the drainage system as a whole, a number of machine experiments are conducted, during which a management strategy is determined, which is aimed at increasing productivity and improving the ecological and reclamation condition and the efficiency of DLs use, as well as the results of short- and medium-term forecasting of productivity and the ecological and reclamation condition of the object of management.

Forecasting calculations performed for DLs "Stubla" of Rivne region indicate that the mathematical model of FACDL, along with the analytical capabilities of assessing the condition of DLs and their territorial subsystems, provides the possibility of reliable forecasting of future conditions of evaluated objects at the development level of operational, short-term and medium-term forecasts of the ecological and reclamation condition and the level of productivity of drained lands.

Exemplified by five typical DLs, after conducting a series of machine experiments on the FACDL model, a strategy for optimal management of DLs was determined in order to increase productivity and improve the ecological and reclamation condition. It has been found that the developed mathematical model of FACDL is an effective means of managing drained lands, including under conditions of uncertainty.

On the basis of the main blocks of the FACDL model, the assessment method of drained lands of Ukraine has been improved, which, for the first time, takes into account the soil parameters' significance degree varying in time and space. The method makes it possible to perform a quality assessment and establish the quality of hydromorphic soils as an object of land reclamation and agricultural use with maximum objectivity, and thus to assess their potential fertility. The obtained assessment values of certain soil contours and DLs as a whole make it possible to reasonably proceed to the development of agricultural, reclamation, and agro-reclamation measures to increase the fertility of drained hydromorphic soils, determine ecological priorities and regulations regarding natural and reclamation regimes of drained soils, programming of crop yield on drained lands, and also provide an opportunity to solve the task of economic assessment of land during land management under the conditions of privatization of drained agricultural lands and the development of market relations in the field of drained lands use. On the basis of the methodology described above, a soil quality map of the drainage system "Holovnytsia" (Rivne region) was compiled.

On the basis of the interpretation of of local optima position for each ERM indicator (x_{ij}^*), the position of the global optimum for DLs as a whole, as well as the values of ecological restrictions for each factor (x_{ij}' and x_{ij}''), the author formulated the criteria for the agrogeosystem stability, the methodology principles for assessing the DLs stability and identification of the sustainability form of a drained natural and man-made system in the ASDL model.

The DLs sustainability as a set of natural and man-made systems created due to anthropogenic modification of the natural structure of hydromorphic (including swamp) geosystems, as well as geosystems of another organization (e.g. river basins), is evaluated from the standpoint of the geosystem's ability to maintain the ability to effectively produce phytomass under the conditions of preservation of its internal (modified) structure. Meanwhile, the stochasticity of the studied systems and the fact that the instantaneous value of any indicator of the system condition describes the condition of the system at the *i*-th moment of time, but does not characterize the stability of the vertical structure over time, are taken into account. Therefore, an approach was applied that allows assessing stability both from the perspective of the instantaneous values of the indicators and from the perspective of their average values (close to mathematical expectation). According to this approach, the instantaneous values of indicators x_{ij} are compared with the environmental restrictions at the *i*-th time (x'_{ij} and x''_{ij}), and the average multi-year values of the indicators $\overline{x_j}$ are compared with the average values of environmental restrictions $\overline{x'_{ij}}$ та $\overline{x''_j}$). In addition, the presence of trends revealed during the statistical analysis of time series for individual indicators that characterize the drift of the system state towards the optimum (preserving stability) or away from it (loss of stability) is taken into account. The presence of a "critical zone" near the outer border of normal states is important, the width of which is determined by the sigmoidality of the change in stability according to a separate indicator from the optimum to the ecological limit in the ecological niche, and other general ecological regularities.

When implementing the methodology, the following partial cases arise:

1. If all indicators of the condition of the evaluated system had average multi-year values during the observation period *(t)* within normal conditions (i.e., the inequality is $\overline{x'_j} \prec \overline{x_j} \prec \overline{x''_j}$ and $\overline{x''_j}$ – the average lower and upper environmental limits) and at each moment of their determination in the ERM system after filtering out gross errors there were the values, which satisfied the inequalities $x'_{lf} < x_{t'} < x''_{tj'}$, then regardless of the direction of the drift of indicators (in the presence of time trends of the day x_j), this system is inert and *environmentally sustainable.*
2. If all indicators of the condition of the evaluated system have average multi-year values during the period t in the interval $\left(\overline{x'_j} + 0{,}1\left(\overline{x^*_j} - \overline{x'_j}\right)\right) \prec \overline{x_j} \prec \left(\overline{x''_j} - 0{,}1\left(\overline{x''_j} - \overline{x^*_j}\right)\right)$, and the instantaneous values of the indicators are outside the interval $x'_{ij} \prec x_{ij} \prec x''_j$, then regardless of the direction of the drift of the indicators (in the presence of time trends for x_j), this system is *renewable* (I form of reversibility) and therefore *quite stable.*
3. If at least one indicator of the condition of the evaluated system had an average multi-year value during the period t outside the interval $\left(\overline{x'_j} + 0{,}1\left(\overline{x^*_j} - \overline{x'_j}\right)\right) \prec \overline{x_j} \prec \left(\overline{x''_j} - 0{,}1\left(\overline{x''_j} - \overline{x^*_j}\right)\right)$, but remained within the interval $\overline{x'_j} \prec \overline{x_j} \prec \overline{x''_j}$, the instantaneous values of the indicators exceeded the interval $x'_{ij} \prec x_{ij} \prec x''_j$ and its drift x^*_j is statistically confirmed ($R^2>0{,}25$ for the trend at the level of significance $\alpha=0.05$ according to Student's test), then this system is *renewable* (form II of renewable) and therefore *relatively stable.*

4. If at least one indicator of the state of the evaluated system had a multi-year average value during the period t outside the interval $\left(\overline{x_j^{'}}+0,1\left(\overline{x_j^{*}}-\overline{x_j^{'}}\right)\right)\prec\overline{x_j}\prec\left(\overline{x_j^{''}}-0,1\left(\overline{x_j^{''}}-\overline{x_j^{*}}\right)\right)$, but within the interval $\overline{x_j^{'}}\prec\overline{x_j}\prec\overline{x_j^{''}}$ and there is a statistically confirmed drift in the direction from x_j^{*}, this system is *plastic* (it moves to another area of normal conditions) and therefore *ecologically unstable* .
5. If at least one indicator of the state of the evaluated system had a multi-year average value during the period t outside the interval $\overline{x_j^{'}}\prec\overline{x_j}\prec\overline{x_j^{''}}$, then, regardless of the variation of the instantaneous values of the indicators and the direction of the drift of the indicators (in the presence of time trends for xj), this system is at the stage of *successional* or evolutionary transformations and therefore it is *ecologically unstable*. Succession of the system can be a partial case of plasticity (in the studied time scale) or evidence of the geosystem destruction with the formation of a new one.

Taking into account the sigmoidal nature of the stability decline from optimum to pessimum, the task of measuring stability on this scale is reduced to a certain type of dependence $\omega=f\left(x_{ij},x_{ij}^{*},x_{ij}^{'},x_{ij}^{''},y_e(\tau),y_b\right)$ and establishing the phase coordinates of the system. In this case, it is proposed to use the exponential unimodal Grynchenko model, which makes it possible to connect the phase coordinates of the system with the relative index of its stability. Thus, the relative indicator of the stability of the vertical structure of the drained geosystem according to the level of phytoproductivity (the yield of each main agricultural crop) is determined by the dependence

$$\omega_k=\exp\left[-k_y\left(\frac{y_{ek}-y_{bk}}{y_{bk}}\right)^2\right],\ \omega_k=\overline{0.1}, \tag{15}$$

where κ_y is a parameter of the curve shape, which depends on the agricultural crop, the type of land, the set of model variables, etc.; y_{ek}, y_{bk} are ecological and biological productivity of the k-th agricultural crop, respectively.

The final value of the relative stability index according to the level of productivity of the drained geosystem ω_1 is determined as the result of averaging ω_k by the relevant stations and, as a whole, for the evaluated agrogeosystem, drainage system, field, crop rotation, etc.

The relative indicator of the stability of the drained geosystem vertical structure according to the degree of compliance with the accepted environmental standards for each i-th value of the j-th variable is calculated according to the dependence

$$\omega_{ij2}=\begin{cases}\exp\left[-k_j\left(\frac{x_{ij}-x_{ij}^{*}}{x_{ij}^{*}-x_{ij}^{'}}\right)^2\right], x_{ij}\le x_{ij}^{*}; \\ \exp\left[-k_j\left(\frac{x_{ij}-x_{ij}^{*}}{x_{ij}^{*}-x^{'ij}}()^2\right)[]_{ijij}^{*}\right]\end{cases} \{_{ij2}\overline{0,1} \tag{16}$$

where k_j is a parameter of the sigmoidal curve shape, which depends on the physical content of the j-th variable, the type of terrain, etc.; x_{ij}, $x_{ij}*$, x'_{ij}, x''_{ij} are, respectively, the actual, optimal, lower critical and upper critical values of the j-th variable at the i-th step of discretization of the ASDL model (the moment of sample selection or measurement during the observationperiod in the ERM system).

After calculating all values of ω_{ij2}, we determine the value of the relative indicators of stability as the arithmetic mean of the partial values of stability for each variable and according to the principle of comparison (ecological niche) as a whole:

$$\omega_{j2} = \frac{1}{n}\sum\nolimits_{i=1}^{n}\omega_{ij2},\ \omega_{j2} = \overline{0,1},\ \omega_2 = \frac{1}{m}\sum\nolimits_{j=1}^{m}\omega_{j2},\ \omega_2 = \overline{0,1}, \tag{17}$$

where m is the total number of variables used in the assessment of the environmental sustainability of the studied system according to the mathematical model of ASDL. After final averaging, we have:

$$\omega = \frac{1}{2}(\omega_1 + \omega_2),\ \omega = \overline{0,1} \tag{18}$$

Based on the calculations of ω according to the ASDL model, considering the sigmoidal nature of the dependence of stability on ERM indicators, as well as taking into account the Coleman principle described above, the following scale was used to interpret the ω indicator: stable (0.75-1.00), fairly stable (0, 50-0.75), relatively stable (0.25-0.50), unstable (0.00-0.25).

The formulated structure of the ASDL model, the algorithm for its implementation presented in the form of a block diagram in Fig.6, as well as the developed computer program compiled on the basis of *Microsoft Excel*, make it possible to evaluate, forecast and manage the ecological sustainability of DLs in the most reasonable way possible, based on the data of ecological and reclamation monitoring of drained lands of Ukraine.

ILLUSTRATIVE EXAMPLE

In the process of identifying the model of productivity and assessment of the state of drained lands, a list of climatic variables of the model of productivity and assessment of the reclamation state of drained lands was determined. It was established that the main climatic indicator for the studied system is the sum of active air temperatures higher than 10°C. This indicator makes it possible to determine, with high reliability (r=0.998; R2=0.996-0.997), the intensity of energy input to drained agrogeosystems based on the obtained regional empirical models. To ensure the possibility of statistical modelling of the average decadal temperature of the soil based on data on the average decadal air temperature, an empirical model was developed, which has the form of *te=1.07611n+0.6017* and a high level of reliability: *r = 0.94; R2 = 0.88 ; Syx=1.69.*

The application of a system approach and measurement theory methods made it possible to substantiate the optimal set of indicators from the point of view of conducting ERM and the needs for performance modelling, which includes: 1) the content of organic matter *(G)* for mineral soils or ash content *(Z)* for peat soils, 2) the content of mobile phosphorus *(P)*, 3) the content of exchangeable potassium *(K)*, 4) reserves of productive moisture in the calculated soil layer *(W)*, 5) the level of groundwater *(h)*, 6) the density of the soil *(d)*, 7) the power of humus horizon $(H_г)$ for mineral soils, or mineralized stratum $(H_м)$ for peat soils, 8) reaction (pH_{KCl}), 9) redox conditions (*Eh* index).

As a result of the use of the energy-material method of decomposition of drained agrogeosystems in S_{O3} according to the Churchman-Akof method from the theory of operations research, the binary relations of preference $g>l>b>t$ and the weighting coefficients of the functional y_a: $a'_g=0.43$; $a'_l=0.29$; $a'_b=0.14$; $a'_t=0.14$ were determined.

Figure 6. Generalized block diagram of the model of assessment of sustainability of drained lands (ASDL)

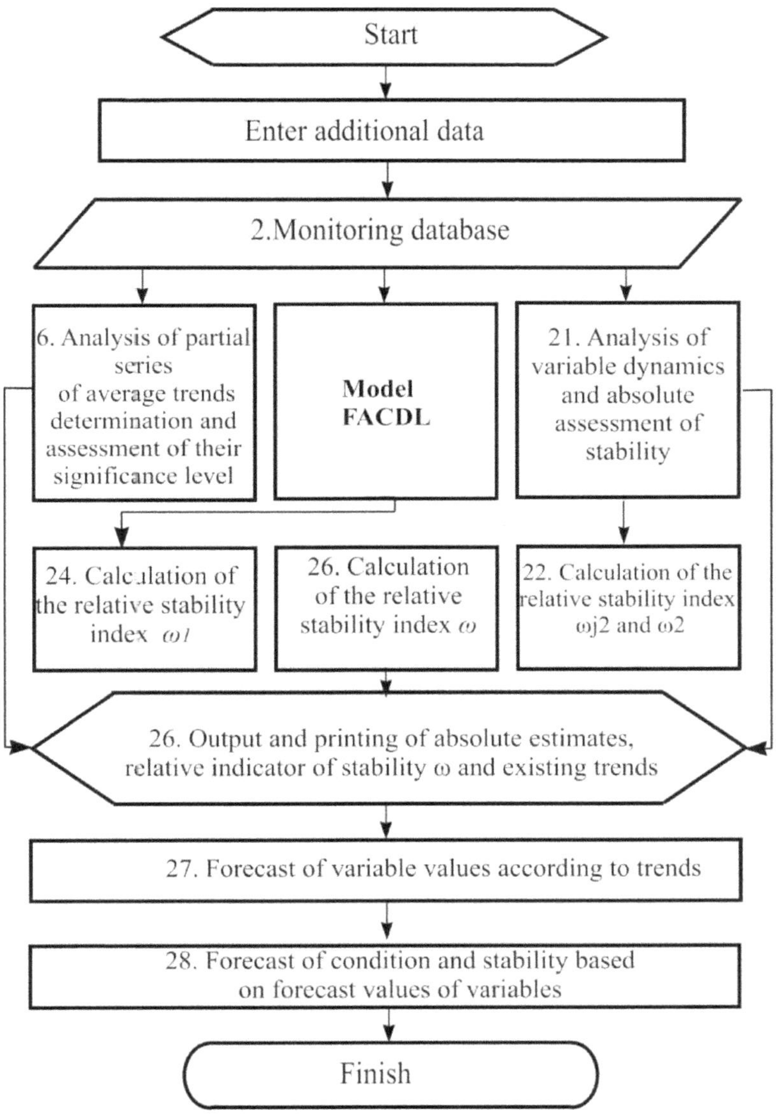

Based on the analysis of the self-regulation graph, the parameters of phytoproductivity forming the environment of the green plant can be described by the binary preference ratio $l > g > t$. Under the condition $l \sim (g+t)$, which derives from the assumption that solar radiation in natural systems is the main source of energy for all thermal and chemical processes in the soil, the value of the weighting coefficients for each of the parameters of phytoproductivity (yield) in the functional y_a was calculated, which describe the level of their significance ("strength") in the process of self-regulation of the **plant-environment** module:

$$\alpha_l^{''} = 0.50;\ \alpha_g^{''} = 0.45;\ \alpha_t^{''} = 0.05;\ \alpha_b^{''} = 0\,.$$

Since the S_{O3} system has four regulatory inputs (Fig. 3), which are considered as connections of the external anthropogenic factor Л with the internal elements of S_{O3}, it is necessary to establish the numerical values of the "weight" of S_{O3} components from the point of view of external regulation. As a result of the statistical analysis, it was found that the yield increases obtained from the use of "strong" and "weak" measures differ significantly according to the Fisher and Student tests ($p_f=3.5\cdot10^{-7}$, $p_t=4.6\text{-}10^{-8}$), which forces us to reject the hypothesis that they belong to one general totality and confirms the fundamental correctness of the accepted division of measures. On the other hand, the conducted Fisher tests show a fairly high one-sided probability of commonality of the samples of gains obtained on peat and mineral soils, respectively, from weak ((p_f =0.77) and from strong (p_f=0,46) measures. Therefore, there is every reason to average the values of gains obtained separately from strong and weak measures (management alternatives) regardless of soil type to establish the overall superiority of the former over the latter. Experimental evidence shows a four-fold advantage of strong measures in terms of the degree of managerial influence.

Having the value of the weight φ_{fv} for a specific management strategy, the variable values of the weighting coefficients $\alpha_f^{'''}$ are calculated. In the case of a full set of management measures,

$$\alpha_l^{m} = 0.074,\ \alpha_g^{'''} = 0.463,\ \alpha_t^{'''} = 0.315,\ \alpha_b^{'''} = 0.148$$ are applied.

The final (averaged) values of weighting coefficients for productivity factors in the case of the most complete set of management measures are $a_g=0.448$, $a_l=0.288$, $a_t=0.168$, $a_b = 0.096$.

Therefore, regulatory measures of direct influence on elements of the S_{O3} system are four times more effective than measures of indirect influence. By changing the values of the variables, as well as the set and quality of the implementation of measures carried out on DLs, we change the weight of each of the phytoproductivity parameters and, thus, the result of the simulation is the value of the phytoproductivity (yield) at a given point.

For the first time, the weighting values of individual productivity parameters obtained on DLs indicate that almost half of the volume of agricultural crop yields and the productivity of quasi-natural phytocenoses on drained lands depend on soil conditions, therefore, on the reclamation state of the lands. The element "Soil fertility and soil conditions" is 4.7 times stronger than "botanical and biological features", 2.7 times stronger than "plants supply with heat", and also 1.6 times stronger than the element "plants supply with light". On the basis of such ratios, it can be asserted that when conducting agriculture and carrying out environmental protection measures on DLs, the greatest attention should be paid to the regulation of water and other soil regimes and the formation of a favourable reclamation (ecological and reclamation) situation. The ERM system should be directed to the collection of actual data mainly on the state of the soil cover and the territorial control of indicators of soil condition to ensure the possibility of managing the productivity of agricultural crops.

The application of measurement theory and graph theory to establish relationships between fertility factors and indicators of the drained soils condition made it possible to substantiate a set of the most representative soil indicators, rank the indicators and parameterize the soil fertility functional (9), calculating respective weighting factors γ'_p:

$$\gamma'_{G(Z)} = 0.182;\ \gamma'_W = 0.152;\ \gamma'_{Eh} = 0.151;\ \gamma'_h = 0.121;\ \gamma'_d = 0.121;$$
$$\gamma'_{pH} = 0.091;\ \gamma'_{H_{G(M)}} = 0.061;\ \gamma'_K = 0{,}061;\ \gamma'_P = 0.06$$

From the point of view of the importance of soil indicators (variables) when describing the fertility and state of the soil cover, the following values of the weighting coefficients were obtained γ''_p:

$$\gamma''_{G(Z)} = 0.167;\ \gamma''_K = 0.061;\ \gamma''_p = 0.154;\ \gamma''_W = 0.128;\ \gamma''_{pH} = 0.115;$$
$$\gamma''_d = 0.090;\ \gamma''_{Eh} = 0.077;\ \gamma''_h = 0.051;\ \gamma''_{H_{G(M)}} = 0.051$$

According to the assessment of the weight of soil indicators according to the degree of their external (anthropogenic) controllability, the value of the weighting coefficients was calculated in the case of the most complete set of management measures for external (anthropogenic) regulation of the S_{O3} system:

$$\gamma'''_K = 0.189;\ \gamma'''_p = 0.176;\ \gamma'''_W = 0.135;\ \gamma'''_{G(Z)} = 0.122;\ \gamma'''_{Eh} = 0.081;$$
$$\gamma'''_h = 0.081;\ \gamma'''_d = 0.081;\ \gamma'''_{pH} = 0.068;\ \gamma'''_{H_{G(M)}} = 0.068$$

The final average values of the weighting coefficients for the functional variables of soil fertility in the case of the most complete set of management measures for external (anthropogenic) regulation of the S_{O3} system are:

$\gamma_G(Z)$=0.157; γ_K=0.139; γ_W=0.138; γ_P=0.130; γ_{Eh}=0.103; γ_d=0.097; γ_{pH}=0.091; γ_h=0.084; $\gamma_{H_{G(M)}} = 0.060$.

In the real conditions of drained lands functioning, the weighting coefficients $\gamma_p \neq$const (*since* $\gamma'''_p \neq const$) and their values depend on the selection, execution time and the quality of agrotechnical, agrochemical, agroreclamation, etc. measures. Such a set of measures and their quality are set by the user at the input of the mathematical model of productivity and assessment of drained lands condition.

The quantitative relationship between the amount of evapotranspiration of agricultural plants and the volume of produced phytomass makes it possible to apply the value of the biological coefficient of total evaporation $k_I = ET / E^0$ (where ET is the amount of evapotranspiration for a certain segment of the growing season, mm; E is evaporation from the surface of the soil devoid of vegetation, mm) and in this way, to parameterize the functional of taking into account the prehistory of the development of plants in ontogenesis (4) by determining the weighting coefficients of the phenophases of the main agricultural crops development. In terms of physical content, $k_б$ shows how many times evapotranspiration differs from evaporation in given soil and climate conditions, characterizing the ability of plants to regulate transpiration, and therefore, the need for life factors to accumulate crop phytomass. Taking into account the value of $k_б$ for the phases of this agricultural crop development, the weighting factor for the i-th phase of development is determined according to the dependence:

$$v_i = k_{bV} / \sum_{i=1}^{n} k_b. \tag{19}$$

Based on the analysis of experimental data, principles and an algorithm for determining the coefficient of crop losses during harvesting have been developed in order to model the size of the stored crop. Since for most types of cereals, natural shedding of grain is the main way of yield loss, and technically and technologically determined losses when using worn-out grain harvesters can exceed the total losses from natural and climatic factors, there is a need to develop and implement an information and advisory system of support and management of economic activity within the ERM framework on drained lands of Ukraine.

To assess the ecological stability of Rivne region drained lands, the ASDL model was applied, the results of which in terms of drainage systems are shown in Table 6. A mapping scheme of assessment and forecasting of ecological sustainability was drawn up on the example of the drainage system "Yazvynka" (Rivne region).

One of the main modern degradation processes that determine the plasticity of the vertical structure, and, accordingly, the loss of stability of the reclamation sub-regions of the Western Forest-Steppe, composed of drained peat layers, is the mineralization of peat organic matter, which is accompanied by the compaction of the peat layer. We found time trends for ash content and density of peat in the form of regression equations of the form:

$$Z= 0.2943\ t + 51.849,\ R^2=0.359,\ a = 0.05; \tag{20}$$

$$d= 0.2338\ t + 0,2897,\ R^2=0.651,\ a = 0.05, \tag{21}$$

where t is the number of decades from April 20, 1988, taking into account the segments from April 20 to October 30 of each year of the observation period; α is the level of significance for the Student's test when assessing the convergence of calculated and empirical values of indicators.

The regression equations obtained as a result of correlation and regression analysis of time series, which describe the trends of individual hydrochemical and radiological indicators at a sufficient level of significance and were used to assess the ecological stability of the Horyn and Styr river basins, are summarized in Table 7.

The results of the evaluation of the environmental sustainability of the basins of the Horyn and Styr rivers are given in Table 8.

Table 6. Ecological sustainability assessment results of Rivne region drainage systems

Drainage system	Absolute assessment		Relative indicators of stability			
	form of stability	degree of stability	ω_1	ω_2	ω	ω_{pr}
Vorobyno	renewable (II)	repatively stable	0.355	0.312	0.336	0.343
Holovnytsia	renewable (I)	fairly stable	0.792	0.768	0.780	0.780
Derazhne-Postiine	plastic	unstable	0.323	0.348	0.336	0.333
Ikva	plastic	unstable	0.357	0.317	0.337	0.325
Stubelka	plastic	unstable	0.478	0.437	0.458	0.455
Stubla	plastic	unstable	0.248	0.224	0.237	0.231
Yazvynka	successional	unstable	0.224	0.333	0.279	0.272

Table 7. Trends in hydrochemical and radiological indicators that determine ecological stability of the Styr river basin and Horyn river basin

Diameter	Indicators	Regression equation C = $f(t)$	Approximation reliability indicator, R^2	Level of correlation significance, α
the Horyn river Vysotsk	Nitrates	$C = 0.1858t^2 - 1.3716t + 2.9225$	0.702	0.01
	Chlorides	$C = -0.5433t + 25.882$	0.570	0.05
	Sulfates	$C = 60.163\ t^{-0.3733}$	0.807	0.01
	BOD	$C = -0.1099t^2 + 0.872t + 5.0923$	0.834	0.01
the river Viliya	Cs-137	$C = 2.0579t^{0.152}$	0.465	0.05
the Styr river Kuznietsovsk	Suspended matter	$C = -0.4282t^2 + 1709.1t - 210^6$	0.451	0.05
	Nitrates	$C = -0.0611t^2 + 0.8182t - 0.5391$	0.883	0.01
	Chlorides	$C = -1.1211t + 2258.3$	0.859	0.01
	Sulfates	$C\text{-}0.4459t^2 - 1785.2t + 210^6$	0.863	0.01
	BOD	$C = 0.0717t^2 - 286.97t + 286975$	0.748	0.01
the Styr river Sopachiv	Suspended matter	$C = -0.5902t + 1198$	0.660	0.01
	Chlorides	$Q = -0.8367t + 1691.9$	0.600	0.01
	Sulfates	$C = -3.462t + 6951.1$	0.823	0.01

Table 8. Environmental sustainability assessment results of the basins of the Horyn and Styr rivers ccording to the monitoring observations data

Diameters and basins	Absolute assessment			Relative indicators of stability	
	Form of stability	Share of indicators	Stability degree	ω	ω_{pr}
the Viliya river, Ostrog	inert	1.00	stable	0.871	0.871
the Horyn river, Vysotsk	inert	0.69	stable	0.376	0.355
the Horyn river basin	relatively stable			0.624	0.613
the Styr river, Kuznietsovsk	inert	0.69	stable	0.578	0.621
the Styr river, Sopachiv	inert	0.69	stable	0.533	0.584
the Styr river basin	relatively stable			0.556	0.603

CONCLUSION

1. The modern state of agriculture and ecological and reclamation monitoring necessitates the development of fundamentally new methods of assessment, forecasting, and management based on mathematical models, which would make it possible to systematically take into account the content of processes, the effect of the basic laws in agriculture, calculate the value of productivity, assess the state of anthropogenically altered natural systems, predict future states, carry out machine experimentation in order to rationally manage drained lands. Since the existing methods and models do not meet the requirements regarding the level of complexity and artificiality of the structure, taking into account the combined effect of variables, compliance with tasks and monitoring databases, etc., there is a need to develop a special mathematical model of productivity and assessment of the condition of drained lands (FACDL), as well as a mathematical model of environmental assessment sustainability of drained lands and natural and man-made systems (ASDL).
2. Taking into account the properties of the modeled system, the use of the Lebesgue measure functional is justified as the main operator of the FACDL model, which has a universal character and can be used for the mathematical modeling of the studied system. To describe the law of tolerance (minimum, optimum, and maximum), a new type of tolerance function of plants and plant associations to the action of internal and external factors has been developed. It has been proven that the dependence, developed and parameterized to describe the law of tolerance, approximate experimental data regarding the properties of natural systems with high accuracy, and therefore can be used in the simulation mathematical model of productivity and assessment of the state of drained lands and other simulation mathematical models of natural and naturally technogenic systems. Based on experimental data, a cumulative action block was developed for the first time, which simulates the cumulative effect of the variables that determine the state of the system under study. For the first time, for natural systems under conditions of agricultural use, the type and parameters of the empirical block of the submodel of the aggregate action of variables were established, and the values of the weighting coefficients were determined, which describe the degree of participation of the effect of the aggregate action of factors in the process of a crop formation.

3. As a result of the conducted theoretical studies, the general structure of the simulation mathematical model of FACDLwas determined, which makes it possible to determine and forecast the productivity of drained lands based on monitoring data with the use of meteorological observation data, and to develop measures for the rational management of drained lands.
4. To create a productivity model and assess the condition of drained lands, a rational set of climatic variables and indicators of ecological and reclamation monitoring has been defined. It was established that the main climatic indicator for the studied system is the sum of active air temperatures higher than 10°C, which makes it possible to determine, with high reliability, the intensity of energy input to the studied agrogeosystem based on the obtained regional empirical models. For the first time, the obtained values of the weight of individual parameters of the productivity on DLs indicate that 0.45 of the agricultural crop volume harvested on drained lands is determined by soil conditions, and therefore, the land reclamation condition. On the basis of the obtained weighting ratios, it can be stated that the greatest attention should be paid to the regulation of soil regimes and the formation of a favorable reclamation (ecological and reclamation) situation when conducting agriculture and carrying out environmental protection measures on reclaimed lands, and the system of ecological and reclamation monitoring should be directed to the collection of soil-reclamation indicators of drained lands to ensure the possibility of managing their productivity.

REFERENCES

Aleksyeyevskii, V., Skrypnyk, O., & Riabtseva, H. (1992). Assessment and control of changes in natural complexes under the influence of drainage. K: UkrNYYHyM.

Organization and management of ecological and reclamation monitoring on drained lands. (1997). *Drained lands*. K: Derzhvohosp Ukrainy.

Duque, M., Zapico, I., Bugosh, N., Tejedor, M., Delgado, F., & Nicolau, M. (2021). A Somolinos quarry land stewardship history: From ancient and recent land degradation to sensitive geomorphic-ecological restoration and its monitoring. *Ecological Engineering*, *170*(November), 106359. doi:10.1016/j.ecoleng.2021.106359

Hadzalo, Ya., Stashuk, V., & Rokochynckyi, T. (2017). Land reclamation and development of the Ukrainian Polissia: a collective monograph. OLDI-PLUS.

Isaeva, S., & Shevchenko, V. (2019). Principles of Ecological Monitoring of Agricultural Lands. *IOP Conference Series. Earth and Environmental Science*, *272*(2), 022230. doi:10.1088/1755-1315/272/2/022230

Klymenko, M., Pryshchepa, A., & Vozniuk, N. (2006). Environmental monitoring. Vydavnychyi tsentr Akademia.

Matsniev, A., Protsenko, S., & Sablii, L. (2000). Monitoring and engineering methods of environmental protection. VAT Rivnenska drukarnia.

Moshynskyi, V. (1995). Monitoring and assessment of the ecological and reclamation condition of drained lands of Rivne region. National University of Water and Environmental Engineering.

Moshynskyi, V. (2005). Methods of managing productivity and ecological sustainability of drained lands. Rivne: NUWM.

Moshynskyi, V., & Sasyuk, Z. (2010). Spatial extrapolation in monitoring data reproduction tasks. Rivne: NUWM.

Moshynskyi V., Tsvyetova O. (2004). Methodology and prospects of ecological and reckamation monitoring in Ukrainian Polesie. *Reclamation and Water Management, 90*, 33-39.

Moszynski W. (2002). *Dynamika stezenia jonow w odwodmonych glebach norfowych terenow lesnostepowych zachodniej Ukrainy*. Roczniki Gleboznawcze. – Warszawa.

Novokhatin, V., Osipova, N. (2020). *Monitoring of agrophysical properties of the active layer of drained long-frozen peat deposits in Western Siberia* / E3S Web of Conferences 164, 07016 Pages 6.

Richling, A., & Solon, J. (1998). Ekologia krajobrazu. Warszawa: PWN.

Schilling, K. E., & Dinsmore, S. (2018). Monitoring the Wildlife, Hydrology and Water Quality of Drained Wetlands of the Des Moines Lobe, Northern Iowa: Introduction to Special Feature. *Wetlands*, *38*(2), 207–210. doi:10.100713157-017-0989-y

Truskavetskyi R. (1992). Negative phenomena in soil formation on drained lands. *Soil fertility. Monitoring and management*. K.: Urozhai.

Tsuman, N., Zinchenko, V. (2021). Agroecological problems of intensive agriculture on drained lands. *Plant Varieties Studying and Protection 17*(4).

Mitrofanov, E., & Mitrofanova, O. (2021).*Tile drainage monitoring in agricultural fields with an unmanned aerial system. IOP Conference Series: Earth and Environmental Science*.

Ukhurebor, K., Aigbe, U., Onyancha, R., & Ndunagu, J. (2022). An Overview of the Emergence and Challenges of Land Reclamation: Issues and Prospect. Hindawi. Applied and Environmental Soil Science. doi:10.1155/2022/5889823

Untenecker, J., Tiemeyer, B., Freibauer, A., Laggner, A., & Luterbacher, J. (2017). Tracking changes in the land use, management and drainage status of organic soils as indicators of the effectiveness of mitigation strategies for climate change. *Ecological Indicators*, *72*, 459–472. doi:10.1016/j.ecolind.2016.08.004

Chapter 7
Long Term Modeling and Predictive Assessment of Climate Change of Ukraine Polesie Zone

Petro Kovalenko
Institute of Water Problems and Land Reclamation, Ukraine

Nataliia Prykhodko
https://orcid.org/0000-0003-1424-2628
National University of Water and Environmental Engineering, Ukraine

Vasyl Turcheniuk
National University of Water and Environmental Engineering, Ukraine

Jerzy Jeznach
Warsaw University of Life Sciences, Poland

ABSTRACT

The analysis of current state of weather and climatic conditions and evaluation of their predicted changes for immediate and distant prospect in the drained areas of Ukrainian Polesie region was carried out. The main trends in changes of meteorological characteristics and their possible effect on the conditions of functioning water management and ameliorative objects; complexes also on the natural and ameliorative state of drained areas were identified. The research uses a method of predictive-simulation modeling with used predictive assessment models of normalized distribution of the basic meteorological characteristics in long-term and one-year vegetation context. It was established that, for today, a high variability in meteorological characteristics can lead to significant deterioration of operation conditions of water management and ameliorative objects and units, as well as natural and ameliorative conditions of drained lands. Core measures regarding the adaptive potential enhancement and development in the region under the conditions of climate change were examined.

DOI: 10.4018/978-1-6684-8248-3.ch007

BACKGROUND

Global warming is one of the numerous ecological, social, and economic problems, which mankind started to face at the end of the last century.

Current climate change may have significant natural, economic and social aftermath. In view of the changes in climate conditions taking place even today, a vitally important part is assigned to the forecasting of possible climate change in future and solving of a number of crucial and complicated tasks, related to the development and implementation of the strategies of farther mankind survival both on a world-wide basis and at a regional level. Therefore, for today a need of predicted evaluation of climate change and development of relevant adaptive decisions arise (Gorzhyk, Yemelyanova, Trofimova et al., 1998; Romashchenko, Sobko, Savchuk et al., 2003; Romashchenko, Rokochinskiy, Halik et al., 2007; Romashchenko, Khvesik, Mikhailov et al., 2015).

Ukraine belongs to the number of regions influenced by ongoing climate change that is rather essential. On the base of numerous meteorological factors and indices the domestic scientists (climate researchers) come to the point, that for the last 10…25 years a new climate is being formed over the territory of Ukraine (Boichenko, Voloshchuk, Doroshenko, 2000). These changes are very tangible in Polesie region of Ukraine where the drained lands are instrumental for the country economic development and provision of food security.

Thus, the evaluation of the predicted changes in weather and climate patterns under the conditions of climate change and their impact on the conditions of water management and ameliorative both objects and complexes as well as on the natural-ameliorative state of Polesie region of Ukraine is of a great concern.

METHODS AND MODELS

To solve the task a large-scaled computer experiment on PC was planned and performed based on long-term retrospective and current observational data obtained in Polesie region (Mazhayskiy, Rokochinskiy, Volchek, et al., 2017; Rokochinskiy, 2010).

The following variants of study were planed and implemented for the average conditions of Western Polesie:

- **Variant 1 – «Base»**: characterization of the meteorological characteristics over a period of vegetation (months IV-X), obtained on the base of long-term retrospective data (1891-1964) (Hydrometioizdat, 1990);
- **Variant 2 – «Transitional»**: characterization of standardized average long-term values of the variables of basic meteorological characteristics and their distribution over the vegetation period, obtained under current conditions (1947-1990);
- **Variant 3 – «Recent»:** characterization of the dynamics and standardized average long-term values of the variables of basic meteorological characteristics and their distribution over the vegetation period, obtained under recent conditions over the years of 1991-2017;
- **Variant 4a – «CCCM»** and **variant 4b – «UKMO»**: characterization of standardized average long-term values of the variables of basic meteorological characteristics characteristics and their distribution over the vegetation period, obtained in view of current and predicted climate change, in accordance with the recommendations (Romashchenko, Sobko, Savchuk et al., 2003), by the

models of Canadian Climate Center (CCCM) – as a more favourable forecast, and of the United Kingdom Met Office (UKMO) – as a less favourable forecast, which foresee an increase in average annual temperature up to 4° C and 6° C relatively – provided that the doubling CO_2 in atmosphere occurs (Romashchenko, Sobko, Savchuk et al., 2003; Shevchuk, Trofimova, Trofimchuk, 2001).

Practicability of the application of «CCCM» and «UKMO» models is proved by their capabilities to involve both less and more critical scenarios of changing weather patterns, and they perfectly comply with the predictive estimate of the rationing of basic meteorological characteristics in the long-term and one-year vegetation context (Rokochinskiy, Halik, Frolenkova et al., 2008).

As a base variant of study were taken the retrospective long-term data (1891-1964 years) according to official reference data (Hydrometioizdat, 1990), because in this period the climate of the studied region had a pronounced stable character.

The forecast was made for five typical groups of vegetation periods of target years in view of general heat and moisture provision (very wet – 10%, wet – 30%, middle – 50%, dry – 70%, very dry – 90% according to Table 1) by such basic meteorological characteristics as air temperature, precipitation, relative humidity and air humidity deficit, photosynthetically active radiation (*PAR*) and water availability coefficient (precipitation/evapotranspiration ratio) by the given diagram (Figure 1) (Rokochinskiy, Halik, Frolenkova et al., 2008).

Table 1. Estimated provision of meteorological factors according to typical groups (estimated years) of vegetation periods in view of general heat and moisture provision

Provision P (%) by P_j, $\bar{H}_j$	**Typical groups (estimated years)P = {p},** $p = \overline{1, n_p}$ **(n_p = 5)**	**Provision P (%) by $\bar{T}_j, \bar{D}_j$**
0...20%	P = 1, very wet	80...100%
20...40%	P = 2, wet	60...80%
40...60%	P = 3, middle	40...60%
60...80%	P = 4, dry	20...40%
80...100%	P = 5, very dry	0...20%

Scientific and methodological basics of machine experiment are the approaches and models for forecasting of typical meteorological regimes on a long-term basis, the use of which is regulate by the relevant industry standards (Rokochinskiy, 2010). The generalized block-scheme of stages of implementation the model of typical distribution of meteofactors in the case of availability of long-term database presented in Figure 2.

Figure 1. Diagram of weather and climate characteristics assessment under retrospective (Base), transitional (Transitional), recent (Recent) and forecasted (CCCM, UKMO) conditions; CCM = model of Canadian Climate Center, UKMO = model of United Kingdom Met Office, PAR = photosynthetically active radiation
Source: own elaboration

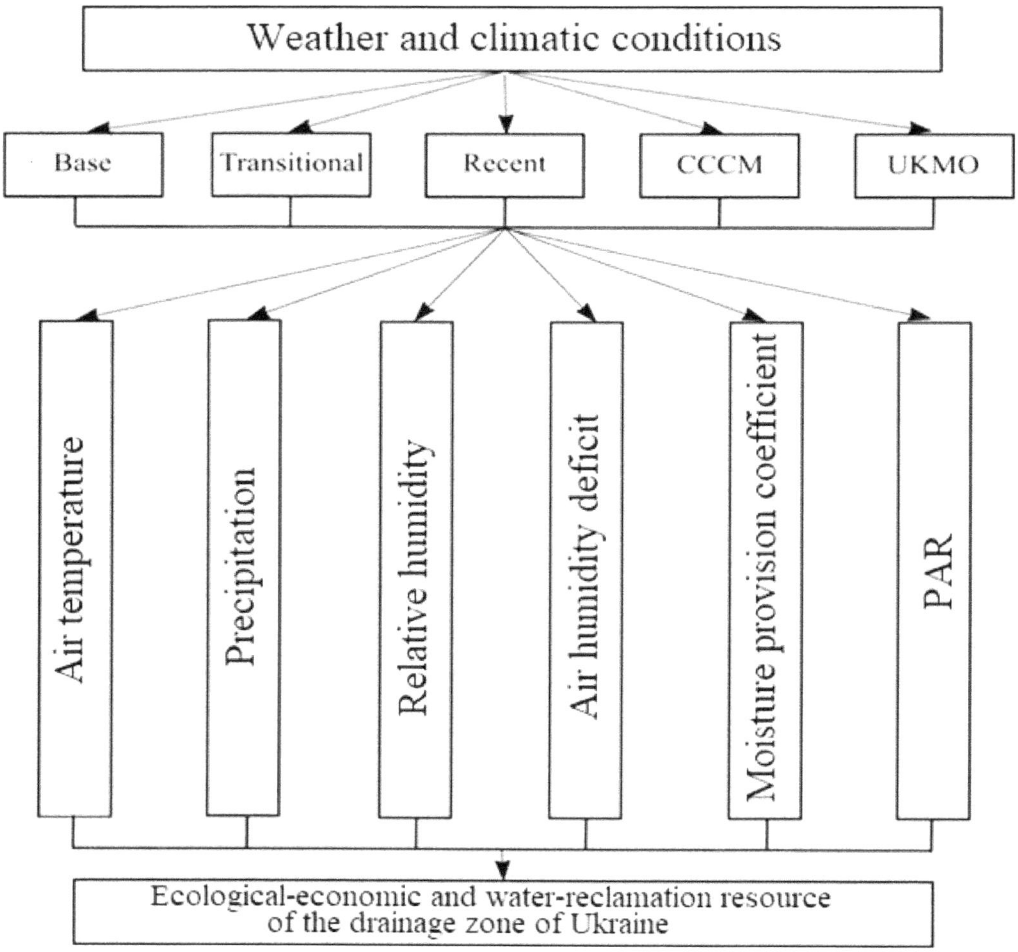

RESULTS AND DISCUSSION

The summarized results of the calculation of the values of basic meteorological characteristics (precipitation, temperature, air humidity deficit and relative air humidity) and their derivate characteristics (*PAR*, water availability coefficient) over the target years were given in the Table 2 as the percentage change of normalized vegetation values of the basic meteorological characteristics by the variants of study (Transitional, Recent, CCCM, UKMO) relative to their baseline values (Base).

The obtained results on the comparative assessment of changes in current and predicted values of basic meteorological characteristics over the target years and on the average for the years enable to make following conclusions (Mazhayskiy, Rokochinskiy, Volchek et al., 2017):

Figure 2. The generalized block-scheme of stages of implementation the model of typical distribution of meteofactors in the case of availability of long-term database

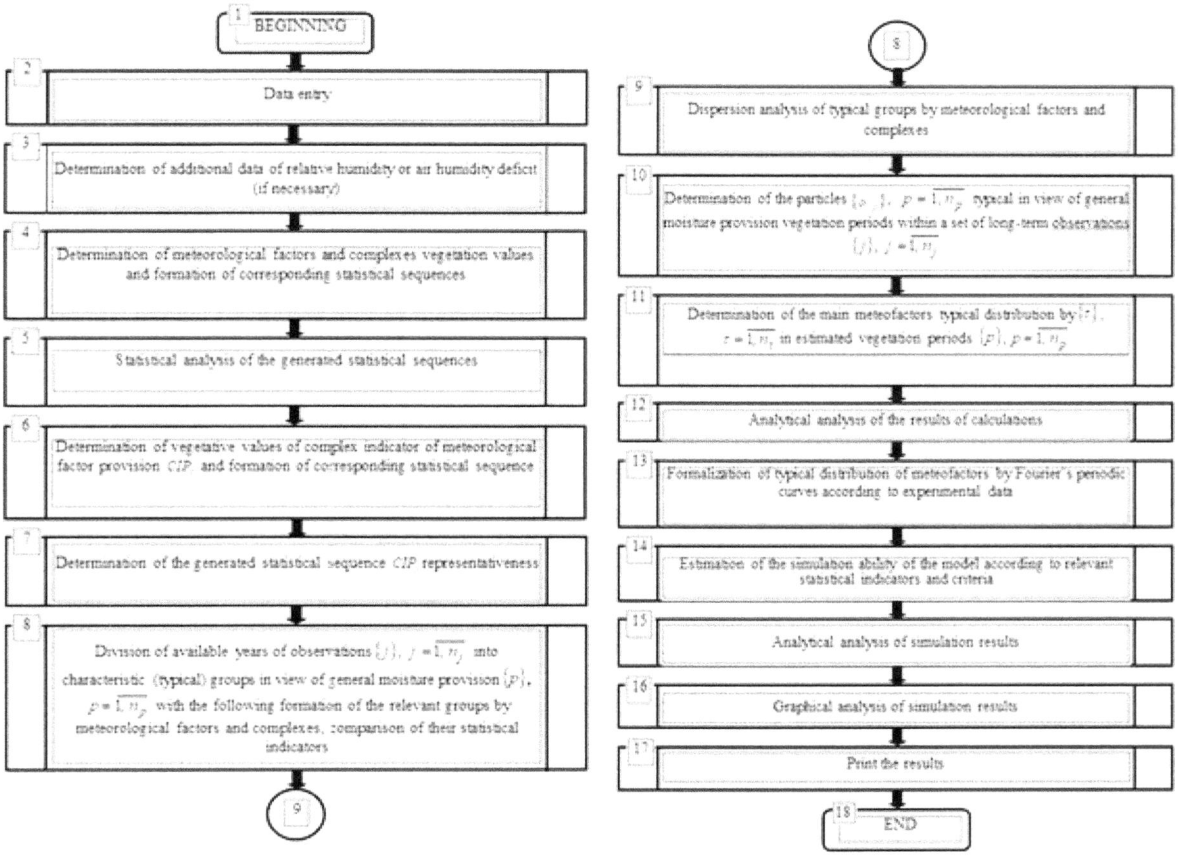

- **On precipitation**: in transitional conditions (Transitional) as compared with the base variant (Base) the precipitation decreases over the target years (from 0,47% in dry year (70%) up to 17,9% in wet year (30%), with an average of 5,30%. As to the current conditions (Recent), as a whole some of the increase in precipitation can be observed (it varies from 2,80% in very wet year (10%) up to 11,87% in very dry year (90%), with an average of 4,07%; respectively by the forecasting variants some of the increase in precipitation can be observed over all target years, with an average of 1,69% for CCCM and 7,47% for UKMO;
- **On air temperature**: over all examined variants the increase in air temperature was observed both over the target years and average for the years that is less evident in transitional conditions (Transitional) – 1,75%, more evident in current conditions (Recent) – 4,84%, and it increases significantly in forecasting conditions: CCCM – 24,59% and UKMO – 34,93%;
- **On air humidity deficit**: the nature of change of this index is similar to the air temperature change with some ranges over the target years and on the average for the years: Transitional – 0,83%; Recent – 5,44%, CCCM – 22,8% and UKMO – 27,19%;

Table 2. Comparative assessment of changes in the values of basic meteorological characteristics by the variants of study for the conditions of Polesie region

Model indicator		Typical group of target years					Average value (%)
		10%	30%	50%	70%	90%	
Precipitation *P*	transitional	–5,34	–17,90	–1,84	–0,47	–0,96	–5,30
	recent	–2,80	0,33	0,06	10,88	11,87	+4,07
	CCCM	1,29	1,43	1,62	1,87	2,23	+1,69
	UKMO	7,05	7,20	7,39	7,66	8,04	+7,47
Air temperature *T*	transitional	1,89	2,07	2,08	0,95	1,76	+1,75
	recent	4,89	7,27	5,42	6,14	0,49	+4,84
	CCCM	25,71	25,04	24,50	24,21	23,47	+24,59
	UKMO	36,12	35,45	34,82	34,53	33,76	+34,93
Air humidity deficit *D*	transitional	3,44	2,55	4,12	–2,12	0,77	+1,75
	recent	4,44	8,79	7,66	0,32	5,98	+5,44
	CCCM	24,07	24,00	22,97	21,95	21,04	+22,81
	UKMO	27,94	28,03	26,97	28,00	25,00	+27,19
Relative air humidity *H*	transitional	1,49	1,20	–1,75	1,32	1,89	+0,83
	recent	–4,85	–2,47	–2,37	0,61	–1,35	–2,09
	CCCM	–7,99	–6,72	–6,68	–6,41	–6,68	–6,90
	UKMO	–8,94	–7,72	–9,04	–8,74	–8,64	–8,62
Photosynthetically active radiation *PAR*	transitional	0,50	1,11	1,64	0,90	1,29	+1,09
	recent	2,84	4,64	3,77	3,84	0,06	+3,03
	CCCM	16,33	16,09	16,10	15,80	15,60	+15,98
	UKMO	22,82	22,65	22,72	22,40	22,23	+22,57
Water availability coefficient K_w	transitional	–8,49	–19,94	–5,73	1,69	–1,72	–6,84
	recent	–6,94	–7,77	–7,05	10,53	5,55	–1,14
	CCCM	–18,36	–18,20	–17,36	–16,47	–15,54	–17,19
	UKMO	–16,33	–16,27	–15,42	–15,89	–13,57	–15,49

- **On atmosphere relative humidity**: in transitional conditions (Transitional) the value of this index on the average increases – 0,83%, in current conditions (Recent) it decreases by 2,09%, according to the forecast it also decreases: for CCCM by 2,09% and for UKMO by 6,9%;
- **On *PAR***: the nature of change of this index corresponds to the change in air temperature, having a little lower intensity of the increase: Transitional – 1,09%, Recent – 3,03%; CCCM – 15,98% and UKMO – 22,57%;
- **On water availability coefficient**: the nature and rate of the change in precipitation and air humidity deficit both over the target years and on the average for the years determine some of the decrease in transitional conditions (Transitional) – 6,84%, less evident decrease in current conditions (Recent) – 1,14%, and a rather significant decrease by the forecast: CCCM – 17,19% and UKMO – 14,49%.

Comparative assessment of the dynamics of basic meteorological characteristics of the vegetative periods for the period of 1991-2017 along with their retrospective and perspective rates was given in Fig. 3.

The obtained results enable to make the following statements (Mazhayskiy, Rokochinskiy, Volchek et al., 2017):

- **As to precipitation**: it was observed a rather wide range of precipitation for the studied period, namely it was from 200 up to 600 when having an average long-term rate as 443 mm with well-defined peaks in 1991, 1999, 2008 and 2017 and decays in 1991, 2002, 2004, 2015; in the next years after the year 2008 it was observed a steady decline in precipitation in all variants of study along with its drastic increase in 2016 and 2017; meanwhile the predicted rates by the models CCCM and UKMO are already within the current range of total values;
- **As to air temperature**: the opposite is the case; starting from the year it was observed an increase in air temperature along with well-defined peaks in 1999, 2002, 2012 and 2016, which were accordingly 15,1° C, 15,0° C, 15,3° C and 15,2° C; meanwhile the average annual air temperature for the studied period do not reach the predicted values by the models of CCCM and UKMO;
    - **As to air humidity deficit**: the dynamics of changes in air humidity as a whole shows the characteristic features of the change in air temperature range: air humidity deficit reaches first and second peaks in the years 2002 and 2005, which were accordingly 1118 and 1112 mm, when having average values of 5,30 and 5,20 mm res respectively; after the year 2005 the total values of air humidity deficit slightly decreased and manifested insignificant range of them; after that it was observed their third and fourth peaks in the years 2012 and 2016 – accordingly 1193 and 1250 mm, when having average values of 5,60 and 5,83 mm respectively; meanwhile, the value of air humidity deficit by the model Base is lower than the average annual rates by the models Transitional and Recent, and its corresponding rates by the models CCCM and UKMO are already within its current range;
    - **As to air relative humidity**: the opposite is the case regarding the dynamics of the change in air relative humidity, it was observed two peaks (about 78%), one – in 1991 and another – in 1998, after that a drastic decrease in the values (near 68%) up to the first and second peaks was recorded in the years 2005 and 2012; meanwhile the average long-term value of air relative humidity by the model Base is higher than the average annual rates by the models Transitional and Recent, and its corresponding rates by the models CCCM and UKMO were within the current range of average annual values.

In whole the forecasting values of studied meteorological characteristics by the models CCCM and UKMO in Polesie region, excepting air temperature, are already within a current range and even higher for some indices, that proves the existence of a steady trend towards the change in climatic conditions.

Rising air temperatures, increase in climate deficit and aridity in the region will inevitably lead to the increase in evaporation and general water consumption while growing crops, including the land with regulated water regime (drained territories).

For the drained land with shallow ground waters, where the most of drainage systems in Ukraine are located, whether conditions have an effect on the formation of water regime of soils and ground waters, influencing the ways and behaviour of ground processes.

Figure 3. Comparative assessment of normalized values of basic meteorological characteristics by the variants of study regarding the dynamics of their valid values in conditions of Polesie region: a) precipitation, b) air temperature, c) air humidity deficit, d) relative air humidity; base, transitional recent CCCM, UKMO as in Figure 1
Source: own study

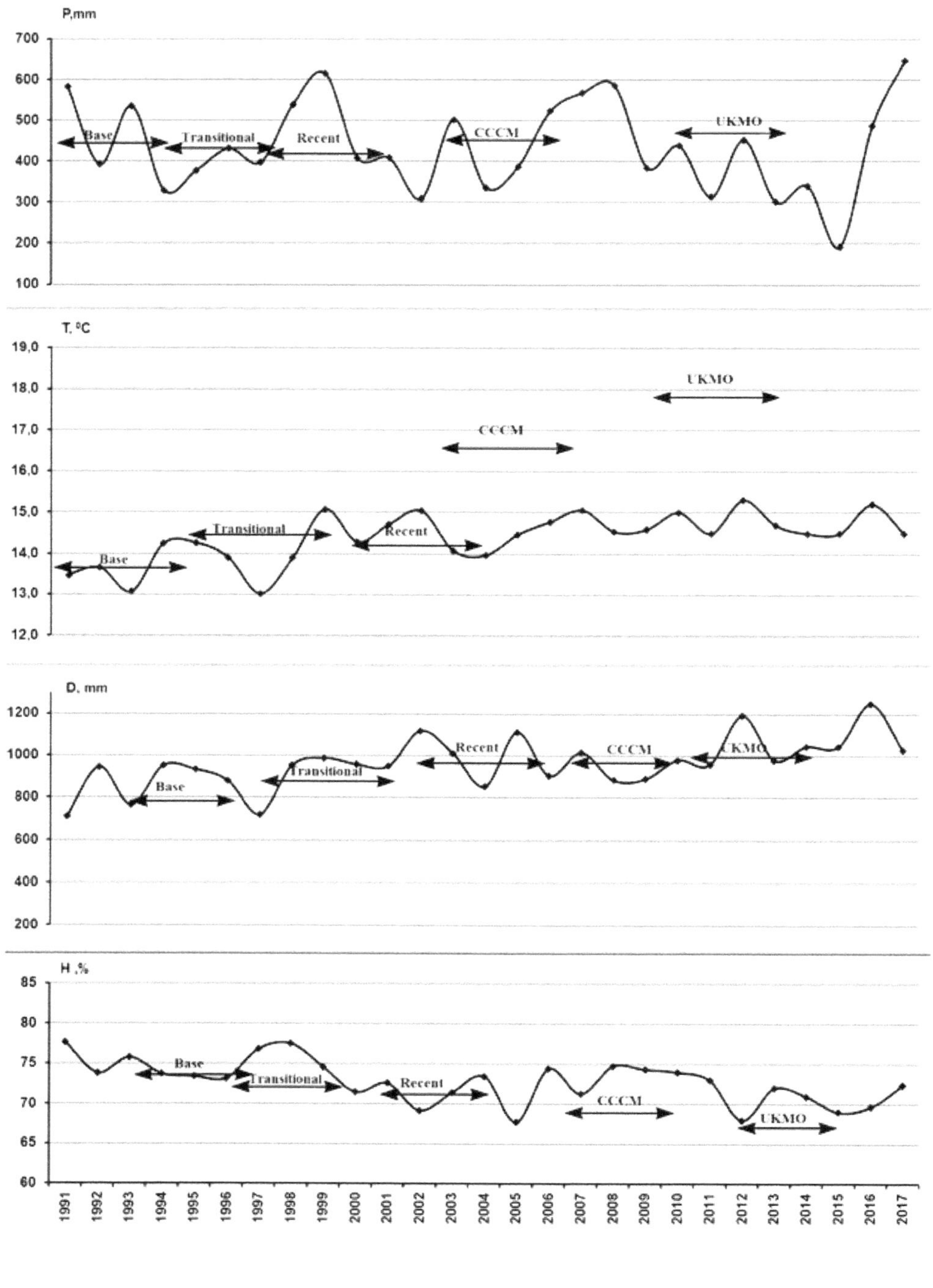

Climatic variability directly or indirectly influences soils, both mineral and organic. It increases the hazard of peatlands burning, their mineralization and deterioration as a result of their excessive draining, worsening of their water-physical properties, which leads to the loss in soil productivity.

Changes in soil and climate conditions necessitate some work to be performed so that crops to benefit. First of all it comes to the replacement of standard crops for the crops requiring a rather large sum of positive temperatures. It is also pressing issue about the possibility of growing new drought-tolerant crops with a rather short vegetation period.

Therefore, climate changes require the soils to be adaptive to these conditions. In this regard one of the most important issues of agrarian and ameliorative science for today is the enhancement of the adaptive capacity of agricultural lands with regulated water regime to benefit most.

Adaptation of agricultural production in conditions of sustainable development to the changes in soil and climatic features calls for the introduction of new technologies of operational water regulation, maximum accumulation of precipitation, reconstruction and renewal of existing ameliorative facilities as well as the construction of the most modern ones in the drained areas.

FUTURE RESEARCH DIRECTIONS

That in turn will have an effect on functioning water management and ameliorative objects and units, as a result of above mentioned changes in natural and ameliorative resources that requires developing some adaptive technical and technological measures for proper managing these objects based on relevant scientific state and international researches and programs.

Under such conditions the adaptation to climate change means an accommodation in natural or social systems as a response to actual or expected climate impacts or their consequences that enables to reduce their environmental risks and damages and take some advantages.

Agricultural lands located within water management and ameliorative objects and units should be considered as agricultural ecosystems. Efficient use of these lands is possible provided the evaluation of their adaptive capacities. In this it must be taking into account: firstly – their nature and exposure to essential climate change; secondly – their capabilities to contend with external effects and damages, which arise as a result of environmental changes (Adger, 2006); thirdly – risk and frequency of natural disasters and their impacts (Schmidt-Thome, 2007).

Based on the studied experience of European countries (Massier, 2012) the adaptive measures on drained lands should involve a range of management, operational and agro-technical, construction and project activities (Table 3).

There are various activities on adaptation as applied to agriculture (Massier, 2012; Paris Agreement ratified by the Law N° 1469-VIII of 14.07.2016), which involve different levels of decision-making: international, national, regional and local and are aimed at both prevention and elimination of consequences. Introducing the adaptive activities should be accompanied by raising awareness of public and land users.

Table 3. Main activities on adaptation to climate change

Group of activity	Activity
Management activities	– establishment of organization and consulting centres – development of reporting, planning and prospective water balance by the basin principle – modelling of river ecological conditions in view of water quality indices in agro- ecosystems
Operational and agro-technical activities	– regulation of water regime – prevention of soil degradation – improvement of plant breeding
Construction activities	– construction, reconstruction and renaturalization of water management and ameliorative objects and units – construction and reconstruction of flood control hydrotechnical structures
Project activities	– ensuring ecological and safe conditions for environmental management – design of closed farming production cycles

Source: (Massier, 2012)

CONCLUSION

Thus, for today a high variability in meteorological characteristics can lead to the significant deterioration of operation conditions of water management and ameliora-tive objects and units, as well as natural and ameliorative conditions of drained lands in Polesie region as a whole. The proposed adaptation activities have their own explicit objectives, which are closely related, therefore to develop a proper program of adaptation activities the potential impacts of climate change, risks and vulnerability as its results need to be better understood. This is the integrative approach to solving the problems will enable to enhance and develop the adaptive capacity of Polesie region.

REFERENCES

Adger, W. (2006). Vulnerability. *Global Environmental Change, 16*(3), 268–281. doi:10.1016/j.gloenvcha.2006.02.006

Boichenko, S., Voloshchuk, V., & Doroshenko, I. (2000). Global warming and its consequences on the territory of Ukraine. *Ukrainskyi heohrafichnyi zhurnal, 2*, 59–68.

Gorzhyk, I., Yemelyanova, Zh., Trofimova, I., & Shereshevskiy, A. (1998). Ukraine and Global Greenhouse Effect: Book. 2. Vulnerability and adaptation of environmental and economic systems to climate change. Kiev, Ukraine: Vydavnytstvo Ahentstva z ratsionalnoho vykorystannia enerhii ta ekolohii.

Massier, E. (2012). The experience of the European Union in adapting to climate change and its application in Ukraine. *OSCE Coordinator for Economic and Environmental Activities*. Biuro Koordynatora z ekonomichnoi ta dovkilnoi diialnosti OBSIE.

Mazhayskiy, Y., Rokochinskiy, A., Volchek, A., Meshyk, O., & Jeznach, J. (2017). Environmental management of Polesie. VNIIGiM.

Paris Agreement ratified by the Law N° 1469-VIII of 14.07.2016. https://ec.europa.eu/clima/policies/international/negotiations/paris_en

Rokochinskiy, A. (2010). *The scientific and practical aspects optimization of water regulation drained lands on environmental and economic grounds.* NUVGP.

Rokochinskiy, A., Halik, O., Frolenkova, N., Voloshchuk, V., Shalai, S., Zubyk, Ya., Bezhuk, V., Zubyk, L., Pokladnov, Ye., Savchuk, T., Koptiuk, R., Nesteruk, L., Volk, P., & Kotiai, L. (2008). Guide to DBN V.2.4-1-99. Reclamation systems and structures. VAT.

Rokochinskiy, A., Volk, P., Pinchuk, O., Mendus, S., & Koptyuk, R. (2017). Comparative evaluation of various approaches to the foundation of parameters of agricultural drainage. *Journal of Water and Land Development*, *34*(1), 215–220. doi:10.1515/jwld-2017-0056

Romashchenko, M., Khvesik, M., & Mikhailov, Yu. (2015). Water strategy of Ukraine for the period till 2025 (scientific basis). Kiev, Ukraine

Romashchenko, M., Rokochinskiy, A., Halik, O., Savchuk, T. & Kolodich, O. (2007). Recent climate change and their manifestations from global to regional manifestations. *Hydromelioration and hydromelioration construction, 32,* 65–79.

Romashchenko, M., Sobko, O., Savchuk, D., & Kulbida, M. (2003). *About some problems of agrarian science in connection with climate change*. Institute of Hydrotechnics and Reclamation UAAN.

Schmidt-Thome, P. (2007). *The spatial effects and management of natural and technological hazards in Europe*. ESPON.

The Ukrainian SSR. (1990). Series 3. Perennial data. Scientific and Applied Handbook on the Climate of the USSR. Russia, Lenynhrad: Hydrometeoyzdat.

Shevchuk, V., Trofimova, I., & Trofimchuk, O. (2001). *Problems and strategy of Ukraine's implementation of the UN Framework Convention on Climate Change*. FAO.

Chapter 8
Increase the Adaptive Potential of Dried Land in Changeable Climatic Conditions

Pavlo Volk
National University of Water and Environmental Engineering, Ukraine

Anatoliy Rokochinskiy
National University of Water and Environmental Engineering, Ukraine

Nataliia Prykhodko
https://orcid.org/0000-0003-1424-2628
National University of Water and Environmental Engineering, Ukraine

Liubov Volk
https://orcid.org/0000-0003-1033-6715
National University of Water and Environmental Engineering, Ukraine

ABSTRACT

Based on the analysis and generalization of the data of domestic and foreign scientists and specialists, as well as the results of the authors' respective research, an assessment of the current state of changes in weather and climatic conditions in the Polesie zone and their impact on agricultural production, fertility, and moisture supply of soil; the conditions of functioning of water management and reclamation facilities, the natural and reclamation state of the drained lands was also given. The general recommendations for the development of adaptive organizational and economic, operational, agrotechnical, construction, and project activities that have their own clear goals and are closely interrelated were considered. On this basis, a complex of adaptive agromeliorative, agrotechnical and hydrotechnical, measures has been developed and proposed, aimed at a gradual transition to the cultivation of new varieties and types of agricultural crops, effective regulation of the water regime of drained lands and the functioning of drainage systems in variable climatic conditions.

DOI: 10.4018/978-1-6684-8248-3.ch008

BACKGROUND

Both at the global and regional levels, climate change has become an indisputable fact, the presence of which has posed the problem of solving several important and complex tasks related to the development and implementation of a strategy for its practical existence and adaptation to new climatic conditions. At the same time, the ongoing climate changes threaten a significant decrease in productivity in crop and livestock production, as well as the degradation of ecosystems, which causes significant risks to the food security of the population and the well-being of future generations. The solution to this problem requires the development and implementation of adaptation strategies that can increase the productivity, efficiency, and profitability of agricultural production (Romashchenko et al., 2020; Kovalenko et al., 2019; Moore 2014; Ciscar et al., 2019).

In agricultural production, reclamation traditionally plays a leading role in ensuring its sustainable development in adverse climatic conditions. Taking into account the expected climate changes, an extremely important role in the adaptation of agriculture to it and, first of all, crop production, is played by water, hydrotechnical, agrotechnical, and other types of reclamation. Therefore, there is a need to develop and implement adaptive measures for agricultural production on lands with a regulated water regime, including drained ones, in changeable climatic conditions to effectively counter modern challenges - energy, food, and water crises, which are intensifying due to ongoing climate change (Kolokytha et al., 2020; Rokochynskiy et al., 2019).

MATERIALS AND METHODS

The analysis and generalization of the content of the program international and national documents regarding the need and general recommendations for the development of adaptive measures related to climate change, the results of research and a general assessment of domestic and foreign scientists and specialists, as well as the results and our respective research were carried out. An assessment of the current state of changes in weather and climatic conditions, including in the Polesie zone, and their impact on the efficiency of agricultural production, fertility and moisture supply of soil, the conditions for the functioning of water management and reclamation facilities, the natural reclamation state of drained lands was given and general recommendations for the development of adaptive organizational and economic, operational and agrotechnical, construction and project activities that have their own clear goals and are closely interrelated were considered (Kolokytha et al., 2020; Rokochynskiy et al., 2019; Mazhaiskiy 2017).

The methods of our research on climate change, its impact on different aspects of agricultural production on drained lands in order to develop appropriate adaptation measures are based on the use of an appropriate set of predictive and simulation models, which includes a terrain climate model, a model of water regime and technologies for water regulation of drained lands, the model of development and formation of the yield of cultivated crops and are implemented on the basis of a long-term forecast. The basic methods of the general theory of experiment were also used: analytical, experimental and statistical; mathematical modeling and machine experiment using modern information and computer technologies. The processing of the results of field research and simulation data was carried out using the methods of mathematical statistics, probability theory and correlation analysis.

RESULTS AND DISCUSSION

The fact of global climatic changes is recognized by the world community and is beyond doubt today (Climate Change and Land, 2020). The problem of global warming arose at the end of the last century, and subsequently, due to its increasingly tangible negative impact, the relevance only increases. In 1988, the Intergovernmental Panel on Climate Change (IPCC) was established by the World Meteorological Organization and the United Nations Environment Program to assess the state of global warming and the associated risks. Recognition of global climate change on the planet and their negative consequences for humanity was declared by the international community in 1992 by the adoption of the UN Framework Convention on Climate Change (UN FCCC) - the agreement was signed by more than 180 countries of the world. In Paris (France) on June 17, 1994, the UN Convention to Combat Desertification and Drought was adopted and opened for signature. In 1997, the UN Framework Convention on Climate Change was supplemented by the Kyoto Protocol.

Later, in connection with an increase in the rate of warming, in 2015 the Paris Agreement was adopted, the main goal of which is to keep the growth of global temperature in the 21st century within 2°C and the possibility of its reduction to 1.5°C.

The solution to this problem requires the development and implementation of adaptation strategies that can increase the productivity, efficiency and profitability of agricultural production.

According to the United Nations Framework Convention on Climate Change, ***adaptation*** means adjustments in natural, social or economic systems in response to actual or expected climate changes, as well as their consequences (Eric, Massey, 2012).

To clarify and expand the applicability of the concept of adaptation to climate change conditions, the following categories were also considered and clearly defined: *adaptation potential; tendency to influence; sustainability; risk; sensitivity; vulnerability* (Stepanenko, Poloviy, 2011).

In accordance with the programmatic international documents devoted to this issue in the field of climate diplomacy in Ukraine and the EU as a whole, as well as a strategy based on real research according to the IPCC criteria on climate change in Ukraine until 2100 and pursuant to the Decree of the President of Ukraine "On the decision of the Council of the National security and defense of Ukraine" of September 14, 2020 and "On the National Security Strategy of Ukraine", etc.) Ukraine, like other participating States, has undertaken to allocate funds for adaptation to climate change and preparation for their implementation (Cook, 2016).

Along with the energy, food and water crises, climate change is one of the main challenges of our time. Thus, the unpredictability of weather conditions, which endangers food production, provokes a rise in sea level, increases the risk of natural disasters and anomalous phenomena that manifest themselves in various forms (drought, waterlogging, flooding of territories and etc.) and are the consequences of climate change, have a global character and unprecedented scale. However, most of the impacts of climate change are expected to persist for several centuries. Therefore, if decisive action is not taken today, then the subsequent adaptation to climate change will require a lot of effort and expense.

To assess the possible consequences of climate change as a whole for the current period, 12 scenarios of the impact on its changes were developed: for each of the three global climate models (GCM) (GISS, GFDL, UKMO), four scenarios were identified (GCM without a direct influence of CO_2 on yield, with direct impact of CO_2 on yield, with direct effects of CO_2 and adaptation of the 1st level, with direct effects of CO_2 and adaptation of the 2nd level). The IBSNAT model was used to assess how climate change and rising carbon dioxide levels could change the world's crop production at 112 sites in 18 countries.

Ukraine is also among the regions of the planet where the ongoing climate changes are tangible. According to the generalized data of the research results of specialists and scientists of Ukrhydromet, the Institute of Water Problems and Reclamation of the National Academy of Agrarian Sciences of Ukraine (NAAS), as well as the Odessa State Ecological University (Romashchenko et al., 2020) over the past thirty years, the average annual air temperature has increased by 1.2 °C in the country as a whole.

According to the results of assessing the regional warming rates, it was found that in the south and north-east of the country in 1991-2019 the average annual air temperature turned out to be by 1.0-1.1 °C higher than for the standard period (1961-1990), in the west - by 1.2-1.3 °C, and in the north and in the central regions - by 1.4-1.5 °C. That is, the growth rate of air temperature for 1975-2019 ranges from 0.61 to 0.82 °C / 10 years, while in the neighboring post-Soviet countries (Russia, Moldova, Belarus) - 0.47-0.59 °C / 10 years, and in the northern hemisphere and Europe - 0.34 and 0.47 °C / 10 years, respectively. These data indicate that the rate of air temperature rise in Ukraine is much higher than the European and global scales.

Usually, an increase in air temperature increases the moisture holding capacity of the atmosphere. With an increase in air temperature by 1 °C, the water-holding capacity of air increases by about 7%, which leads to an increase in potential evaporation (Romashchenko et al., 2020).

According to the calculations of scientists from the Institute of Water Problems and Land Reclamation of the NAAS (Romashchenko et al., 2020)., an increase in the average monthly air temperature by 1 ° C increases the potential average monthly total evaporation (excluding other factors) by 9%. In this regard, there are six main typical zones in Ukraine: *excessive moisture supply* - occupies 4.5% of the territory; *humid* - 30.0%; *insufficient moisture supply* - 16.0%; *arid* - 20.0%; *dry* - 22.0%; *very dry* - 7.5%.

In the period from 1991 to 2015, compared to 1961-1990, territories with a significant deficit of natural moisture supply (dry and very dry zones) increased by 7% and cover a total of more than 29.5% of the area of Ukraine or 11, 6 million hectares (37%) of the country's arable land. At the same time, the territory of the country with excessive and sufficient atmospheric moisture, on the contrary, decreased by 10% and occupies only 22.5%, including 7.6 million hectares of arable land.

According to the forecast until 2050, according to the scenario based on the A1B model (Rokochynskiy et al., 2020), the general trend of an increase in the temperature regime by 1.24-1.48 ° C is expected, and with an increase in precipitation in general by 8%, the climatic water balance of the territory of Ukraine may decrease by 45-115 mm, and its deficit in the Steppe zone will be 560 mm or more. Taking this into account, the territory of the country with an insufficient level of moisture will increase to 56%, and only 28% of the territory will meet humid and excessively humid conditions.

In the long term, by the end of 2100, a part of the country's territory with an insufficient level of moisture can reach 71% against 50% in 1991-2015 with a simultaneous decrease in the area of arable land with sufficient moisture to 5.5-1.8 million hectares.

In general, on the territory of Ukraine already in the twentieth century, there was a general trend towards an increase in air temperature and an increase in the amount of atmospheric precipitation: so the annual temperature increased by 0.3-0.7 °C, and precipitation - by 50-100 mm, the frequency of significant anomalies of both indicators also increased significantly (in 1.5 times). Their intraseasonal variability also increased. In all scenarios, the amount of precipitation will also increase. In some seasons, this increase may exceed the existing level by 20% (Rokochynskiy et al., 2020).

In the warm season, as a result of a sharp increase in air temperature, most of the precipitation will be spent on evaporation. The main feature of the warming period is the constant unevenness of precipitation in the middle of the year and in certain years. At the same time, the greatest influence, due to global warming, is experienced by the Ukrainian Polesie zone during the winter and spring seasons.

On reclaimed agricultural lands with a close occurrence of the groundwater level, weather and climatic conditions are directly involved in the formation of the water regime of soil and groundwater, determining the direction of the soil processes both in the natural state and in certain technological periods of growing crops. In this regard, the forecast assessment of weather and climatic conditions is an essential condition for the implementation of the assessment of the overall efficiency of use and the ecological and reclamation state of drained lands.

To assess retrospective, modern and expected climate changes in the Western Polesie zone of Ukraine and its impact on the efficiency of agricultural production, soil fertility and moisture supply, the conditions for the functioning of water management and reclamation facilities, as well as with the aim of developing appropriate adaptation measures, we have planned and implemented a large-scale machine experiment using modern information and computer technologies. The research is based on the use of an appropriate complex of forecasting and simulation models and will include a model of the local climate, a model of the water regime and technologies for water regulation of drained lands, a model of the development and formation of the yield of cultivated crops, which are implemented on the basis of a long-term forecast (Kovalenko et al., 2019).

With regard to our assessment of climate change, then, according to (Rokochynskiy et al., 2019)., the forecast was carried out:

- according to basic meteorological characteristics: air temperature, precipitation, relative humidity and air humidity deficit;
- for five groups of vegetation periods of estimated years in terms of total heat and moisture supply: very humid - 10%; humid - 30%; medium - 50%; dry - 70%; very dry - 90%;
- for the following levels and corresponding periods of assessment: retrospective - 1945-1990; recent -1991-2015; forecast - for the climate model UKMO - the model of the United Kingdom Met Office, provides for an increase in the average annual air temperature by 6 ° C when the CO_2 content in the atmosphere doubles. In forecasting regime calculations, this model takes into account more critical scenarios of changes in weather and climatic conditions and is in better agreement with the models that we used for predictive assessment of the normalized distribution of the main meteorological characteristics in the long-term and intra-growing section.

The generalized results of the established average annual norms of vegetation values of the indicators of the main meteorological factors, as well as the possible nature of their change according to the considered levels and assessment periods and the last real years (2015-2020) for the conditions of Western Polesie of Ukraine, are shown in the Figure 1.

The obtained results convincingly indicate that the predicted values of almost all the main meteorological indicators in the zone of Western Polesie, except air temperature, are already within the limits of their current fluctuations and even exceed them in individual positions, which indicates a stable trend of changing climatic conditions in this region.

At the same time, for volatility, as the initial component of total evaporation, the nature of the change in this indicator for modern conditions relative to retrospective ones seems to be insignificant (an increase by 5.44%), and for forecast conditions according to the UKMO model it is very significant - 27.2%.

Figure 1. Comparative assessment of the average annual norms of vegetation values of the indicators of the main meteorological factors for the conditions of Western Polesie by the considered levels and periods

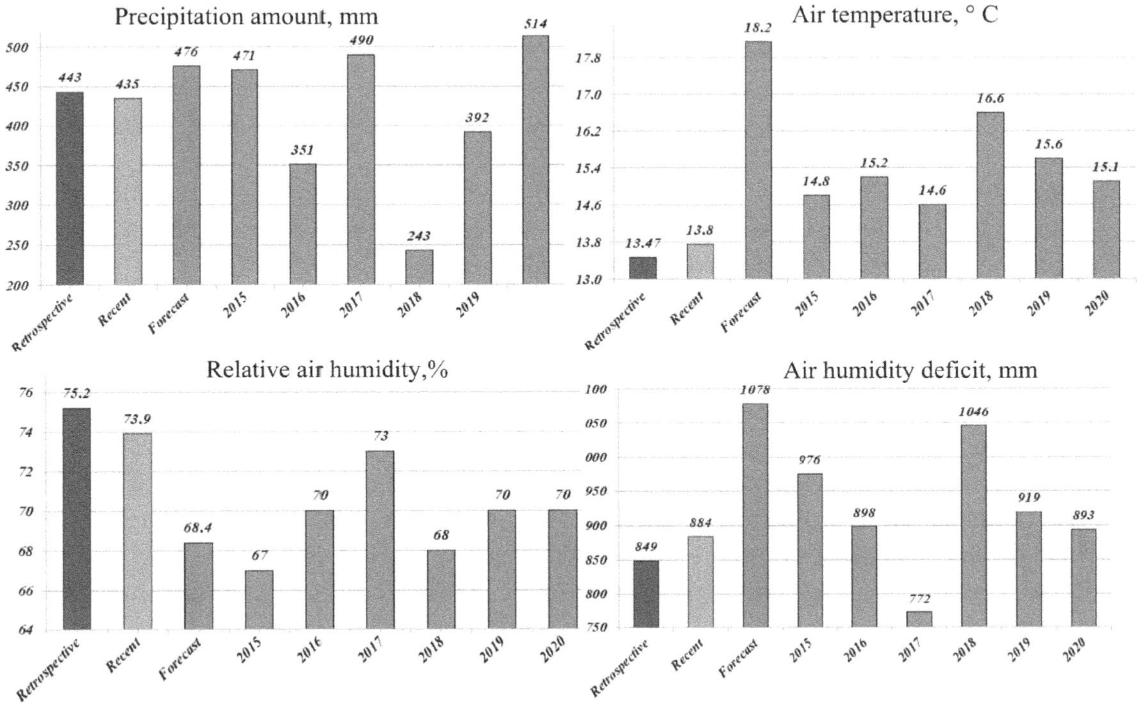

A detailed analysis by FAO shows that climate change affects the agricultural sectors in many different ways, which vary from region to region. Some regions are expected to face prolonged droughts and water scarcity. The widespread melting of glaciers and snow cover in large mountain ranges, especially in Asia, will affect the volume and timing of water flows, ultimately reducing the availability of water for irrigation downstream. The increase in temperatures will lead to changes in the geography and frequency of disease outbreaks and pest outbreaks. In low-latitude regions, even a slight warming will lead to lower yields.

According to many studies (Moore 2014; Ciscar et al., 2019), covering a wide range of regions, the negative impact of climate on crop yields will be more common than positive.

Attention should be paid to the expansion of the boundaries of agriculture in many countries, since it can not only be caused by global changes in the region (population growth, changes in the landscape and natural and climatic conditions), but also cause other changes.

Currently, the agricultural industry in Ukraine is not extremely vulnerable to climate change yet. However, changes in weather conditions (an increase in air temperature, uneven distribution of precipitation, which are stormy in the warm season, ineffective accumulation of moisture in the soil, etc.) lead to an increase in the number and intensity of arid phenomena. Together with other negative factors of anthropogenic impact, this can lead to the expansion of the zone of risky agriculture and desertification in the southern regions of Ukraine. Due to intense warming over the past decades, there have been changes in the structure of agricultural production, the area under agricultural crops and levels of their yield.

According to (Rokochynskiy et al., 2020), the climate plays a decisive role in the formation of agro-ecological conditions for agricultural production. One of the most important tasks that must be solved when studying it is the need for knowledge: what kind of agricultural crops are profitable to grow in a particular area; how the climate affects the soil on which these crops are grown; what methods are best used when growing them.

Climate has played and still has a decisive role as one of the leading factors in soil formation, vegetation, and productivity of agricultural crops.

Solar energy absorbed by the soil is spent on processes such as heating, evaporation, transpiration, photosynthesis, humus synthesis, and so on. The sum of active temperatures is traditionally used for agronomic and soil assessment of the territorial thermal regime.

An increase in the average annual temperature, the value of the radiation balance and the sum of active temperatures per year leads to an increase in the intensity of weathering, the synthesis of organic matter, the activation of the vital activity of animals and microorganisms, an increase in the intensity of soil-forming processes, as well as to a reduction in the interphase periods and the vegetation cycle and, accordingly, the onset of earlier terms of ripening and harvesting of cultivated zoned agricultural crops.

One of the elementary processes of soil formation is the evaporation of soil moisture, which, first of all, depends on temperature. An increase in precipitation as a result of increased leaching of nutrients (mineral and organic) substances from the soil will also negatively affect soil fertility. The regions with sufficient and excessive moisture can suffer the most from this, which occurs in the north-western regions of Ukraine and can lead to a decrease in soil fertility in this territory by 20% or more.

Nowadays in Ukraine there are about 1 million hectares of peatlands drained as a result of reclamation, the bulk of which is located in the right-bank Polesie within the Kiev, Volyn, Rivne and Zhytomyr regions. Peat bogs, as one of the most vulnerable wetlands, are threatened with degradation on a global scale today.

In especially dry years in the spring and summer months, protracted fires in the peatlands of Polesie continue for several weeks, or even more than a month. In recent years, long-term fires have been recorded annually and have led to local environmental disasters.

Thus, it is expected that in the near future climate change in Ukraine will have both positive and negative impact on agriculture. Moreover, the balance of these influences will be different in different agroclimatic zones. A decrease in the duration of the growing season will be beneficial for agriculture in the northern half of the country, and it will lead to an increase in droughts for the southern half of the country.

Agro-climatic resources can increase and significantly improve the efficiency of agriculture. The implementation of this is possible in the case of the rise and radical adaptation of agriculture to climatic conditions synchronized with the pace of climate change, that is, it is necessary to optimize natural reclamation conditions with the help of specially developed reclamation measures. Otherwise, climate change can lead to a drop in the average yield and instability of agricultural production in general.

The focus of reclamation activities is determined, first of all, by the required level of productivity of the agricultural land used with a regulated water regime of soil, in accordance with the level of requirements of agricultural crops to the conditions of their cultivation. Therefore, it is of interest to assess changes in weather and climatic conditions and their impact on the general natural reclamation regime and productivity of reclaimed lands at the regional level. That is, it is necessary to identify general tendencies and the nature of the relation in the system of *weather and climatic conditions ⇒ natural reclamation regime ⇒ productivity of drained lands* on the example of the zone of Western Polesie of Ukraine.

The generalized results of the predictive assessment of the impact of changes in weather and climatic conditions on the productivity of the main cultivated zoned agricultural crops on drained lands under the considered recent and forecast conditions in relation to retrospective ones for the most indicative dry (p = 70%) and very dry (p=90%) vegetation periods obtained as a result of the obtained simulation modeling on ECM are presented in Table 1.

Table 1. General trends towards a decrease in the productivity of the main cultivated zoned agricultural crops on drained lands in changeable climatic conditions

No.	Agricultural crops	Deviations of the main indicators from the normalized average annual favorable conditions, %							
		Recent conditions				Forecast conditions			
		$\Delta\tau_T$	ΔY_Y	ΔY_W	ΔY	$\Delta\tau_T$	ΔY_Y	ΔY_W	ΔY
Dry growing season, p = 70%									
1.	Winter cereals	8.55	13.3	17.0	15.2	16.7	34.6	40.4	42.6
2.	Potatoes	7.70	22.2	22.0	29.3	15.4	45.2	43.9	46.8
3.	Perennial grasses	5.30	10.8	34.1	36.8	15.8	46.8	48.6	48.2
Very dry growing season, p = 90%									
4.	Winter cereals	25.0	37.4	38.8	38.5	33.3	48.2	43.3	43.6
5.	Potatoes	15.4	43.6	48.0	48.0	23.1	54.4	52.4	56.7
6.	Perennial grasses	26.3	38.8	43.8	43.8	31.6	60.8	59.2	62.4

The note. Designations used in the table: $\Delta\tau T$ a decrease in the duration of the growing season of cultivated crops with an increase in temperature relative to its normalized average long-term one; ΔYT ΔYW a decrease in the yield of agricultural crops when the temperature and water regimes deviate, respectively, from the normalized average long-term favorable conditions; ΔY- a general decrease in the yield of the main cultivated zoned crops on drained lands in changeable climatic conditions.

The presented data Table 1 convincingly indicate that the existing recent and expected predicted climate changes, associated primarily with an increase in the temperature regime and a decrease in the moisture supply of the territories Figure 1, in general, worsen conditions and can lead to a decrease in the yield of traditional zoned varieties and types of agricultural crops grown on drained lands on average from 15 to 60%. It should be noted that this level of yield decline, and even lower, takes place in the last observed years with the manifestation of abnormal phenomena in comparison with the retrospective period Figure 1, when growing crops on drained lands without additional moisture in real production conditions.

The projected increase in air temperature and increased aridity under climate change, which are already observed today, will inevitably lead to a decrease in the natural moisture supply of the territory and an increase in potential evaporation and, accordingly, the total water demand when growing crops on drained lands. In turn, the lack of natural moisture requires additional moistening of the grown crops through the implementation of appropriate water regulation technologies to moisten the drained lands.

Therefore, by analogy with described above, a predictive assessment of changes in water demand was made when growing crops on the drained lands of Western Polesie of Ukraine in changeable climatic conditions to substantiate the appropriate adaptive solutions.

The main results of a comparative assessment of water demand according to the main indicators of the regime and technique of subsurface moistening (irrigation and moisture rates, the number of irrigations, etc.), as the most common technology for moistening drained lands, for the existing and forecast conditions under consideration are presented Figure 2 in the form of diagrams, which in a comparable form reflect the values of the maximum norms of water supply for the most indicative very dry (p = 90%) year in the ten-day section when growing the main crops on peat soil.

Figure 2. Dynamics of the maximum average ten-day values of water demand of cultivated crops according to water supply rates during dry growing seasons (p = 90%) with subsurface moistening of drained lands

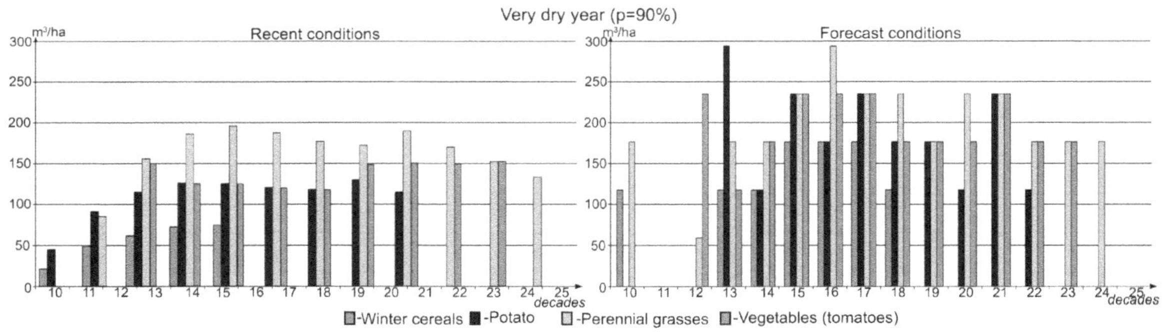

The results obtained to determine the total evaporation, water demand and technological efficiency of humidification of drained lands in the changeable climatic and agromeliorative conditions of Western Polesie of Ukraine indicate a possible increase in the water demand of grown crops by 1.5-2.0 times. This determines the need to reassess the functionality and technical condition of the existing drainage systems based on the transition to humidification of drained lands on an ongoing basis.

Thus, at the existing rates and levels of changes in weather and climatic conditions, a deterioration in natural reclamation conditions in general, both in the Polesie zone and in Ukraine as a whole should be expected. This will inevitably affect the functioning of water management and amelioration facilities and complexes as a result of corresponding changes in the ecological and economic resource, which determines the need to develop adaptive regime, technological and technical measures for the management of these facilities on the basis of relevant research, comprehensive scientific industry, state and interstate programs.

The modern concept of adaptation to climate change is presented in the Fifth Assessment Report of the IPCC and Working Group II and was further developed in the cycle of the Sixth Assessment Report (2015-2022).

At the same time, it is stated that one of the most important components of playing a leading role in the adaptation process is a clear understanding of why they are carried out, and what adaptation measures can be introduced. As basic, various measures are considered to address different aspects of adaptation Figure 3.

The implementation of measures to adapt agriculture to climate change must necessarily provide, first of all, in accordance with generally accepted European practice, a set of activities to raise awareness of the public and land users.

Figure 3. General characteristics of adaptation measures
(Paris Agreement/ United Nations. New York, 2016)

Activities aimed at the formation of adaptive potential	Activities aimed at reducing risk and sensitivity	Activities aimed at increasing capacity to cope with the consequences of emergency events	Activities to benefit from climate change
Promote national and public awareness of the climate change process, its consequences and the ability to respond to them	Preparatory measures aimed at increasing the degree of resilience and protection from short-term, medium-term and long-term climate changes	Activities during and after extreme events (floods, fires, hurricanes) to reduce the consequences and curb natural disasters	Activities to benefit from climate change; climate change is beneficial for someone
Examples: the research of the impact of climate change, disaster response plans, etc.	**Examples:** early warning systems, new varieties of crops, etc.	**Examples:** cooling areas during extreme heat, a dedicated emergency fund, special teams on hand to clear roads, etc.	**Examples:** benefit from longer growing seasons of new commercial crops.

The general methodology for carrying out such actions can be presented in the form of the following algorithm Figure 4.

Figure 4. Algorithm for making decisions on adaptation in an agroecosystem

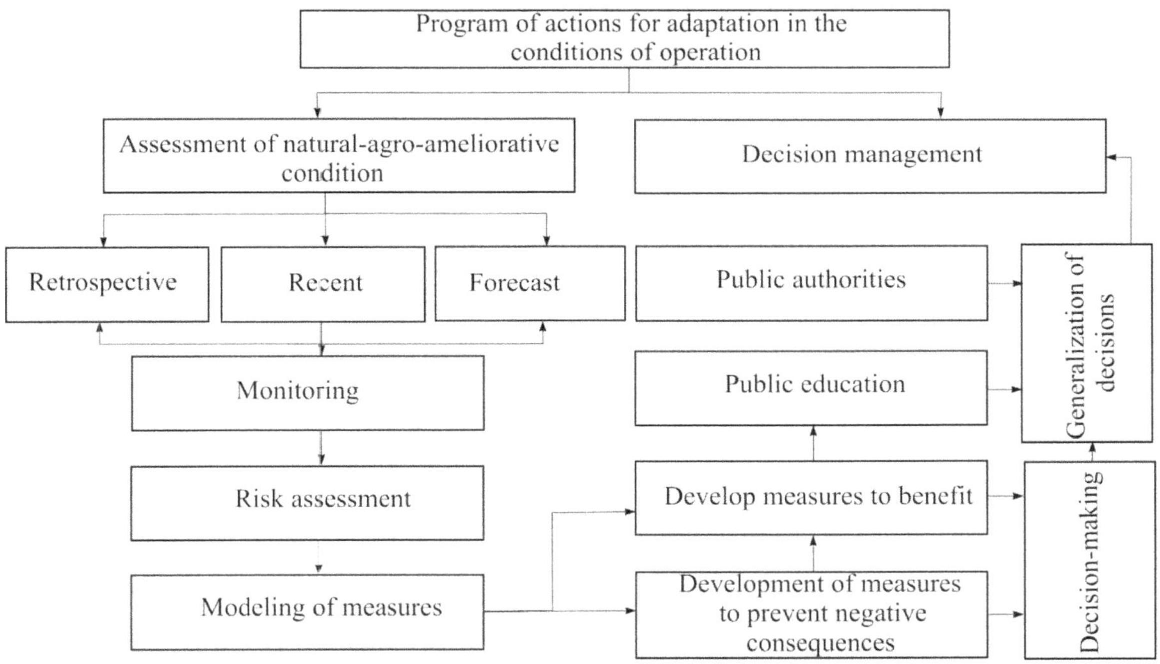

As for adaptive measures for reclaimed agricultural land, then, according to (Romashchenko et al., 2020; Kovalenko, Rokochynskiy et al., 2019), for their development it is necessary to carry out much larger, in comparison with the minimum necessary, comprehensive research. Among other important research on adaptation of agricultural production to climate change in general, this research, including for the drainage reclamation zone, should primarily be aimed at developing new energy, water and resource-saving methods of integrated land reclamation, the latest technical means of irrigation, water regulation, application of fertilizers and chemical ameliorants, irrigation and fertilization regimes for agricultural crops in accordance with the projected changes.

Therefore, based on the analysis and generalization of the content of international and national program documents regarding the need and general recommendations for the development of adaptive measures related to climate change, research results and general assessment of domestic and foreign scientists and specialists, as well as the results of our relevant research to assess the current state of changes in weather and climatic conditions, including in the Polesie zone, and their impact on the efficiency of agriculture, moisture supply and soil fertility, a set of measures has been developed to adapt agricultural production on drained lands, which will make it possible to effectively resist recent challenges associated with ongoing and expected climate changes Table 2.

Table 2. General recommendations on the types of measures for the adaptation of agricultural production on drained land in changeable climatic conditions

Organizational and economic activities	*Agrotechnical activities*	*Agro-ameliorative activities*	*Hydraulic engineering activities*	
Operational			*Project*	*Construction*
Creation of organizational and advisory centers to raise awareness of the public and land users.	Changes in varieties and the introduction of crops that are resistant to biotic and abiotic threats associated with climate change (primarily temperature and water regimes); growing hybrid crops; use of crops of later ripening.	Improvement of technologies and tillage tools of drained soil aimed at improving their water-physical properties and storage capacity on the basis of improvement, first of all, technologies and means of deep continuous loosening of soil, which will make it possible to effectively accumulate moisture in the soil profile and on the drained massif as a whole.	Improvement of technologies and water regulation regimes on drained lands, their transfer to regular humidification, regulation and water accumulation within the system; improvement of types and design of hydroameliorative systems on drained lands; improvement of methods of design and calculation of drainage systems based on the use of optimization methods in compliance with modern economic and environmental requirements, the widespread use of information and computer technologies.	New construction, reconstruction and modernization of existing drainage systems to effectively meet modern challenges associated with ongoing and expected climate change.

CONCLUSION

Thus, in order to effectively counter current challenges associated with ongoing and expected climate changes, a set of adaptive agromeliorative, agrotechnical and hydraulic engineering measures has been developed and proposed, aimed at a gradual transition to the cultivation of new varieties and types of crops, effective regulation of the water regime, regulation and accumulation of moisture in the soil profile and within the system, the transition from the traditional periodic to the implementation and provision of regular moistening of drained lands, improvement of water regulation technologies, types and designs of drainage systems and their technical elements, methods of their design and calculation, corresponding to modern economic and environmental requirements.

REFERENCES

About approval of the Concept for the implementation of state policy in the field of climate change until 2030: Order of the Cabinet of Ministers N° 932-r of December 7, 2016 https://www.kmu.gov.ua/ua/npas/249573705 (access date 29.04.2020)

Ciscar, J. C., Rising, J., Kopp, R. E., & Feyen, L. (2019). Assessing future climate change impacts in the EU and the USA: Insights and lessons from two continental-scale projects. *Environmental Research Letters*, *14*(8), 084010. doi:10.1088/1748-9326/ab281e

IPCC. (2020). *Climate Change and Land: An IPCC Special Report on climate change, desertification, land degradation, sustainable land management, food security, and greenhouse gas fluxes in terrestrial ecosystems.* IPCC. https://www.ipcc.ch/reports/

Consensus on consensus: a synthesis of consensus estimates on human-caused global warming / Cook J. et al. // *Environmental Research Letters*. 2016. Vol. 11. 4. P. 1-7. https://doi.Org/10.1088/1748-9326/ll/4/048002

Eric E. Massey, 2012. The experience of the European Union in adapting to climate change and its application in Ukraine. Office of the Coordinator for Economic and Environmental Affairs of the OSCE P-40.

Global temperature [Data set]. (n.d.). NASA. https://climate.nasa.gov/vital-signs/global-temperature/

Zacon. (2020*). Irrigation and drainage strategy in Ukraine until 2030.* Cabinet of Ministers of Ukraine. https://zakon.rada.gov.ua/ laws / show /

Kolokytha, E., & Malamataris, D. (2020). Integrated Water Management Approach for Adaptation to Climate Change in Highly Water Stressed Basins. *Water Resources Management*, *34*(3), 1173–1197. doi:10.100711269-020-02492-w

Kovalenko, P., Rokochynskiy, A., Jeznach, J., Koptyuk, R., Volk, P., Prykhodko, N., Tykhenko, R. (2019). Evaluation of climate change in Ukrainian part of Polissia region and ways of adaptation to it. *Journal of Water and Land Development*, *41* (IV-VI), 77-82. . doi:10.2478/jwld-2019-0030

Moore, F., & Lobell, D. (2014). Adaptation potential of European agriculture in response to climate change. *Nature Climate Change*, *4*(7), 610–614. doi:10.1038/nclimate2228

United Nations. (2016). *Paris Agreement.* UN. https://treaties.un.org/doc/Treaties/2016/02/20160215%2006-0 3%20PM/Ch_XXVII-7-d.pdf

Rokochinskiy, A., Volk, P., Jeznach, J., Turcheniuk, V., Frolenkova, N. & Koptiuk, R. (2019). Reclamation projects development improvement technology considering optimization of drained lands water regulation based on BIM. *Scientific Review Engineering and Environmental Sciences*, *3*(85), 432-443. doi:10.22630/PNIKS.2019.28.3.40

Rokochynskyi, A., Volk, P., Turcheniuk, V., Tokar, L., Volk, L., Mazhayskiy, Y., & Chernikova, O. (2020). Construction and agricultural drainage parameter optimization considering economic and environmental requirements. *Engineering for rural development*. Jelgava. http://www.tf.llu.lv /conference/proceedings 2020/Papers/TF0 10.pdf

Romashchenko M., Gusev Yu., Shatkovskiy A., Saidak R., Yatsyuk M., Shevchenko A. and Matiash T. (2020). Impact of climate change on water resources and agricultural production. *Reclamation and water resources management*, (1), 5 - 22. doi:10.31073/mivg202001-235

Romashchenko, M. I., Saydak, R. V., & Matyash, T. V. 2019. Development of irrigation and drainage as the basis of sustainable agriculture in Ukraine in climate change. *IX International scientific and technical conference*, 22-27. IX.

Supit, I., van Diepenb, C. A., de Wit, A. J. W., Wolf, J., Kabat, P., Baruthc, B., & Ludwig, F. (2012). Assessing climate change effects on European crop yields using the Crop Growth Monitoring System and a weather generator. *Agricultural and Forest Meteorology*, *164*, 96–111. doi:10.1016/j.agrformet.2012.05.005

United Nations Convention to Combat Desertification. (n.d.). *Documents*. UNCCD. https://www.unccd.int/official-documents

UNFCC. (2020). United Nations framework convention on climate change. *Kyoto Protocol*, Kyoto. https://unfccc.int/ resource / docs / convkp / kpeng.pdf (access date 30.04.2020).

UN. (1992). *United Nations framework convention on climate change*. United Nations. . https://unfccc.int/resource/docs/convkp/ conveng.pdf

Chapter 9
The Improvement of the Design Technology of Reclamation Projects for the Optimization of the Drained Lands Water Regulation Based on BIM Technologies

Anatoliy Rokochinskiy
National University of Water and Environmental Engineering, Ukraine

ABSTRACT

Improvement principles of development reclamation projects technologies considering drained lands water regulation optimization based on BIM were observed. Modern methodological basis was used, and, at the same time, a universal technical tool that allows the improvement of the practice of designing complex objects and systems that is successfully used and developed in almost all fields of science, technology, and industry is computer aided design building information modelling (BIM). In structural terms, BIM is an organizational and technical system consisting of a large number of interrelated and interacting components. The main function of BIM is to carry out object aided design and their components through the use of mathematical models and other automated design procedures and computer technology. So, transition to optimization design methods will help to improve the overall ecological and economic efficiency of drainage reclamation, especially in working out new construction, reconstruction, and modernization of irrigation and drainage systems.

DOI: 10.4018/978-1-6684-8248-3.ch009

INTRODUCTION

Irrigation and drainage sector formed as an independent one in Ukraine in the last 35-40 years and is positioned on the intersection of environmental management, agriculture and hydrotechnical construction (Shumakov, 1996; Aydarov et al., 1990; Romashchenko et al., 2003; Rokochinskiy, 2010; Karuk, 1992; Rex, 1995). The same problems we observed in Poland (Pierzgalski, Jeznach, 2012). The most intensive development of land reclamation in Ukraine occurred in 1966-1990, but in 1991 the state capital investment in the water management sector in general and specifically in Meliorative construction was sharply reduced. Gradually sharp decline over the years of construction and reconstruction of drainage systems and complete shutdown of their state funding occurred in 2002 (Rex, 1995).

This situation leads to inadequate supply of material and technical resources, fertilizers, lack of land improvement measures, disruptions in energy supply, reducing the amount of work on maintain and repair of reclamation equipment, permanent loss of skilled personnel, incomplete use of scientific development, insufficient information support of farms and so on. Existing meliorative systems (MS), especially drainage, do not correspond to progressive demand of agricultural production, they are also obsolete outdated and require renovation and modernization. This shooed be corresponded to the «National Program of Water Management and Environmental Rehabilitation of Dnieper river basin till 2020» and other standard legal acts which allow water melioration functioning.

In addition, today a very acute problem of climate change arises (Rokochinskiy et al., 2019). Therefore, a search for alternative ways of development of hydraulic variants of melioration, optimal types and designs of drainage systems, more efficient, reliable and environmentally friendly ways of reduction the material input, energy, fund, job and water capacity of agricultural production, also increase of food production and agricultural raw materials is an urgent position.

Reclamation development in Ukraine in, general, and particularly, in the drained area inherent all objectively common both as positive and negative trends. First of all, it concerns the fact, that the large-scale reclamation development requires significant investment, but the resulting economic effect of this is at least 60...70% from the project (Shumakov, 1996). Failure to achieve efficiency from reclamation project is due to lack of knowledge about the results of interaction of human reclamation activity with natural processes and taking them into account in the construction of reclamation facilities. Therefore, ways to improve drainage reclamation lies in the development and improvement of methods of optimization of design and calculation of drainage systems and their components on environmental and economic bases. In particular, transition to multivariance in the substantiation of design solutions is the search for the optimal design solution from selected possible alternatives. The main objective of comparative evaluation is to choose more economical and environmentally friendly investment options, that is, to ensure maximum impact from reclamation activities at the lowest cost for their realization (Shumakov, 1996).

As in previous years, and now, in practice, procedure of development of alternatives design solutions (DS) and selection of the best ones is absent, although in state building codes (SBS V. 2.4-1-99) «Melioration systems and structures» fixed that the technical solutions as for schemes and structures of basic reclamation constructions should be taken on the comparison of key technical and economic indexes of different options. However, in practice, on the initial stage, after the necessary studies and research, project organization developed a technical draft of a project and later, it just clarifies, adds and substantiates its feasibility and effectiveness. Although the final choice of water regulation methods and their related type constructions and projects of MS can be done only considering technical-economical substantiation (TES) of optimal decisions by comparing their alternative variants (ways and projects of water regulation) on the object, which is under discussion.

METHODS AND TECHNIQUES

The need for this approach may be explained by the lack of reasonable and effective methods of research and selection of the optimal variant reclamation project. Therefore, substantiating design solution on environmental and economic basis needs improvement of the design process, especially when considering clarifying or drafting stages (Shumakov, 1996; Mazhayskiy et al., 2017; Mioduszewski et al., 2010).

Modern methodological basis and at the same time, universal technical tool that allows to improve the practice of designing complex objects and systems that is successfully used and developed in almost all fields of science, technology and industry is computer aided design – Building Information Modelling – (BIM) (Rahman et al., 2013; Juszczyk et al., 2015).

In structural terms BIM is an organizational and technical system consisting of a large number of interrelated and interacting components. The main function of BIM is to carry out object aided design and their components through the use of mathematical models and other automated design procedures and computer technology (Mazhayskiy et al., 2017).

One of the major design goals is to find the optimal design solution from the chosen possible options. The essence of optimization thus is fixed on finding the best (possible) design decision based on climate, landscape, soil, geological, agronomic conditions, which gives the minimum (maximum) of some aimed function, that characterizes the overall (complexed) effectiveness of the designed object. So for the implementation of optimization the main requirement is to possess the alternatives, of options totality of design decisions as for natural agro-reclamation conditions and formation of a real object terrain.

RESULTS AND DISCUSSION

Therefore, at the design stage of new agricultural production system (APS) and hydromeliorative system (HS) according to (Mazhayskiy et al., 2017) design, technical, technological solutions (DS) are accepted only once and subsequent correction put in them the object management strategies according to changing weather conditions is impossible or impractical. Through practical inability to forecast the real nature of change within the design lifetime of the MS (30…50 years), this strategy can be implemented as follows.

Regarding the (Rokochinskiy et al., 2010; Mazhayskiy et al., 2017) integrated optimization model the optimal DS in general can be defined as

$$\begin{cases} U_0 = \underset{\{i\}}{extr} U_i, i = \overline{1, n_i}; \\ Z_{0j} = \min\limits_{\{i\}} \left| Z_{ji} - \widehat{Z}_j \right|, j = \overline{1, n_j}; i = \overline{1, n_j} \end{cases} \tag{1}$$

Where U_0 – extreme values accepted by economic optimality criterion chosen U corresponding to the optimum set for DS options $I=\{i\}$, $i=\overline{1,n_i}$ – set, $j=\overline{1,n_j}$ criteria (physical indicators) relevant ecological efficiency DS variants; $\widehat{Z}_j$ – according to the accepted limiting environmental indicators.

DS variants project set $I=\{i\}$, $i=\overline{1,n_i}$ composed by different technologies of drained land water regulation and appropriate design solutions on MS and their technical elements.

Then, as for to (1), the value criteria of economic and environmental optimization for each project solution, taking into account climatic management strategy in the long year period of the object is determined by the formulas:

$$U_i = \sum_{p=1}^{n_p} u(I,P)\cdot\alpha_p, \tag{2}$$

$$Z_{ji} = \sum_{p=1}^{n_p} z_j(I,P)\cdot\alpha_p,\ j=\overline{1,n_j};\ i=\overline{1,n_i}, \tag{3}$$

Calculation base optimal level defines the parameters of the project design solutions of the specific object, it can be determined, unlike the accepted practice of rationing, also on environmental and economic grounds, taking into account existing natural and agro-reclamation conditions considering complex optimization model:

$$\begin{cases} P_0 = \underset{\{p\}}{extr}\, U_0(P), i=i_0; p=\overline{1},n_p; \\ Z_0 \gtrless \widehat{Z}_j, j=\overline{1},n_j, \end{cases} \tag{4}$$

where p_0– the typical optimal pattern of object climatic conditions considering calculation base of the growing season of a combination $\{p\}$, $p=\overline{1,n_p}$, which is defined by the extremum of the function $U_0(P)$ composed for economically optimal DS $i=i_0$; Z_{0j}– set $\{j\}$, $j=\overline{1,n_j}$, criteria of environmental efficiency in $i=i_0$.

Realization of complex optimization models require definition of indicators and criteria of economic and environmental efficiency of DS which are formed, primarily, within the reclaimed field. Their integral evaluation by the number of fields with project or planned rotation in the structure of agricultural use system and based on technical elements of MS will help to characterize the ecological and economic effects for MS in general.

All components of the general optimization model, such as, technical and economic indicators (TEI), that make up the economic and mathematical models (investment, value obtained crop production, current agricultural and reclamation costs, etc.) as well as environmental indicators (criteria) water and general nature reclamation regimes of drained lands, are defined by accepted methods and water regulation schemes they are changeable and are dependent on many factors, the main of them are climatic, soil reclamation, agronomic and other object conditions. They can be schematically represented as baseline data for formulating and solving optimization problems due to multiple relevant indicators:

- meteorological stations or posts $\Omega\{\omega\}$, $\omega=\overline{1,n_\omega}$, their shares f_ω or areas $F\omega_s$ervice within the system;
- drained soil types G={g}, $g=\overline{1,n_g}$, their shares f_g or areas F_g of dissemination;
- species of cultivated crops $Q=\{k\}$, $k=\overline{1,n_k}$, their shares f_k or areas F_k in the design crop rotation structure and their average calculated values (project or planning) yields $\overline{Y_k}, k=\overline{1,n_k}$, on drained lands;
- possible ways of water regulation on drained lands $S=\{s\}$, $s=\overline{1,n_s}$ and possible areas f_s their usage within the system;
- calculated (typical) in terms of heat and moisture periods during the growing season $P=\{p\}$, $p=\overline{1,n_p}$ and their shares αp, $p=\overline{1,n_p}$ within the designed lifetime of the system or the usage of drained land;
- structural elements of the system for the specific terrain, soil reclamation differences etc $L=\{l\}$, $l=\overline{1,n_l}$ their shares f_l or areas F_l of dissemination;
- specific cost parameters for the determination of technical and economic indicators (TEI), which are part of the economic and mathematical models and so on.

Determining the required values of common optimization models components is possible only on the basis of solving complex and diversified, multiparametered, multifunctioned task using mathematical modeling computers and is based on creation the complex of hierarchical connected simulation sub-models with forecasting on long-term basis of water and general reclamation nature modes of drained lands, their impact on the grown crops and created environmental impact.

Generalized structure of forecast-simulation and optimization calculations assessing the effectiveness of drained lands water regulation on the long term based due to various natural reclamation conditions is given in Fig. 1.

Characteristic features implementation of designed structure is block construction. As well, it can be pointed out such relatively independent generalized units:

- data block of the initial data sets by major affecting nature-climatic, soil reclamation and other factors (block 2, Fig. 1);
- block of forecast calculations on long-term basis based on a set of relevant models: meteorological regimes, water regime and water regulation technology, productivity of drained land for schematized natural agro-reclamation conditions of a real object (blocks 5, 8, 10, 12, Fig. 1);
- block of optimization calculations, which provides definition adopted in accordance with the structure of their operation (single or n-level), adopted criteria of economic and environmental efficiency of DS, that are considered according to the tasks formulated under accepted conditions of optimization as at the level of each reclaimed fields (grown culture), so as on the fields totality within the system (block 14, Fig. 1). Due to them general optimization condition and final selection of the management decision are implemented, which is done by informal way through expert assessment of the relevant expert at the design stage or routine operation of drainage systems, that take into account, on the one hand, economic efficiency, and on the other - environmental acceptability of its implementation (block 15, Fig. 1);

– block of formation and output of any intermediate (if required) and final results of forecast-optimization calculations at any stage of their execution (block 16, Fig. 1).

Figure 1. The generalized block-chart of forecast-optimization calculations with substantiation of design decisions on environmental and economic grounds

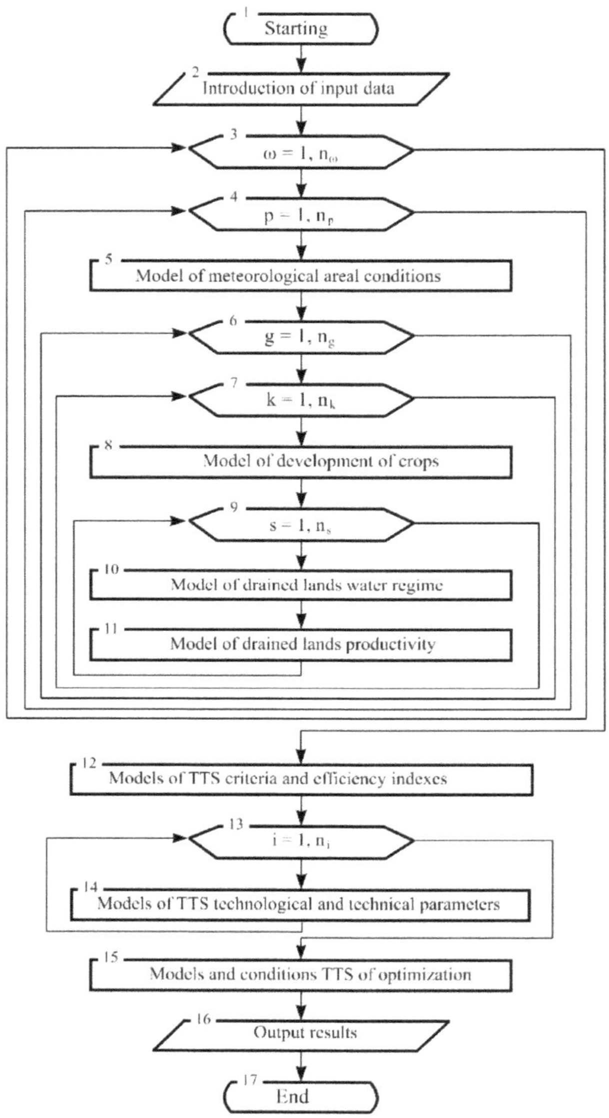

Certainly, shown in Fig. 1 general-purpose structure of forecasting, optimization calculations will vary in each case of its application depending on the optimization problem level that is solved, and accepted for consideration according to formulated task, leveled structure of their implementation and specific nature conditions and other reclamation facility management.

Practical realization of forecast-imitation major amount and optimization calculation is possible only by using corresponding fools, which corresponds to BIM.

So, for all the characteristic features, BIM refers to complex systems and methodological basis of its establishment and operation, as when creating a forecast-optimization methods and models are: system approach and system analysis.

Therefore, according to (Shumakov, 1996) the overall technology design of complex objects, processes and systems on a multiple basis at principle stages of their development can be represented as a block diagram (Fig. 1). This model illustrates general principles and approaches to the process of ecological and economic assessment of alternatives reclamation projects of construction and reconstruction, having the choice of optimal substantiate investment in the future (Fig. 2).

This model illustrates general principles and approaches to the process of ecological and economic assessment of reclamation projects construction and reconstruction alternatives, having the choice of optimal investment and its substantiation. Consider the basic steps in detail.

The given chart depicts the principles of transition from traditional mostly multivariate approach to substation Of DS in creation of water melioration objects, to possible transition on optimization methods choosing DS optimal an multivariate base.

Considering all this, corresponding to modern demands, it is necessary to do ecology-economic evaluation of meliorative project, general model of which is given on (Fig. 3).

Its implementations include the following studies:

Design initiation involves, primarily, the formation of idea and goal of the project, setting goals, forming its main characteristics and preliminary feasibility assessment of investments for its implementation.

Design searching project. At the search pre-project stage collection and analysis of the necessary data are conducted, survey project works are carried out and, necessary studies to further definition of technical solutions possible options, thus to select the most economically and environmentally sound.

Identification of DS potential options. An essential principle of reclamation design effectiveness and optimization is multi-variants of examined DS, aiming to select the best of them.

CONCLUSION

The main condition to implement optimization is to have alternatives, of the set of all possible variants of design decisions considering possible variable nature agro-reclamation conditions of the real object. When designing reclamation activities, as a rule, a considerable number of different technical and technological solution options appear, that, accordingly, affect the economic and environmental efficiency of their implementation.

It is necessary to substantiate and determine the optimal design solution of the type, construction, and parameters of the reclamation system which are accepted by taking water regulation mode (scheme) on the system, considering existing natural and economic conditions. Therefore, at this stage, all potential options for the project are examined to decide choosing the best one in the final stages of development. These forecasts, firstly, water regulation technology determination mostly suitable for implementation in terms of the goals and objectives of the project and terms of the analysis of local conditions (climate, landscape, soil, geology, agronomy, etc.), availability and cost of raw materials, energy, and labor. The results of the performed research determine all technologically possible this object's methods of melioration and corresponding to them meliorative systems.

Figure 2. Typical logic chart of design process on a multiple basis by the principle design procedures

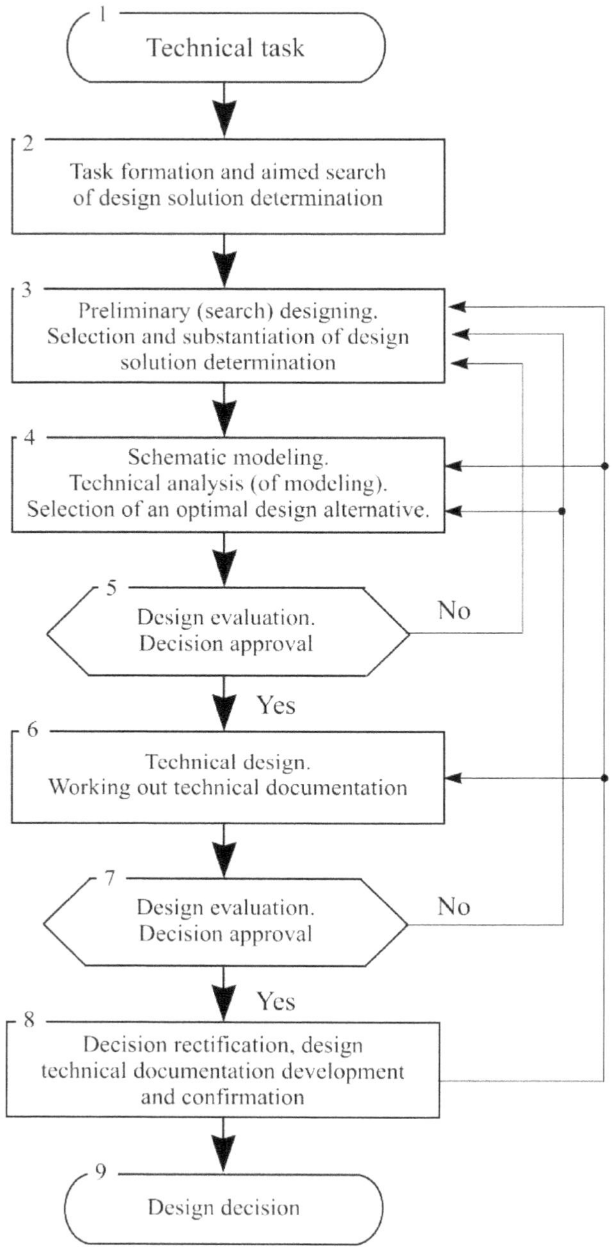

Variants of technical and technological solutions are composed of different water regulation technologies of reclaimed land and appropriate design solutions for reclamation systems and their technical elements. After all, water regulation technology (or methods) determines the necessary structural decisions as for the type of adjusting and central network, hydrotechnical stations, etc., and thus the type and design of melirative system on reclaimed lands.

Figure 3. The general model of ecological and economic assessment reclamation project

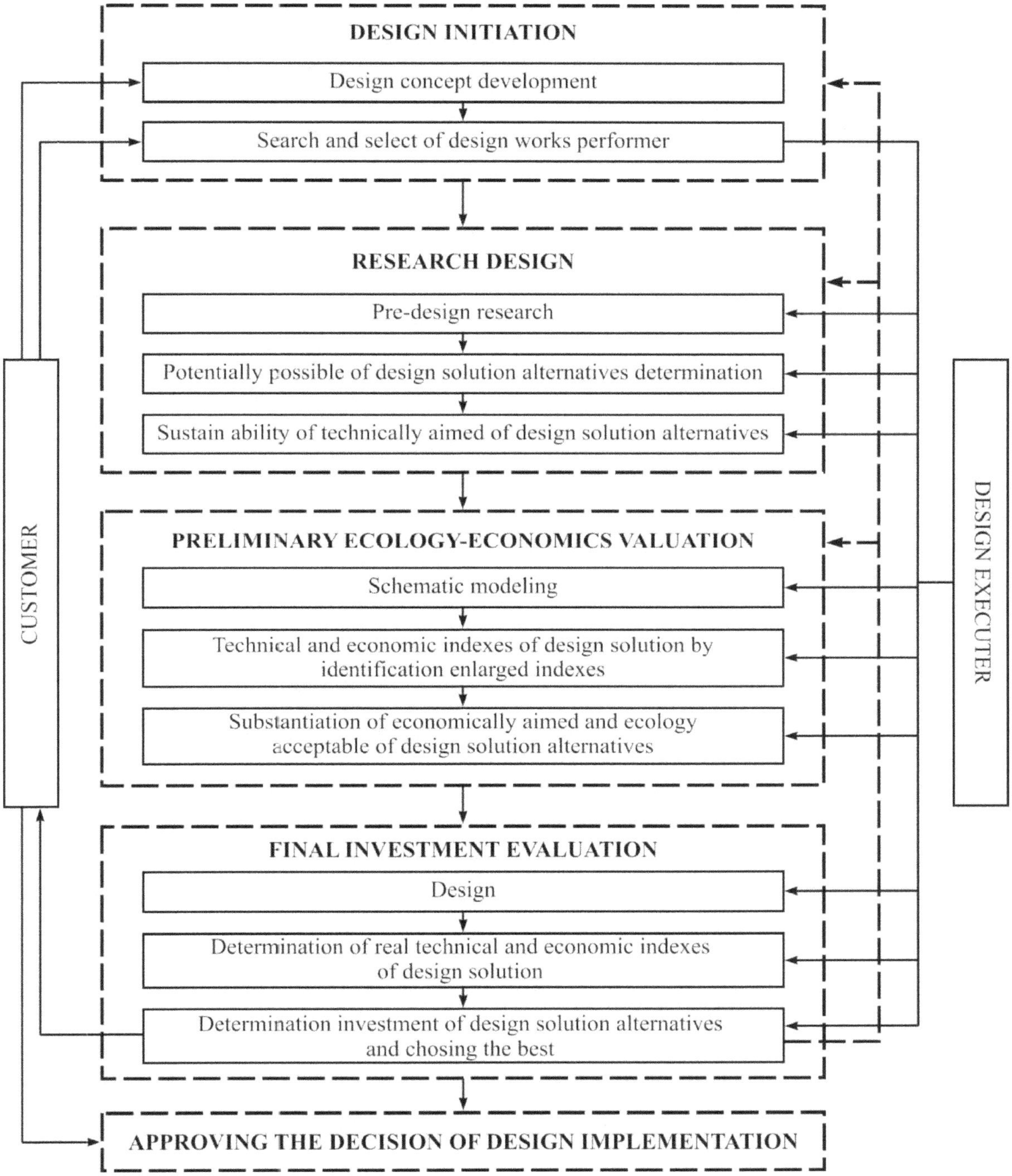

To substantiate optimal DS on a multivariate basis there is a need of chose for further conceptual design composed of a set of potential technical and technological solutions identified in the previous stage, the most suitable for implementation in production in natural and agro-ameliorative conditions of a real object. From a practical point of view, their rational number may be 2...6 versions of DS. At this stage, options of DS for further research and evaluation can be selected on comprehensive criteria considering technical, technological, and environmental conditions and restrictions on the use of expert opinion. The real interest from the number of possible water regulation schemes options primarily attracts those which forecast water regulation on schemes when to all crops rotation the only one way of water regulation is used also on the scheme which is a combination of water regulation ways composed of optimal modes for each planned crop field 8.

Previous environmental and economic assessment is carried out at the stage of the feasibility study, which includes a schematic design and definition of technical and economic indicators for project options.

In general, schematic designs are preliminary plans of problem solutions in specific conditions and usually, they cover the development of the major issues with a small number of indicators. The schematic design stage is developed for the conceptual definition of territorial requirements, functional, and environmental object solutions, fundamental confirming the possibility and feasibility of creating it. At the stage of schematic design preliminary development is carried out and the main types and volumes of work for each of the alternatives DS analyzed are determined.

Subsequently, based on the schematic design, calculation of all required technical, technological, operational, environmental, and economic parameters chosen to consider for draft options based on enlarged and standard indicators, including power facility, the total number of employees, productivity, cost of major products, estimated cost of construction, cost of fixed assets, participation share of investors, annual demand for raw materials, energy, water, basic building materials costs, field area, used reclaimed lands directions and so on is done. These figures are output data for further economic evaluation and comparison of reclamation project alternatives.

After that, the comparative environmental and economic efficiency of the project developed variations and environmental conditions of comparable DS options are performed. A common approach to selecting the best DS in the field of land reclamation is focusing on their economic feasibility and environmental acceptability.

From the set of pre-selected technical and technological parameters options for the project by environmental and economic criteria, choose one or more similar technical and economic indicators of environmentally acceptable for DS for their furthermore detailed development and evaluation.

For selected, in the previous stage, promising environmental and economic project option needed to acting requirements working documentation is developed, acceded constructional decisions and their technical parameters are clarified, the actual volume of the main construction and installation work, equipment, structures, material and energy needs, labor force, duration, and the term of construction work and others are determined.

So, transition to optimization design methods will help to improve the overall ecological and economic efficiency of drainage reclamation, especially in working out new construction, reconstruction and modernization of Irrigation and drainage systems.

REFERENCES

Aydarov, I., Golovanov, A., & Nikolsky, Yu. (1990). *Optimization of meliorative regimes of irrigated and drained agricultural land*. Agrometeoizdat.

Juszczyk, M., Výskalab, M., & Zimaa, K. (2015). Prospects for the use of BIM in Poland and the Czech Republic – Preliminary Research Result. *Procedia Engineering*, *123*, 250–259. doi:10.1016/j.proeng.2015.10.086

Karuk, B. (1992). System approach to assessing the impact on the environment of drainage reclamation facilities in Ukraine. *Theory and practice of ecological and land reclamation monitoring in the Ukrainian Polesie,* 22–32.

Mazhayskiy, Y., Rokochinskiy, A., Volchek, A., Meshyk, O., & Jeznach, J. (Eds.). (2017). Environmental management of Polissya. VNIIGiM of A. Kostiakov.

Mioduszewski, W., Kowalewski, Z., Zurawski, R., & Stankiewicz, J. (2010). Drainage system in the Kampinos National Park. *Journal of Water and Land Development*, *14*(1), 83–95. doi:10.2478/v10025-011-0007-3

Pierzgalski, E., Jeznach, J., Baryła, A., Brandyk, A., Stańczyk, T., Szejba, D., & Wiśniewski, S. (2012). *Weryfikacja systemów melioracyjnych pod kątem znaczenia dla bezpieczeństwa powodziowego w Regionie Wodnym Wisły Środkowej*. Maszynopis SGGW Warszawa.

Rahman, Md., Suwal, S., & Jäväjä, P. (2013). *Diverse approach of BIM in AEC industry: a study of current knowledge and practice.* The 30 th CIB W78 International Conference, Beijing, China.

Rex, L. (1995). *System research of meliorative processes and systems.* Aslan.

Rokochinskiy, A. (2010). *The scientific and practical aspects optimization of water regulation drained lands on environmental and economic grounds.* NUVGP.

Rokochinskiy, A., Volk, P., Pinchuk, O., Mendus, S., & Koptyuk, R. (2017). Comparative evaluation of various approaches to the foundation of parameters of agricultural drainage. *Journal of Water and Land Development*, *34*(1), 215–220. doi:10.1515/jwld-2017-0056

Rokochynskiy, A. (2016). System optimization of water regulation on reclaimed lands. *Problems of rational use of natural resources and sustainable development of Polissya.* 111–114.

Rokochynskiy, A., Volk, P., Frolenkova, N., Prykhodko, N., Gerasimov, I., Pinchuk O. (2019). Evaluation of climate changes and their accounting for developing the reclamation measures in Western Ukraine. *Scientific Review Engineering and Environmental Sciences*, *1*(83), 3–13.

Romashchenko, M., Sobko, O., Savchuk, D., Kulbida, M. (2003). *About some tasks of ag-ricultural science in relation to climate change.* The scientific report-information NAAS Kyiv.

Shumakov, B. (1996). Melioration in the twenty-first century. *Reclamation and water management, 3*, 4–6.

Chapter 10
Scientific and Practical Principles of System Optimization of Various Parameters of Drainage Systems

Pavlo Volk
National University of Water and Environmental Engineering, Ukraine

ABSTRACT

Approaches to the systemic optimization of technological and design solutions for the creation and operation of drainage systems have been determined. It seems that the drainage system is a complex natural and technical ecological and economic system. Finding the general optimum in such a system based on the system optimization consists in substantiating local optima for all its main components of heterogeneous elements in the system effect - mode - technology - design in their interconnection. The practical implementation of a complex of predictive-simulation and optimization calculations in projects for new construction, reconstruction and modernization of drainage systems based on the developed scientific, methodological, information and software can be carried out using the appropriate toolkit, which is CAD and modern BIM technologies.

BACKGROUND

In modern conditions, a significant amount of agricultural products in order to improve food security is grown on lands with a regulated water regime using various types and designs of reclamation systems (RS). The aggravation of the greatest challenges of our time concerning the energy, food and water crises under the influence of changes in climatic and anthropogenic factors both at the planetary and regional levels necessitates further improvement of scientific and methodological approaches to the creation and operation of water management and reclamation facilities that are adapted to these changes, including for drainage systems (DS) in the drainage reclamation zone based on an assessment of their overall

DOI: 10.4018/978-1-6684-8248-3.ch010

efficiency in compliance with modern economic and environmental requirements (Várallyay, 2010, Kovalenko et al., 2019).

The beginning of large drainage activities in the western part of the former tsarist Russia on an area of more than 8 million hectares was organized by the Western Expedition to drain swamps under the leadership of the general J. I. Zhilinskiy in 1873. These works in the Polesie zone were the first large state program for land drainage with open canals and closed drainage, the parameters of which were taken mainly constructively. At the same time, locks, dams, and cofferdam were installed on many canals to control (regulate) the water regime.

In the future, larger-scale work on the expansion and drainage of agricultural land in the zone of the Ukrainian Polesie (1.5 million hectares of drained land, including more than 700 thousand hectares with moisture), as well as in the Polesie territories now adjacent to Poland, Belarus and Russia, took place during the Soviet period of development of this region, starting from the second half of the 60s of the last century. Traditionally, these issues were resolved on the basis of the use of hydrodynamic methods for calculating the movement of groundwater to dehumidifiers (water receivers) and hydraulic calculations of the parameters of drainage lines, canals, rivers, etc., which were used to determine the position of the groundwater level during periods critical for waterlogging conditions.

In the current practice, this problem was solved by performing predictive engineering and reclamation calculations to substantiate the type and design of DS, including in the conditions of Ukraine, first according to the water balance model (Yangol et al., 1970), and later - according to its improved interpretations.In the current practice, this problem was solved by performing predictive engineering and reclamation calculations to substantiate the type and design of DS, including in the conditions of Ukraine, first according to the water balance model (A.M. Yangol, 1970), and later - according to its improved interpretations. However, their implementation made it possible to obtain a virtually uncontested result regarding the assessment of the effectiveness of the application of possible technologies for water regulation of drained lands in specific natural-agro-reclamation conditions of a real object. In addition, the parameters of the constituent technical elements and the system as a whole were determined on the basis of the standardization of the level of estimated security, which reduced the objectivity of the validity and the overall effectiveness of these decisions.

Thus, according to a generalized assessment (Shumakov et al., 1996), it was determined that the development of reclamation in the world in general and in the drainage zone of the Polesie region in particular is characterized by all general objectively determined, both positive and negative trends. First of all, this concerns the fact that the large-scale development of land reclamation is associated with significant capital investments, very tangible for the economy of any country, but the resulting effect is, at best, 60 ... 70% of the planning one. The failure to achieve the planning efficiency of land reclamation was explained by the lack of our knowledge about the results of the interaction of human land reclamation with natural processes and their consideration in the design of the RS.

Therefore, in the design and operation of RS, including DS, at all stages of decision-making in time (1 - project, 2 - planned operation,3 - operational management of the facility), it is extremely important to determine the parameters of the system and its constituent technical elements (regulatory network, conducting network of canals, regulating hydraulic structures, pumping stations, etc.) correctly (objectively). It is they, who largely determine the cost of the system, its economic and environmental efficiency. In turn, this will increase the requirements for the quality of assessment, forecasting and optimization of management of water and general natural reclamation regimes of drained lands, as a prerequisite for solving this problem (Rokochinskiy et al., 2010, 2017).

General conceptual approaches to the optimization of the reclamation regime based on the formulation of general principles and the definition of indicators, criteria and the creation of optimization models are considered in the recommendations (Aydarova et al., 1990, Rex et al., 1995).

The optimization approach has been very successfully applied to determine the parameters of RS and their operating modes (Downey, 1972; Doorenbos, 1979; Kassam, 1979; Rabochev, 1984; KK.T.hommik, 1985; Minaev, 1986; Likhatsevich, 1993; N.A.Lazarchuk, 2009). As the most promising, they adopted an economic and mathematical method that combines the advantages of traditional hydromechanical and empirical methods, based on the implementation of a complex of forecast and optimization calculations.

But the optimization methods and models developed at that time were considered and used mainly to substantiate local single solutions for individual elements of the system or water regulation technologies, in particular, the optimal parameters of the main canal, hydraulic structures, drainage, etc. (Lazarchuk et al., 2009).

At the same time, it turned out that in the modern conditions of the transition to market relations in the country, this method, as a simplified optimization method, in the form in which it was implemented, considered only the economic component of optimization and does not take into account environmental efficiency when determining the optimal technical and technological solutions and their parameters, that does not correspond to modern requirements (Mazhaiskiy et al., 2017).

In this regard, there is a need to search for new approaches, methods and models based on the further development of the general optimization theory for its application to substantiate the optimal type, design and parameters of DS and its constituent technical elements on ecological and economic principles, taking into account changes in climatic conditions.

METHODS AND TECHNIQUES

Research methods are based on the application of systems theory with the basics of systems analysis and modeling in the development of modern approaches to the system optimization of technical, technological and constructive solutions for the creation and operation of drainage systems on ecological and economic principles. At the same time, the systematic approach includes: consideration of the drainage system is not like a technical one, but like a complex natural-technical system; research of their elements, patterns of functioning and development; decomposition of complex goals and objectives of a predominantly hierarchical nature inherent in such systems; application of the methodology of the descending hierarchy of analysis and the ascending hierarchy of synthesis in the development of forecast-optimization models and methods of their implementation.

This is the most modern methodological basis, with the complex of optimization and forecasting-simulation models developed by us, scientific-methodological and information support for their implementation in the structure of the corresponding BIM technologies (Kovalenko et al.,, 2019) are used to further improve scientific and methodological approaches to the design and operation of water management and reclamation facilities based on a multivariate assessment of the overall technical, technological, economic and environmental efficiency in accordance with changing conditions.

RESULTS AND DISCUSSION

In the development of these approaches, we (A.N. Rokochinskiy et al. 2010, 2017) presented that such a system as part of an agricultural production system (APS) and hydro-reclamation systems (HRS) should be considered as ***a complex natural and technical, ecological and economic system (CNTEES).*** A necessary condition is to find a general (global) optimum in such a system only on the basis of the application of system optimization, the essence of which is the sequential determination of local (intermediate) optima for all its main components of heterogeneous elements *(in the subsystem effect - mode - technology - design)* in their relation. At the same time, for the current level of development of water management, reclamation and agricultural science, system optimization to substantiate optimal design decisions regarding the type, design and parameters of the HRS and its constituent technical elements is carried out at the existing (certain or specified) level of economic (crop yield) and ecological efficiency of functioning of the investigated object.

The solution of such a problem requires, first of all, the development of a model of the system that allows for any element or level of its functioning to consistently determine the optimal total ecological and economic effect created from the action and interaction of interconnected heterogeneous components of the system as a whole with the environment.

The optimal total (integral) ecological and economic effect, which corresponds to the optimal level of technical, technological, economic and ecological efficiency of the DS and is created during its functioning in time (τ) and space (f), in the general case can be determined by a model of the form

$$Y^o = \int_0^{T_{np}} \int_0^{F_s} y_i^o(T_i, F_i) d\tau \, df, \quad i = \overline{1, n_i}, \tag{1}$$

where Y^o– is the optimal integrated ecological and economic effect created by the functioning of the system in accordance with the current requirements; y_i^o – parameters of the optimal overall ecological and economic effect created from the action and interaction of interconnected heterogeneous components of the DS elements with the environment; T_i, T_{pr}–respectively, the project periods of functioning for the *i-th* heterogeneous elements and the system as a whole; F_i, F_s – respectively, the area for the *i-th* heterogeneous elements and the system as a whole.

In its turn,

$$y_i^0 = f_1^*\left(f_2^*\left(f_3^*\left(z_i^0\right)\right)\right), \; i = \overline{1, n_i}, \tag{2}$$

where f_1^*– is the function of optimizing the parameters of the natural reclamation regime θRi, $i = \overline{1, n_i}$, within the system; f_2^*– the function of optimizing the parameters of water regulation technologies θSi, $i = \overline{1, n_i}$, on the system; f_3^*– the function of optimizing the parameters of design solutions for the system and its technical elements θKi, $i = \overline{1, n_i}$ ensuring the implementation of appropriate technologies for water regulation on drained land.

The search for the optimal parameters of the components of a complex function (2) and, first of all, the parameters of the modes and associated technological solutions regarding the technologies of water regulation and technical solutions by the type, design and parameters of the system, as well as the constituent technical elements that provide them, depending on the created the general ecological and economic effect, can formally be implemented sequentially through the corresponding inverse functions:

- regarding the optimal parameters of water regulation modes z_2^0,

$$z_2^0 = f_1^{*-1}\left(\hat{y}_i\right),\ i=\overline{1,n_i}\ ; \tag{4}$$

- regarding the optimal parameters of technologies,

$$z_3^0 = f_2^{*-1}\left(f_1^{*-1}\left(\hat{y}_i\right)\right),\ i=\overline{1,n_i}\ ; \tag{5}$$

- regarding the optimal design parameters,

$$z_4^0 = f_3^{*-1}\left(f_2^{*-1}\left(f_1^{*-1}\left(\hat{y}_i\right)\right)\right),\ i=\overline{1,n_i}\ , \tag{6}$$

where the given or adopted parameters of the overall ecological and economic effect for the corresponding heterogeneous elements of the system.

To implement the optimization functions (2) – (6), it is necessary to use the results of the research of the regularities of interconnected processes of water movement both in the main constituent elements of subsystems and in the system as a whole, as well as existing approaches to solving optimization problems in CNTEES.

In this formulation, models (1) and (2) in a general implicit form make it possible, at least theoretically, to substantiate the possibility of setting optimization problems and to determine the optimal parameters of operating, technological and design solutions consistently for all heterogeneous constituent elements in their interconnection and the system as a whole at least at the empirical or empirical-functional level of determining the relation between them.

Based on the results of the considered prerequisites and the performed theoretical analysis on the system optimization of technological and design solutions in CNTEES, which includes DS, the management optimization model underlying the implementation of the optimization approach based on the economic and mathematical method in general form is represented as (A.N. Rokochinskiy et al. 2010, 2017)

$$Z_i^0 = \underset{\{m_i\}}{extr}\left(Z_{mi}\right),\ m_i = \overline{1,n_{m_i}},\ i=\overline{1,n_i}\ , \tag{7}$$

where Z_i^0 – the extreme values according to the accepted conditions of the selected optimality criteria, corresponding to multiple optimal technical and technological solutions for a set of dissimilar elements $\{i\}$, $i=\overline{1,n_i}$ and the system as a whole; Z_{m_i} – the values of the optimality criteria for a set of possible alternative options for the implementation of the *i-th* element within the system $\{m_i\}$, $m_i = \overline{1, n_{m_i}}$, $i = \overline{1, n_i}$.

On the basis of the considered approaches, we have improved methods and models of optimization to substantiate the optimal type, design and parameters of DS on ecological and economic principles, taking into account climate change for the short and long term by: 1) changing the established practice of considering the concepts of reclamation objects; 2) determining the presence in such a system of a structural connection between the parameters of its heterogeneous elements; 3) the development of principles for the construction and implementation of an integrated model of system optimization of regime, technological and design solutions for water regulation on drained lands: it includes a model of economic optimization, built on the basis of the traditional economic and mathematical approach, and its ecological component, as a limitation, determines the ecological acceptability of the optimal economic solution; 4) substantiation of the criteria for economic and environmental optimization for different levels of management decision-making over time; 5) development of a complex of predictive and simulation models for predictive assessment on a long-term basis of variable conditions calculated for heat and moisture supply of growing periods: the corresponding sets of schematized meteorological regimes, water regime and water regulation technologies, productivity (yield) of crops grown.

In turn, the transition to optimization methods determines the need to change the design technology for water management and reclamation facilities based on the use of a multivariate approach, modern information and computer technologies. The practical implementation of a complex of predictive-simulation and optimization calculations in projects for new construction, reconstruction and modernization of DS based on the scientific, methodological, information and software developed by us can be applied with the use of appropriate tools, which is CAD and modern BIM technologies (Rokochinskiy et al., 2019).

On their basis, a number of optimization and forecasting tasks for water regulation of drained lands were solved, including in the Polesie zone of Ukraine (Mazhaiskiy, et al. 2017, Kovalenko et al., 2020.):

1) in projects for the construction and reconstruction of drainage systems:
    - substantiation of the choice of possible ways to regulate the water regime and the corresponding types and structures of HRS on drained lands;
    - optimization of the parameters of the control network;
    - optimization of the estimated supply and parameters of main canals and regulated water receivers;
2) in projects for the operation of drying and moisturizing systems:
    - complex automation of planned water regulation of drained lands; - development of systemic plans for water regulation on a long-term basis;
    - creation of informational and advisory systems for operational planning of water regulation;
    - implementation of planned water regulation in production conditions;
    - substantiation of rational water consumption for moistening the drained lands of Ukraine; - substantiation of ecologically optimal regimes of reclaimed lands of the humid zone of Ukraine.

A characteristic feature of the implementation of system optimization in projects for the creation and operation of DS is a kind of research based on a computer experiment with the widespread use of modern computer and information technologies.

As an example, we consider the application of system optimization, based on improved design technologies (P.I. Kovalenko, A.N. Rokochinskiy, E. Eznakh, P.P. Volk, 2019) when justifying the type, design and parameters of DS, as well as its basic technical elements, in the project of their reconstruction on the drained lands of the “Rassvet” farm of the Rokytnovskiy district of the Rivne region in the Polesie zone of Ukraine.

The soil of the researched area with a total area of 410 hectares are represented by sod-medium-podzolic gley-sandy soil on sands with a filtration coefficient (k_f= 1.2*m*/*dey*) and share participation in the system $\left(f_{g_m} = 0.1\right)$, soddy-gley sandy loam soil (k_f= 1.0*m*/*dey*, $f_{g_m} = 0.3$) and medium-thick medium-decomposed peatlands (k_f= 0.4*m*/*dey*, $f_{g_m} = 0.6$). The crop rotation on the array is represented by: oats with equity participation (f_k=0.12), perennial grasses (f_k=0.25), winter wheat (f_k=0.12), corn for silage (f_k=0.13), potatoes (f_k=0.13).

Table 1 presents a comparative characteristic of the main indicators and their parameters, established by traditional methods and approaches in a real project, as well as by improved methods using system optimization in the reconstruction project of this facility.

Table 1. Comparative characteristics of the main indicators and their parameters established by traditional and advanced methods and approaches

N°	**Key indicators and parameters**	**Using:**	
		Traditional methods and approaches in a real project	**Improved methods based on system optimization in a renovation object project**
1.	Regulating network	Open network of canals with a distance between them B_r= 60 m, calculated module of drainage flow q_r=0.5 *l/s·ha*	Closed systematic drainage with a distance between drains B_r= 20m, optimal drainage flow module q_r^0 =0.6 *l/s·ha*
2.	Conducting and protective network of open canals, hydraulic structures (gateways - regulators)	Structurally	Structurally
3.	Main canal (main collector and its branches) - water intake	Estimated provision of sowing consumption is 10%, $Q_{10\%}$ = 0.73 m³/s; estimated parameters of the main collector and its branches: width along the bottom *b*= 2 m; slope placement coefficient *m*= 2; channel depth *H* = 2m;	The optimal estimated provision of the design flow rate is 15%, $Q_{15\%}$ = 0.505 m³/s; optimal parameters of the main collector and its branches: *b*= 1.5 m; *m*= 2; *H* = 1.5m;
4.	Type, construction of DS	Draining	Draining with preventive locking
5.	Economic efficiency	Land use ratio K_3=0.88; payback period T_0= 10 years	Land use ratio K_3=0.96; discounted payback period $\hat{T}_0$ = 7 years
6.	Environmental efficiency	Not determined	*Acceptable* (permissible) - relative to the formation of the average module of drainage flow in the system, which is permissible according to environmental requirements, q_s= 0.42 *l/s·ha*, which is less than $\hat{q}_{5:>;}$ = 0.46 *l/s·ha*; quite reliable - relative to the coefficient of environmental reliability of the project, k_H=0.54, should be in the range of values ($0.5 < k_H \leq 1.0$).

The note. *Preventive locking* - regulation of a part of the flow within the drainage system during the recession of the spring flood.

CONCLUSION

Thus, new approaches are considered and proposed, they are based on the application of system optimization to substantiate on ecological and economic principles the type, design and parameters of drainage systems and their constituent technical elements in projects of new construction, reconstruction and modernization in accordance with recent changing conditions.

REFERENCES

Aydarov, I. P., Golovanov, A. I., & Nikolskiy, Y. N. (1990). *Optimization of meliorative regimes of irrigated and drained agricultural land*. Agrometeoizdat.

Doorenbos, J., & Kassam, A. H. (1979). Yield respons to water. FAO Irrigation and Drainage Paper. FAO.

Downey, L. A. (1972). Water-Yield relations fornonforage crops. *Journal of the Irrigation and Drainage Division*, *98*(1, IR1), 107–115. doi:10.1061/JRCEA4.0000853

Khomik, K. T. (1985). *Basics of calculating drainage systems*. Bureau of Scientific and Technical Information of the Ministry of Agriculture of the Estonian SSR.

Kovalenko, P., Rokochinskiy, A., Jeznach, J., Volk, P., Koptyuk, R., & Prykhodko, N. (2019). Evaluation of climate change in Polissie region and ways of adaptation to it. *Journal of Water and Land Development*, *41*(1), 72–82. doi:10.2478/jwld-2019-0030

Kovalenko, P., Rokochinskiy, A., Mazhayskiy, Yu., Volk, P., Volk, L., & Chernikova, O. (2020). *Construction and agricultural drainage parameter optimization considering economic and environmental requirements*. Engineering for Rural Development. Jelgava. doi:10.22616/ERDev.2020.19.TF237

Lazarchuk, M. O., Cherenkov, A. V., & Rokochinskiy, A. M. (2009). Optimization of drainage systems calculation and management. Monograph, Rivne- NUWREM.

Likhatsevich, A. P. (2001). Land reclamation of Belarus. Minsk: BelNIIMiL

Mazhaiskiy, Y. A., Rokochinskiy, A. N., Volchek, A. A., Meshik, O. P., & Eznakh, E. (2017). Environmental management of Polesie. Ryazan: Meshcher. Branch of All-Russian Research Institute of Hydrotechnics and Melioration Named after A.N. Kostyakova.

Minaev, I. V. (1977). *Design principles of drainage and drainage-humidification systems with optimal parameters*. Water Industry of Belarus.

Rabochev, I. S., & Pyagay, E. T. (1984). Optimization of drainage parameters on floodplain soil. Bulletin of agricultural science, 103-115

Rex, L. M. (1995). *System research of meliorative processes and systems*. Aslan.

Rokochinskiy, A., Frolenkova, N., Jeznach, J., Koptiuk, R., Turcheniuk, V., & Volk, P. (2019). Reclamation projects development improvement technology considering optimization of drained lands water regulation based on BIM. *Przeglad Naukowy Inzynieria i Ksztaltowanie Srodowiska*, *28*(3), 432–443. doi:10.22630/PNIKS.2019.28.3.40

Rokochinskiy, A., Korobiichuk, I., Kuzmych, L., Volk, P., & Kuzmych, A. (2020). *The System Optimization of Technical*. Technological and Construction Parameters of Polder Systems., doi:10.1007/978-3-030-40971-5_8

Rokochinskiy, A. M. (2010). *The scientific and practical aspects of optimization of water regulation of drained lands on ecological and economic bases: monograph*. NUWEE.

Shumakov, B. B. (1996). *Melioration in the twenty-first century*. Reclamation and Water Management.

Várallyay, G. (2010). The impact of climate change on soils and on their water management. [Special Issue II]. *Agronomy Research (Tartu)*, 385–396.

Yangol, A. M. (1965). *Humidification recommendations for drained lands*. Urogay.

Chapter 11
Drainage Systems Project Development Taking Into Account Drained Lands Topography

Roman Koptyuk
https://orcid.org/0000-0002-7086-3608
National University of Water and Environmental Engineering, Ukraine

ABSTRACT

The results of assessing the nature and degree of influence of the terrain on drained lands are presented in making a design decision on water management technology, type and design of drainage system based on the indicator of land reclamation efficiency using BIM-technologies, forecasting and optimization calculations. The schematization of the relief conditions of drained lands according to the slopes and differences of the surface of the earth is considered. Schematization was performed and the classification of the drained lands was developed according to the index of reclamation efficiency of the relief, which depends on the degree of development of the relief by the slopes and differences of the surface of the land. The generalized results of the estimation of the land reclamation efficiency of the terrain of a separate section of the drainage system, the zonal conditions of the Rivne region and the Polesie of Ukraine within their administrative limits are presented.

BACKGROUND

Drainage systems design is performed on waterlogged lands by selected technologies of water regulation depending on climatic, relief, soil, hydrological, farming, and other conditions of the object which are evaluated by the results of the project organization survey. However, the terrain is one of the main factors influencing water regulation technology and design reclamation system acceptance.

DOI: 10.4018/978-1-6684-8248-3.ch011

In this regard, as a rule, in substantiation of reclamation activities a considerable number of different techniques and technology solution options that decisively influence, first of all, the economic and environmental efficiency of their implementation tend to occur.

As the research results show also experience and practice in the creation and functioning of water management and reclamation facilities in the humid zone, even at slopes greater than 0,002 the required level of moisture supply for crops when using warning and moisture locking on a large part of the drained lands is not reached, that, in its turn, influences the yield reduction and the overall efficiency of reclamation.

According to (Koptyuk et al., 2008) the territory with developed relief given the level of groundwater (GWL) is supported only on a small part of the reclaimed area. In the lowlands groundwater outlet to the surface is possible, but in the highlands, the depth of the groundwater is much higher than the drainage rate. In the area of excess moisture upper elements of land relief (watersheds, steep slopes) are less waterlogged, from them precipitation flows down in the form of runoff, thus moisturizing the located area below. Most swampy are stagnant, slow-flow lowlands and plains where surface water stagnates, especially when there is insufficient natural territory drainage.

Therefore, at the design stage estimation of drained areas and uneven relief should be performed, which, in its turn, affects the water regime of drained lands and the object efficiency in general. Based on the water regime features formation it should differentially determine the drained areas, on which the reclamation system will work in various modes: drainage, warning locking through regulation of a part of the flow during spring flood decline, underground moisturizing by periodic retention or lifting GWL during the growing season - correspondingly apply different water regulation technologies and irrigation and drainage systems constructions.

A characteristic feature of water regime formation on drained land with developed relief is also a runoff formation, uneven distribution of moisture, and depth of GWL in the system, this should be taken into account in the model of the long-term forecast of water regime and water regulation technologies on drained lands (Koptyuk et al., 2014; Rokochinsky et al., 2011).

The actuality of solving this problem increases significantly as a result of climate change both on the planetary and regional levels. Recent changes in environmental conditions are due, above all, to climate change and bring changes in requirements for the rational use of water and land resources in the reclamation areas, working out the complex adaptive measures aimed at regulation and accumulation of precipitation, the nature and mode of which varies considerably (Kovalenko et al., 2019).

METHODS AND TECHNIQUES

Under the existing morphometric classification drained lands relief considering the size of individual forms within the specified mezorelief can be positioned as a combination of microform and Nanoform (Rychagov, 2006). Therefore, the basic linear indicators of the relief are slope of the land (i) and differences in the surface of the earth: general (ΔH_{gi}), by a slope (ΔH_i) and the local high lands or lowlands ($\pm\Delta h_g$).

Based on the analysis, generalization and arrangement of the relief project conditions of real objects located in sufficient and unstable moisture Ukraine zones, four basic schemes of their formation by slopes and by differences in the surface of the earth are pointed out and positioned to consider (Fig. 1) (Koptyuk et al., 2011)

1) $i \neq 0$, $\Delta H_i \neq 0$, $\Delta h_g \neq 0$ - availability of slopes, differences in the surface of the earth by slopes and local differences in the surface of the earth;
2) $i \neq 0$, $\Delta H_i \neq 0$, $\Delta h_g = 0$ - availability of slopes and differences in the surface of the earth by slope, lack of local differences in the surface of the earth;
3) $i = 0$, $\Delta H_i = 0$, $\Delta h_g \neq 0$ - lack of slopes, availability of local differences in the surface of the earth (local downgrade or rise);
4) $i = 0$, $\Delta H_i = 0$, $\Delta h_g = 0$ - lack of slopes and differences in the surface of the earth [basic conditions]

Figure 1. The main calculated schemes of drained land relief variability

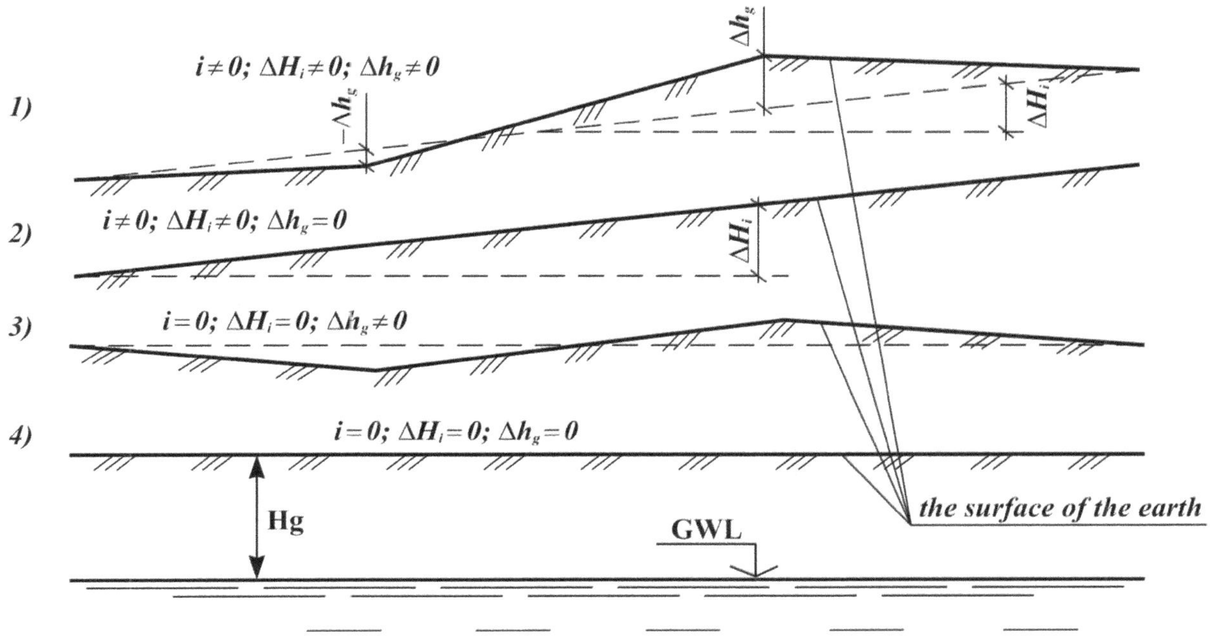

Thus, schematization of drained lands relief conditions allows for improving the existing model of long-term forecasting water regime and water regulation technologies accounting for the balance of groundwater with different levels of their occurrence within each selected relief unit on the drained area (Koptyuk et al., 2011).

This task to implement a large-scale experiment on the computer machine in an accelerated time was planned and carried out. As the basis for the experiment implementation complex of the forecast, simulation models were considered accounting for major structural and technological variable parameters of drainage systems and climatic conditions of the area, water regime, water regulation technologies and drained land efficiency for schematized natural, farming, and reclamation conditions. Models were developed at the department of water engineering and water technology, at the National University of Water and Environmental Engineering (Mazhayskiy et al., 2017; Rokochinskiy et al., 2011). Their practical application corresponds to standards regulated by the State Water Agency of Ukraine.

To evaluate the nature and level of the terrain relief influence on the formation of water regime and yield of crops grown considering various water regulation technologies, natural and land improvement conditions of 32 real objects in sufficient and unstable moisturizing zone of Ukraine were summarized and schematized by:

- **climatic zones** $\{\omega\} = \overline{1, n_\omega}$ (Carpathian, Polesie, forest-steppe);
- **calculated period of growth** $\{p\} = \overline{1, n_p}$ (humid, wet, average, dry, very dry);
- **complex of cultivated crops** $\{k\} = \overline{1, n_k}$ (Cereals, technical, herbs);
- **levels of productivity of cultivated crops** $\{\beta_Y\} = \overline{1, n_{\beta_Y}}$ (Low, medium, high);
- **species and soil permeability** $\{g\} = \overline{1, n_g}$ (Sand, k_f=1.2*m/day*, Sandy, k_f=0.7*m/day*, Loam, k_f=0.2*m/day*)
- **water regulation technologies** $\{s\} = \overline{1, n_s}$ (Drainage, locking alarm, underground wetting);
- **differences in the surface of the earth** $\{\Delta h\} = \overline{1, n_{\Delta h}}$ (-0.4 m, -0.2 m, 0.0 m, 0.2 m, 0.4 m, 0.6 m, 0.8 m, 1.0 m, 1.2 m)
- **slopes** $\{i\} = \overline{1, n_i}$ (0, 0.002, 0.005, 0.01).

14,580 calculated variants taken by the appropriate implementation schemes considering multiple natural and land improvement conditions, using complex of the forecast-imitation models correspondingly were realized.

As for the conducted research on machine experiment it was found that the terrain profile impact on the water regime formation and the yield of crops grown considering various water regulation technologies has a pronounced optimum, which is differentially forms depending on the multiple nature and land improvement conditions of a real object (Fig. 2).

Figure 2. The level of productivity βY different types of crops, depending on the drained land developed relief $\Delta Hg_{i:}$ 1 - winter wheat; 2 - potatoes; 3 - grasses for green mass

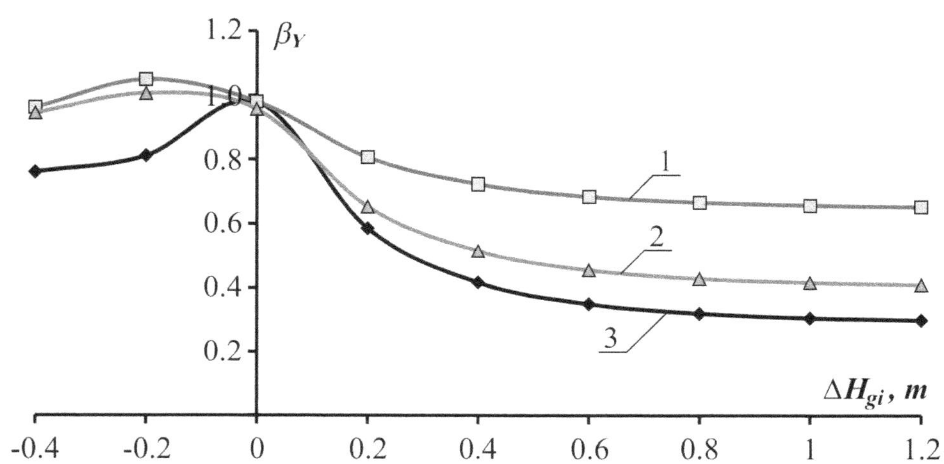

Thus, the productivity of reclaimed land in a developed relief in comparison with flat terrain can vary within a large limits(up to ± 60%).

In this regard, schematization was performed and classification of drained land in terms of reclamation efficiency of relief was depeloped, which depends on the degree of developed relief considering slopes and differences in the surface of the earth.

This index is a relative value, which is based on machine experiment it was determined by the ratio of land productivity (yields) with varying degre es of developed relief ($i \neq 0$, $\Delta H_{gi} \neq 0, m$) to its values in plain conditions ($i=0$, $\Delta H_{gi}=0m$) and it reflects possible level of productivity of reclaimed land, depending on the degree of the terrain developed relief:

$$k_{\Delta h} = {}^{Y_{\Delta h \neq 0}}\!\big/_{Y_{\Delta h=0}}, \tag{1}$$

where $k_{\Delta h}$ - an indicator of the efficiency of reclamation; $Y_{\Delta h \neq 0}$ - yields on drained lands with developed relief kg / ha; $Y_{\Delta h=0}$ - yields on drained lands with flat topography, kg / ha.

Table 1 shows the classification of drained lands relief reclamation efficiency as for its ranked quantitative characteristics.

Table 1. Reclamation effectiveness of drained land relief

The degree of relief variety	Quantitative indexs of relief		Reclamation efficiency index	Reclamation effectiveness of relief
	i	$\Delta H_{gi}, m$		
Poorly developed	<0.001	<0.2	1.0 ... 1.2	***High***
Medium developed	0.001 ... 0.004	0.2 ... 0.6	0.8 ... 1.0	***The average***
Highly developed	> 0.004	> 0.6	0.4 ... 0.8	***Low***

Specified values of ameliorative efficiency in certain natural and land improvement conditions of real object can be determined by the relevant forecast-imitation calculations.

According to common practice at the stage of design decisions as for the types, structures and parameters of drainage system, such work as routing channels and commutators their construction of longitudinal profiles from which it is possible to obtain data on slopes, on differences in the surface of the earth, on the whole area of the system and its hierarchical integrated levels (at reclaimed fields for crop rotation design, soil level within the system, at the structural elements of the system by the characteristic relief conditions, etc.) is performed.

At the present stage of development of science and technology providing a highly informative computer technology in all spheres of life, computer-aided systems as a universal technical tool that allows to improve the practice of designing complex facilities and systems, primarily of natural –technogenic character and also to solve related series of scientific and economic problems are intensively use.

Considering the complexity of water management and reclamation facilities design, for design automation and optimization BIM-technology should be used (Rokochinskiy et al., 2019). Powerful software package is Autodesk Civil 3D, where workflows are based on the technology of information modeling.

This software package enables to work in the field of geodesy, topography, master plan, geology, vertical planning and ordering area, new construction and reconstruction linearly extended objects (roads and railways, canals, dams, pipelines, pressure and non-pressure) using GIS base (geographic information system), to perform analysis on the ground slope and elevation marks, to model complex watersheds and determine the direction of flow and provides tools for planning earth's surface.

Therefore, on the pre stage we offer to perform a preliminary analysis of the ground and determine the reclamation efficiency of drained lands using software complex Autodesk Civil 3D, that allows you to examine the nature of the surface, to determine the required land surfaces indicators and their values (height, variations, deviations, area specific areas), to show the surface both as by given and by automatically specified range of values of marks and slopes and to arrange in a tabular form marks, differences in the surface of the earth, slopes and specific areas of land.

Average estimated value of slope of the land in the whole system is proposed to define by the formula:

$$\overline{i}^{(cuc)} = \sum_{n=1}^{n_i} \left(i_n \cdot f_n\right) / \sum_{n=1}^{n_i} f_n, \tag{2}$$

where f - characteristic surface areas of land, ha; i - slope of the characteristic areas of the surface.

The corresponding average calculated value of the difference in the surface of the earth in the whole system is defined similarly

$$\overline{\Delta H_{gi}}^{(cuc)} = \sum_{n=1}^{n_i} \left(\Delta H_{gi_n} \cdot f_n\right) / \sum_{n=1}^{n_i} f_n, \tag{3}$$

where ΔH_{gi} - overall characteristic difference in the surface of the earth, m.

FUTURE RESEARCH DIRECTIONS

To determine the reclamation efficiency of relief plot of 330 hectares of drainage system "Ikva" Dubno district, Rivne region the model of a relief recieved by surveying was made, the character of the surface was identified, the required parameters and their values were fixed: height varieties, slopes, specific areas (Fig. 3, tab. 2).

Performed topography analysis on the examined reclamation system showed that the average slope is 0.004, and the overall average difference in the surface of the earth is - 0.34 m. Thus, the drainage system relief considering the developed relief degree equals medium developed and reclamation effectiveness of relief - the average, its index may be within 0.8 ... 1.0.

Defined average slopes and difference in the surface of the earth are the output data to perform calculations on the long-term forecasting water regime and water regulation technologies of drained lands model considering the relief terrain, which is based on the implementation of the equation of the water balance of the active root layer of soil, taking into account the formation of surface runoff and equation of ground water balance considering the surface differences (ΔH_{gi}) (Koptyuk et al., 2014; Rokochinskiy et al., 2011).

Figure 3. Identification of specific relief areas of drainage system "Ikva" Dubno district, Rivne region

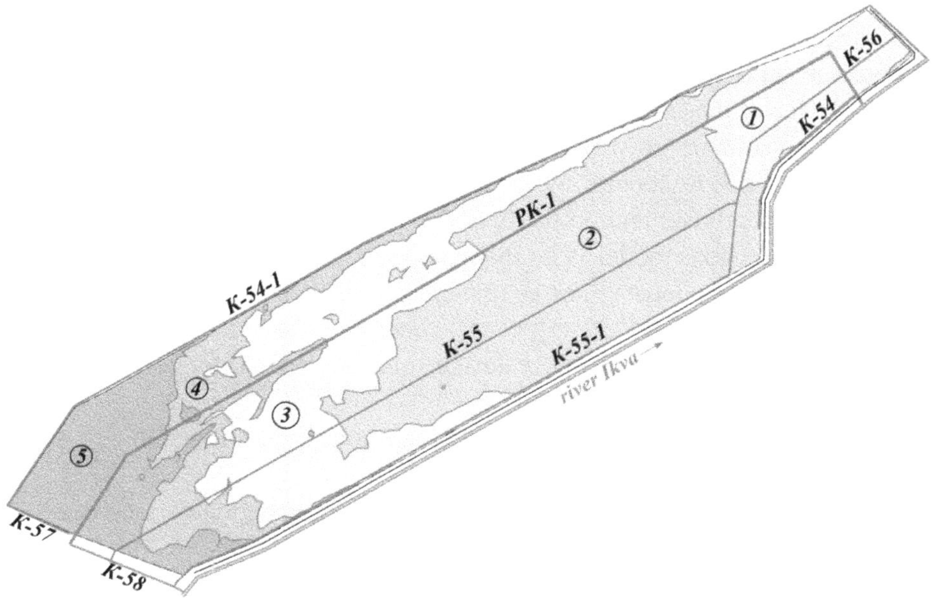

Table 2. Summarized results of the evaluation of reclamation terrain efficiency in the drainage system "Ikva" Dubno district, Rivne region

Field-number	Quantitative indexes of relief		Square		The degree of developed relief	Reclamation efficiency index	Relief Reclamation effectiveness
	Slope, *i*	ΔH_{gr}, *m*	*Ha*	%			
For specific areas of relief							
1	0.004	0.5	31.8	9.6	***develop medium.***	0.96	***average***
2	0.001	0.2	143.2	43.4	***develop poorly.***	1.02	***high***
3	0.002	0.3	78.4	23.8	***develop medium.***	0.92	***average***
4	0.003	0.4	35.7	10.8	***develop medium.***	0.87	***average***
5	0.018	0.9	40.9	12.4	***strongly develop.***	0.64	***low***
Totally in the system							
	0.004	0.34	330	100	***develop medium.***	0.93	***average***

The need for taking into account number of influence factors while substantiating both as rational water regulation technology of drained lands and corresponding technologies to the type and design of drainage system that grant to achieve project economic and environmental effect indicators according to the project after modern requirements predetermine the necessity of using the optimization approach (Rokochinskiy, 2010).

In addition, the presence of a pronounced optimum at different stages of developed relief on drained lands considering multiple natural and agromeliorative conditions also predetermine the necessity of using the optimization approach to the design study decisions as for water regulation technology and related to them the type and design of the system on drained lands with developed terrain.

Considering the hierarchical approach to forming and implementing the optimization model of constructive decisions substantiation as for types, constructions and options of drainage system, slopes and differences in the surface of the earth it should be pointed out within the reclaimed fields related to the regulatory networks and soils for grown crop rotation and design decision options by water regulation way (Mazhayskiy et al., 2017; Rokochinskiy, 2010).

Based on the above mentioned and on the optimization model forming principles the output constructive substantiation decisions model as for the types, structures and parameters of the whole system and its technical elements components must be implemented in accordance with the hierarchy levels:

- at project crop growing rotation level, for design solutions options by water regulation way;
- Ground level within the system;
- system structural elements by the specific relief conditions level, for variable parameters of slopes (i) and differnces in the surface of the earth ($\Delta \boldsymbol{H}_{gi}$);
- the whole system level for design solutions for selected water regulated schemes

As a criteria of water regulation optimality on drained lands under drainage systems construction and reconstruction, it is appropriate to use given costs index based on weather and climate risk (Mazhayskiy et al., 2017; Frolenkova et al., 2007)

$$ZP_0 = \min_{\{i\}} \left\{ \left[\left(A_i + C_i^{C2} + C_i^{M} \right) + E_H K_i + R_i \right] / V_i \right\},\ i = \overline{1, n_i}\,, \tag{4}$$

where C_i - operating costs for getting production according to variants in technical solutions that consist of agricultural C_i^{A3} and operational C^e costs. The latter include deductions for depreciation and repair A_i and reclamation costs C_i^M to maintain the system;

E_H - regulatory factor of economic efficiency of capital investments K_i for the corresponding options of technical solutions together $\{i\},\ i = \overline{1, n_i}$;

R_i - weather and climat risk, which is determined by such statistical model

$$R_i = \sqrt{V_i - \widehat{V}_i},\ i = \overline{1, n_i}\,, \tag{5}$$

where V_i - gross output value for the actual yield, got by i-those variant of design solution; and $\widehat{V}_i$ - the value of gross output for the potential yield of the object.

Generalized results of the optimal variant substantiation as for the type and system construction based on the water regulation appropriate technology on the drained lands (DS - drainage system, WL - warning locking, ML - moisturing locking) considering the terrain (i=0.004, Δh=0.34m) for the object being studied are presented in Table 3.

Table 3. Basic indicators of overall design decision options economic efficiency

Number s/n	Indicator	Water regulation technology		
		DS	WL	ML
i=0.004, Δh=0.34m				
1.	Total operating expenses C_i UAH. / Ha	14330	15392	17388
2.	Weather and climate risk R_i UAH. / Ha	12628	11154	10424
3.	Capital expenditures K_i UAH. / Ha	26526	28592	37126
4.	Given cost index based on weather and climate risk ZP_i	1.66	**1.53**	1.59

Investment evaluation of the optimal variant design decision by the basic parameters used in calculating the investment attractiveness of the reclamation project net profit value (NPV), profit index (PI), discounted payback period (DPP) and internal rate of return (IRR) of investment (Mazhayskiy et al., 2017; Frolenkova N.A. et al., 2007)

$$\mathrm{PI}_0 = \max_{\{i\}} \sum_{j=1}^{m} \mathrm{PI}_{V_j} \cdot \alpha_{p_j}, \; i = \overline{1,n}, \tag{6}$$

considering that

$$\begin{cases} \mathrm{NPV}_V \geq 0; \\ \mathrm{IRR}_V \geq d_i; \\ \mathrm{DPP}_V \leq '', \end{cases} \tag{7}$$

where d_i - discount rate of the project; T - suitable for investors discounted payback period of investment.

Generalized results of the investment efficiency evaluation of considered design solutions for the object being studied are presented in Table 4.

Obtained results illustrate that the level of developed relief affects both on technological and onr the overall water regulation economic efficiency on drained lands that is resulted on the selection of optimal design solutions corresponding to the type and design of drainage systems.

Table 4. Basic indicators of economic efficiency of investments

Number s/n	Indicator	Technology vodorehulyuvannya		
		DS	WL	ML
i=0.004, Δh=0.34m				
1.	Profit index of investment PI	1.56	**1.57**	1.36
2.	Net profit value NVP, UAH. / Ha	14787	**16369**	13186
3.	Discounted payback period DPP, years	8	**7**	9

This approach as for the evaluation of the terrain slopes and differences in the surface of the earth can also be implemented using geographic information systems (GIS), allowing to evaluate the relief on the macro level.

Thus, using software ArcGIS, we have built a digital terrain models for zonal conditions of Rivne region and Ukraine Polesie, in general, based on radar topographic satellite mapping data SRTM (Shuttle radar topographic mission) and the analysis of the earth's surface slopes has been done (Fig. 4, 5) and their areas have been defined (tab. 5).

Figure 4. The elevation marks and slopes of the earth's surface in Rivne region

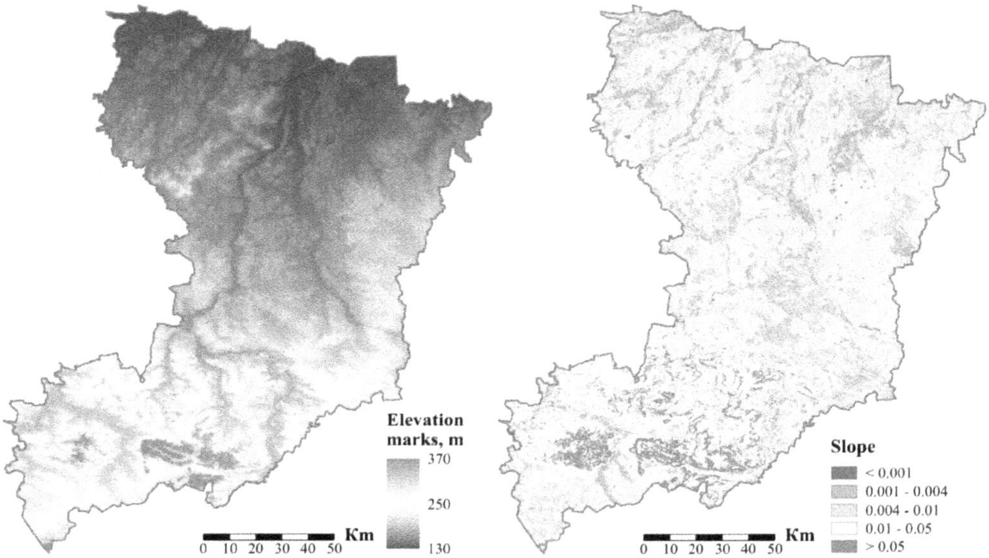

By the analysis results of relief elevation marks in Ukrainian Polesie it was determined that the absolute maximum marks are up of 330 m, the minimum are- 88 meters, heights differences are- 242 m. For the Rivne region absolutely maximum marks are up of 370 m, the minimum are - 130 m heights differences are - 240 meters. Poorly and middle developed relief distribution area for the Ukrainian Polesie zone is 18.12%, for Rivne region is - 16.38%, the average slope of the land correspondigly is 0.022 (1.26°) and 0.027 (1.55°).

According to (Rychagov, 2006), the relief of zonal conditions studied in Rivne region and Ukrainian Polesie, as a whole, by the steepness of the earth's surface to the plain-hilly terrain. But from the ameliorative point of view, namely to subsoil drainage systems, which in climate change terms are the most matter-of-fact in the area of Ukrainian Polesie, the relief with existing slopes is already well developed, but its reclamation efficiency is low.

Developed relief evidence brings to inefficient rainfall use due to the surface runoff occurance and soil erosion threat, uneven GWL formation and water regime in general on drained lands.

Figure 5. The elevation marks and slopes of the earth's surface of Ukraine Polesie

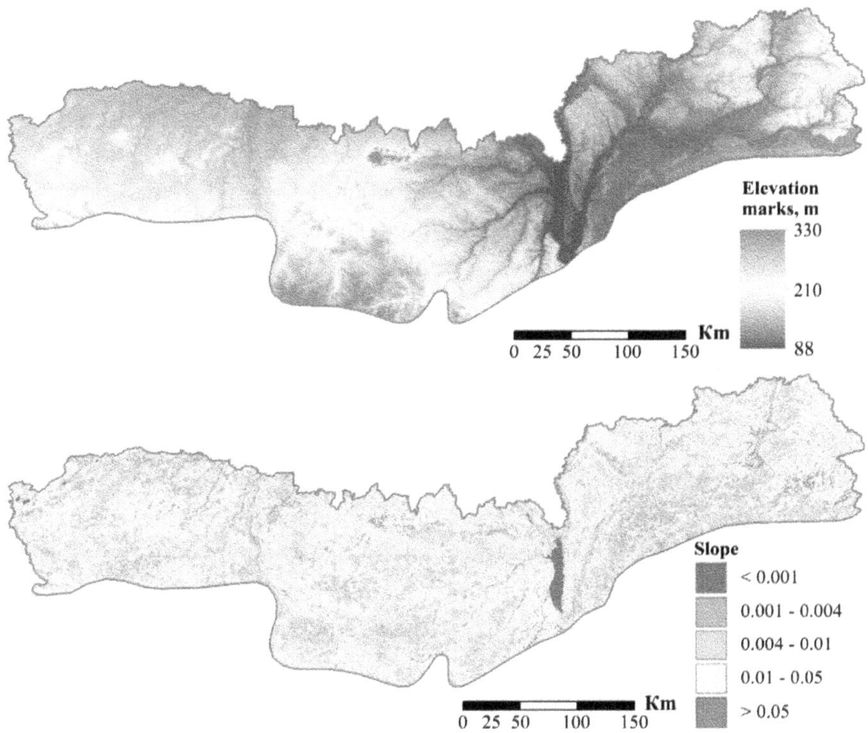

Table 5. Analysis of the Earth's surface slopes on studied objects

Number s/n	Slope, *i*	Area		The degree of developed relief	Reclamation effectiveness of relief
		km²	%		
By special relief plots in Rivne region					
1	<0.001	317	1.58	*poorly developed*	*high*
2	0.001 - 0.004	2967	14.8	*medium developed*	*average*
3	0.004 - 0.01	5532	27.59	*highly developed*	*low*
4	0.01 - 0.05	10348	51.62	*highly developed*	*low*
5	> 0.05	883	4.41	*highly developed*	*low*
Overall average value in Rivne region					
	0,027	20047	100	*highly developed*	*low*
By special relief plots in Ukraine Polesie					
1	<0.001	2121	2.21	*poorly developed*	*high*
2	0.001 - 0.004	15259	15.91	*medium developed*	*average*
3	0.004 - 0.01	29932	31.22	*highly developed*	*low*
4	0.01 - 0.05	46558	48.56	*highly developed*	*low*
5	> 0.05	2010	2.1	*highly developed*	*low*
Overall average value in Ukraine Polesie					
	0.022	95880	100	*highly developed*	*low*

It should be noted, that the heights exactness of relief general digital model SRTM corresponds to the heights gained from topographic maps of 1: 100,000. Considering constant mistake it is possible to increase the accuracy of high and ultrahigh resolution SRTM matrix, which can be used to create maps of 1: 25,000 and 1: 10,000 scales for areas of flat and hilly terrain (Postelnyak, 2013).

Thus, microrelief due to its size levels in creating large-scale maps based on digital relief model SRTM. So to consider microrelief in the design of drainage systems is possible on the basis of surveying areas of the object construction, which according to (Rokochinskiy et al., 2010) is to meet the following requirements: topographical (engineering and topographical) plan at a scale of 1: 2000; relief intersection is 0.5 m for plain areas and for areas with micro-relief is 0.25 m; square leveling is 20x20.

CONCLUSION

Thus, modern methods of ecological and economic substantiation of investment projects, which are widely implemented in current practice in most cases are not adapted to the functioning of each individual field. Evaluation of reclamation projects requires taking into account the large number of diverse natural and technical factors that significantly affect the final result. Together with considering the terrain to solve a number of complex forecasting, optimization models it should be taken into account soil types, levels of natural water availability, agronomic conditions, practice of crop rotation patterns and so on.

REFERENCES

Koptyuk, R., Budz, M., & Rokochinskiy, A. (2011). Influence of irregularity of relief on construction and reconstruction of drainage systems. *Agrarian Science*, (99), 98–102.

Koptyuk R., Rokochinskiy A. (2008). Substantiation of necessity and modern approaches to optimization of structures of drainage systems taking into account the terrain. *Bulletin of the National University of Water Management and Environmental Management*, *1* (41), 476-483.

Koptyuk, R., & Rokochinskiy, A. (2014). Model forecasting of the water regime and technological water regulation of the drained lands with the terminal of the vestry. *Pour overlapping of the efficiency of the irritated earthlings*. Novocherkassk ROSNIPM.

Kovalenko P., Rokochynskiy A., Jeznach J., Koptyuk R., Volk P., Prykhodko N., Tykhenko R. (2019). Evaluation of climate change in Ukrainian part of Polissia region and ways of adaptation to it. *Journal of Water and Land Development*, *41* (IV-VI), 77-82.

Mazhayskiy, Y., Rokochinskiy, A., Volchek, A., Meshyk, O., & Jeznach, J. (Eds.). (2017). Environmental management of Polissya. VNIIGiM of A. Kostiakov.

Postelnyak A. (2013). Evaluation of the accuracy of the heights of digital models of the relief SRTM and Aster GDEM. *Bulletin of geodesy and cartography*, *4*, 17-21.

Rokochinskiy, A. (2010). *The scientific and practical aspects optimization of water regulation drained lands on environmental and economic grounds*. NUVGP.

Rokochinskiy, A., Antonov, O., & Shalai, S. (2010). *Engineering surveys for water management and environmental construction*. NUVGP.

Rokochinskiy A., Jeznach J., Volk P., Turcheniuk V., Frolenkova N., Koptyuk R. (2019). Reclamation projects development improvement technology considering optimization of drained lands water regulation based on BIM. *Scientific Review Engineering and Environmental Sciences*, *3*(85), 432-443.

Rokochinskiy, A., Stashuk, V., Duplyak, V., Frolenkova, N., & Koptyuk, R. (2011). Temporary recommendations for the forecast evaluation of the water regime and technologies of water regulation of drained land in projects of construction and reconstruction of reclamation systems. Rivne, Ukraine: NUVGP. .

Frolenkova, N., Rokochinskiy, A., & Kozhushko, L. (2007). *Ecological and economic evaluation in the management of reclamation projects*. NUVGP.

Rychagov, G. (2006). General geomorphology. Publishing House of Moscow University: Science.

Chapter 12
Calculation of Groundwater Flow and Drainage Under Regulation of Water–Physical Conditions on the Reclaimed Lands of Polesie

Vadym Poliakov
Institute of Hydromechanics NAS of Ukraine, Ukraine

ABSTRACT

The urgency of improving the methods of calculation of the groundwater flow regulation due to global warming as applied to the soil-climate conditions of the Polesie zone of Ukraine is discussed. The calculation schemes of drainage, which are typical for Polesie, are systematized. Recommendations on accounting for the hydrodynamic resistance in the calculations of imperfect drainage are given. A number of engineering methods are presented. They are intended, firstly, for prediction of the water table (WT) dynamics against the background of a single and systematic horizontal drain laid in the uniform ground or in the upper or lower layers of the two-layer ground, secondly, theoretical substantiation of the drainage design parameters (depth of laying drain, distance between regular drains). The dependencies for the calculation of regular drainage systems operating in the draining or subsurface wetting regimes are given. Proposals on additional consideration of influence of open conductive network and molehills system on the WT regime of temporarily waterlogged lands.

BACKGROUND

The changes in the climate conditions that have being taken place over the past decades, expressed primarily in a steady increase in temperature indicators, have led to irreversible changes in the aquasphere. In Polesie, they primarily affected the expenditure items of the water balance. Thus, water exchange between the atmosphere and the soil cover has being intensified, especially physical evaporation, and the need for

DOI: 10.4018/978-1-6684-8248-3.ch012

plants in water increased. As a result, the soil moisture reserves were generally reduced. And therefore, the water-physical conditions which are unfavorable for growing crops are more and more common on the agricultural lands. But if earlier, when monitoring the natural situation, developing measures and technical means in order to optimize the water regime of reclaimed land, special attention was paid to the early spring and its characteristic excess of soil moisture, now conceptual approaches to improving water regulation are being revised here. It should also be noted the overdrying and overwetting of the soil in the late spring and summer, which used to happen sporadically. To eliminate their negative consequences, it was enough to take minimal measures. Now the imbalance between the amount of moisture available and the needs of the agricultural crops in it is observed more often and is more significance. As a result, the management of the water conditions of temporarily overdried and waterlogged lands in the Polesie zone not only remains relevant, but also needs to be adapted to new realities.

Drainage is the most important engineering tool for regulating the water-physical conditions in natural and artificial porous structures. Therefore, it plays a key role in water construction in Ukraine. Generally, drainage has great potential in managing the water-physical conditions in agricultural land within the region under consideration. Mathematical modeling of the groundwater flow and moisture transfer in porous media in the presence of concentrated and distributed effluents and sources helps considerably to fully realize this potential.

The solutions of correctly formulated mathematical problems of drainage (using analytical methods and the hydrodynamic and hydraulic statements) serve as the basis for the development of engineering methods for determining its design and technological parameters. Reliable data on the water-physical properties of the aquifer, the patterns of water exchange between it and the atmosphere are necessary for the successful practical application of these methods. It is important that such information obtained during a complex of the targeted experimental studies in natural and laboratory conditions allows, first of all, to schematize the natural environment at the land reclamation site and choose an appropriate drainage scheme on this basis, and then provide a calculation method with a set of values of all the model parameters contained in it. Therefore, the value of such techniques consists not only of predicting the spatio-temporal changes in the desired characteristics, in substantiating the parameters of drainage, but also in the rational planning of the aforementioned studies, and processing of their results in order to specify the initial data.

The engineering methods presented in this chapter are developed thanks to the theoretical analysis in the hydraulic approximation of the groundwater flow occurring against the background of two types of drainage in a variety of soil, climate and hydrogeological conditions. Obviously, they can be successfully applied only if the coefficients characterizing the structure and permeability of the water-saturated soil, as well as its water exchange with the adjacent head horizon and the surface layer of the atmosphere are preliminarily selected. Special attention was paid to the horizontal drainage. It is possible to more flexibly and efficiently control the water-physical conditions in the soil and, in particular, the groundwater flow.

The main provisions of the schematization are briefly described below and indicate the need for the correct consideration of drainage hydrodynamic imperfection, that is, imperfection in the degree of opening of the aquifer, and additionally in the character of its opening. As to the natural reclamation sites in the Polesie zone, the imperfection of the drainage is mainly due to the hire power of the groundwater flow in comparison with the depth of the drainage. Therefore, we specifically dedicated one section to the establishment of the hydrodynamic component of the total hydrodynamic resistance. The separate accounting of the effect on the inflow to the drain of the upper worse permeable (peat) layer and the lower more permeable (mineral) layer is provided when calculating the hydrodynamic resistance. At the same time, a virtually layered porous medium is treated as conditionally uniform in calculating the hydraulic characteristics at a sufficient distance from the drains.

A number of important regime-forming factors, which are characterized by significant variability in space and time, essentially complicates the natural situation at the Polesie reclamation facilities. It is currently unrealistic to take into full account the complex of water-physical, hydrogeological, climate conditions for predicting the development of this situation due to the limited information available on them and the high cost of replenishing it. However, this is not particularly necessary. So, some of the mentioned factors are secondary for regulating the water-physical conditions in agricultural lands.

DRAINAGE SCHEMES AND CALCULATION METHODS

The schematization of the hydrogeological conditions is of paramount importance for the drainage calculations. At the same time, the upper section (layer) of thick aquifers deserves attention due to a diverse soil biota and, above all, root systems of plants, fluctuations in water table (WT) and intensive gas exchange within it. Therefore, the drainage schemes intended for the subsequent various engineering calculations are selected based on an analysis of the mentioned conditions. They share in common the large thickness of the soil layer to a partial or absolute waterproofing with slight differences in permeability inside it. Therefore, it is justified to average the WT, the hydraulic conductivity and the thickness of the aquifer considering it uniform in theoretical developments.

Three varieties of the aquifer complex are widespread in Polesie in accordance with the accepted classification. It combines seven characteristic hydrogeological schemes and requires an individual approach to the drainage calculations. Complex A is a geostructure formed by three layers of different thickness and permeability. It is advisable to consider it in the first approximation as uniform with the following weighted average hydraulic conductivity k_e

$$k_e = \frac{k_1 m_1 + k_2 m_2 + k_3 m_3}{m_1 + m_2 + m_3} \quad (1)$$

where k_i, m_i (i=1,2,3) are the hydraulic conductivity and power of the i-th layer; when laying the drains in the upper layer, m_1 is the either more roughly their height above the nearest interlayer boundary, or more precisely the same height supplemented by the averaged excess of the groundwater level above them in case of the drains laying in the upper layer. The corresponding drainage schemes are presented in Fig.s 1 - 5. Here, the dashed line indicates the possible boundary between the layers. It is possible to slightly increase the reliability of the calculations for the hydrodynamic section AI^1 if we select the upper layer, the permeability of which is noticeably worse than the two underlying. The latter again should be considered as a conditionally uniform layer. Its hydraulic conductivity should be calculated by the formula (1) having previously set k_1=m_1=0. Thus, the complex A can be treated in the drainage calculations either as a two-layer porous structure with a less powerful and worse permeable upper layer or with a slightly lower accuracy as a uniform structure in the absence of hydraulic connection with the underlying head horizon (Figure 1 - 3). The decoding of the designations adopted on these and other design drawings is given in the corresponding sections of the chapter.

The peculiarity of the hydrogeological conditions in the complex Б is due to the presence in the aquifer of an interlayer (zone of colmatage), which can occur at a relatively small depth and have a very low permeability. Therefore, water exchange between groundwater and a thick layer of watery marl located below is extremely difficult and is not able to significantly affect the drainage operation. The groundwater flow in the hydrodynamic section BII3 is complicated by the presence of swamp deposits here. The permeability of the upper layer of peat of low power is several times less than the permeability of the underlying layer of sandy loam. In this case it is preferable to use the drainage scheme with two-layer soil. A simpler scheme can also be used (after carefully averaging the hydraulic conductivity) (Figure 1 - 5). It is important to note that the thickness of the peat layer is still quite great, so that both of the horizontal drains and channels of the conducting network (molehills) will be hanging in it.

Figure 1. The calculation scheme of unsteady systematic imperfect horizontal drainage

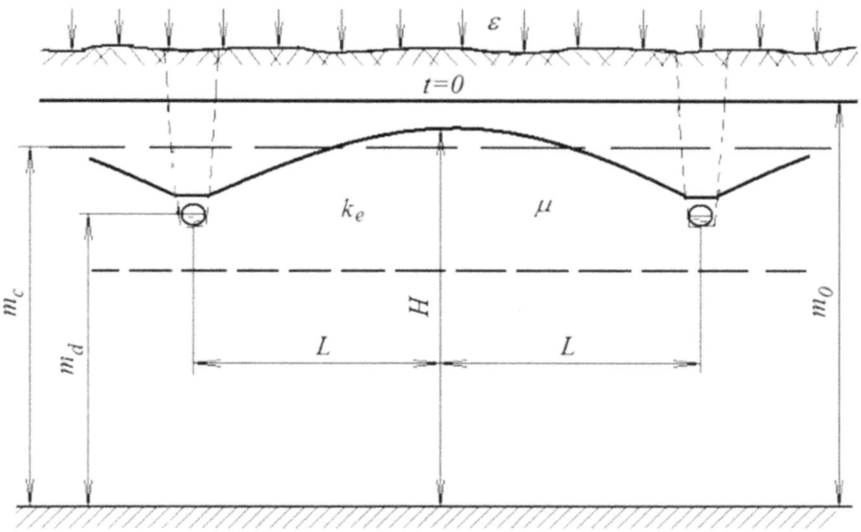

The most complicated hydrogeological conditions are characteristic of the complex *B* due to the possible hydraulic interconnection between ground and underground waters. The relatively high permeability of the loam layer as a dividing bulkhead, even with its considerable thickness, can cause a significant inflow of head water into the upper layers and, as a result, a noticeable decrease in the drainage effect of drains during critical periods of water deficit of the year. If the head difference at the borders of bulkheads is equal to Δh, then the intensity of the water flow through it under the rigid conditions of groundwater flow will be $k_s \Delta h/m_s$, where k_s, m_s are the hydraulic conductivity and the thickness of the low-permeable interlayer. In such cases it makes sense to use drainage calculation schemes with a uniform or two-layer structure of the groundwater flow taking into account head supply additionally. A corresponding to a uniform or two-layer structure with such power diagram is shown in Figure 4. The two-layer hydrogeological scheme is better due to the specific water-physical properties of the upper peat layer for the hydrodynamic section VIII3. The specified layer has sufficient power so that the drains and shallow channels are within it.

Figure 2. The calculation scheme of subsoil wetting by systematic imperfect horizontal drainage

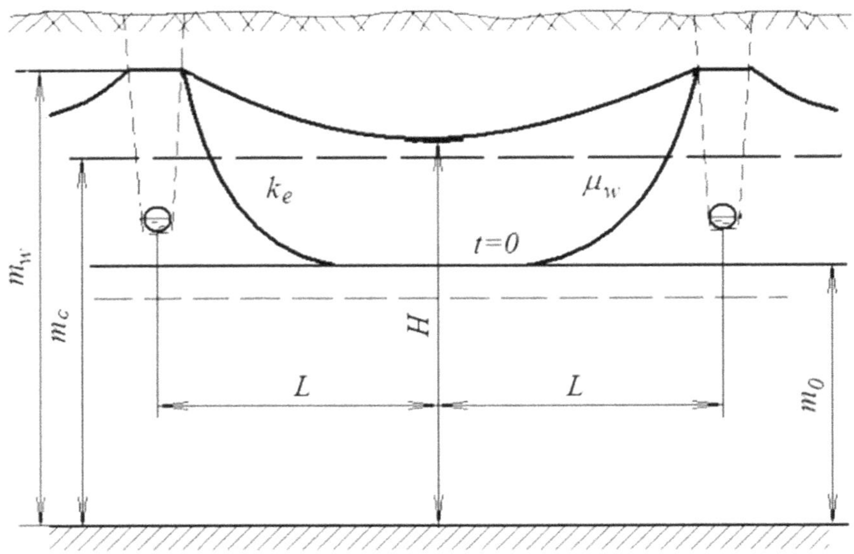

Figure 3. The calculation scheme of a single imperfect horizontal drain

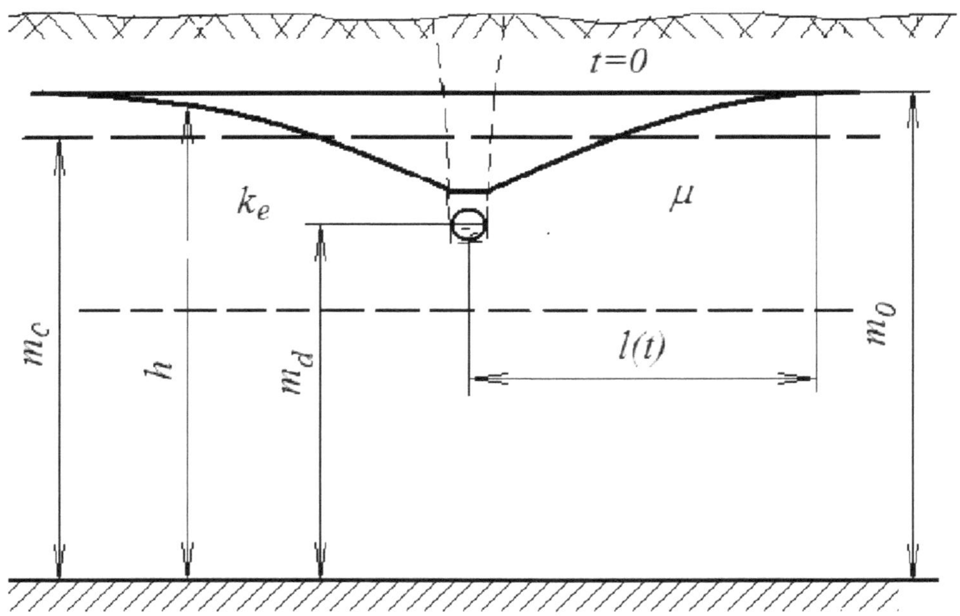

Figure 4. The calculation scheme of the steady-state systematic imperfect drainage

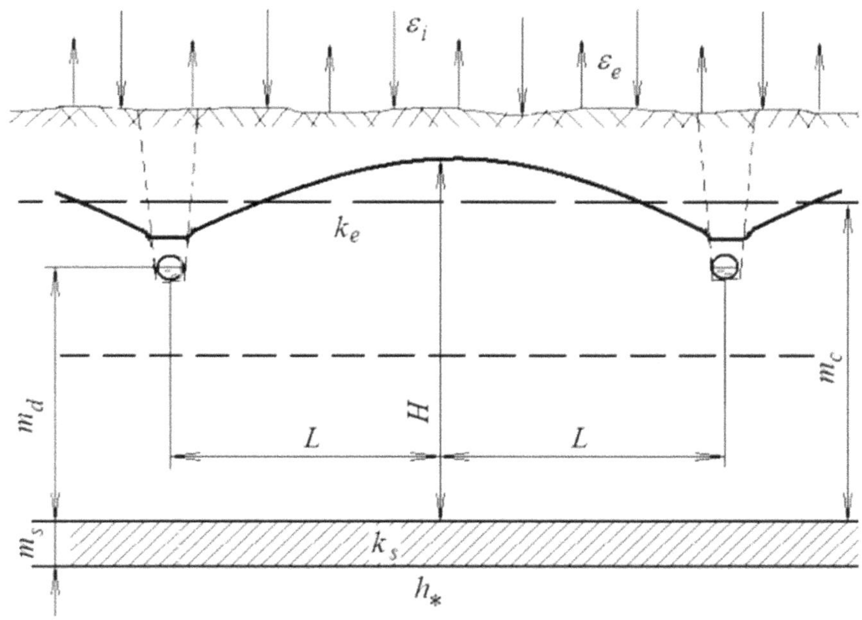

To achieve a significant drainage or wetting effect in large areas due to the localization of the influence of a single drain, drainage systems with the same spacing 2*L* can be used to ensure identical regulation of the water-physical conditions. Systematic drainage schemes are successfully applied to calculate the hydraulic characteristics on these systems Figure 1, 2, 3, 4. Their main design parameter, namely spacing *L*, should be substantiated theoretically, is spacing *L*. The influence of the drainage system on adjoining territories is mainly due to the action of boundary drains (channels) and is unsteady. Therefore, a single drainage scheme is suitable for its assessment in a layer that is unlimited in plan Figure 3. Sometimes the draining effect of the drainage system can be noticeably increased by the influence of the conductive channels. The drainage effect due to the action of the channels can be estimated in the first approximation using the planned scheme of systematic drains and the corresponding mathematical models Figure 5. In this case, the drained area is divided into many identical sections of approximately rectangular shape. Drainage of each of them is ensured by drains (channels, molehills) located on all four boundaries.

The developed calculated techniques must first be provided with initial information in full to carry out practical calculations and obtain reliable results. The composition of such information is dictated by the drainage scheme and the type of formulas and equations used in the calculations. The specified scheme is selected after processing the entire available array массива of empirical data, taking into account the purpose of the drainage. The hydraulic conductivities and powers are set (taking into account the initial or stable position of the WT) in accordance with the scheme, first of all, for a two-layer or uniform soil stratum. Then, information is formalized depending on the adopted scheme. It must quantitatively characterize the water exchange of groundwater with ground, surface waters and the atmosphere including the biological factor, and also reflects the design features of the drainage and the requirements for its operation.

Figure 5. The calculation scheme of unsteady both systematic imperfect horizontal drainage and the network of conductive channels

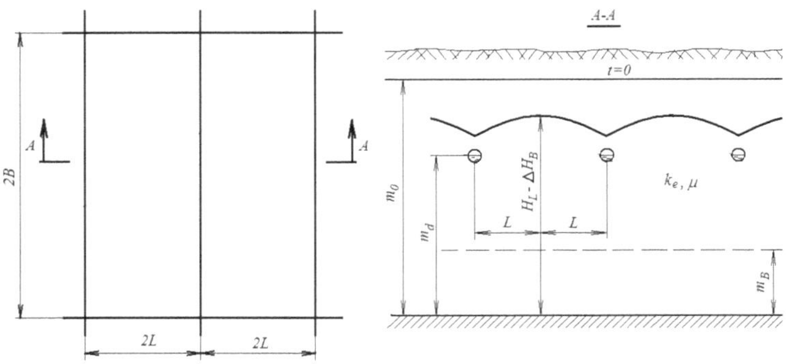

The spring and summer periods (early, late) are responsible for the reclamation drainage in the region under consideration. Strong waterlogging of the soil is often observed in the early spring, and sometimes its flooding in lowered areas due to intensive snowmelt is accompanied by thawing of the upper frozen soil layer. Therefore, the first period is usually the main period to assess the effectiveness of drainage. It should ensure the removal of excess moisture from the soil in the standard time during this period. Regular changes in time of hydraulic characteristics and, at the same time, rational parameters of drainage, are established using non-stationary hydraulic models of the groundwater flow. The groundwater flow is calculated due to the relative short duration of the spring calculation period, low temperatures and precipitation without taking into account water exchange with adjoining media and at the highest position of the WT (Figure 1, 3). The initial water level depends on the previous meteorological conditions, the terrain, and sometimes it is believed in the calculations that the soil is completely saturated with water at the beginning of the drainage operation. It should be noted that even the formation of a water layer on the soil surface is possible in special cases. Then the drain at the first, short stage of drainage works as one under riverbed дрена на первом, коротком этапе осушения работает как подрусловая. The modeling of the groundwater flow here becomes more complicated not only due to its hydrodynamic nature, but also, possibly, due to the dynamics of the surface runoff, which is strongly influenced by the slope of the terrain.

There are often drastic changes in the weather conditions during the crops growing season. As a result, excessive overwetting and overdrying of the soil alternate. Head supply entering the groundwater through the underlying low-permeability layer together with abundant precipitation, intensive physical evaporation, and transpiration can worsen the water-physical state of the soil (Nasedkin, 1981).

It is often sufficient despite the significant non-uniformity in time and the unpredictability of precipitation to introduce a generalized constant value into the initial unsteady or steady-state model of drainage due to the low intensity of the groundwater flow. The rate of infiltration ε has to accurately reflect the volume of water received from the soil surface into the ground for time considered for a representative year (Figure 5).

Prolonged evaporation from the surface of the groundwater flow can cause an excessive decrease in the WT and, as a result, in excessive depletion of moisture reserve in the root zone of the soil. The intensity of such evaporation εe depends to a large extent on the depth of the WT (Averyanov, 1982; Shih, 1983). It often changes significantly with progressive drainage of the soil, as well as along the groundwater flow. This fact is usually taken into account in the calculations of the drainage or wetting effect of drains using the following power dependence:

$$\varepsilon_e = \varepsilon_s(h - h_k)^\delta \tag{2}$$

where ε_m,δ are empirical constants, M is the total thickness of the soil stratum (from waterproofing to the soil surface), hk is the critical depth (evaporation ceases). If the coefficient ε_s is adjusted accordingly, then it can be taken into account evapotranspiration using evaporation (2). Subsoil wetting performed by a regular system of the wetting drains (Figure 4) (Dmitriev et al., 1992; Maslov et al.,1981; Tishenko, 2014; Yangol, 1970) can quickly eliminate serious overdrying of the soil or prevent it altogether.

The position of the WT is controlled, on the one hand, by the groundwater flow, and on the other hand, by moisture transfer in the aeration zone. The direction of water exchange between saturated and unsaturated zones of the soil is determined by the regulation mode of its water-physical state. The rate of water exchange, as a first approximation, is justified to be considered proportional to the velocity of the WT movement. The proportionality coefficient also depends on the regulation mode, but to an even greater extent on the depth of groundwater. Although the values of the indicated coefficient differ during draining and subsoil wetting, this difference, as a rule, must be neglected due to the limited reliable information about the moisture properties of the soil and the laws of moisture transfer in it. The relationship between the coefficient of averaged water loss (saturation deficiency) μ and the hydraulic conductivity k, depth of the WT is more significant (Ivitsky, 1988; Oleinik and Poliakov, 1987). If no suitable experimental data are available, then generalized empirical formulas are used for evaluating calculations. It is convenient to present them in the following single form:

$$\mu(h) = \alpha_k\beta(M - h)^\gamma \tag{3}$$

Empirical values are recommended: for mineral soils (G. D. Erkin) α=0.056, β=0.5, γ=0.333; for lowland peat (A. I. Ivitsky) -α=0.116, β=0.375, γ=0.75. The formula of V. P. Nasikovsky and V. Ya. Shapran is also suitable for the conditions of Polesie (Nasikovsky and Shapran, 1991).

When performing practical calculations of systematic drainage and the initial (h0) and final (h) positions of the WT set in the middle between the drains, it is first necessary to calculate the corresponding value of the averaged water loss (or saturation deficiency) according to the formula:

$$\mu_c(h) = \frac{1}{3}\left[\mu(m_d) + \mu(h_0) + \mu(h)\right] = \frac{\alpha k^\beta}{3}\left[(M - m_d)^\gamma + (M - h_0)^\gamma + (M - h)^\gamma\right]. \tag{4}$$

Only then the time it takes for the WT to reach in the soil the mark h can be determined. If the estimated time is known in advance, then the corresponding WT is necessary to determine by selection due to the monotonous increase or decrease in the current value μ (without its additional averaging). Obviously, this coefficient has physical meaning only in the case of unsteady groundwater flow.

Recently, numerical and analytical methods have been carried out to study water exchange between saturated and unsaturated zones of the soil on the basis of models of combined groundwater flow and moisture transfer. Then the dependences of water loss μ_d and saturation deficiency μ_w on the depth *M–h* and velocity of the WT dH/dt were substantiated in the following form (Poliakov and Kalugin, 2016):

$$\mu_{d,w}\left(H,\frac{dH}{dt}\right)=\lambda_{d,w}\left(\frac{dH}{dt}\right)\mu_{dm,wm}\left(H\right).$$

The second factor is the key factor for calculating the effect of drainage. Such expressions were adopted for it in accordance with (Poliakov, 2016; Van Genuchten, 1980)

$$\mu_{dm}\left(H\right)=\left(\theta_s-\theta_r\right)\left\{1-\left[1+\alpha^n\left(M-H\right)^n\right]^{\frac{1-n}{n}}\right\},$$

$$\mu_{wm}\left(H\right)=\sqrt{\theta_s^2-\theta_r\theta_s}\cdot\sqrt{1-\left[1+\alpha^n\left(M-H\right)^n\right]^{\frac{1-n}{n}}},$$

where 'θ_s,θ_r' are the residual and total (minus trapped air) moisture; α,n,m are the empirical coefficients. The following representation is recommended for the first factor:

$$\lambda_{d,w}\left(\frac{dH}{dt}\right)=\left(1+a_{d,w}\sqrt{\pm\frac{dH}{dt}}\right)^{-1},$$

where the sign "+"and a_w correspond to wetting, the sign "-"and ad - to drainage. The calculations of representative values of the coefficients are based on the international classification of finely dispersed mineral soils FAO. The variety of such soils in nature is divided according to FAO into 12 classes depending on the content of sandy, dusty, silty particles (Shein et al., 2001). Table 1 gives the information on the initial and calculated parameters of the first three classes, which differ in their lighter mechanical composition. There is similar information for all twelve classes at the present time too.

Table 1. The values of the initial and calculated parameters for the three classes of fine mineral soils

Soil class	$\alpha, <^{-1}$	n	θ_s	θ_r	*k,m/day*	a_d, $day^{0.5}m^{-0.5}$	a_w, $day^{0.5}m^{-0.5}$
Sand	14.5	2.68	0.43	0.045	7.13	0.1837	0.0904
Loamy sand	12.4	2.28	0.41	0.057	3.5	0.3056	0.1457
Sandy loam	7.5	1.89	0.41	0.065	1.06	0.5799	0.2688

Accounting for Hydrodynamic Imperfection of Drainage

The modeling of the processes of draining and wetting of the soil using drainage is significantly complicated due to the spatial nature of the groundwater flow near it. The horizontal and vertical drainages in Polesie are characterized by a double compression of the saturation flow near them. This happens because of the small size in comparison with the flow power, not only the water-intake elements, but also the drains themselves, as well as the depth of their laying. Usually, stronger compression and the serious additional head losses resulting from it are observed in the near-drain zone of the flow with a length approximately equal to its power. They can be reliably determined theoretically only using the models of the hydrodynamic saturation flow. The effect of the primary flow compression is usually taken into account in the calculations of the WT by means of a special hydrodynamic resistance Φ_h. The value Φ_h is practically independent of time in view of the rapid stabilization of the groundwater flow near the drain. Numerous formulas have been proposed to determine it in relation to the horizontal drainage. They are obtained as a result of exact solving stationary two-dimensional (profile) groundwater flow problems against the background of the drainage. In this case, the operation of a single (one-level) drainage system was considered in uniform and layered aquifers with various types of groundwater supply (Oleinik, 1981). It is enough to use to two of the many existing formulas in relation to the specific natural conditions of Polesie. Since the aquifers related to all three soil-head porous complexes allow their formal replacement by a uniform geostructure when schematizing the water-physical state, the main formula is the well-known formula by V. V. Vedernikov (Vedernikov, 1939). It looks after adaptation to groundwater flow with free surface for horizontal drain as follows

$$\Phi_h = \frac{m_c}{2\pi} \ln \frac{m_c}{2\pi r_d \sin \pi \left(\Delta m_c + 0.5 r_d \right) / m_c} \tag{5}$$

where m_c is the average power of the groundwater flow, $\Delta m_c = (m_0+H_p-2m_d)/4$, m_0 is the initial WT mark for calculations, m_d is the height of the drainage above the (partial) waterproofing, r_d is the radius of the drain taking into account the filter. If the groundwater flow actually begins at a considerable distance from the drain and the head here does not change much over time, then its value m_0 is used in the corresponding unsteady mathematical problem under the initial and boundary (at infinity) conditions. Then, m_c is justified to determine as $(m_0+m_d)/2$, and Δm_c as $(m_0-m_d)/2$. In the case of the systematic drains, the groundwater flow is formed starting in the middle between them, where the head (WT) H is gradually reduced. Let it fall from the mark m_0 to the mark H_p by the reference time. In this case, m_c is recommended to calculate taking into account the dynamics of the head as $(m_0+m_d+H_p)/3$, and Δm_c as $(m_0+H_p-2m_d)/4$. The value m_c should be found taking into account the control head in the drain m_w for wetting and the unsteady operation of the regular drain system. Then it will be equal $(m_w+m_0+H_p)/3$ and Δm_c becomes equal $(2m_w-H_p-m_0)/4$. The installation of well-permeable drain filling allows to reduce the hydrodynamic resistance of the drain. Suitable recommendations are given in the literature (Oleinik, 1981; Oleinik and Poliakov, 1987).

The effect of soil heterogeneity on the groundwater flow is more significant near drains. Therefore, it is reasonable for the calculation of Φ_h to take into account the presence of an upper thin less permeable layer in the aquifer in relation to hydrodynamic sections *БII*[3] and *BIII*[1]. When horizontal drains are located in the upper layer, the following formula allows to determine with high reliability Φ_h (Oleinik, 1981):

$$\Phi_h = \chi \frac{k_2 \left(m_1 + m_2 \right)}{2\pi k_1} \ln \frac{m_1 + m_2}{2\pi r_d \sin \dfrac{\pi \left(m_1 + m_2 + 0.5 r_d \right)}{m_1 + m_2}} + \frac{m_1 \left(k_1 - k_2 \right)}{2\pi k_1} \ln \frac{m_1}{2\pi r_d \sin \dfrac{\pi \left(m_1 + 0.5 r_d \right)}{m_1}}. \tag{6}$$

Here $\chi = 1 - \left[0.21\sqrt{\frac{k_2 - k_1}{k_2 + k_1}} + \frac{r_d\left(k_2 - k_1\right)}{m_1\left(k_2 + k_1\right)} \right] \lg\frac{m_2}{m_1}$, m_1 is the average power of the saturation flow in the upper layer.

Sometimes other factors negatively affect the drainage effect. The phenomenon of colmatage is distinguished among them. It has a physical, chemical or biological nature. A noticeably greater draining effect can be achieved in the cohesiveless (sandy) soils due to their suffusion. There are different phenomena initiated by drainage which actively influence it and can simply be taken into account in the groundwater flow calculations using the concept and method of the hydrodynamic resistances (Oleinik, 1978; Shestakov, 1965). Then the formulas derived with respect to perfect (ideal) drains should be the basis of such calculations for regulating the water-physical conditions in soils. These formulas have to be adjusted taking into account additional interferences noise, which can significantly reduce the intensity of the water-physical process

The basic formula for determining the flow discharge of imperfect drainage per unit of drain length q (uniform soil, the saturation flow with and without free surface, other important hydraulic characteristics) is the following formula:

$$q = km_c \frac{\Delta h}{\Phi_0 + \Phi_h + \Phi_d + \ldots} \tag{7}$$

Here k is the hydraulic conductivity, Δh is the head loss (level) between the drain and the external boundary of the drainage area, Φ_0 is the hydrodynamic resistance of a perfect drain, Φ_d is the hydrodynamic resistance due to imperfection of the drainage design (Pivovar and Bugai, 2000; Murashko, 1982; Lennoz-Gratin, 1989; Skaggs, 1978; Willardson and Walker, 1979). The remaining resistances, if necessary, are simply introduced into the formula (13.7) (Alekseev and Kommunar, 1974; Poliakov, 2014; Tugay et al., 2001; Pinck-Pfeiffer et al., 2000). The eight techniques outlined below provide reliable calculation of the operation and parameters of imperfect drains in various environmental conditions. The degree of imperfection is characterized by a general additional hydrodynamic resistance Φ. Formally, Φ includes all possible types of additional resistance. Since Φ_h most often makes the main contribution, the two formulas that are suitable for reclamation objects of Polesie with a water mode regulated by horizontal drainage are presented above for its determination.

If drainage of a waterlogged area is carried out by imperfect wells, it is recommended that the corresponding component Φ_h of their total hydrodynamic resistance Φ_c be determined using the known formulas. They are presented in the monographs together with supporting figures and tables (Oleinik, 1978; Shestakov, 1965). The formulas were obtained for a single well based on expression (7) and the exact solutions of spatial groundwater flow problems for point and linear sinks. In particular, it was proposed that the approximate value Φ_h be determined using Fig. 6 when the well filter was located near the roof or the bottom of the head reservoir. Fig. 7 allows to clarify this value, taking into account the head loss within drain. To find it, A. Ya. Oleinik, A. M. Tugai have used their own results of in-depth hydrodynamic studies of the imperfect well effects. In the case of a regular system of such wells, the additional resistance due to their mutual influence should be neglected, since the ratio of the distance between the wells to the saturation flow power is much more than 1 (Bochever et al., 1969).

Figure 6. Dependence $\Phi_h(l_w/m_c)$: 1–m_c/r_w=10, 2–m_c/r_w=30, 3–m_c/r_w=100, 4–m_c/r_w=200, 5–m_c/r_w=500, 6–m_c/r_w=1000

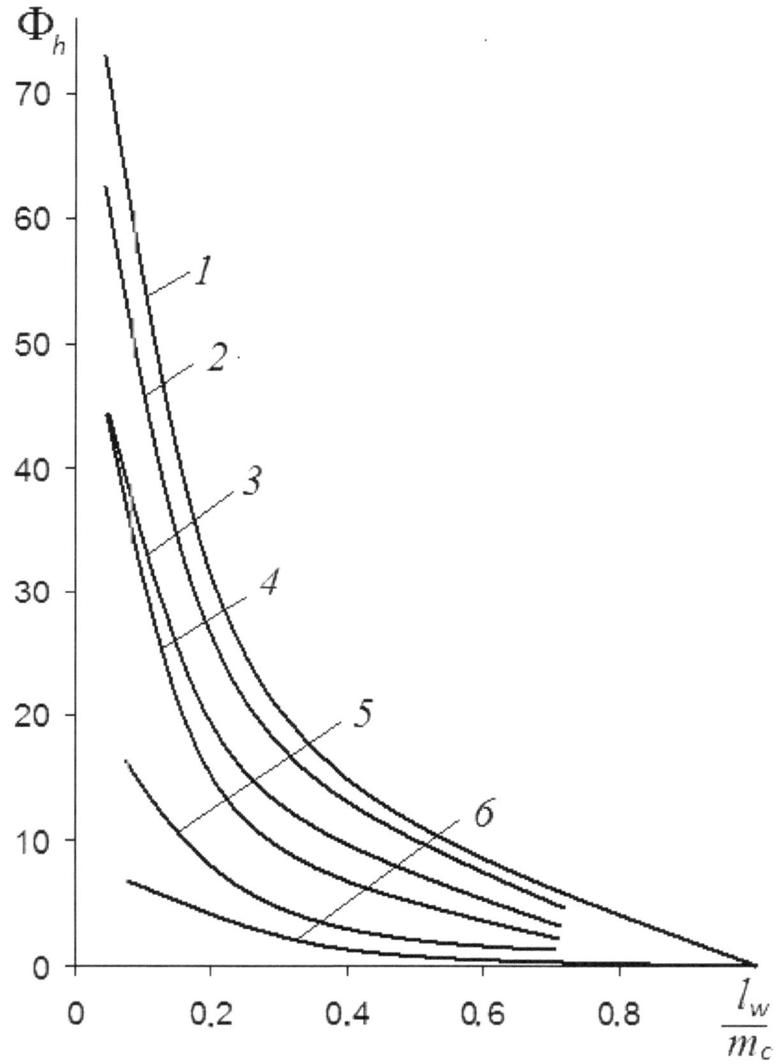

Here l_w is the height of the well filter, αw is the correction factor, which should be multiplied by the approximate value Φh according to Fig. 6, 7 in order to obtain its more reasonable and already final value.

Unsteady Inflow to Systematic Horizontal Drainage

A special place among the eight calculation methods presented in this chapter is occupied by the technique described below and illustrated in Fig. 1. It is intended to predict the development of the unsteady inflow to the systematic horizontal drainage and substantiate their design parameters in relation to the main calculation period. The soil water - physical conditions are especially unfavorable for agricultural

Figure 7. Dependence $\alpha_w(l_w/m_c)$: 1–r_w/m_c=0.01, 2–r_w/m_c=0.015, 3–r_w/m_c=0.02, 4–r_w/m_c=0.05

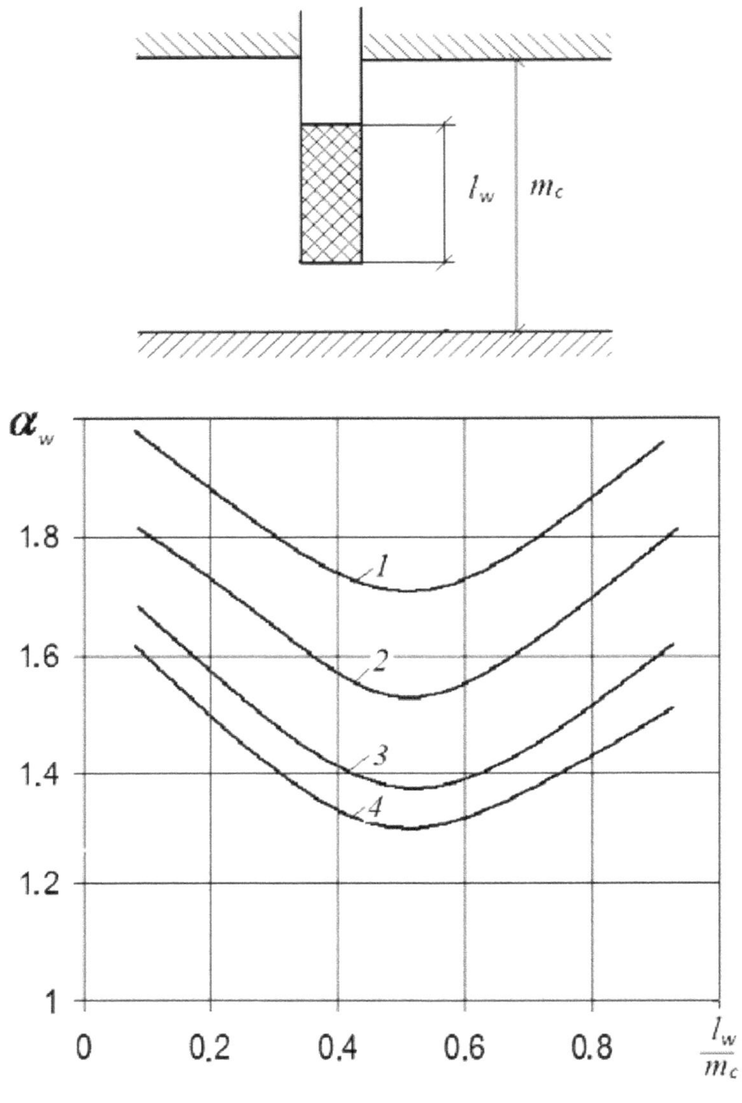

work most often early in the spring and drainage requires the fast removal of a large amount of excess water. The initial degree of the soil overwetting initial for the operation of drains is characterized by the initial WT height above the waterproofing m_0. The value m_0 is traditionally assumed to be constant when formulating unsteady groundwater flow problems against the background of the drainage. In this case, the process of draining the soil will formally proceed in two stages. Primary drainage will continue until the influence zones of neighboring drains close. It is determined by the formula (8) and will be equal to

$$t_1 = \frac{\mu_c}{k_e m_c}\left(\frac{L^2}{12} + \frac{2}{3}\Phi L - \frac{8}{3}\Phi^2 \ln\left(1 + \frac{L}{4\Phi}\right)\right) \qquad (8)$$

where L is the half of the drain spacing.

However, the question of choosing the initial form of the WT cannot be unambiguously resolved from a physical point of view due to its dynamism in the spring. Therefore, it is important to set the value m_0 only for the middle between the drains when calculating the WT dynamics. The time t_1 can be neglected, especially since the duration of the first and, in fact, auxiliary stage of drainage is much less than the duration of the second stage. The most important calculation dependences are given below just for the main stage. The dependence for determining the time to decline the WT is of special importance, namely:

$$t = \frac{L^2 + 6L\Phi}{3k_e m_c} \mu_c(h) \ln \frac{\Delta m \left(4\Phi L + 2Lx - x^2\right)}{\left(L^2 + 4L\Phi\right)\left(h - m_d\right)} \tag{9}$$

where $\mu_c(h)$ is calculated according to (4). A formula follows from (13.9) for calculating, firstly, a decrease in the WT in the middle between the drains with time, and secondly, the most important parameter of the systematic drainage. The dynamics of the WT here $H(t)=h(L,t)$ is described by the dependence

$$t = \frac{L^2 + 6L\Phi}{3k_e m_c} \mu_c(H) \ln \frac{\Delta m}{H - m_d} \tag{10}$$

An extensive literature is dedicated to the determination of the drain spacing, for example (Bashkis, 1974; Myaskov et.al., 1982; Van der Pboeg et al., 1999; Watson and Whisler, 1976). Here it is proposed to calculate L for a given drainage time t_O and drainage norm S_O as follows

$$L = \sqrt{\frac{3k_e m_c t_O}{\mu_c(M - S_O) \ln \dfrac{\Delta m}{M - S_O - m_d}} + 9\Phi^2} - 3\Phi \tag{11}$$

where $\mu_c(M–S_O)$ according to (4). One-side drainage inflow can be accurately found at any time by the formula

$$q = \frac{2k_e m_c}{L + 4\Phi} \exp\left[-\frac{3k_e m_c t}{\mu_c(h_0)\left(L^2 + 4L\Phi\right)}\right] \tag{12}$$

where $h_0(t)=h(0,t)$. The sequence of calculations taking into account the variability looks like this: first you need to set a value h_0, then from (19) find the values of time t and coefficient μ_c corresponding to it. And finally, substitute this set of values in (15). As a result, the desired value at a specified time q is easily calculated. Calculation of the WT and flow discharge are simplified if it is possible to operate with a constant average value μ_{cc} of variable coefficient μ_c. Then the change of h with time and along the groundwater flow is immediately calculated by the formula

$$h(x,t)=m_d+\Delta m\frac{4\Phi L+2Lx-x^2}{4\Phi L+L^2}\exp\left[-\frac{3k_e m_c}{\mu_{cc}\left(L^2+6\Phi L\right)}t\right] \tag{13}$$

and the discharge q can be calculated directly from (12) assuming $\mu_c(h_0)=\mu_{cc}$.

Noteworthy are the results of comparing formula (11) with a similar well-known formula, which was previously obtained assuming a uniform along the length of the water flow from the unsaturated zone, and with the same intensity as in the middle between the drains (Oleinik and Poliakov, 1987). By the way, the previous formula is quite reliable, as it was substantiated for a variety of soil-reclamation conditions by comparison with other authoritative formulas and with the data of numerical calculations based on the original mathematical model. However, formula (11) is more accurate from a formal point of view, since its derivation additionally took into account the fact of increasing water inflow from the aeration zone to groundwater as it was approaching the drain. With regard to perfect drainage. the distance L calculated using the new formula is 22.5% greater with regard to imperfect drainage.

Unsteady Groundwater Flow When Subsoil Wetting of The Soil

Long evapotranspiration in the absence of rain often leads to depletion of moisture available to plants. Usually, the critical water-physical state of the soil is accompanied by a large depth of the groundwater. As a result, the influx of moisture from the saturation flow into the root layer is too small. The detrimental effects of the soil moisture deficiency on crops can be minimized or even avoided through subsoil wetting. If the design of the drainage system provides for the use of its drains for wetting, then the projection of their operation becomes important. The control effect on the water-physical conditions in the overdrying soil is realized in practice by creating an increased head within the drain. It causes a very rapid rise in the WT over the drain (in the drain backfill). Then the water spreads in both directions, forming tongues on the surface of the groundwater Fig. 2. Thus, the first stage of wetting takes place. Obviously, a certain amount of time must pass before the neighboring tongues come together. But since this time, as a rule, is small compared with the calculation time, it is justified to neglect it for engineering purposes. Then, the dynamics of the WT caused by the increased head in the drains m_w, can be calculated with a slight error using the dependence:

$$t=\frac{L_w^2+6\Phi_w L_w}{3k_e m_c}\mu_w(h)\ln\frac{\Delta m_w\left(4\Phi_w L_w+2L_w x-x^2\right)}{\left(L_w^2+4\Phi_w L_w\right)\left(m_w-h\right)} \tag{14}$$

where the presence of the index "*w*" indicates that the characteristic (coefficient, parameter) refers to the wetting process; $\Delta m_w=m_w-m_0$; the average value of the averaged saturation deficiency μ_w in accordance with (4) is:

$$\mu_w(h)=\frac{1}{3}\left[\mu(m_w)+\mu(m_0)+\mu(h)\right]=\frac{\alpha_w k^{\beta_w}}{3}\left[(M-m_w)^{\gamma_w}+(M-m_0)^{\gamma_w}+(M-h)^{\gamma_w}\right]. \tag{15}$$

The substantiation of the drain spacing and during its operation - the control head m_w is of particular interest at the design stage of the drainage system for wetting. The main structural parameter is calculated on the basis of the given time t_w and the wetting norm S_w (optimal depth of WT by the end of the wetting event). The corresponding formula for L_w follows from (14) and has the following form

$$L_w = \sqrt{\frac{3k_e m_c t_w}{\mu_w(S_w)\ln\dfrac{\Delta m_w}{m_w - M + S_w}} + 9\Phi_w^2} - 3\Phi_w \quad (16)$$

The rational value of the parameter m_w is also determined on the basis of (14) for given t_w, S_w and in addition L_w. The required technological parameter is present in this dependence in both explicit and non-explicit (in the expression for Φ_w (15)) forms. Therefore, it is more accurate to find it directly from the dependence (14) assuming that $t=t_w$, $h=M-S_w$ in it and accepting (15) for Φ_w. But since the presence m_w in the last formula has a little effect on the results of its calculations, then it is recommended to calculate m_w taking into account the comments on the formula (15) and (12) by selecting from the equation

$$t_w = \frac{L_w^2 + 6\Phi_w L}{3km_c(m_w)} \cdot \mu_w(m_w)\ln\frac{m_w - m_0}{m_w + S_w - M}. \quad (17)$$

Unsteady Inflow to a Single Horizontal Drain in an Infinite Aquifer

Drains are capable of introducing significant disturbance into the water-physical conditions in aquifers, which, nevertheless, is local in nature. A number of modern techniques for the engineering calculation of reclamation drainage have been developed on the basis of unsteady hydraulic models under the assumption of an even decrease in the WT along the saturation flow. It approximately occurs in the cases of systematic drainage over the entire space between drains and a single drain in the zone of its influence. Either its actual value in some characteristic flow section (in the middle between drains for systematic drainage), or less frequently its average value in the motion area was taken for a constant decrease rate. The second assumption regarding the WT velocity better corresponds to the physics of the groundwater flow and, therefore, allows us to derive more accurate calculation dependences (although it turns out they are somewhat more complicated). It was this assumption that was used in theoretical developments related to the drainage. Specifically their most important outcomes are presented in this chapter. Most of the calculation methods developed above and below contain only elementary functions, which make the calculations as easy as possible. Only one technique has to deal with the special integral functions. The calculation dependences given here allow for a comprehensive assessment of the drainage effect due to a separate horizontal drain in a (conditionally) uniform soil. They correspond to the calculation scheme in Figure 3. and follow from an approximate solution to the problem of unsteady inflow to a single imperfect drain in an unbounded reservoir (Poliakov, 2016). This solution, as well as other basic solutions for the techniques in this chapter, is suitable for the groundwater flow with and without free surface (with a sufficiently large flow power) in principle, since it is based on a uniform linear model. The high accuracy of the indicated solution follows from the results of its comparison with the exact solution (for perfect drainage). The calculation dependences are presented in a dimensionless form for compactness.

The expansion of the drain zone of the influence is described by the equation

$$\bar{t} = \frac{\bar{l}^2}{12} + \frac{2}{3}\bar{\Phi}\bar{l} - \frac{8}{3}\bar{\Phi}^2 \ln\left(1 + \frac{\bar{l}}{4\bar{\Phi}}\right) \tag{18}$$

The change in the relative one-sided discharge $\bar{q}$ over time is recommended to be determined using the following expression

$$\bar{t} = \frac{\left(1 - \bar{m}_d^2\right)^2}{12\bar{q}^2} + \frac{8}{3}\bar{\Phi}^2 \ln\frac{4\bar{\Phi}\bar{q}}{1 - \bar{m}_d^2} - \frac{4}{3}\bar{\Phi}^2 \tag{19}$$

The WT (head) changes over time within the specified zone according to the following dependence

$$\tilde{h}\left(\bar{x}, \bar{l}\right) = \frac{m_0 - h(x,l)}{\Delta m} = \frac{\left(\bar{x} - \bar{l}\right)^2}{\bar{l}^2 - 4\bar{\Phi}l} \tag{20}$$

Accepted here:

$$\bar{t} = \frac{k_e m_c}{\mu_c m_0^2} t,\ \bar{l} = \frac{l}{m_0},\ \bar{\Phi} = \frac{\Phi}{m_0},\ \bar{m}_d = \frac{m_d}{m_0},\ \bar{x} = \frac{x}{m_0},\ \Delta m = m_0 - m_d;$$

k_e, μ_c are the effective hydraulic conductivity and water loss (elastic, averaged), moreover, μ_c equals

$$\alpha k^{\beta}\left[\left(M - m_0\right)^{\gamma} + \left(M - m_d\right)^{\gamma}\right]$$

or slightly less; m_c is the average power of the groundwater flow in the zone of drain influence and equals to $(m_0+m_d)/2$ (or the power of the head horizon); l is the length of the zone of influence; h, m_0, m_d are the marks of the current, initial WT and the height of the drain above the waterproofing.

Dependencies (8) - (10) can be successfully used not only to predict changes in the zone of influence of a single drain, the position of the WT, its discharge, but also for solving engineering problems. In particular, they allow you to rationally choose the depth of the drainage, based on the requirements for it to ensure the WT drop to a given mark at a certain point of time and distance L. These dependencies make it possible to more accurately calculate the systematic drainage as will be shown in the next section.

Unsteady Inflow to Systematic Drainage Adjusted for Infiltration

The calculation procedure presented in the previous section can be used in principle to determine the WT and check the drain spacing also in the case of short-term rainfall, which periodically occurs in late spring and summer (Figure 1). Let w be the entire volume of rain (irrigation) water that enters the soil through each part of its surface of a unit area and the initial WT is at a mark m^0. Then the WT will quickly rise to the mark m_0 that is proposed to be calculated on the basis of the groundwater balance and taking into account (3) according to the following formula:

$$m_0 = M - \left[\left(M - m^0 \right)^{\gamma+1} - \frac{w}{\alpha k^{\beta}} \right]^{\frac{1}{\gamma+1}} \tag{21}$$

It is advisable to carry out the subsequent calculation of drainage of the soil by systematic drainage with the value m_0 given in this way using dependencies (9) - (12). We can also neglect the time during which infiltration water reaches groundwater. The dynamics of the WT and the drain spacing are determined more precisely by solving a similar problem of unsteady inflow to a systematic drainage, but now in the conditions of stable infiltration supply of the groundwater flow. The intensity of such supply ε (can take place during the entire calculation period or its first part) is assumed in this technique to be constant as usual.

The main dependence designed to calculate the time for moving the WT to as far as the given mark h at an arbitrary distance from the drains up to the middle between drains is the following:

$$t = \frac{\mu_c(h)}{\psi} \ln \left[\frac{2\psi \left(L + 6\Phi \right) \Delta m - 3\varepsilon \left(L + 4\Phi \right)}{L + 4\Phi} \times \frac{4\Phi L + 2Lx - x^2}{2\psi \left(L^2 + 6L\Phi \right) \left(h - m_d \right) - 3\varepsilon \left(4\Phi L + 2Lx - x^2 \right)} \right], \tag{22}$$

where $\psi = 3k_e m_c / \left(L^2 + 6\Phi L \right)$. The direction of the WT movement is determined by the ratio between the drainage discharge and the infiltration supply. Therefore, the WT rise immediately begins if

$$\Delta m < \frac{\varepsilon \left(L^2 + 4\Phi L \right)}{2k_e m_c}$$

If the condition

$$\Delta m > \frac{\varepsilon \left(L^2 + 4\Phi L \right)}{2k_e m_c}$$

is met, the WT will monotonically decrease. In both cases, it will asymptotically strive to achieve a stable position, which corresponds to the mark

$$m_d + \frac{\varepsilon\left(L^2 + 4\Phi L\right)}{2k_e m_c}$$

The change in one-side discharge over time is described by the expression

$$q(t) = k_e m_c \frac{h_0(t) - m_d}{2\Phi} \tag{23}$$

where the water level above the drain should be found by solving such equation

$$t = \frac{\mu_c(h_0)}{\psi} \ln\left[\frac{4\Phi}{L + 4\Phi} \cdot \frac{2\psi(L + 6\Phi)\Delta m - 3\varepsilon(L + 4\Phi)}{2\psi(L + 6\Phi)(h_0 - m_d) - 12\varepsilon\Phi}\right] \tag{24}$$

Also, the selection allows to establish a key design parameter. It is recommended to calculate the distance L for the given time t_O and the drainage requirement S_O from the equation

$$t_O = \frac{\mu_c(S_O)}{\psi} \ln \frac{2\psi(L)(L + 6\Phi)\Delta m - 3\varepsilon(L + 4\Phi)}{2\psi(L)(L + 6\Phi)(M - S_O - m_d) - 3\varepsilon(L + 4\Phi)} \tag{25}$$

Obviously, the calculation dependences from section 13.3 are in fact a particular case of equations (23), (14), (25). Then the dependences can formally be deduced from the equations if ε goes to 0.

Unsteady Inflow to Drainage Subsystems of Conductive Channels, Molehills

The subsystem of conducting channels (molehills) along with drains takes part in the drainage process in the reclaimed area diverting a part of the excess water from the waterlogged soil Figure 5. Their contribution to the regulation of the water-physical conditions is significantly smaller due to the significantly greater distance between the channels even despite a deeper their laying (Ivitsky, 1988). Therefore, there is no need for a theoretical analysis of the planned groundwater flow when taking into account the influence of the channels. It is enough to consider the operation of the channel subsystem independently of drains, and add up the draining effects due to the operation of both subsystems. Then, it is justified to use the first technique, i.e., the formulas from section 13 in the calculations of the hydraulic characteristics and, first of all, the WT. The same formulas are also suitable for the case of the combined operation of systematic drains and molehills, which are located perpendicular to the drains and contribute to the accelerated removal of water from the upper (arable) soil layer. And here a simplified approach is appropriate due to the proximity of molehills to the soil surface.

The procedure for calculating the effects of systematic drainage, taking into account the influence of a regular channel network, is as follows. First of all, we should choose the cross-section of the groundwater flow and the WT in it, for example, take a value H in the middle between the drains. The corresponding coefficient value $\mu_c(H)$ and time point tp are calculated according to (19). Here the WT will drop by the value ΔHB for a given period of time under the influence of channels. It is proposed to be determined with $\mu_c=\mu_c(H)$ by the formula similar to (13)

$$\Delta H_B = \Delta m_B \cdot \exp\left[-\frac{3k_e m_c t_p}{\mu_c(H)\left(B^2+6\Phi_B B\right)}\right], \tag{26}$$

where $\Delta m_B=m_0-m_B$, m_B is the water level in the channels, B is half the distance between adjacent channels, Φ_B is the hydrodynamic resistance of the channels. As a result, at the time t_p the WT decreases in the middle between the drains approximately to the mark $H-\Delta H_B$ at the point of time t_p.

It is recommended to calculate the adjusted value L at a given value B using the formula (8) with the adjusted drainage requirement $\tilde{S}_O$ and coefficient μ_c. For this, according to formula (18), for $t=t_O$ and $\mu_c=\mu_c(S_O)$ a correction Δ_{HB0} is found. Then the corrected values are substituted in (11).

The regular subsystem of molehills is usually located close to the surface of the earth and serves to accelerate the removal of excess water from the surface zone of the soil (Eggelsman, 1984). The drainage process proceeds in two stages in this situation. The WT decreases rapidly during the first stage due to the combined draining effect of both subsystems. It ends at the point time t_k of the decrease in the WT to the level of molehills mk. To determine the time t_k, the decrease in WT is monitored due to inflows to drainage and molehills that are not dependent on each other. Based on the drainage effect achieved at the required moment (in the center of each drainage area the WT at this point is at the mark m_k), the following equation has been derived

$$\Delta m \cdot \exp\left[-\frac{3k_e m_c t_k}{\mu_c(m_k)\left(L^2+6\Phi L\right)}\right] + \Delta m_k \cdot \exp\left[-\frac{3k_e m_c t_k}{\mu_c(m_k)\left(B_k^2+6\Phi_k B_k\right)}\right] = \Delta m_k\,, \tag{27}$$

where $\Delta m_k=m_0-m_k$, Φ_k is the hydrodynamic resistance of molehills, B_k is a half of the distance between them. Subsequent calculation of the WT dynamics (at $t>t_k$) is performed according to the dependencies from section 3 since only main subsystem operates. At the same time m_k is accepted as m_0.

Steady Groundwater Flow Against the Background of Drainage Taking Into Account Water Exchange with Adjacent Environments

A general and correct idea of the wide possibilities of drainage in the long-term management of the water resources of the reclaimed lands of the Polesie zone allows us to perform a theoretical analysis of its effect on the basis of the steady-state groundwater flow models. Obviously, the concepts of time and the calculation period lose their meaning with this approach. The total evaporation, head and infiltration supply of groundwater (Figure 4), together with surface runoff and other less significant factors take an active part in the formation of the resources. The position of the WT is of particular importance in modeling the groundwater flow against the background of drainage during the crops growing season, since it depends on: directly - water exchange of groundwater with a head horizon and the aeration zone, and indirectly - water exchange with the surface layer of the atmosphere. Drainage is able to provide reliable control over the stabilized water-physical conditions in soil in principle. Moreover, it usually is unable to completely smooth out the fluctuations of the hydraulic characteristics due to sharp changes in weather conditions during the mentioned period. Nevertheless, it is able to significantly reduce the amplitude of such oscillations and thereby keep the water-physical conditions in the soil in the permissible range.

At the same time, the use of drains leads to uneven drainage of the soil and the curvature of the free surface, which makes it difficult to regulate its water-physical conditions, in general, and the WT, in particular. As previously noted, the evaporation and head supply intensities are reliably described by the power-law functions on the WT mark (12). Thus, drainage can cause a pronounced unevenness of water exchange in the soil stratum, so that its rates near the drain and at a distance from it will noticeably differ (Poliakov, 2012). Disregarding this circumstance leads to an underestimation or overestimation of the drainage effect and serious errors in the calculations of the WT. It should be noted that a long-term favorable water-physical situation is achieved at the drainage systems only with a scientifically substantiated drain spacing (Wenyan et al., 1994). It is recommended that such substantiation be applied to drainage systems using the calculation dependences given in this section. They follow from an exact solution of the corresponding steady-state problem in linear approximation containing the equation of water balance with a variable free surface along the flow (Poliakov, 2016). The indicated solution has an integral form if the exponent *p* in expression (12) does not take integer values, which complicates engineering calculations. However, it is often assumed in calculation practice $p=1$ (Vasiliev et al., 1970) and other coefficients in (2) are adjusted if necessary. Just at this value *p*, the dependencies and equations are derived and given below that take into account the unevenness of evaporation and head supply over the drainage area with a stable water-physical conditions on it. In addition, approximate formulas are given for calculating the WT and distance *L*. They are simpler and suitable for any value *L*.

Therefore, the position of the WT is recommended to be determined at $p=1$ based on the relationship:

$$h(x,H)=\frac{1}{2a\phi}\left[(\phi-b)^2+a^2H^2+2abH\right]. \tag{28}$$

Here the curvature of the free surface between drains and the changes in the intensity of evaporation and head supply due to the curvature are taken into account,

$$\phi(x,H)=(aH+b)\exp\left(-\sqrt{\frac{a(x-L)^2}{k_e m_c}}\right),$$

the water exchange function $G(h)=a_h+b$, $a=\varepsilon_e+k_s/m_s$, $b=-(k_sH_s)/m_s-\varepsilon-\varepsilon_e h_k$, $m_c=(m_d+H)/2$ (in the case when thick aquifers are present, mc i_s acceptable to be equal md), H i*s* the level of the WT in the middle between drains, which should be calculated by selection from the equation:

$$\frac{km_c}{4\Phi^2}\left[h_0(H)-m_d\right]^2=ah_0^2(H)+2bh_0(H)-aH^2-2bH \tag{29}$$

In this case, the WT over the drain h_0 is expressed as follows:

$$h_0(H)=\frac{1}{2a\phi_0}\left[(\phi_0-b)^2+a^2H^2+2abH\right], \tag{30}$$

where $\phi_0(H)=(aH+b)\exp\left(-\sqrt{\frac{aL^2}{k_e m_c}}\right)$. It is sufficient to solve the equation (29) by selection to establish the position of the WT in the middle between drains. The head function over the drains is described here by the expression (30). Essentially, you need to find one of the four roots of an algebraic equation of the fourth degree. Obviously, only the root has physical meaning whose value is in the range from m_d to M (draining). The WT mark is simply determined in any other cross-section of the groundwater flow with the known H by the formula (20). Finally, it is easy to establish using (29) and (30) the distance L, at which the WT in the middle between drains will be stably maintained at the mark $H_O=M-S_O$. The value LI is exactly calculated within the framework of the accepted mathematical model by the formula

$$L=-\sqrt{\frac{k_e m_c}{a}}\ln\left[\frac{G(h_0(H_O))}{G(H_O)}-\sqrt{\frac{G^2(h_0(H_O))}{G^2(H_O)}-1}\right] \tag{31}$$

where

$$h_0=\frac{2\Phi^2}{4a\Phi^2-k_e m_c}\left[\sqrt{\left(2b+\frac{k_e m_c m_d}{2\Phi^2}\right)^2+4\left(a-\frac{km_c}{4\Phi^2}\right)\left(aH_O^2+2bH_O+\frac{k_e m_c m_d^2}{4\Phi^2}\right)}-2b-\frac{k_e m_c m_d}{2\Phi^2}\right]$$

The calculation formulas are greatly simplified if we assume that the value G is constant and, in addition, equal to its value in the middle between drains, i.e. $G(H)$. Then the WT will be with a certain error at $x=L$ at the actually the lowest (or greatest) intensity of the total water exchange

$$H=\frac{2k_e m_c m_d-b\left(L^2+4L\Phi\right)}{2k_e m_c+a\left(L^2+4\Phi L\right)} \tag{32}$$

The distance L is simply determined at a given drainage requirement H_O as follows

$$L=\sqrt{4\Phi^2-\frac{2k_e m_c(H_O-m_a)}{G(H_O)}}-2\Phi \tag{33}$$

Unsteady Inflow to a Single Self-Flowing Well

Vertical drainage is capable of strongly influencing the flow in aquifers with relatively low capital and operating costs (Vasiliev et al., 1970; Oleinik, 1978). In view of the considerable resistance that a fluid moving in a porous medium experiences, the perturbation of the water-physical conditions due to the operation of a vertical drain, as well as a horizontal drain, is localized near it. It is possible to significantly intensify the inflow because of pumping with a large discharge, the creation of a small head in the well filters. However, at the same time, the unevenness of groundwater reserves distribution in the drainable territory increases in a similar way due to the sharp curvature of the free surface. It is preferable to use self-flowing wells to regulate the water mode of agricultural lands. Really, the decrease in the WT over the area occurs more evenly due to the constancy of the head within the wells. A single well and a large group of regularly located wells (systematic drainage) are described and the calculation dependencies are given in dimensionless form here and in the next section (Poliakov, 2014). The imperfection of the wells is taken into account using the general hydrodynamic resistance. The WT (piezometric surface), well discharge, and besides extent of the zone of influence were the object of the calculations in the first technique.

Therefore, it is proposed to calculate the relative decrease in the WT from the initial mark m_0 and the decrease in the well discharge over time depending on the relative radius of the zone of influence $\bar{R}$ Figure 8 according to the following formulas

$$\bar{S}\left(\bar{r},\bar{R}\right)=\frac{\bar{r}-\bar{R}-\bar{R}\ln\dfrac{\bar{r}}{\bar{R}}}{\bar{R}\ln\bar{R}+\left(1-\Phi_c\right)\left(1-\bar{R}\right)}, \tag{33}$$

$$\bar{q}\left(\bar{R}\right)=\frac{\bar{R}-1}{\bar{R}\ln\bar{R}+\left(1-\Phi_c\right)\left(1-\bar{R}\right)} \tag{34}$$

In fact, $\bar{R}$ presents in expressions (33), (34) as a parameter that formally changes in time from 1 to ∞, since the zone of influence of the well can, in principle, expand infinitely. The corresponding regularity $\bar{R}\left(\bar{t}\right)$ is deduced in the form of an inverse integral function, namely:

$$\bar{t}=\frac{1}{12}\int_1^{\bar{R}}\frac{P_1(\xi)+\Phi_c P_2(\xi)}{\xi\left[\xi\ln\xi-\left(1-\Phi_c\right)\left(\xi-1\right)\right]}d\xi, \tag{35}$$

where $P_1(\xi)=2\left(\xi^3-1\right)\ln\xi-3\left(\xi^2-1\right),\qquad P_2(\xi)=2\xi^3-3\xi^2-6\xi+7,$

and the relative variables and parameters in (33) - (35) are introduced as follows:

$$\bar{S}=S/S_d=\left(m_0-h\right)/S_d,\ \bar{r}=r/R_d,\ \bar{R}=R/R_d,\ \bar{t}=\left(k_e m_c t\right)/(\mu_c R_d^2),\ \bar{q}=q/\left(2\pi k_e m_c S_d\right),$$

S_d is the constant decrease in head within the well with radius R_d, comments on the coefficient μc are given in the second section. To determine the specific position of the WT and the well discharge, you should first establish the value $\bar{R}$, and then use Mathcad or other similar software packages to calculate the corresponding time using formula (35). At the same time, the WT and the discharge are found by formulas (33), (34).

Figure 8. The calculation scheme of a single imperfect self-flowing well

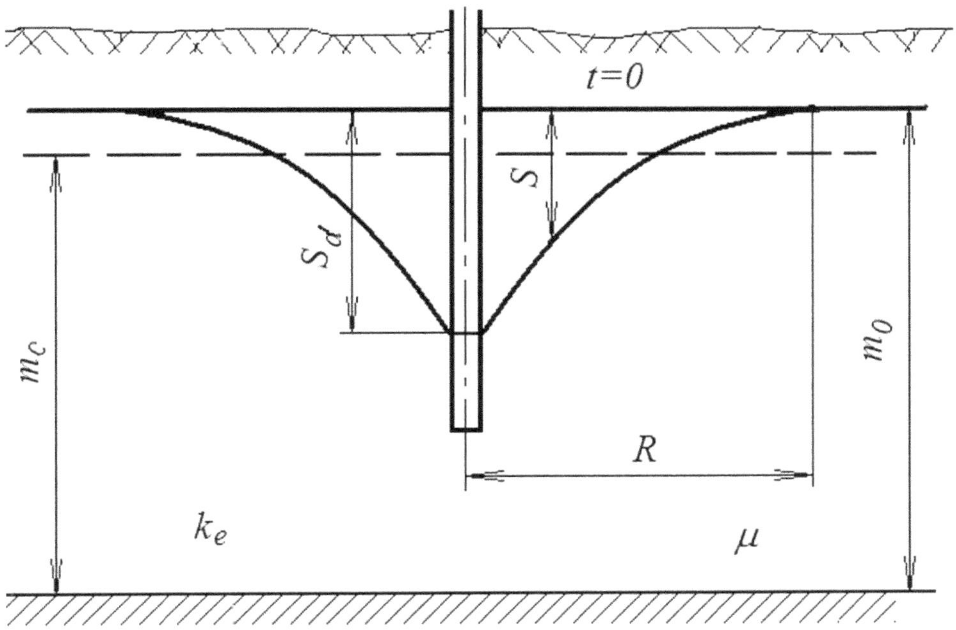

Unsteady Flow to The Regular System of Self-Flowing Wells

Regular systems of the self-flowing wells actively influence the water-physical conditions of the agricultural lands Figure 9. The necessary effect of soil drainage can be achieved, as a result of the operation of the self-flowing wells due to the rational choice at the design stage of the drain spacing and during operation – the head within the well. The last technique presented below primarily allows you to predict the movement of the free surface and the decrease in the discharge over time. At the same time, it allows to theoretically substantiate the distance between the wells and the head in them needed to fulfill the specified requirement S_0 and drainage time t_0.

Suppose that during the calculation period, water with intensity ε flows to the WT. First of all, a fixed influence zone of an arbitrary internal well is highlighted. If the wells are located regularly with a distance 2L between adjacent ones, then this zone is a square in plan. It is conditionally replaced by a circle with the same area, the radius of which Re is associated with L by the ratio

$$R_e = 2L / \sqrt{\pi} \quad (36)$$

and appears in all the calculation dependencies.

The first of them allows you to accurately determine the movement over time of a free groundwater surface using the following formula

$$\bar{S}(\bar{r},\bar{t})=1-\frac{\bar{\varepsilon}}{2}\theta(\bar{r})-\left[1-\frac{\bar{\varepsilon}}{2}\theta(\bar{R}_e)\right]\frac{\bar{R}_e\ln\bar{r}-\bar{r}+\Phi_c\bar{R}_e-\Phi_c+1}{\bar{R}_e\ln\bar{R}_e-\bar{R}_e+\Phi_c\bar{R}_e}e^{-\frac{\bar{t}}{\psi}}, \tag{37}$$

where

$$\bar{\varepsilon}=\frac{R_d^2\varepsilon}{k_e m_c S_d},\ \theta(\bar{r})=\bar{R}_e^2\ln\bar{r}-\frac{\bar{r}^2}{2}+\Phi_c\bar{R}_e^2,\ \bar{R}_e=\frac{R_e}{R_d},\ \psi=\frac{\bar{R}_e^2}{2}\left(\ln\bar{R}_e+\Phi_c-\frac{7}{6}\right).$$

The decrease in the relative well discharge with time is described by the dependence

$$\bar{Q}(\bar{t})=\frac{Q(t)}{2\pi k_e m_c S_d}=\frac{\bar{\varepsilon}\bar{R}_e^2}{2}+\left[1-\frac{\bar{\varepsilon}}{2}\theta(\bar{R}_e)\right]\frac{e^{-\frac{\bar{t}}{\psi}}}{\ln\bar{R}_e+\Phi_c-1} \tag{38}$$

Figure 9. The calculation scheme of a regular system of imperfect self-flowing wells

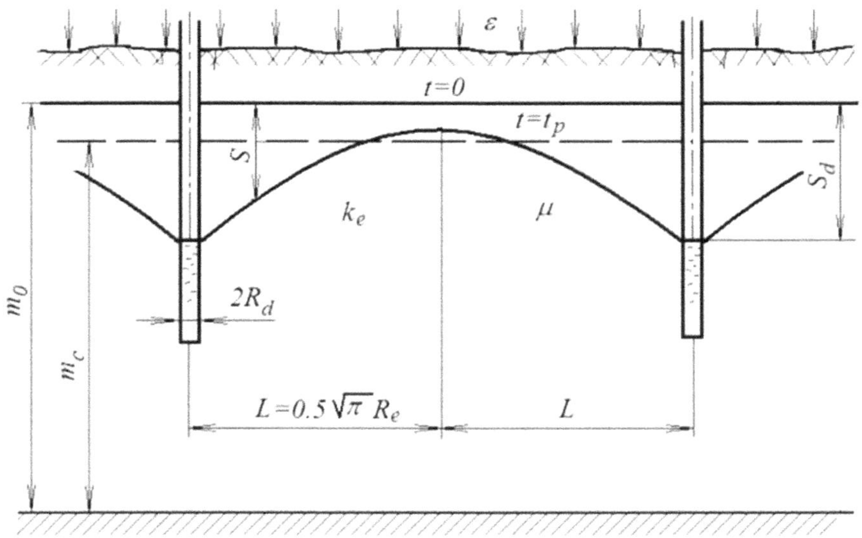

An equation is derived with respect to the radius $\bar{R}_O$ from (37). The solution of the equation allows to determine its rational value by selection for given value $\bar{t}_O,\bar{S}_O$, namely:

$$\bar{S}_p = \bar{S}_O - \frac{M - m_O}{S_d} = \left[1 - \frac{\bar{\varepsilon}}{2}\theta\left(\bar{R}_O\right)\right]\left[1 - e^{-\frac{\bar{t}_O}{\psi\left(\bar{R}_O\right)}}\right]. \tag{39}$$

If infiltration supply can be neglected, then formulas and equation (37) - (39) will be simplified to this form

$$\bar{S}\left(\bar{r},\bar{t}\right) = 1 - \frac{\bar{R}_e \ln \bar{r} - \bar{r} + \Phi_c \bar{R}_e - \Phi_c + 1}{\bar{R}_e \ln \bar{R}_e - \bar{R}_e + \Phi_c \bar{R}_e} e^{-\frac{\bar{t}}{\psi}}, \tag{40}$$

$$\bar{Q}\left(\bar{t}\right) = \frac{e^{-\frac{\bar{t}}{\psi}}}{\ln \bar{R}_e + \Phi_c - 1}, \tag{41}$$

$$\frac{\bar{R}_O^2}{2}\left(\ln \bar{R}_O - \frac{7}{6} + \Phi_A\right) = -\frac{\bar{t}_O}{\ln\left(1 - \bar{S}_p\right)} \tag{42}$$

It is expedient to continue theoretical researches of drainage operation on the reclaimed lands taking into account serious changes in the climate conditions. In order to provide them with initial information, it is necessary to collect and generalize field data on water exchange between atmosphere and soil, in particular, on strengthening of physical and biological evaporation, intensification of precipitation during the vegetation period. Special attention will be paid to surface runoff formation, the distribution of the storm precipitation between it and infiltration, deterioration of the permeability of the near-surface layer. It is supposed to develop special measures for accumulation of surface runoff with its subsequent use in subsoil wetting.

CONCLUSION

Therefore, a number of engineering techniques have been developed for a comprehensive theoretical analysis of the development of the water mode under the influence of two types of drainage применительно in the specific natural conditions of Polesie. They make it possible to perform a reliable projection of spatio-temporal changes in the most important hydraulic characteristics, first of all, the WT and the discharge, to substantiate the design parameters, namely, the depth of the drainage, and, also the drain spacing for the regular drainage systems based on requirements to optimize the water-physical conditions in the soil in a given point of time. The techniques for calculating the operation and the key design parameter of the systematic imperfect horizontal drainage are of fundamental importance. Really, they are intended to be applied to the time intervals that are critical for the drainage operation. The basic techniques are supplemented by the techniques that allow us to analyze the dynamics of the WT in the areas with a regulated groundwater flow and special, more complicated from a formal point of view conditions.

A system of self-flowing wells can be an alternative to horizontal drainage when regulating the water-physical conditions on the agricultural lands. To identify the advantages and disadvantages of their use during the engineering process, as well as to operate them with the greatest return, the last two techniques make it possible. In conclusion, it should be emphasized that all of the presented techniques, except one, are based on the approximate solutions of one-dimensional (axisymmetric) problems of transient groundwater flow against the background of drainage. Therefore, much attention was paid in the conducted theoretical studies to assessing the accuracy of these solutions. It was found that all the dependences and equations derived in the framework of the linear hydraulic theory and recommended for engineering calculations allow the desired characteristics and parameters to be determined with a minimum (of the order of a few percent) error.

The linearization of the initial groundwater flow models can lead to serious errors in the calculations if the limits of the WT change are comparable with the thickness of the aquifer and the relationship between the water exchange intensity between the saturated and unsaturated zones of the soil and the position and velocity of the WT is clearly nonlinear. Then, it is impractical to use the calculation formalisms presented in this chapter and important to rely on the nonlinear theory of drainage, in particular, attracting new developments in this direction (Poliakov 1, 2016; Poliakov 2, 2016).

REFERENCES

Alekseev, V. S., & Kommunar, G. M. (1974). Colmatage of filters and filter zones of water wells. *Water Resources*, *5*, 170–177.

Averyanov, S. F. (1982). *Outflow from canals and its influence on the mode of groundwater*. Kolos.

Bashkis, K. K. (1974). *On the issue of analysis of calculation methods of horizontal drainage*. Transactions of Lit. NIIGiM. Issues of land drainage using pottery drainage. -. *Jelgava*, *IX*, 5–24.

Dmitriev, A. F., Bezusyak, A. V., & Khlapuk, N. N. (1992). *Improvement of drainage and wetting systems*. Svit.

Ivitsky, A. I. (1988). *Fundamentals of design and calculation of drainage and drainage-humidification systems*. Science and Technology.

Maslov, B. S., Stankevich, V. S., & Chernenok, V. Ya. (1981). *Drainage and humidification systems*. Kolos.

Murashko, A. I. (1982). *Agricultural drainage in the humid zone. Methods of calculations of irrigation and drainage systems*. Kolos.

Nasedkin, I. (1981). Nutrition of waterlogged lands of Polesie with pressure waters and its study. *Reclamation and water management. 53*, 25–29.

Oleinik, A. (1978). Hydrodynamic calculations of vertical drainage. - Kjiv: Science Dumka, 202 p.

Oleinik, A. (1981). Geohydrodynamics of drainage. Science Dumka.

Pinck-Pfeiffer, S., Ragusa, S., Sztajnbok, P., & Vandevelde, I. (2000). Interrelationships between biological, chemical and physical processes as an analog to clogging in aquifer storage and recovery (ASP) wells. *Water Research*, *34*(7), 2110–2118. doi:10.1016/S0043-1354(99)00356-5

Pivovar, N. G., & Bugai, V. L. Friedrichson [et al.]. (2000). Drainage with fiber filters to protect territories from flooding. Institute of Hydromechanics.

Poliakov, V. L. (2012). Calculation of the steady-state action of the drain-humidifier system under conditions of intensive evaporation. *Problems of Water Supply, Water and Hydraulics. -. Kjiv: KNUBA.*

Poliakov, V. L. (2014). Hydrodynamic deformations in drained soils: theory and applications. Kjiv: Agrar Media Group.

Poliakov, V. L. (2016). On the modeling of controlled water regime in the soil column. *Reports of NAS of Ukraine*, *9*, 51–60.

Poliakov, V.L. (2016). Calculation of the action of a single imperfect horizontal drain in an unlimited headless reservoir taking into account the aeration zone. *Hydrotechnical construction, 5*, 21–31.

Poliakov, V. L. (2016). *Calculation of the established pressure-free groundwater flow for systematic drainage during water exchange with adjacent media.* Problems of Water Supply, Water and Hydraulics - Kjiv. *KNUBA*, *27*, 291–301.

Poliakov, V. L., & Kalugin, Yu. I. (2016). Mathematical modeling of water exchange between saturated and non-saturated zones of drained soil. *Reports of NAS of Ukraine*, *9*, 43–49.

Shih, S. F. (1983). Soil surface evaporation and water table depth. *Journal of Irrigation and Drainage Engineering*, *109*(4), 366–376. doi:10.1061/(ASCE)0733-9437(1983)109:4(366)

Skaggs, R. W. (1978). Effects of drain tube openings on water – table drawdown. *Journal of the Irrigation and Drainage Division*, *104*(1), 13–21. doi:10.1061/JRCEA4.0001183

Tishenko, O. (2014). Regulation of the water regime of soils on the two-way systems in the Polesie of Ukraine. *Water Statehood of Ukraine*, *6*(114), 35–39.

Tugay, A. M., Oliynik, O. Ya., & Tugay, Ya. A. (2001). Productivity of water inlets in the minds of colmatage. Kharkiv: KHNAMG.

Van der Pboeg, R. R., Kirkham, M. B., & Marquardt, M. (1999). The Golding equation for soil drainage: Its origin, evolution and use. *Soil Science Society of America Journal*, *63*(1), 33–39. doi:10.2136s saj1999.03615995006300010006x

Van Genuchten, M. Th. (1980). A closed-form equation for predicting the hydraulic conductivity of unsaturated soils. *Soil Science Society of America Journal*, *44*(5), 892–898. doi:10.2136ssaj1980.0361 5995004400050002x

Vasiliev, S. V., Verigin, N. V., & Glaser, B. A. (1970). *Methods of calculations of irrigation and drainage systems.* Kolos. [et al.]

Vedernikov, V. V. (1939). *The theory of groundwater flow and its application in the field of irrigation and drainage*. Gos-stroizdat.

Wenyan, W., Bing, S., & Zhilu, L. (1994). Drain-spacing calculation considering influence of evaporation. *Journal of Irrigation and Drainage Engineering*, *120*(3), 563–572. doi:10.1061/(ASCE)0733-9437(1994)120:3(563)

Willardson, L. S., & Walker, R. E. (1979). Synthetic drain envelope-soil interactions. *Journal of the Irrigation and Drainage Division*, *105*(4), 367–373. doi:10.1061/JRCEA4.0001274

Yangol, A. M. (1970). *Two-side regulation of moisture during drainage*. - Moscow: Kolos, 136p. Lennoz-Gratin, Ch. (1989). Effect of envelopes on flow pattern near drain pipe. *Journal of Irrigation and Drainage Engineering*, *115*(4), 626–641.

Chapter 13
Improvement of the Calculation of the Closed Collector–Drainage Network of Drainage Systems

Liubov Volk
https://orcid.org/0000-0003-1033-6715
National University of Water and Environmental Engineering, Ukraine

Maksym Haponiuk
National University of Water and Environmental Engineering, Ukraine

ABSTRACT

The necessity and scientific and methodological approaches to improving the calculation of a closed collector-drainage network of drainage systems are considered. Based on the application of the system methodology, the structure of the hierarchical and hydraulic connection between the water regime of the field and the operation mode of the closed collector-drainage network has been established. On the basis of the theoretical and experimental studies performed, relatively new scientific provisions have been proposed. This will allow, based on the application of the obtained universal equations, to construct a distribution profile of the total turbulent kinematic viscosity and the average flow velocity in the collector pipe. It becomes possible to evaluate the efficiency of the flow movement both in the constituent elements (collector pipes) and in the closed collector-drainage network as a whole. And it also becomes possible to improve the methods of designing and calculating their technological and design parameters.

BACKGROUND

Climatic conditions are changing in the world, including changes in the amount and intensity of precipitation, soil moisture and rising air temperatures, total evaporation, and water demand of crops. They determine the need to ensure the productivity of drained lands through the application of adaptive

DOI: 10.4018/978-1-6684-8248-3.ch013

measures for the regulation and accumulation of moisture in the soil and within the reclamation system, as well as the transition from periodic to regular soil moisture (Watkiss 2015, Labędzki et al. 2021).

At the same time, the technical condition of drainage systems built 40-50 years ago has deteriorated due to wear and tear and failure to perform the necessary set of operational measures. This has led to the deformation and siltation of the closed collector-drainage network, as the main regulating element of such systems that operate in the regime of drainage and soil moisture. As a result, their parameters deviated from the design. The regime of operation of the closed collector-drainage network was also disrupted and its capacity was reduced. Therefore, the overall efficiency of such systems and the productivity of drained lands have decreased by 25… 50% of the design system (Rokochinskiy et al. 2019a, Kovalenko et al. 2020).

Therefore, the issue of changing approaches to the creation and operation of water reclamation facilities on drained lands is extremely important. The methodologies for substantiation in projects of construction and reconstruction of drained systems, their optimal constructive decisions (type, design, parameters of systems, and components of their technical elements) should change. First of all, this is relevant for a closed collector-drainage network. It is a regulating element of the drainage system and can work in the regime of drainage and in the regime of moisture.

At the same time, when designing and operating drainage systems at all stages of time management decisions (1-project, 2-planned operation, 3-operational facility management) (Hebbink et al. 1999, Rokochinskiy et al. 2020), it is extremely important to correctly (objectively) determine the parameters system and its components technical elements. These issues concern first of all the regulating closed collector-drainage network, as well as the leading network of canals, regulating and supporting hydraulic structures, pumping stations, etc. They determine the cost of the system, its overall technological, economic, and environmental efficiency.

In the 70-80s of the last century, the issues of automation of water regulation management and production processes of water distribution on drainage and humidification systems were solved by means of hydroautomatics to increase the efficiency of drainage systems with bilateral regulation of the water regime of lands (Kamra et al. 2019, Rokochinskiy et al. 2019b, Kovalenko et al. 2021).

These developments on the methodology of creation and operation of drainage systems were of a high scientific level. They were included in industry standards and applied in practice. But, as practice and experience have shown (Kamra et al. 2019, Kovalenko et al. 2021, Seyedmohammadi et al. 2016, Rokochinskiy et al. 2021), unfortunately, the required level of economic and environmental efficiency of land reclamation has not been achieved. Also, the reason for this was the lack of perfection of the methodology of design and calculation of such objects and its non-compliance with the requirements. This methodology did not take into account the relationship between the mode of operation of the closed collector-drainage network and the water regime of drained lands.

In this regard, the purpose of our research is to develop new approaches to improving methods for calculating a closed collector-drainage network of drainage systems based on the justification of the relationship and taking into account the impact of its efficiency on the efficiency of water management of drained lands.

MATERIAL AND METHODS

The analysis and generalization of existing researches and generally accepted methods on the substantiation of type, design, and parameters of the closed collector-drainage network are executed. At the same time, the closed collector-drainage network is the main regulatory element of drainage systems during drainage and subsoil wetting.

A systematic approach and systematic analysis were used to determine the structural relationship between the regime of operation of the closed collector-drainage network and the water regime of the reclaimed field. The drainage system is studied as a complex natural and technical system.

In performing theoretical and experimental research, we used methods of mathematical modeling of the hydrodynamic structure of turbulent flow in pressure pipes using generally accepted Navier-Stokes differential equations (Bar-mei 2022, Malecha 2022).

To confirm the adequacy of the obtained analytical models, generally accepted methods of statistical processing of the results of experiments and generally accepted experimental data of such scientists as Nikuradze I., Shevelov F.O., Altshul A.D. were used (Peiqing 2021). These scientists studied the regularities of the distribution of the averaged velocities of turbulent flows in pressure pipes.

Based on the application of the methodology of the system approach and system analysis in the creation and operation of water management and reclamation facilities, drainage systems are complex natural and technical ecological and economic systems. These systems have a structural relationship of the form *effect*⇔regime⇔technology⇔design. This determines the need for more detailed studies of the dependences of water processes in the main elements and in the system.

In the development of such an approach, it is advisable to identify and consider the subsystem *soil*⇔groundwater level regime⇔closed collector and drainage network. The elements of such a system have a structural hierarchical and hydraulic relationship. In this case, the peculiarity of drainage systems with two-way control is that the processes of water flow in the elements when working in the drainage regime and in the regime of subsoil humidification are similar and opposite.

The connection between the groundwater level regime and the closed collector-drainage network in such a subsystem is provided by the transformation of the filtration flow into the structure of the set of water turbulent flows in hierarchically and hydraulically connected pressure drainage pipes. These drainage pipes are components of a closed collector-drainage network with different conditions for the formation of pressure gradients and velocity. These networks operate in drainage regime and in soil humidification regime. This can be reflected by the corresponding subsystem of the form - *soil flow regime*⇔flow regime in a closed collector-drainage network⇔flow regime in a collector pipe as its main regulating element of a closed collector-drainage network.

According to the general theory of water flow in a pressure pipe, the efficiency of the flow regime in a closed collector-drainage network is determined by the parameters of the hydrodynamic structure of the velocity distribution in the cross-section of the flow in the collector-drainage pipe. This pipe is the main element of the network and determines the efficiency of the closed collector-drainage network.

Based on the analysis and generalization of research, a structural and logical scheme of hierarchical and hydraulic interconnection and the influence of the regime of operation of the closed collector-drainage network on the groundwater level of the reclaimed field are developed. The efficiency of the network is determined by the hydrodynamic structure of the velocity distribution in the cross-section of the flow in the drainage pipe Figure. 1.

However, in the case of hydraulic engineering problems such as the design and parameters of pressure pipes in the presence of imperfect theory, difficulties arise due to the inability to determine the distribution of total turbulent kinematic viscosity, average pipe velocities, tangential stresses, and angular velocity. It is also impossible to determine sufficiently scientifically sound technological and design parameters of pressure pipes, including drainage pipes. This determines the objective need for further development of a complete theory of turbulent flow.

Accordingly, many scientists worked on the solution of this problem. They proposed semi-empirical, in their view universal, theories of turbulent flows, in which the unknown relationships between turbulent stresses and averaged strain velocities are concretized on the basis of relevant hypotheses, physical conjectures, theories of dimensions obtained by generalizing both own and materials of other researchers.

But the semi-empirical theories developed and proposed by them cannot be universal, as they are limited by the range of conditions for their receipt and application, which necessitated the adoption of appropriate assumptions.

According to these semi-empirical theories, models based on power and logarithmic profiles have been proposed. The logarithmic profile has become widespread, despite the fact that this dependence does not correspond to the boundary conditions on the axis and on the inner surface of the pipes. In order to ensure the boundary conditions on the inner surface of the pipe, scientists have developed two-layer and three-layer models, but they also do not meet the boundary conditions along the flow axis of the pipe (Rubenstein *et al.* 2021).

Due to the fact that the existing dependences of the distribution of averaged velocities in pipes do not provide boundary conditions near the pipe wall (power equations), and the proposed logarithmic equations do not provide boundary conditions not only near the wall but also on the pipe axis, there is an objective the need for further development and deepening of this theory.

Figure 2a - 2d show graphs of the dependence of the friction factor of the pipes λ and the Reynolds number R*e*. These graphs show the flow regimes in the pipes. The experimental points shown in the graphs were obtained on the basis of unique experimental studies by I. Nikaradze and F. Shevelyov (RAffel e*t al.* 2007, GIroud e*t al.* 2000).

In a turbulent flow, the friction factor depends not only on the Reynolds number but also on the pipe inner surface. Therefore, there are many dependencies that are found in the reference literature. All equations are recommended only for specific conditions (material, type of surface roughness, the height of roughness, the distance between roughnesses, their shape and location on the surface). Adequate dependence of the change in the friction factor on all the main factors in force can only be determined experimentally (Farahnak-Ghazani *et al.* 2021).

We have hypothesized that the Navier-Stokes differential equation can be used to calculate the averaged velocities also for the turbulent regime. To do this, it is necessary to replace the molecular kinematic viscosity v, in the Navier-Stokes equation, with a total turbulent kinematic viscosity v_{tot}, which includes kinematic viscosity on the inner surface of the pipe v_s and the kinematic turbulent viscosity v_t of mass transfer from one layer to another layer, recommended by J.V. Boussinesq (Rodriguez *et al.* 2022, Speight 2017, Gooch 2011, Volk *et al.* 2021). Then equation will have the form

$$u_{xt} = \lambda \operatorname{Re}^2 \frac{v^2}{64 v_{tot} r_0^3}\left(r_0^2 - r^2\right) \tag{1}$$

where ν_{tot} - the total turbulent kinematic viscosity; r_0 - the pipe radius; r - the coordinate along the radius of the pipe; ν - the molecular kinematic viscosity; λ - the friction factor of the pipes; Re - the Reynolds number.

Figure 1. Structural and logical scheme of hierarchical and hydraulic influence of the closed collector-drainage network operation regime on the groundwater level of the reclaimed field, the efficiency of which is determined by the hydrodynamic structure of velocity distribution in the flow cross-section in the collector pipe as an element of the control network

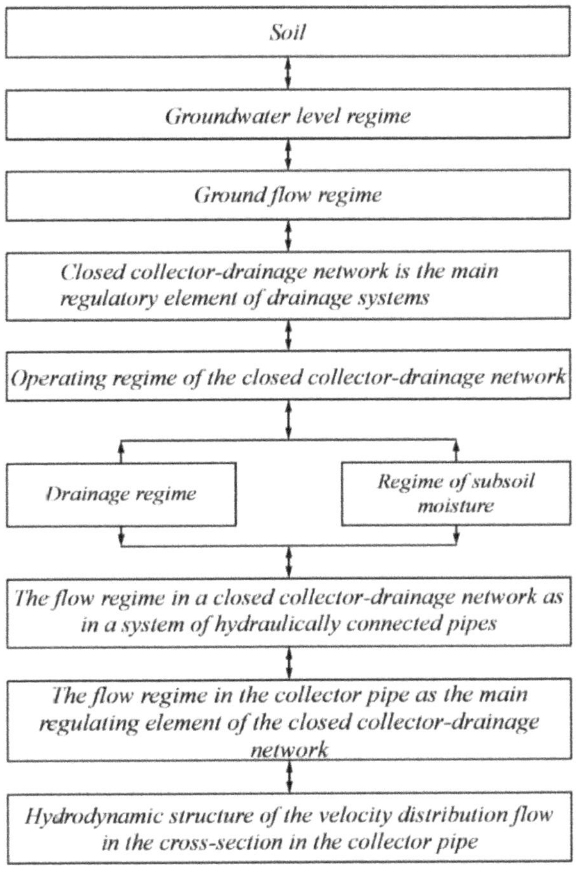

Turbulent kinematic viscosity between flow layers is also not a constant physical quantity, and its determination requires the necessary research. Therefore, it is advisable, given the canonical equation of the ellipse, to assume that the sum of the relative kinematic turbulent viscosity at the point of flow and the relative distance to this point in appropriate power is equal one.

$$\left(\frac{\nu_t}{\nu_{tmax}}\right)^{1/m}+\left(\frac{r}{r_0}\right)^{n}=1 \tag{2}$$

Figure 2. Graphs of the dependence of the pipe friction factor on the Reynolds number: a – pipes with uniform granular roughness (according to I.Nikuradze experiments); b – smooth pipes (according to I.Nikuradze experiments); c – new steel pipes (according to F.O. Shevelyov experiments); d – new cast-iron pipes (according to F.O. Shevelyov experiments); 1 – laminar flow ; 2 – transition region; 3 – region of hydraulically smooth turbulence (H. Blasius); 4 – almost quadratic resistance region; 5 – quadratic resistance region

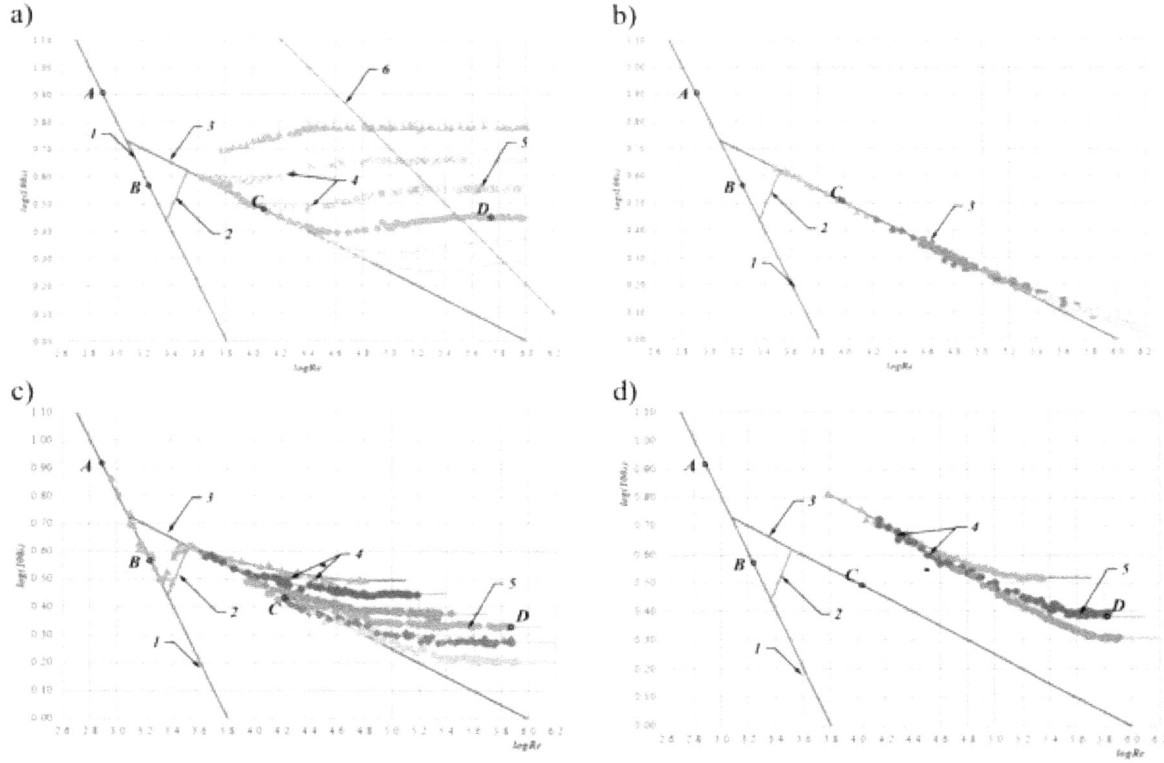

Given that the flow interacts with the inner surface of the pipe, it is necessary toconsider the kinematic viscosity on the inner surface of the pipe $v_s = \lambda v$ in equation (2). Then equation (2) has the form

$$\left(\frac{v_{tot}}{v_{tmax}}\right)^{1/m} + \left(\frac{r}{r_0}\right)^{n} = 1 + \left(\frac{v_s}{v_{tmax}}\right)^{1/m} \tag{3}$$

The equation of total kinematic turbulent viscosity will take the form

$$v_{tot} = \frac{1}{r_0^{nm}}\left(v_{tmax}^{1/m}\left(r_0^n - r^n\right) + v_s^{1/m} r_0^n\right)^m \tag{4}$$

where unknown parameters are determined by the system of equations

$$\left.\begin{aligned} \nu_{tmax} &= k\nu\lambda Re \\ k &= algRelg(100\lambda)+b \\ m &= clgRelg(100\lambda)+d \\ n &= 2/m \\ \nu_s &= \lambda\nu \end{aligned}\right\} \tag{5}$$

where ν_{tmax} – the maximum value of the kinematic turbulent viscosity at the pipe axis; *k*, *m* i *n* - the constant parameters of some flow regime; (*lgRe*; *lg*(100*λ*)) - coordinates on the Nikuradze graph, which depend on the region; a,*b,c,d*- constant coefficients based on the experiments data.

If we take into account in equation (4) equation (5) we obtain

$$\nu_{tot} = \frac{\nu\lambda}{r_0^{nm}}\left(\left(kRe()^{1/m}\left(r_0^n - r^n\right)_0^n\right)()^m\right). \tag{6}$$

The original design parameters *k* and *m*, determined by equation (5), include unknown coefficients. They obtained from experimental points using the least-squares method with a probability of 95% and, accordingly, and are equal *a*= -0.000675; *b*=0.0273; *c*= -0.0195; *d*=0.915.

RESULTS AND DISCUSSION

To analyze theoretical studies based on the theory of planning experiments, a plan for the implementation of experimental studies on a laboratory hydraulic installation was created. As a result, distributions of averaged water flow velocities in various types of pipes were obtained. Statistical processing of the data was carried out and the variance and main deviations were determined. The analysis of the conducted experimental and statistical studies confirmed the correctness of the analytical models with a confidence level of 95%.

On the basis of experimental data, graphs of parameters dependence on factors are constructed. Figure 3 shows the graph of the parameter *k* on the multiplication of coordinates *lgRe*, lg(100*λ*).

The graph shows that the parameter *k* is linearly dependent on the multiplication of coordinates *lgRe*, lg(100*λ*) that determine the points corresponding to the hydraulic flow regime on the Nikuradze graph. As multiplication increases, the parameter value k decreases linearly.

The parameter is determined by the equation

$$k = 0.0273 - 0.000675\ lgRelg(100\lambda) \tag{7}$$

Figure 4 shows the graphs of the parameter dependence *m* on the multiplication of coordinates *lgRe*, lg(100*λ*).

Figure 3. Graph of dependence of parameter k on the multiplication of coordinates (lgRe, lg(100λ)): 1 - value of parameter k on the basis of experiments; 2 - by equation (7)

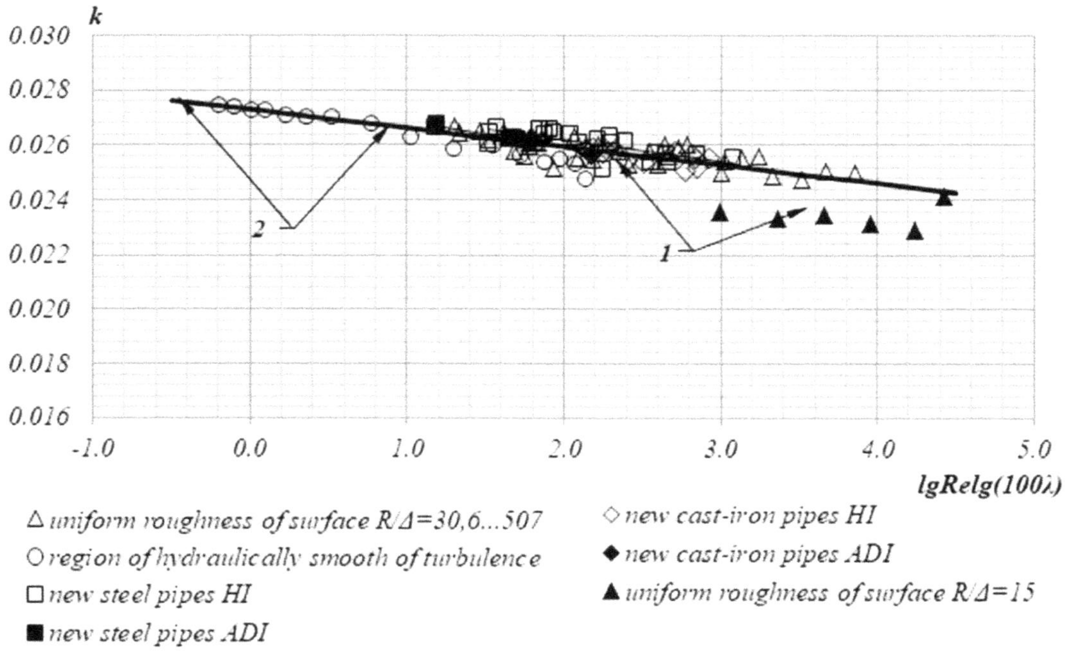

Figure 4. Graph of the parameter dependence im on the multiplication of coordinates (lgRe, lg(100λ)): 1 – value of parameter m on the basis of experiments; 2 - by equation (8)

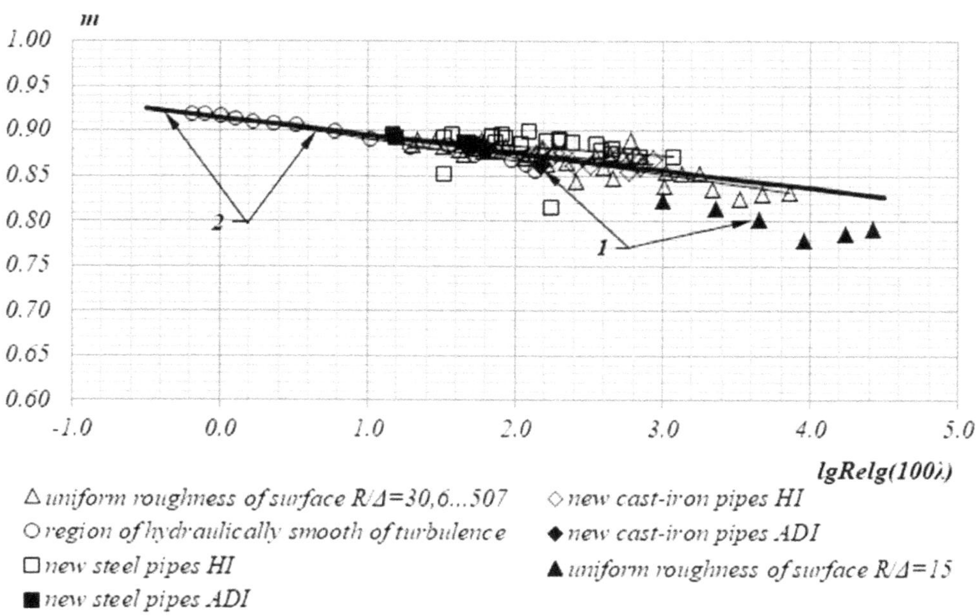

The graph shows that the parameter m is linearly dependent on the multiplication of coordinates $lgRe$, $\lg(100\lambda)$. As the multiplication of coordinates increases, the parameter m decreases. The parameter m is determined by the equation

$$m= 0.915 - 0.0195\, lgRelg(100\lambda) \tag{8}$$

The unknown coefficients was obtained by the method of least squares at the accepted values of $n=2/m$ and $v_s=\lambda v$. The totality of experimental points, which was adopted in determining the constant coefficients, included experiments in pipes of different diameter, different roughness, and pipe material, type of liquid and its temperature, measuring equipment, as well as hydraulic and aerodynamic installations. Such conditions are associated not only with random errors but also with systematic errors. Despite this, the experimental values of the parameters k and m differ from their regression equations by no more than 2-3%. Experimental pipe points with a smoothness of 15.6 and a diameter of 2.42 cm deviate from the equations by 7%.

For example, Figure 5 shows the experimental points and graphs of the relative total kinematic turbulent viscosity v_{tot}/v on $lgRe$, $\lg(100\lambda)$ for smooth pipes with a diameter of 10 cm, determined by the equation

$$\frac{v_{tot}}{v} = \frac{1}{r_0^{nm}}\left(\left(\frac{v_{tmax}}{v}\right)^{1/m}\left(r_0^n - r^n\right) + \left(\frac{v_s}{v}\right)^{1/m} r_0^n\right)^m \tag{9}$$

Figure 5. Graphs of relative total kinematic turbulent viscosity distribution v_{tot}/v by the Reynolds number and the friction factor in smooth pipes (10 cm in diameter): 1 – value v_{tot}/v based on experiments is indicated by markers; 2 - by equation (9)

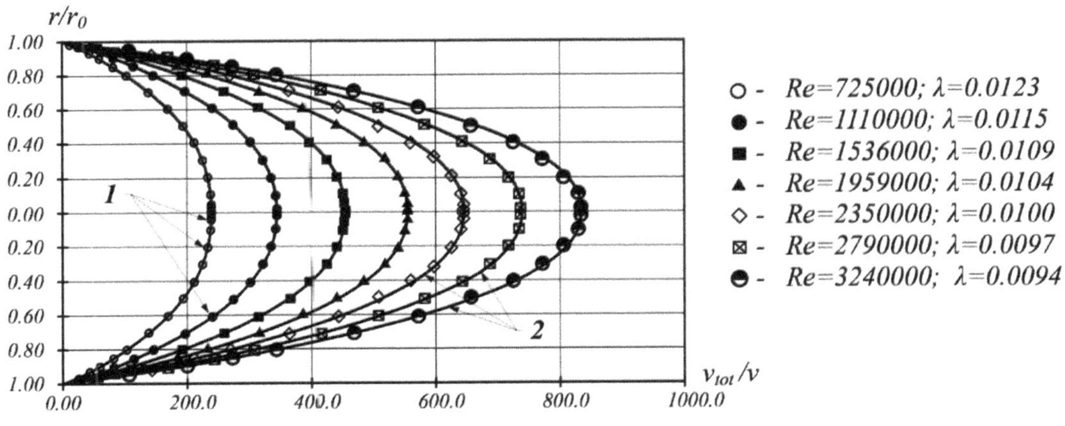

If we take into account the equation of the total turbulent kinematic viscosity (9), the dependence of the distribution of the averaged velocity of the turbulent flow will be:

$$u_{xt} = \frac{\nu \, \mathrm{Re}^2 \left(r_0^2 - r^2\right)}{64\left(\left(kRe()^{1/m}\left(r_0^n - r^n\right)_0^n\right)()^m{}_0\right)} \tag{10}$$

Figure 6 shows the profiles of the average velocities of the fluid for laminar flow, for the region of hydraulically smooth turbulence, almost quadratic resistance region, and quadratic resistance region of turbulent flow.

These dependencies, in contrast to existing semi-empirical theories, make it possible to construct an averaged velocity distribution profile and compare the change in averages of averaged fluid velocities for different regimes and their regions by revealing the complete hydrodynamic flow structure for all turbulent regions Fig. 6. For example, Fig. 7. shows the averaged velocity profiles for a pipe with a diameter of 10 cm.

Figure 6. Distribution of the averaged velocity of the flow: 1 – laminar flow; 2 – the region of hydraulically smooth turbulence; 3 – almost quadratic resistance region and quadratic resistance region of turbulent flow

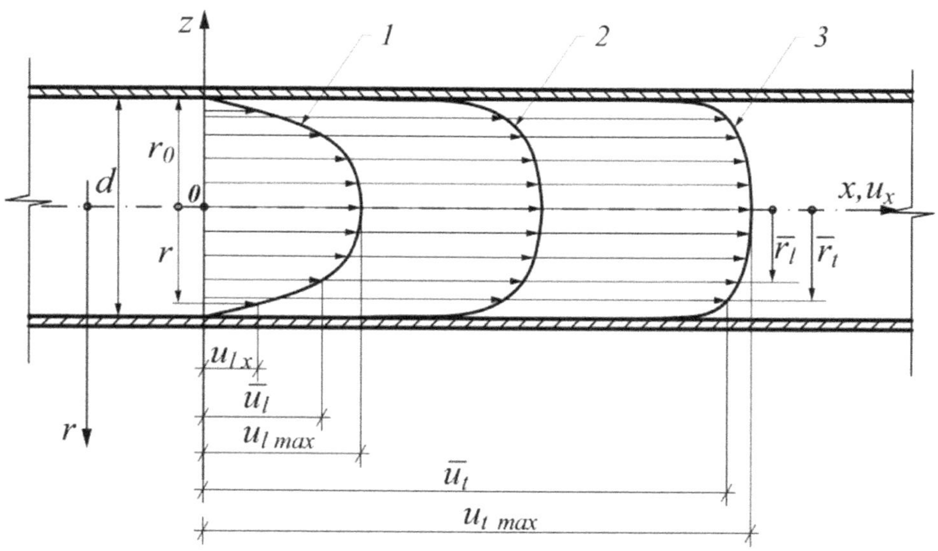

Also, the adequacy of the established and proposed new theoretical positions and analytical models obtained on their basis is confirmed by almost complete coincidence (with 95% confidence) of the main parameters of turbulent flow calculated by them with the results of experimental studies of leading and recognized scientists in this field Nikuradze I., Shevelyov F., and Altshul A.D. by statistical processing of them by general methods using a computer.

Figure 7. Profiles of averaged velocities of turbulent flow in the region of hydraulically smooth turbulence for a pipe with a diameter of 10 cm: 1 - values based on experiments are given using markers; 2 - according to equation (10)

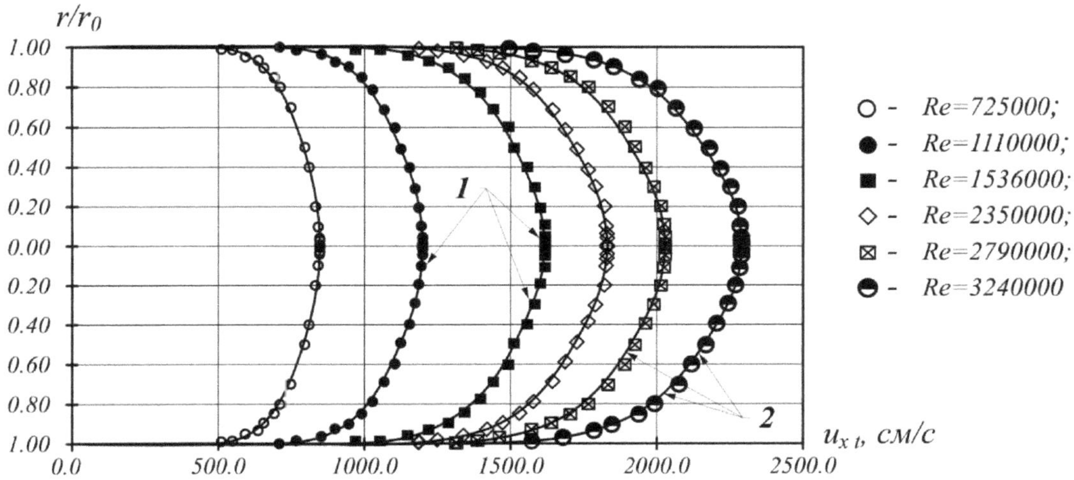

CONCLUSION

Thus, improved theoretical provisions for the further development of the general theory of turbulent flow in pressure pipes are obtained. These provisions, in contrast to existing semi-empirical theories, describe the complete hydrodynamic structure of the flow for all areas of turbulent flow. Based on the application of the obtained universal equations, construct a distribution profile of the total turbulent kinematic viscosity and averaged velocity. The development of a general theory of turbulent flow in pressure pipelines, and as a result of this, the obtained average velocity distribution profile will increase the accuracy of calculations when determining the average flow velocity and pipe throughput by 7-10%. The considered approach evaluates the flow efficiency in the components of the collector pipes of the network and in the closed collector-drainage network. This makes it possible to further improve the methods of design and calculation of its technological and design parameters. Thus, it can increase the overall technical, technological, economic, and environmental efficiency of drainage systems in accordance with modern conditions.

REFERENCES

Bar-Mei, G. (2022). *Basics of Fluid Mechanic*. Potto Project Publication. doi:10.5281/zenodo.6462400

Farahnak-Ghazani, M., & Mirmohseni, M. Masoumeh Nasiri-Kenari. (2021). On Molecular Flow Velocity Meters. IEEE Transactions on Molecular, Biological and Multi-Scale Communications (pp. 224 – 238). IEEE. doi:10.1109/TMBMC.2020.3044772

Giroud, J., Palmer, B., & Dove, J. E. (2000). Calculation of Flow Velocity in Pipes as a Function of Flow Rate. *Geosynthetics International*, *7*(4-6), 583–600. doi:10.1680/gein.7.0183

Gooch, J. W. (2011). Kinematic Viscosity. In: Encyclopedic Dictionary of Polymers. Springer. doi:10.1007/978-1-4419-6247-8

Hebbink, A. J. (1999). Land drainage. In *Environmental Geology. Encyclopedia of Earth Science*. Springer. doi:10.1007/1-4020-4494-1_196

Kamra, S. K., Kumar, S., Kumar, N., & Dagar, J. C. (2019). Engineering and Biological Approaches for Drainage of Irrigated Lands. In J. Dagar, R. Yadav, & P. Sharma (Eds.), *Research Developments in Saline Agriculture*. Springer. doi:10.1007/978-981-13-5832-6_18

Kovalenko, P., Rokochinskiy, A., Mazhayskiy, Y., Volk, P., Volk, L., & Chernikova, O. (2020). Construction and agricultural drainage parameter optimization considering economic and environmental requirements. *Engineering for Rural Development.*, *19*, 1009–1017. doi:10.22616/ERDev.2020.19.TF237

Kovalenko, P., Rokochinskiy, A., Volk, P., Turcheniuk, V., Frolenkova, N., & Tykhenko, R. (2021). Evaluation of ecological and economic efficiency of investment in water management and land reclamation projects. *Journal of Water and Land Development*, *48*(1–3), 81–87. doi:10.24425/jwld.2021.136149

Łabędzki, L., Kaca, E., & Brandyk, A. (2021). Irrigation and Drainage in Polish Agriculture: State, Problems and Needs. In M. Zeleňáková, K. Kubiak-Wójcicka, & A. M. Negm (Eds.), *Quality of Water Resources in Poland. Springer Water*. Springer. doi:10.1007/978-3-030-64892-3_5

Malecha, Z. (2022). Turbulence and Fluid Mechanics. *Energies*, *15*(3), 1116. doi:10.3390/en15031116

Pasaribu, K., Lambert, L., Lambert, D., English, B., Clark, C., Hellwinckel, C., Boyer, C., & Aaron Smith, S. (2021). Profitability of irrigating for corn, cotton, and soybeans under projected drought scenarios in the Southeastern United States. *Irrigation Science*, *39*(3), 315–328. doi:10.100700271-020-00707-x

Peiqing, L. (2021). Experimental Fluid Mechanics. A General Theory of Fluid Mechanics. Springer. doi:10.1007/978-981-33-6660-2_5

Raffel, M., Willert, C., Wereley, S., & Kompenhans, J. (2007). Examples of Application. In *Particle Image Velocimetry. Experimental Fluid Mechanics* (pp. 259–388). Springer. doi:10.1007/978-3-540-72308-0_9

Rodriguez S., Fathi N., Pourghasemi P. (2022). Theoretical Approach for the Fast Estimation of the Turbulent Kinematic Viscosity for Internal Flows. *Journal of Nuclear Engineering and Radiation Science*. . doi:10.1115/1.4054342

Rokochinskiy, A., Frolenkova, N., Turcheniuk, V., Volk, P., Prykhodko, N., Tykhenko, R., & Openko, I. (2021). The variability of natural and climatic conditions in investment projects in the field of nature management. *Journal of Water and Land Development*, *48*(1–3), 48–54. doi:10.24425/jwld.2021.136145

Rokochynskyi, A., Volk, P., Turcheniuk, V., Tokar, L., Volk, L., Mazhayskiy, Y., & Chernikova, O. (2020). The drainage module is an important factor in the design of drainage systems reconstruction and construction projects in the polesia region. *Engineering for Rural Development.*, *19*, 36–47. doi:10.22616/ERDev.2020.19.TF010

Rubenstein, D. A., Yin, W., & Frame, M. D. (2021). Fundamentals of fluid mechanics. In *Biofluid Mechanics: An Introduction to Fluid Mechanics* (pp. 17–70). Macrocirculation, and Microcirculation., doi:10.1016/B978-0-12-818034-1.00002-5

Seyedmohammadi, J., Esmaeelnejad, L., & Ramezanpour, H. (2016). Land suitability assessment for optimum management of water consumption in precise agriculture. *Modeling Earth Systems and Environment*, *2*(3), 162. doi:10.100740808-016-0212-9

Speight, J. G. (2017). Kinematic Viscosity. Rules of Thumb for Petroleum Engineers. Wiley. doi:10.1002/9781119403647.ch209

Volk, L., Bezusyak, O., & Volk, P. (2021). Improving the dimensioning of closed collecting and drainage network of drainage systems. *Land Reclamation and Water Management*, (1), 98–106. doi:10.31073/mivg202101-269

Watkiss, P. (2015). The Cost of Climate Change in Europe. In K. Steininger, M. König, B. Bednar-Friedl, L. Kranzl, W. Loibl, & F. Prettenthaler (Eds.), *Economic Evaluation of Climate Change Impacts. Springer Climate*. Springer. doi:10.1007/978-3-319-12457-5_2

Chapter 14
Improvement Deep Loosening as an Effective Adaptive Agromeliorative Practice

Olexander Lukyanchuk
National University of Water and Environmental Engineering, Ukraine

Vasyl Turcheniuk
National University of Water and Environmental Engineering, Ukraine

Roman Koptyuk
https://orcid.org/0000-0002-7086-3608
National University of Water and Environmental Engineering, Ukraine

ABSTRACT

This section analyses the technologies and means of traditional deep loosening as an agro-ameliorative measure, and their inconsistency with the requirements for improving the technology of water regulation of ameliorated lands in changing climatic conditions is determined. Therefore, it is advisable to switch from the traditional and most widespread technologies of slot and strip loosening to continuous loosening in profile and area, as well as differentiated loosening of deep soil horizons. It is proved that improved deep loosening can be the most effective means of ensuring the necessary filtration and simultaneous accumulation of excess soil moisture to ensure maximum productivity of ameliorated land throughout the entire growing season. Its implementation in the conditions of Polesie of Ukraine contributed to the improvement of water-physical properties, water permeability, and accumulating ability of the most common drained sod-podzolic soil, which provided an increase in the yield of cultivated crops by 20 ... 40%

DOI: 10.4018/978-1-6684-8248-3.ch014

BACKGROUND

In today's world, environmental problems from the man-made use of water and land resources are becoming more acute, among which food, energy and water are relevant. All of them are interconnected and are currently one of the global problems of mankind in terms of its further normal environmental development. In particular, this applies to the rational use of water and land resources of agricultural lands of the water complex and technical means of their processing.

Polesie is a unique physical and geographical region, which is located on the territory of the Polesie lowland almost in the center of Europe, is a peculiar landscape region of the Eastern European physical and geographical region, has an area of about 300 thousand km^2.

Polesie of Ukraine is a flat sandy-swampy plain with close occurrence to the daily surface of groundwater and with insignificant runoff. The region is characterized by a fairly temperate climate and uneven rainfall at different times of the year. The soil background of the region is sod-podzolic clay-sandy or sandy soils, sod gley clay-sandy, as well as meadow gley sandy soils. In the Polesie area in the summer due to changes in weather and climatic conditions of the area waterlogging alternates with a lack of moisture. The impact of drainage on mineral wet soils is ambiguous. After reducing excessive moisture, the arable layer due to regular tillage acquires favorable properties, and deeper - gley and compacted soil layers, when drained retain and, often, even worsen their existing properties.

Polesie is a unique natural complex. Amelioration has become a significant factor in the socio-economic transformations of this region and plays an important role in the development of economic sectors. Ameliorated waterlogged lands are important in the humid zone of Ukraine, and they are the only guarantors of stable farming and the creation of a reliable forage base to meet the needs of animal husbandry, regardless of weather and climatic conditions in many regions. Therefore, the Polesie zone of Ukraine is the main area for drainage amelioration.

The water management complex is a fairly capital-intensive sector of agricultural production. To maintain the high productivity of waterlogged soil, it is recommended to carry out a number of agromeliorative practices simultaneously with the drainage device, aimed at increasing the accumulating capacity of the active soil layer, improving the surface and subsoil runoff, and enriching the soil with nutrients.

In today's variable climatic conditions, when theamount and intensity of precipitation, temperature conditions change, the implementation of existing water management technologies on ameliorated lands can further increase the inefficiency of their use. This is manifested in the fact that the unevenness of precipitation and the traditional haphazard drainage of excess water into the open network by drainage systems create significant fluctuations in the groundwater level and the moisture content of the root layer of drained lands during the growing season that leads to inefficient use of natural water reserves in the soil. In this case, the lower soil layers become almost waterproof under conditions of long-term operation (Ward, Flower, Cordingley, Weeks, Micin, 2012; Iglesias, Garrote, 2015; Hamzaa, Anderson, 2005).

Despite the fact that the average rainfall over the growing season is 250...400 mm, due to a change in established climatic regimes, crops suffer either from waterlogging with an increase in the intensity and amount of rainfall over a short period of time, or from overdrying of the soil with excessive dry season evaporation. As a result, this leads to crop losses of up to 30...40%. Amelioration and further agricultural land use change the direction and pace of development of soil processes, as well as the water-physical properties of soil. In general, their productivity increases, but at the same time, they do not completely eliminate the negative soil properties in relation to cultivated crops (Iglesias & Garrote, 2015; Medvedev, Laktionova, & Pocheptsova, 2004; Havrysh et al., 2013).

Current climate changes require urgent actions to adapt to them, including industrial agricultural production.

In accordance with the programmatic international and national documents on this issue (the UN Framework Convention on Climate Change, the Kyoto Protocol and the Paris Agreement, documents in the field of climate diplomacy in Ukraine and the EU as a whole, etc.), states, including Ukraine, took commitments to allocate funds for adaptation to climate change and preparation for their implementation. Based on the analysis and generalization of the data of domestic and foreign scientists and specialists, the content of program, international and national documents, it becomes necessary to develop and implement adaptive agromeliorative practices, including on lands with a regulated water regime, to effectively counter modern challenges to energy, food and water crises, which are exacerbated by ongoing climate change.

To ensure sustainable development of agricultural production in the conditions of expected adverse climate changes, deep loosening plays an extremely important role as the most effective adaptive agromeliorative practice, and, at the same time, an effective alternative to expensive reconstruction of existing reclamation systems (Ward, Flower, Cordingley, Weeks, Micin, 2012; Iglesias, Garrote, 2015; Lukyanchuk, Turcheniuk, Prykhodko etc., 2019).

Modern changes in natural conditions, caused primarily by climate changes, lead to a change in the requirements for the rational use of water and land resources in the drainage amelioration zone (Kovalenko, etc., 2019). As the result, the question of changing the principles for the formation of approaches to water regulation on drained lands and the development of a number of measures to adapt water management complexes to these changes becomes extremely urgent. Instead of the main principle of traditional drainage, which consists in maximally accelerating the transfer of surface runoff to groundwater, and then draining it through the drainage beyond the boundaries of the site, it is more expedient to carry out its maximum accumulation and regulation in the soil and the boundaries of the drained site, but only to discharge the residues through the drainage.

In this regard, the issue of rational use of water and land resources of ameliorated lands defines the following tasks:

- the need to differentiate the degree of drainage of the soil to improve the regulation of surface and subsoil runoff, which is related to the direction and intensity of soil processes,
- simultaneously with the use of drainage, carrying out the agromeliorative practices aimed at increasing the accumulating ability of the active layer of the soil and enriching it with nutrients and the drainage system as a whole.

To minimize flooding and loss of deficient moisture by evaporation, it is necessary to ensure the fastest transfer of surface runoff obtained during floods and heavy rains in a productive soil mass (up to 1 m deep) with maximum accumulation in it for use in dry periods.

Increasing the accumulative capacity of productive layers to the maximum can be done by dumpless continuous tillage of the entire soil mass in depth and area, with controlled production of the desired soil structure, differentiated by horizons of the massif according to their purpose (surface protective, seed, root and filter).

Several types of agromeliorative practices are known as special methods of tillage for regulating its moisture by draining excess water from the surface and from the arable layer of soil. In variable climatic conditions, deep loosening can be the most effective adaptive measure and an alternative to expensive reconstruction of existing amelioration systems, including systems of European Polesie.

Traditional deep loosening plays an important role in regulating the water-air regime of drained soils: in periods of waterlogging, it helps to accelerate the release of arable land from excess moisture, accelerating its movement to lower layers, and in dry periods - the accumulation of moisture in loosened soil. subsequent more efficient use of its plants. At the same time, it significantly improves the operating conditions of closed systematic drainage.

Depending on hydrological, soil and other conditions, continuous loosening, continuous loosening-taming, strip loosening, strip loosening-taming, slot loosening are used.

Continuous loosening - the main type, aimed at loosening the entire thickness of the soil in the treated layer on wet soils with surface slopes up to 0.01.

Continuous loosening-mowing ensures, in addition, the formation of moles below the loosening zone on resistant to mowing soils.

Continuous deep loosening of stony soils is not carried out in the presence of stones in the layer of loosening with a thickness of 0…80 cm and a maximum diameter of more than 30…40 cm.

If for organizational reasons it is impossible to perform these measures, it is advisable to use strip loosening (after 4…8 m) with a subsequent transition as the cultivation of soils to continuous deep reclamation loosening.

Strip loosening, carried out mainly by single-hull rippers, creates narrow separate loosening strips (usually after 4 m). Uncultivated areas serve as a kind of support for tractor engines during subsequent passes. This increases the duration of the effectiveness of fluffy strips.

Strip loosening-mowing creates narrow strips of loosening, supplemented at a depth of 0.7…0.75 m molehills. It is used in areas with a large number of "saucers". The distance between the strips is taken to be 2…2.5 m for clay soils and 3…4 m for loamy soils.

Slitting is an agro-meliorative measure aimed at destroying compacted waterproof layers of subsoil horizons to a depth of 0.40… 0.45 m, increasing water permeability and internal soil runoff due to mechanical formation of vertical cracks and fissures 0.25…0.4 m wide.

Deep loosening at humidity below the optimum leads to a deterioration in the quality of loosening and increased traction, which increases the energy consumption of loosening the soil.

The main parameters of deep loosening include depth of loosening; loosening intervals; quality of loosening.

Deep loosening of wet soils is recommended to be carried out only in combination with drainage in compliance with the relevant work technology.

In changing climatic conditions, within the global water, food and energy challenges of today, to mitigate the negative effects of soil droughts and evaporation of greenhouse gases from the soil, the question arises increasing the level of efficiency of soil moisture regulation for further high-performance functioning of agricultural lands that encourages to changes in approaches to obtaining soil structure of functional soil layers through necessity their adaptation to a high level of moisture accumulation and differentiation of spatial redistribution of soil runoff based on the methodology of improving existing agro-ameliorative technologies and creating energy-efficient technical means on ecological and economic principles in accordance with modern conditions and requirements.

Improving the level of efficiency of soil moisture regulation is possible due to the controlled production of soil structure of different moisture accumulation capacity and differentiation of spatial redistribution of soil runoff

The use of deep cultivation as an effective adaptive measure in variable climatic conditions determines the need for its improvement.

Traditional deep loosening involves crevice loosening of the soil by rack-mounted working bodies (Laziuta, Kabulova, & Gasanov, 2016). In this case, vertical channels are formed in the soil profile to transfer the entire surface runoff to the subsoil and then drain it through the drainage. In the current conditions of alternating periods of drought and excessive humidity, this technology can cause significant negative environmental consequences due to the increased leaching regime of the land. In case of excess moisture, the drainage system, which is originally designed to be more gentle, will be heavily overloaded. In case of a dry period the soil will be unproductive.

It is advisable to change the principle of traditional deep loosening - the transition from technologies of local crevice loosening of the vertical soil profile and strip loosening over the area to a continuous area and differentiated loosening of the soil profile depth.

The essence of the improved continuous loosening is that the upper layer of the soil, the plow sole and the lower layer of the soil are loosened separately. Loosening of each layer is carried out continuously both on height of a profile of each layer, and on the field area, the top layer to a plow sole is loosened to formation of average and small lumpy structure in its top part and granular and small lumpy structure in the lower part. the layer is loosened to form a large- and medium-lumpy structure to a depth less than the depth of drainage. The formation of granular and fine-grained structure makes it possible to make the most efficient use of soil moisture accumulation potential due to the formation of a large surface area of soil particles and small gaps between them in the area above the plow sole where potential root systems are located. cultures.

The formation of large and medium-lumpy structure to the depth of drainage makes it possible to quickly remove excess fluid due to increased gaps between soil particles simultaneously with some accumulation in the soil due to the formation of sufficient surface area of soil particles. The formation of a medium-lumpy structure in the upper quarter of the upper layer protects the accumulated moisture slightly below from excessive evaporation from the surface and weathering. As a result, in each of the soil layers a moisture-accumulating structure is obtained, which is suitable for the purpose of each of them: with maximum moisture accumulation potential in the upper part and sufficient water capacity with simultaneous accumulation of some moisture in the lower part of the soil profile.

METHODS AND TECHNIQUES

The practice and the accumulated experience of agricultural production, according to which the vertical soil profile of farmland should have an anti-erosion top layer (0 ... 0.05 m), a root-inhabited layer (0.05 ... 0.4 m) and a lower filtration layer (0.4 ... 0.8 m), additionally speaks in favor of this. At the same time, the optimal soil structure (the percentage by weight of the groups of lumps of "valuable" and other sizes) of each of these layers is different and should be appropriate for each of them (Havrysh et al., 2013).

The recommended parameters of the treated soil layer for crops are determined by agronomic requirements. Generalized indicators which are as follows (Medvedev, Laktionova, Pocheptsova, 2004; Havrysh et al.., 2013; Panchenko, 1999):

- the structure of the soil for sowing should be fine-grained with a predominance of lumps with a diameter of 1 (3) ... 10 mm; the composition of the upper layer (0…5 cm) of units of 10…20 mm;

- the number of units more than 2 mm in the surface layer - 22 ... 40%, to ensure erosion resistance; the content of aggregates more than 1 mm - not less than 50%, aggregates smaller than 1 mm are considered erosive;
- the content of lumps up to 5 mm - not less than 75%; the content of lumps in the size of 1 ... 25 mm - not less than 80%; the content of lumps up to 50 mm – 70…90%;
- density of the surface layer - 1.2…1.3 g / cm^3 (to prevent unproductive evaporation of moisture); soil density below the seed bed up to 1.1 ... 1.3 g / cm^3; the кdensity of the upper root-containing layers above the seeds 0.9 ... 1.0 g / cm3; density of root-containing layers 1.2 g / cm^3;
- volume of air wells (porosity) - about 70%; height of ridges on the surface of the field - up to 3 ... 4 cm;
- water permeability - more than 30 mm / h; filtration coefficient - 5 ... 10 dm^3 / day;
- soil hardness - up to 3 MN (up to 300 t / m^2); specific energy consumption of the process - up to 33…35 kW / m of capture width.

Quality of loosening of the soil is a complex concept, which characterizes the degree of approximation of the actual quantitative indicators of the physical condition of loosened soil to the relevant normative (or recommended) quantitative indicators determined by industry requirements.

On the basis of industry loosening of soils can be construction, reclamation and agricultural. Naturally, the requirements for the quality of loosening the soil in each industry are specific and related to the purpose of loosening. At the same time, there is a group of indicators that characterize the effectiveness of loosening, regardless of industry requirements.

According to A.N. Zelenin, loosening of soils is characterized by the coefficient of loosening k_p, which is determined by the ratio of the volume of loosened soil V_p to the volume of the same soil that he occupied in the natural deposit V (Zelenin, 1969):

$$k_{@}=\frac{V_p}{V} \qquad (1)$$

The average values of kp are in the range of 1.08...1.35, and the residual loosening coefficient is in the range of 1.01...1.09 (Chernyonok, Brusilovsky, 1983).

Smaller values correspond to sandy soils, larger – clay. Over time, loose soil "coagulates", compacts and retains only residual looseness. The completeness of loosening is characterized by the coefficient of completeness of loosening k_n, which is the ratio of the area of loosened soil in the cross section of the gripping strip of the working bodies of the ripper F_p to the sum of the areas of loosened and un loosened soil in the same cross section $F_p + F_n$ (Chernyonok, Brusilovsky, 1983):

$$k_n = \frac{F_p}{F_p + F_n}. \qquad (2)$$

An important indicator of the quality and energy consumption of such working bodies is the coefficient of grinding of the developed soil (Panchenko, 1999). Indicators of quality and energy intensity directly depend on the ratio of size and quantity of formed structural soil aggregates. The grinding coefficient of the whole volume will be proportional to the multiplicity of grinding of the cross section of the soil slice loosened by the working element, which can be defined as the ratio of the initial cross-sectional area of the soil slice S to the cross-sectional area of the averaged soil aggregate S_k:

$$K_n = \frac{S}{S_k}. \tag{3}$$

To determine the multiplicity of grinding through technological parameters, we assume that the formed averaged soil aggregates are of regular square shape as the most resistant to further destruction by grinding. We can also assume that in the future the tops of such soil aggregates will still be destroyed due to the interaction of soil aggregates with each other and they will be approximately similar to the real thing. If this assumption is correct then it can be written down,

$$K_n = \frac{S}{S_k} = \frac{h \cdot L}{l_{k.A5@}^2}, \tag{4}$$

where h, L – respectively, the height and width of the cross section of the soil slice; l_k ser is the linear size of the averaged soil aggregate. Additional moisture capacity characterizes the increase in the accumulation capacity of the active root layer of the soil h (0.5...0.6 m) by reducing its density and the corresponding increase in porosity during deep loosening and is defined as

$$\Delta Wh_p^0 = Wh_p^0 - Wh_n^0, \tag{5}$$

where ΔWh_p^0 is the indicator of additional moisture content of the soil layer h as the difference between the change in the maximum reserve of productive soil moisture after Wh_p^0 and before Wh_n^0 its loosening. In turn, the indicators of the maximum stock of productive soil moisture are determined by generally accepted experimental or calculated methods.

For deep loosening of the soil, depending on the conditions and type of soil, different types of tillers are used: with active and passive working bodies (Lukyanchuk, Turcheniuk, Prykhodko etc., 2019; Kravets etc., 2018; Pivovar, 2011).

Active deep tillers belong to pulsed-power machines. During loosening the soil, they use various types of vibration and shock loads on the soil. The working body of an active deep-tiller develops the soil due to the flow of energy coming from an autonomous source that allows using the additional vibrational energy to reduce the energy intensity of soil loosening. But this is justified only for surface tillers and at low speeds of the tiller moving. With an increase in the speed of movement of the traction vehicle and an increase in depth, the effect of the vibrating working bodies decreases significantly, and the energy consumption increases significantly compared to passive working bodies (Kravets etc., 2018).

The working body of the passive deep tiller develops the soil and carries out its longitudinal movement due to the pulling force of the base tractor only.

Active and passive-active deep-tillers are much more complicated than relatively passive one in design and operation. They are effective only on hard (rocky) and dense soil, and on relatively soft non-frozen soil they have practically no advantages over passive ones.

Therefore, the most widespread tillers are passive deep-tillers. They are reliable, simple in design and in operation, but require more powerful tractors (Fig. 1).

Figure 1. The designs of passive deep tillers (a), c), d), e), f) - rack-mount; b), e), f) - the perimeter; a) - RU.65.2.5: 1 - frame; 2 – wheeled mover; 3 - drawbar; 4 - clip; 5 - depth regulator; 6 – sod knives; 7 - removable racks; 8 – mole plow; 9 - knives; 10 - plows; 11 - rack knives; 12 - shield; 13 - tubular beam; b) - RG-0.8A: 1 - bit; 2 - a place of fastening of a basic wheel; 3 - auxiliary device; 4 - load-bearing beams; 5, 6 - mounting brackets to the rear linkage of the tractor; 7 - side knives)

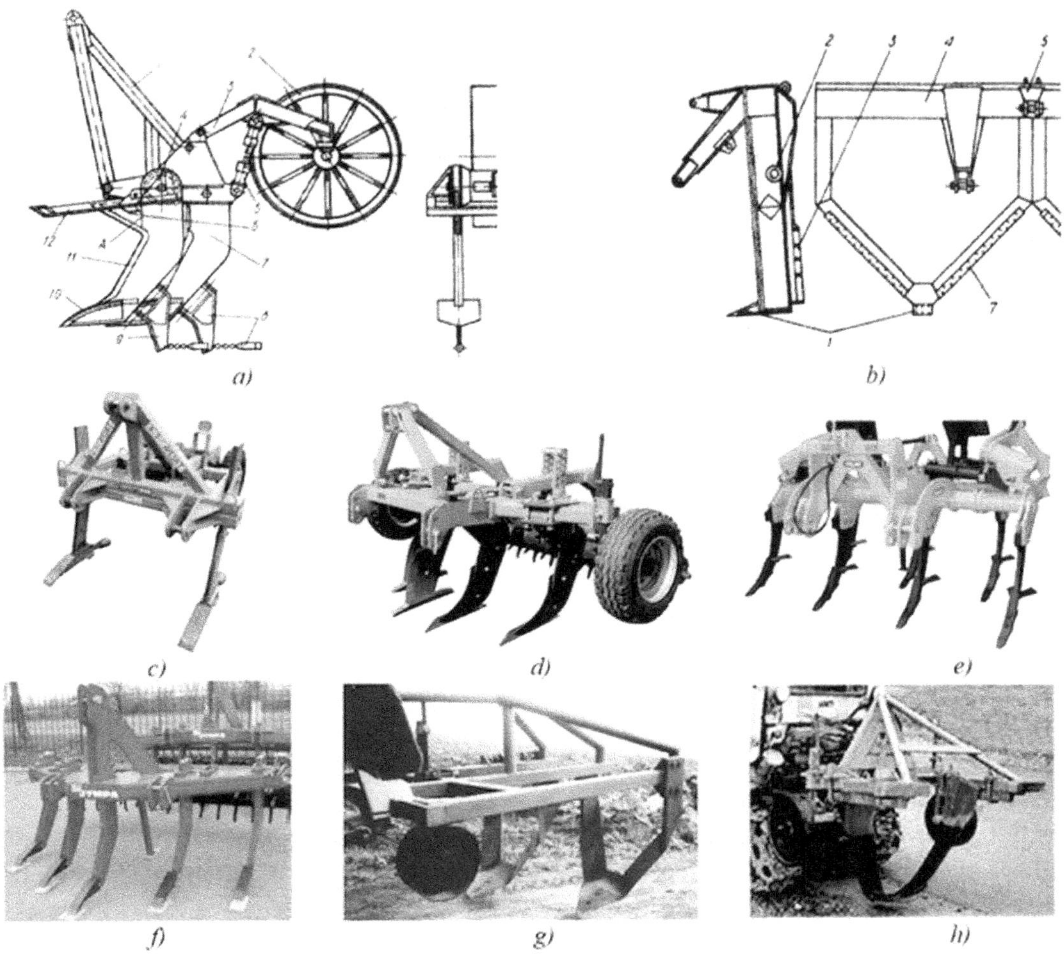

In many countries, for deep loosening, quite simple in design and reliable in operation tools are used, including a frame on supporting wheels or a hinged one with several ripping or shear working elements mounted on it.

Agricultural and agromeliorative deep tillers of a traditional and modern design are rack-mount or perimeter working bodies that loosen the soil to a maximum depth of mainly 0.30 ... 0.35 m - agrarian, and up to 0.45 ... 0.65 m - agromeliorative.

The working bodies of rack-mounted deep-tillers consist of a straight or curved rack where a flat, figured or combined bit is fixed. To increase the area of loosening, expanders are additionally mounted on racks. For laying drainage (moling), a drainer is installedbehind the rack. The working bodies are arranged on the frame in a row and in a checkerboard pattern, V- or Δ- alike (Fig. 1, a, c-f), loosen the soil massif with local crack formation with cutting.

The working bodies of the perimeter deep-tillers are made in the form of a frame that has V-, U-, П -shaped and other profiles (Fig. 1, b, f, h) that loosen the soil massif by volumetric crack formation with cutting.

The design of deep-tillers requires a lot of energy for loosening the soil to place the racks of the deep-tiller in one row (Fig. 1, c), since each working body works with blocked cutting of the soil. The advantage of this type of tiller is its relatively small size. Deep tillers of this design include the tiller of the Austrian company Lemken, the French company Quivogne and the Italian company Badalini.

Placing the racks of the deep-tiller at an angle (Fig. 1, e, f) leads to an increase in its sizes, however, it requires less energy for the loosening process, since only the front working body works in case of blocked soil destruction. The following working bodies, placed at a certain angle to the front working body, work with semi-blocked soil cutting. Deep tillers of this design are manufactured by the Austrian company Hatzenbicher, the Spanish company Jympa, the Ukrainian company KET Ukraine, and the French company Quviogne (Kravets etc., 2018).

In case of a staggered arrangement of racks (Fig. 1, e), the energy intensity of the process of loosening the soil is reduced due to the fact that the front working bodies work with blocked cutting of the soil, and the next two working bodies are placed behind with semi-blocked or almost free cutting of the soil (depending on location step). Tillers of this type are manufactured by the United States company John Deer, the French companies Vogel Noot and Gregorie Besson, the Spanish company Jympa, and the Italian companies Machio Gaspardo and Badalini (Kravets etc., 2018).

Rack-mounted deep-tillers require significant energy consumption when loosening, since during loosening the soil to a depth of 0.45 ... 0.65 m, the width of the working elements of the bit is from 30 to 120 mm, with a corresponding critical cutting depth - 0.1 ... 0.5 m. Therefore, these deep tillers partially work at supercritical depths with plastic deformation of the soil that significantly increases their energy intensity. Deep tillers racks that have lateral wings increase the width of the loosening zone of the soil, resulting in increased critical depth and their effectiveness. Existing disadvantages lead to a significant increase in energy intensity and deterioration in the quality of loosening the soil (Kravets etc., 2018).

Loosening of the soil by perimeter deep tillers is carried out by cutting in the horizontal plane by a sharer, in the vertical plane by side knives and due to compression by the side walls, after that the loosened soil crumbles under the influence of its own weight to the bottom of the cut slot. The designs of this type of deep tiller can improve the quality of the cultivation process, however, at the same time, the energy consumption of cultivation increases due to soil compression between the side knives. Deep tillers of this type have a small working width and, through additional soil compression by side knives, increase the total resistance of soil cutting that negatively affects the energy intensity of soil loosening. Due to the design features, the deep tiller is difficult to adjust for various soil conditions.

The rack-mounted deep-tillers (single and multi-rack) during the passage extend the surface of the soil 10 ... 15 cm upward in zones of 15 ... 20 cm on both sides of the tracks of the risers, and in the spaces between them - by 3 ... 4 cm. The degree of loosening of soil in the zone near the racks - 8 ... 12%, in the zone between the racks - 40 ... 60% (Havrysh etc., 2013; Lukyanchuk, Turcheniuk, Prykhodko etc., 2019).

The distribution of loosening levels by zones indicates that the riser, forming a slot, spreads the soil to the sides, presses it in and rubbing it into the side walls of the void. This leads to deterioration in the quality of loosening and a significant increase in resistance to movement

It is established that when the soil of the perimeters is loosened by the working body, the vertical slots close after it passes, therefore, the soil does not spill from the humus layer into the slot. At the same time, the volume of soil restrained between the side knives is deformed, and the front of deformation directed from the lower sharer and side knives up to the "day" surface. After the passage of the working body, the raised soil settles under the influence of its own weight and at the same time loosens.

Perimeter working bodies are characterized by the same drawbacks as rack-mounted, namely: pressing and grouting the soil into the side walls of the cut slots, great resistance to movement, the inability to predict and obtain the required quality of loosening in the horizons of the vertical profile of loosening the soil.

Therefore, based on the foregoing, the designs of traditional deep-tillers do not guarantee the necessary quality of loosening the soil, their structure, and protection of the soil from technological impact. They condense the soil of the walls of the cut slots, require significant energy consumption when loosening, andhave insufficient completeness of loosening and the duration of its aftereffect. (Mazhayskiy, Rokochynskiy etc. 2017; Simon, Friday, Orok, 2018; Celik, Raper, 2012; Li, Zhang, Zhang, Osman, 2012; Odey, Manuwa, 2016).

Thus, the effectiveness of improving deep cultivation is constrained by the imperfection of existing technologies and tools that do not provide the necessary quality and efficiency of loosening, and therefore require further improvement in accordance with the following requirements (Havrysh etc., 2013; Mazhayskiy, Rokochynskiy etc. 2017):

1) to provide a long after-effect;
2) to improve the water-physical properties and structure of the soil, to increase its fertility;
3) to provide high accumulating and sorption abilities of the soil, to increase the efficiency of its water regulation;
4) to ensure resource conservation, high economic and energy efficiency;
5) to meet modern principles of adaptive nature management.

Obtaining soil structure of different moisture-accumulating capacity of functional soil layers with low energy consumption of the working process is possible due to external mechanical action of tiered working bodies aimed at controlled compaction of the soil mass using low-energy deformations. This causes the formation of a large set of vertical and horizontal microcracks and pores in which moisture will enter and retain soil runoff. The sizes of microcracks and pores are commensurate with similar ones formed from the growth of root systems of plants.

This approach, in addition to the required quality result in soil structure, will also allow to minimize the specific energy consumption in relation to ensuring the operation of technical means and energy efficiency of the land as a whole.

In order to eliminate the shortcomings of traditional deep loosening tools and take into account the necessary technological changes, new working bodies with a multi-level arrangement of risers and an improved working element were developed and tested (Figure 2) (Lukyanchuk, Ryzhyi, Ihnatiuk, 2017).

In general, the working body of such a tiller consists of a frame with load-bearing racks, on which work elements are fixed at different levels (tiers) in the form of a horizontal knife mating with a concave symmetrical grooved loosening surface of variable curvature. Through the shape and parameters of this surface, the required deformation, degree of stress, and, consequently, grinding of the developed soil layer are set. The general arrangement of the working body of the deep-tiller is determined by the spatial spacing of the working elements in three mutually perpendicular directions.

Their use will make it possible to move from ineffective traditional technologies of slotted and strip loosening to highly efficient advanced continuous loosening of soil. Thanks to the new principle of deep loosening, for the first time, it became possible to control the process of loosening the soil and to obtain the desired degree of loosening in each horizon of the developed soil. In the design of the improved subsoiler, the zone of elastoplastic deformations is minimized, it is replaced by less energy-intensive deformations of move, shear and bending due to the belt loosening of the soil by a separate working element at sub-critical depths (non-plastic destruction). The use of such working bodies ensures the creation of favourable conditions for improving the water-physical state of the soil in each layer simultaneously over the entire area and profile (Mazhayskiy etc., 2020).

Figure 2. Working bodies for deep continuous cultivation of the soil with improved racks

The process of tillage is as follows. When the operating unit is moving, the soil particles which are cut off with every horizontal knife, are moving along the surface and are gradually bent from the edges to the middle due to the surface channel shape and are structured by warping bend in two coordinate planes reaching the estimates in the end of the surface (Figure 3).

The use of such working bodies ensures the creation of favourable conditions for the restoration of the water-physical state of the soil in each layer simultaneously over the entire area and profile of the soil.

Soil loosening by multi-tier working bodies is characterized by a decrease in density in the arable layer to 1.1 ... 1.15 t / m^3, in the subsoil - to 1.2 ... 1.3 t / m^3, which corresponds to a loosening coefficient within 1, 23 ... 1.32; the prevalence of soil aggregates up to 25 ... 50 mm across (<25 mm - 65%;

<50 mm - 93%) in the structural composition of the loosened layer; an increase in porosity by 25 ... 35%, an increase in the permeability of the loosened layer by more than 4 times to 1.36 m / day, and an increase by 3.5 times of water loss (Havrysh et al., 2013; Mazhayskiy et al., 2017; Lukyanchuk, Ryzhyi, & Ihnatiuk, 2017).

Figure 3. Operating principle of the tiered operating unit for deep differentiated tillage (R_k – the radius of the surface finite curvature, L - the width of the surface; H – the total tillage depth; h_1 – the depth of tillage by the upper tier; h_2 – the depth of tillage by the lower tier; ψ – the angle of the cutting edges line slope to the horizon; α – the angle of the surface to the horizon)

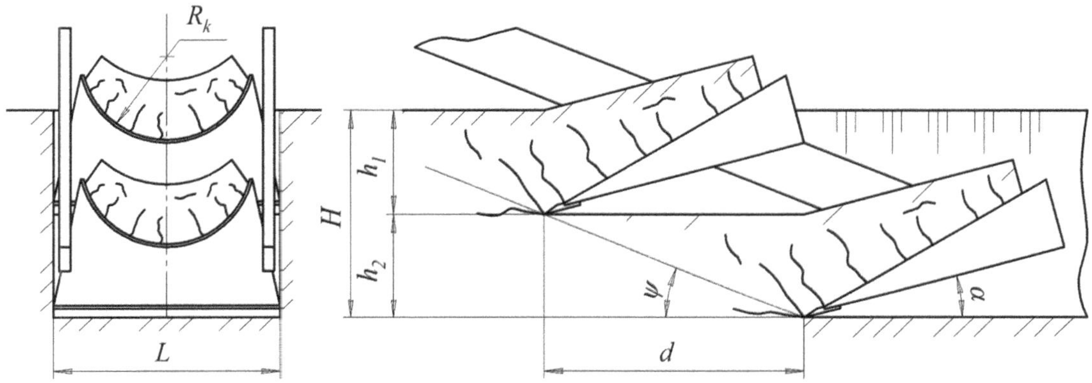

The option of performing the working elements separate and swept from two symmetrical grooved parts with the gradual principle of grinding the soil layer was adopted for subsequent use during the process of improving the design (Figure 4).

If we consider the principles of grinding the soil layer by working elements, then in general they are similar, but there are some differences. In one variant (Figure 4, a), with the cultivator moving, soil shavings cut by each horizontal knife moves along the loosening element and gradually bends from the edges to the middle as a result of the grooved shape of the last one and is structured by bending in two coordinate planes, reaching calculated indicators at the end of the surface. Loosening of the soil occurs evenly and simultaneously across the entire width of the cut soil layer with the rise along the surface of the loosening element to the cutting angle.

In another case (Figure 4, b), due to the different curvature and sweep of the symmetric radial surfaces of the loosening elements, the differentiation of the quality of loosening of the soil in depth occurs with less traction resistance of the working body due to the gradual grinding of the soil layer without its pre-cutting by a horizontal sharer and additional friction from subsequent movement through the loosening element.

Considering the above, the variant with gradual grinding should have an advantage in the towing force of the tractor, since it lacks unproductive transportation and lifting of the soil, although there may be some slight loss in the cutting force of the soil in next tier to the current one.

In qualitative terms, the sequence diagram of grinding is almost the same and should give an identical final indicator of soil loosening.

Figure 4. Variants of the working elements of the deep-tiller (a) grooved; b) - sharp jib)

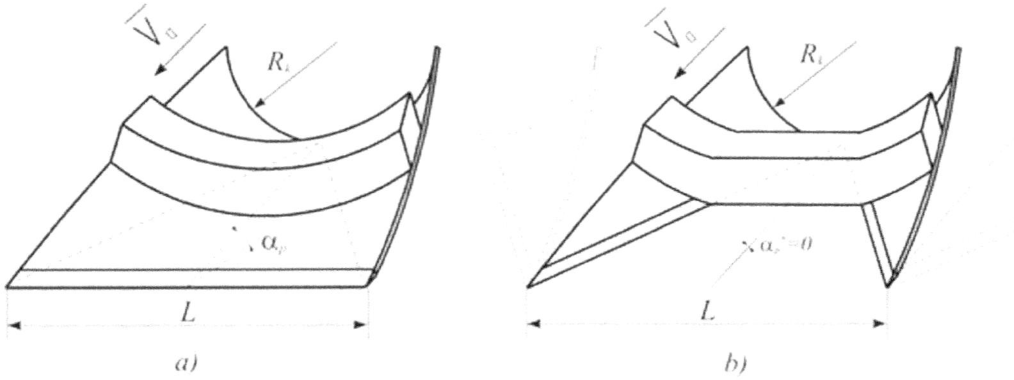

On the basis of the sharp jib variant of the implementation of the loosening elements, working bodies were developed for carrying out continuous differentiated loosening with reduced energy consumption for effective control of the filtration and moisture accumulation for soil of various types that is taken into account in the form of a transverse profile of loosened soil: rectangular - for light soil, trapezoidal - for medium soil and triangular - for heavy soil (Figure 5).

The working element of such deep-tillers consist of a frame with oblique load-bearing racks, on which there are fixed loosening elements in tiers in the form of two concave symmetric arrow radial semi surfaces of equal shift curvature with the smallest radius of curvature at the exit from them, the size of which is proportional to the size of the planned structural element of the soil of a separate tier.

Figure 5. Deep tiller for soil

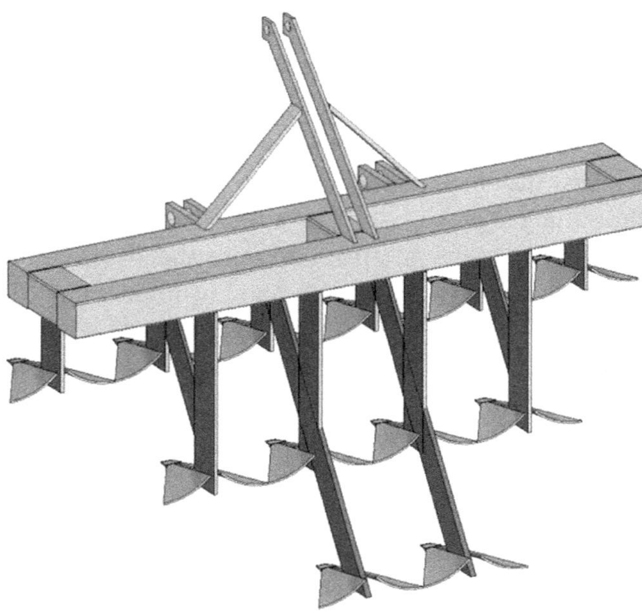

The process of loosening the soil is accompanied by the formation of elemental shavings, which, moving along the surfaces of the working bodies, are crushed into soil aggregates. The latter ones, after descending from the surfaces of the working bodies, move into the space freed by them. At the same time, additional destruction of the formed soil aggregates without mixing the upper tiers with the lower ones is observed (Gaffarov, 2016).

After having been loosened by such a working element, the soil structure is determined by a number of the soil mass grinding stages on the soil mellowing surface, and their number, in turn, depends on the value of the surface curvature. Knowing the parameters of the soil mellowing surface, one can estimate the theoretical soil structure and determine the degree of its loosening due to the multiplicity of crushing.

Given the ratio of the modulus of elasticity of the soil environment in compression and tensile, after mathematical transformations, to simplify calculations and practical use, we can obtain the following dependences on the value of the calculated aggregates of loose soil in a separate tier:

$$l_k^{(n)} = \frac{h \cdot (3.1...3.5)^{n-1}}{(4.1...4.5)^n}, \; n = (3.6...4.0) \cdot \ln\left(\frac{34.333 \cdot h}{R_k}\right). \tag{6}$$

As can be seen from the obtained dependences, with increasing n (stages of grinding) the size of soil aggregates decreases, and their number increases, which is logical in terms of process physics. The number of stages of grinding, and hence the energy consumption of the process, is proportional to the depth of soil development in the tier and inversely proportional to the magnitude of the final radius of the loosening element. The total number of stages of grinding n, for possible indicative design values, is in the range of 10…16 pcs.

Mathematical formulas for calculating the parameters of soil loosening surface are as follows

$$R_k = l_k\left(\frac{1}{2\varepsilon_p} + 1\right), \; L = \frac{\pi}{2} R_k \tag{7}$$

where ε_p – relative warping of the fissure; l_k – the diameter value of the calculated structural soil unit.

The main parametres of tiered operating units for deep soil differentiated tillage depend on soil conditions.

The maximum tillage depth is selected depending on the technological needs or physical and mechanical properties of soil horizons. If the soil is uniform, without any expressed illuvial horizons, the tillage depth of 0.4 ... 0.6 m is adopted. On the reclaimed lands with drainage systems, in order to improve their functioning, the depth of loosening is 0.15 ... 0.20 m lower than the depth of laying drains, counting from the top of drainage tubes.

That is

$$\boldsymbol{H} = \boldsymbol{H}_{dr} - \boldsymbol{d}_{dr} - (0.15 \; \ldots \; 0.2) \tag{8}$$

where H_{dr} – the depth of laying drainage; d_{dr} – the drainage tube diameter.

The angle of soil cutting with a plowshare is determined by the soil conditions

$$\alpha p > 45^\circ - 0.5\varphi^\circ \quad (9)$$

where φ° – is the angle of soil internal friction.

Given that the angle of soil internal friction is 22° ... 30 °, it is considered that α_p > 30...35°.

The general assembly of the deep soil scarifier operating unit is determined by spacing of soil loosening surface in three mutually perpendicular directions.

The value spacing of the tilling tools along the width of the operating unit is within the range limited to the principle of free soil passing between tiers without supporting of the upper tier by the lower one and avoiding the soil contact in the zone of active soil crushing which determines the stand inclination angle ψ.

Vertical spacing is determined by the height of the soil layer that can be efficiently tilled with one surface.

The number of tiers and stands with soil loosening surfaces is determined by the total tilling depth and the width of the cutting unit grip.

The working elements with various defining parameters have different values of the averaged soil aggregate and the grinding multiplicity of the soil mass. And if the value of the averaged soil aggregate has a nearly proportional dependence on the depth of development in the layer and the end radius, then the magnitude of the grinding multiplicity has a rather complex dependence. This difference is due to the fact that the magnitude of the grinding multiplicity is influenced by the width of the soil loosening surface, the value of which in the calculation is variable. It does not affect the magnitude of the formed aggregates. The most effective soil grinding, taking into account the size of the averaged soil aggregate, occurs at a depth of 150 mm and an end radius of 150 mm, and the least effective – at a development depth of 150 mm and an end radius of 200 mm. (Table 1, Figure 6).

Figure 6. The structure of loose soil by the number of soil aggregates (h - tier height, mm; R_k - the radius of the end surface curvature, mm; l_k - soil aggregates, mm)

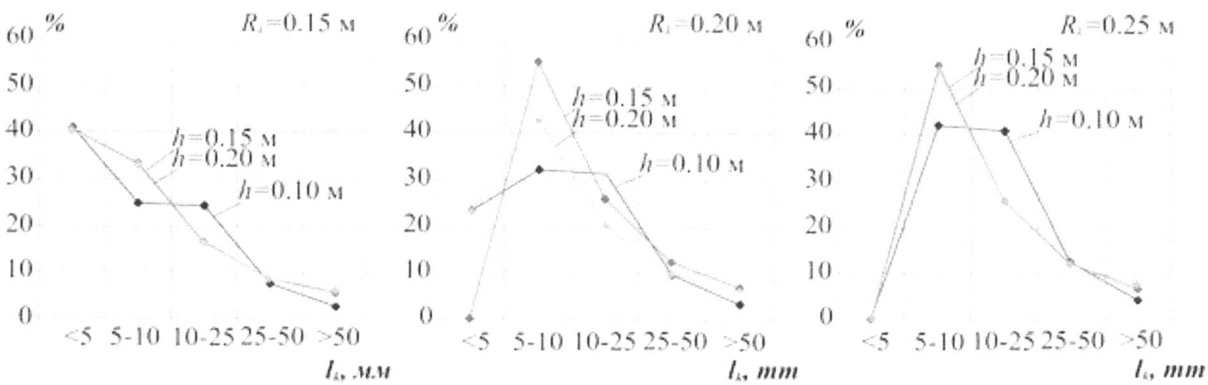

Due to the loose soil theoretical structure, it is possible to determine the necessary parameters of the loosening elements for the upper anti-erosion layer, the roots-like layer and the lower filtration layers when designing the universal layered working parts of the deep loosening devices.

Table 1. Results of the soil grinding structure analytical study

Initial parameters		Structure of loosened soil, l_k							Grinding multiplicity of the soil initial volume
Tier height, *h*, mm	Ende Radius, R_k, mm	Dimensions of formed soil aggregates	<5 mm	5–10 mm	10–25 mm	25–50 mm	>50 mm	Averaged soil aggregate, mm	
100	150	Pcs,%	41.07	24.85	24.12	7.37	2.59	4.92	973
		Capacity,%	2.59	4.28	18.76	33.30	41.07		
	200	Pcs,%	23.37	32.31	31.37	9.59	3.37	5.58	1009
		Capacity,%	1.47	4.33	18.98	33.68	41.54		
	250	Pcs,%	0.00	42.16	40.94	12.51	4.39	6.33	981
		Capacity,%	0.00	4.39	19.26	34.19	42.16		
150	150	Pcs,%	40.44	32.79	15.43	7.26	4.09	5.74	1072
		Capacity,%	1.54	4.39	9.33	19.83	64.90		
	200	Pcs,%	0.00	55.05	25.90	12.19	6.87	7.38	865
		Capacity,%	0.00	4.46	9.48	20.14	65.92		
	250	Pcs,%	0.00	55.05	25.90	12.19	6.87	7.38	1081
		Capacity,%	0.00	4.46	9.48	20.14	65.92		
200	150	Pcs,%	40.23	32.62	15.35	7.22	4.59	6.75	1034
		Capacity,%	1.19	3.40	7.22	15.35	72.84		
	200	Pcs,%	22.75	42.15	19.83	9.33	5.93	7.65	1072
		Capacity,%	0.67	3.42	7.26	15.43	73.23		
	250	Pcs,%	0.00	54.57	25.67	12.08	7.68	8.68	1043
		Capacity,%	0.00	3.44	7.31	15.53	73.72		

From figure 5, it results that in order to loosen the anti-erosion top layer (0 ... 0.05 m), it is expedient to use working surfaces with a finite radius of about 200 mm, for a roots-containing layer (0.05 ... 0,4 m) with a finite radius of about 150 mm, for the lower filtration layers (> 0.4 ... 0.6 m) with a finite radius of about 250 mm.

At the same time, due to the design features, such working bodies can carry out both strip and continuous cultivation, if it is necessary, and such tillers can have up to 40% less specific resistance to movement during cultivation than in traditional existing deep-tillers (Havrysh etc., 2013, Mazhayskiy, Rokochynskiy etc., 2017, Lukyanchuk, Ryzhyi, Ihnatiuk, 2017).

Comparison of the technological efficiency of various technologies and deep loosening equipment was carried out according to the experimental results of 3 years of testing in the "Uyezdtsy" experimental section and in the "Pechalivka" drained massif in the Rivne region of Ukraine (Havrysh etc., 2013; Mazhayskiy, Rokochynskiy etc., 2017). The soil of the studied object is represented by sod-podzolic gley sandy loam sand, formed under conditions of a close occurrence of the groundwater level. The filtration coefficient of the arable soil layer (0 ... 20 cm) is 0.13 ... 0.4 m / day, and at the depth of 20 cm or more – 0.13 ... 0.003 m / day. Soil conditions are typical for the physical and geographical area of the Western (Volyn) Polesie of Ukraine.

Study of various technologies and means of deep loosening on water-physical properties of soils and land improvement status and operation of the existing drainage system in general, conducted to the following options (Figure 7):

1 - slit loosening to a depth of 0.6 m with two-tier deep-ripper type RU -45;
2 - strip loosening to a depth of 0.6 m three-tier deep-ripper of a new type with one riser in each tier;
3 - solid loosening to a depth of 0.6 m two-tier deep-ripper of a new type with three risers in each tier;
4 - control version without loosening.

Figure 7. A comparative diagram of the variants for completeness of loosening the soil by area (in the top) and profile (in the bottom) in a single site (a) - slotted loosening by a single-rack deep-tiller; (b) - strip loosening by a three-rack deep-tiller; (c) - continuous differentiated loosening by a multi-tier deep-tiller
Mazhayskiy et al., 2017

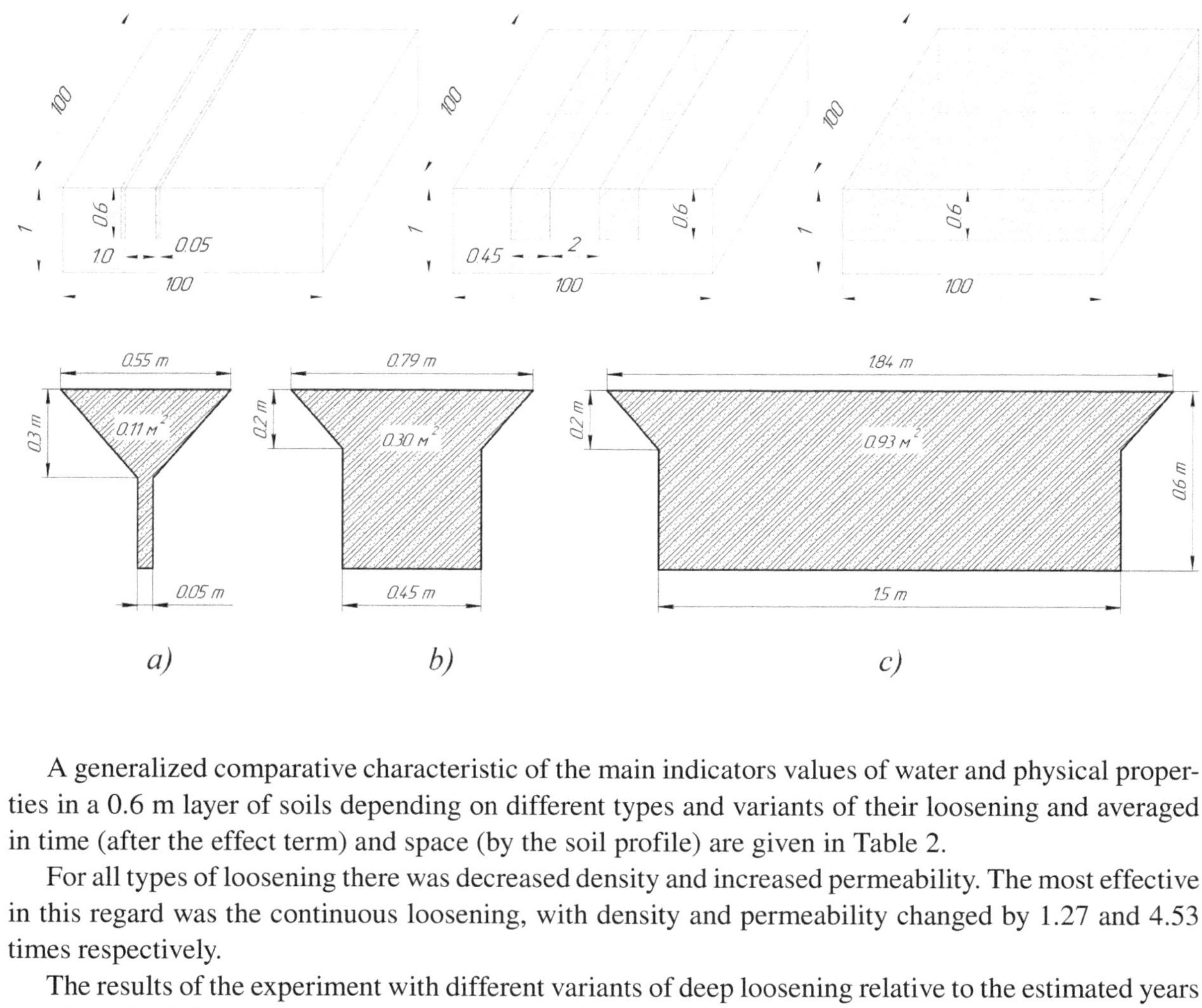

A generalized comparative characteristic of the main indicators values of water and physical properties in a 0.6 m layer of soils depending on different types and variants of their loosening and averaged in time (after the effect term) and space (by the soil profile) are given in Table 2.

For all types of loosening there was decreased density and increased permeability. The most effective in this regard was the continuous loosening, with density and permeability changed by 1.27 and 4.53 times respectively.

The results of the experiment with different variants of deep loosening relative to the estimated years are shown in Table 3.

Table 2. Comparative characteristic of water and physical properties of 0.6 m soil layer in terms of loosening variants

Indicators Variants of loosening	Density, t /m³			Porosity, %			Water permeability, m/day		
	abs value	from control		abs value	from control		abs value	from control	
		t /m³	%		t /m³	%		t /m³	%
Without loosening (control)	1.68	–	–	38.3	–	–	0.30	–	–
Slit	1.63	0.5	2.9	41.8	3.53	9.2	0.65	0.62	206
Strip	1.53	0.5	8.9	43.2	4.9	12	0.82	0.79	296
Continuous	1.32	0.6	21	50.0	11	30	1.36	1.33	443

Table 3. Estimated crop yields of rice crop rotation in terms of loosening variants

No b/o	Loosening variants	Crops	Crop yields in estimated years, %					Predicted crop yields, cwt/ha
			10	30	50	70	90	
1	Without loosening (control)	Perennial herbs	134.8	230.7	251.9	154.1	102.1	183.2
		Spring cereals	12.9	22.0	30.1	27.8	18.6	23.6
		Winter cereals	10.3	20.8	29.5	27.6	18.3	22.7
		Potato	164.2	38.4	216.8	148.3	83.0	136.0
2	Slit loosening	Perennial herbs	151.7	255.2	261.5	177.8	116.7	201.1
		Spring cereals	14.5	24.7	32.4	29.3	20.5	25.6
		Winter cereals	11.6	23.4	31.8	28.5	20.2	24.5
		Potato	184.7	42.4	233.9	156.7	91.8	147.6
3	Strip loosening	Perennial herbs	168.6	287.5	280.2	217.9	142.6	228.7
		Spring cereals	16.1	28.8	36.6	32.4	23.8	29.0
		Winter cereals	12.9	27.2	35.9	31.6	23.6	27.8
		Potato	205.2	50.2	227.7	185.9	110.8	160.8
4	Continuous loosening	Perennial herbs	209.6	331.7	318	263.9	173.1	269.2
		Spring cereals	22.2	33.4	42.5	35.4	26.2	33.4
		Winter cereals	17.8	31.5	41.4	34.9	26.2	32.0
		Potato	262.2	58.7	267.7	231.6	126.5	194.9

Various means and methods of deep loosening provide different levels of environmental reliability of drained mineral soil. In this case, the use of continuous loosening by the multi-tier deep-tiller, in contrast to slot and strip one, increases the ecological reliability coefficient of ameliorated soil from 0.31 –not environmentally reliable value, to 0.52 - ecological reliable value, and thereby ensures the maintenance of a favorable ecological and ameliorative state of soil (Mazhayskiy et al., 2017; Kravets, Kovan'ko, Lukianchuk, 2015).

The summary results of the application of the new technology show the effectiveness of continuous differentiated deep loosening and are presented through graphical dependences of the experimentally obtained physical, energy, and filtration characteristics on the R_c coefficient of completeness of soil loosening in the researched massif with a thickness of 0.6 m, Figure 8.

According to the results presented, the application of deep loosening provides the following increase in crop yields: slit – 5...10%; strip – 10…20%; continuous (improvement) – 20...40%.

Figure 8. The dependence of physical, energy and filtration characteristics on the R_c coefficient of completeness of soil loosening in the researched massif with a thickness of 0.6 m (γ_{zp}^{cep} – average soil density at the experimental site in the layer of 0.6 m; $E_в$ – integral indicator of energy consumption; Q – indicator of the specific energy consumption for loosening 1 m^3; φ – coefficient of energy equivalent (potential) of the obtained agricultural products; K_f – filtration coefficient; Wh° – maximum reserve of productive soil moisture)

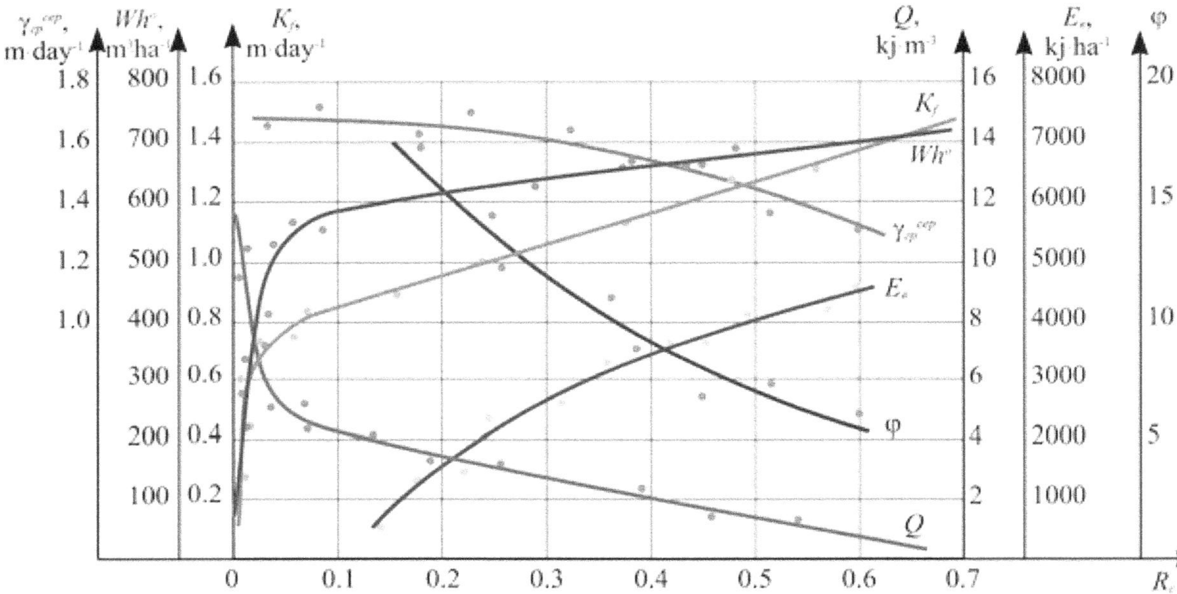

FUTURE RESEARCH DIRECTIONS

For successful practical implementation of the proposed solution, it is necessary to work out the following tasks:

- scientific and practical justification for changing approaches to the use of agro-ameliorative measures to maintain the functioning of agriculture. lands in changing climatic conditions;
- determination of criteria for improvement of deep loosening technologies on ecological and economic principles in accordance with modern conditions and requirements;
- practical implementation of scientific and technical developments and evaluation of their effectiveness based on research results.

CONCLUSION

In today's variable climatic conditions, it is advisable to switch from the traditional and most common technologies of slot and strip loosening to continuous in profile and area, as well as differentiated in deep horizons loosening of the soil.

Improved deep loosening is an effective adaptive tool for the accumulation of free-soil moisture in case of the transition from technologies of slot and strip loosening to continuous in area and differentiated in depth soil loosening.

Conducted continuous layer-by-layer deep loosening according to improved principles allowed to obtain medium- and small-lumpy soil structure in the upper part of the profile and large- and medium-lumpy structure in the lower part, according to agrotechnical requirements and functional purpose of soil layers.

Due to the new principle of deep loosening, for the first time it became possible to control the process of loosening of the soil, namely, to obtain the desired degree of loosening in each horizon of the vertical profile of the soil, that is, to differentiate the loosening structure by depth.

The proposed solution is one of the most effective adaptive means of increasing land productivity in agricultural production and at the same time an alternative measure of expensive reconstruction of existing ameliorative systems in the absence of sufficient funds for this in variable climatic conditions. Its implementation in the conditions of Polesie of Ukraine contributed to the improvement of water-physical properties, water permeability and accumulating ability of drained the most common sod-podzolic soil that provided an increase in the yield of cultivated crops by 20 ... 40%, as well as an increase in the duration of its aftereffect to 3-4 years in case of payback no more than 1 year.

REFERENCES

Celik, A., & Raper, R. L. (2012). Design and Evaluation of ground-driven rotary subsoilers. *Soil & Tillage Research*, *124*, 203–210. doi:10.1016/j.still.2012.06.010

Gaffarov, Kh.R. (2016). What determines the quality and energy performance of machines and implements for deep tillage? *Young Scientist. Technical science.* *5*(109). 31-32.

Hamzaa M.A., Anderson W.K. (2005). Soil compaction in cropping systems: A review of the nature, causes and possible solutions. *Soil and Tillage Research, 82*(2), 121-145.

Havrysh, V. S. (2013). Scientific and methodical recommendations for applying deep loosening on drained mineral soils of the Western Polissia of Ukraine. Department of Defense.

Iglesias, A., & Garrote, L. (2015). Adaptation strategies for agricultural water management under climate change in Europe. *Agricultural Water Management*, *155*, 113–124. doi:10.1016/j.agwat.2015.03.014

Kovalenko, P. etc. (2019). Evaluation of climate change in Ukrainian part of Polissia region and ways of adaptation to it. *Journal of Water and Land Development, 41*(IV-VI). 77-82.

Kravets, S. V., Kovan'ko, V. V., & Lukianchuk, O. P. (2015). Scientific bases of creation of earth-tiered cars and underground-moving devices: monograph. NUWEE.

Laziuta, M., Kabulova, B., & Gasanov, C. (2016). Research of soil tillage by advanced design slit ripper. *Agricultural Engineering. Research papers, 45*(2).

Li, X., Zhang, D., Zhang, R., & Osman, A. N. (2012). Performance of an oscillating subsoiler in reducing resistance. [St. Joseph, Michigan.]. *Presentation of the American Society of Agricultural and Biological Engineers., 12*, 1341191.

Lukyanchuk, O. P., Ryzhyi, O. P., & Ihnatiuk, R. M. (2017). Design of a tiered operating unit for deep differentiated tillage. [Ukraine.]. *Scientific Bulletin of the National Mining University, Is., 4*, 43–48.

Lukyanchuk, O. P., Turcheniuk, V. O., & Prykhodko, N. V. etc. (2019). Necessity and possible approaches to applying deep loosening when cultivating rice. INMATEH - Agricultural Engineering, 57(1),. 199–207.

Mazhayskiy, Y., & Rokochynskiy, A. etc. (2017). Environmental management of Polesie: monograf. Ryazan: Meshcher.f-VNIIIGiM them. A.N. Kostyakova.

Mazhayskiy, Yu., Rokochynskiy, A., Lukianchuk, O., Turcheniuk, V., Volk, P., & Chernikova, O. (2020). Deep loosening as effective adaptive agromeliorative practice on drained mineral soil of european polesie in variable climatic conditions. *Engineering for Rural Development., 19*, 28–35. doi:10.22616/ERDev.2020.19.TF009

Medvedev, V. V., Laktionova, T. M., & Pocheptsova, L. G. (2004). Influence of soil structure on filtration ability. [Ukraine.]. *Herald of Agrarian Sciences., 4*, 5–6.

Odey, S. O., & Manuwa, S. I. (2016). Design steps of narrow tillage tools for draught reduction and increased soil disruption - a review. *Agricultural Engineering International: CIGR Journal, 18*(1), 91–102.

Odey Simon, O., Ovat Friday, A., & Okon Orok, O. (2018). Draughts, Power Requirements and Soil Disruption of Subsoilers. *International Journal of Emerging Engineering Research and Technology., 6*(90), 24–38.

Panchenko, A. N. (1999). The theory of soil pulverization by tillage implements): monograph. DGAU, 140 p. Dnepropetrovsk, Ukraine.

Pivovar, V. (2011). Deep rippers. *Information and analytical newspaper Agro-business today*. http://agro-business.com.ua/agro/mekhanizatsiia-apk/ item/847-hlybokorozpushuvachi.html

Ward, P., Flower, K., Cordingley, N., Weeks, C., & Micin, S. F. (2012). Soil water balance with cover crops and conservation agriculture in a Mediterranean climate. *Field Crops Research, 132*, 33–39. doi:10.1016/j.fcr.2011.10.017

Zelenin, A. N. (1969). Laboratory workshop on cutting soil: textbook. IOP.

Chapter 15
Energy Intensity of Structural Grinding of the Soil by Deep Loosening

Olexander Lukyanchuk
National University of Water and Environmental Engineering, Ukraine

ABSTRACT

In changing climatic conditions, the deep continuous loosening of soils can be the most effective alternative measure for expensive reconstruction of existing reclamation systems, and, at the same time, an effective adaptive means for the accumulation of free-soil moisture. The accumulation efficiency depends on the structure of the loosened soil. The value of the proportionality coefficient of the Bond comminution hypothesis was experimentally established, which makes it possible to determine the energy saturation of technical means for loosening a certain soil mass. It has been established that the required capacity of technical means for deep loosening is approximately 2 times less for sandy soil than for loamy soil. So, with a total working width of 2 m, a working speed of 2 m / s for deep loosening (by 0.75 m) and 4 m / s for the basic tillage (by 0.25 m), it is necessary to have the power of 170- 240 kW of technical equipment for sandy soil and 390-440 for loamy soil

BACKGROUND

In changeable climatic conditions, the deep continuous loosening of mineral soil developed by us can be the most effective alternative measure for expensive reconstruction of existing reclamation systems, and, at the same time, an effective adaptive means for the accumulation of free-soil moisture (Iglesias, Garrote, 2015; Kovalenko etc., 2019; Mazhayskiy etc., 2020).

Deep loosening of ameliorated lands is carried out by special tools to a depth of 0.6...0.8 m or more to ensure water drainage through loosened soil to closed collectors (drains). They are used on heavy mineral soil, as well as on light loam and sandy loam soil with the presence of dense, poorly permeable to water layers in the subsoil horizons (at a depth of 0.3...0.4 m and deeper). Loosening of drained soil to a depth of 0.6...0.8 m against a background of closed drainage improves their water-physical proper-

DOI: 10.4018/978-1-6684-8248-3.ch015

ties and increases the yield of cultivated crops. The influence of the depth of cultivation on the work of drainage is manifested in an increase in drainage runoff and a decrease in surface runoff. The depth of cultivation has the greatest influence in the first year after it is carried out, in the future its influence decreases.

The depth of loosening depends on the depth and strength of poorly permeable soil horizons. It should be 0.2…0.3 m less than the minimum depth of drainage, i.e., 0.6…0.7 m. When loosening to a depth of 0.6…0.7 m, chemical analyses of the soil in the loosening zone are performed. If harmful compounds are detected, it is recommended to reduce the depth of loosening.

On soils with water-resistant layers and moderate waterlogging, located on poorly drained plains and levelled watersheds, the depth of loosening can be 0.5…0.6 m.

Loosening drained soils to a depth of 0.6…0.7 m on the background of closed drainage improves their water and physical properties and increases the yield of cultivated crops.

It was found that increasing the depth of loosening from 0.4 to 0.6 m, as a rule, does not affect the density and overall porosity of the arable soil layer. In the soil layer of 0.25…0.4 m, the density may decrease and the porosity may increase (about 5% in the first year and 2…3% in three years). A more significant effect occurs in the layer of 0.4…0.6 m, in which the density decreases to 10%, and the total duty cycle increases by 15…20%.

Depth of loosening affects the work of drainage. This is manifested in some increase in drainage and decrease in surface runoff. The depth of loosening has the greatest influence in the first year after its carrying out, further influence of loosening decreases. Improving the water and physical properties of the soil and increasing the effect of drainage with increasing depth of loosening has a positive effect on the yield of crops grown on drained lands.

The efficiency of deep loosening is influenced by the design of the working body. The use of tools with passive working bodies requires a lot of traction, and one of the criteria for justifying the depth of loosening is also the choice of the type of tractor-tractor and ripper.

Loosening intervals (distance between the working bodies of the leavening agent) depend on the water-physical properties of the soil and the design of the working bodies of the leavening agent. On soils with a filtration coefficient of less than 0.1 m/day, the distance between the working bodies should be 0.6…0.8 m. It can be increased to 1…1.2 m by reducing the density of the subsoil horizon and the presence of permeable layers.

For this purpose, a number of researchers are developing and evaluating the effectiveness of the working bodies of subsoilers in the destruction of compacted soil layers. At the same time, such works on various types and shapes of subsoilers are presented in terms of thrust and requirements for the energy of destruction of the soil massif, reduction of compaction and ease of operation. However, one of the important points that stands out in this case is the degree of soil destruction or the quality of loosening. Traditional rack-mount subsoilers are very unimportant in this case. In addition, it is highlighted that the development of subsoilers may be preferable to traditional means due to a decrease in the overall demand for specific engine power (Odey, Manuwa, Seth, 2018).

To reduce traction and reduce the use of power for increased soil crushing, some works suggest the use of vibratory (oscillatory) or rotary subsoilers (Celik, Raper, 2012; Li, Zhang, Zhang, Osman, 2012; Odey, Manuwa, 2016).

In addition, this direction has its drawbacks in the form of the need to install additional equipment, low productivity and speed.

In order to eliminate the shortcomings of traditional and active means of deep loosening and take into account the necessary technological changes, new multi-tiered working bodies were developed and proposed. (Mazhayskiy, etc., 2020; Odey, Manuwa, 2016; Laziuta, Kabulova, Gasanov, 2016; Mazhayskiy, Rokochynskiy etc., 2017; Ogbeche, Idowu, Theophilus, 2018). Figure 1.

Figure 1. Subsoilers for soil of various types (design option and completeness of loosening of the transverse profile): a) - for light soil (sandy loam and light loam); b) - for medium soil (medium loam); c) - for heavy soil (heavy loam, clay) 1 - complete structural loosening, 2 - partial loosening by cleavage)

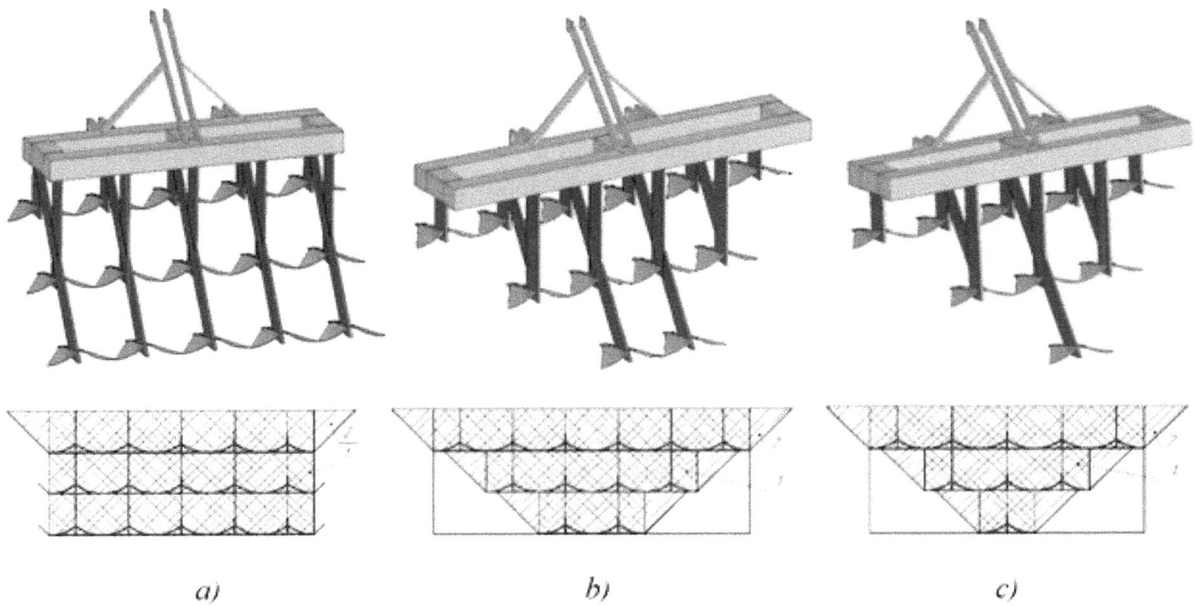

The spatial arrangement of the working bodies and the shape of the transverse profile of the loosened soil should be taken into account when solving the problem of moisture-accumulating loosening with the movement of excess moisture into the subsoil horizons during prolonged dry periods as a result of changes in the precipitation regime and the climate in general.

When performing deep loosening of the soil, it is important to choose the type of ripper and a tractor for it according to the required tractive effort. The choice of the type of the ripper depends mainly on the characteristics of the object: type of soil, type and intensity of water supply, relief, degree of soil acidity and the number of stony inclusions. It is also necessary to take into account the design of the shape of the working elements of the subsoiler, the speed of movement and the depth of tillage (Vetokhin, 2008; Odey, Friday, Draughts, 2018; Kravets etc., 2018).

The qualitative and energy performance of machines and tools for deep tillage depends on the type, shape and parameters of working bodies, their arrangement on the frame, physical and mechanical properties of the soil, etc. One of the most effective ways to reduce the energy consumption of machines and tools for deep tillage is the transformation (transition) of its blocked cutting by the working body into free or semi-free cutting (Gaffarov, 2016).

The traction resistance of the working equipment of a multi-tier subsoiler consists of the resistance of the movement of the working bodies, the excavation and transportation of soil along the surfaces of the working bodies and racks (Kravets, Kovan'ko, Lukianchuk, 2015; Gaffarov, 2016). Because the design of multi-tiered equipment is modular, by determining the traction resistance of one such module, it is possible to estimate the resistance of the working equipment as a whole and, respectively, the required force of the tractor. In this case, it is necessary to take into account the working conditions of the working bodies regarding the water-physical properties of the developed soil. Depending on their placement in each soil horizon, they can perform blocking, semi-blocking or free excavation of the soil. The ratio of power costs between them is taken 2×1.5×1 (Gaffarov, 2016).

According to preliminary research (Kravets, Kovan'ko, Lukianchuk, 2015; Lukyanchuk, Ryzhyi, Ihnatiuk, 2017), one of the main parameters which the traction resistance and soil loosening parameters depend on are the tillage depth and the final radius of curvature of the surface of the working bodies.

The accumulation efficiency depends on the structure of the loosened soil. Due to the significant amount of work with continuous deep loosening and the need for significant comminution of the soil massif, this process is quite energy-consuming (Odey, Manuwa, 2016; Lukyanchuk, Ryzhyi, Ihnatiuk, 2017; Panchenko, 1999; Lukyanchuk, Turcheniuk, Prykhodko etc., 2019).

The process of soil comminution can be viewed from different perspectives. Classical mechanics provides the following: external forces cause the accumulation of internal energy of elastic deformations, stresses in the soil body increase until the tensile strength is exceeded as a result of the concentration of stresses in some local defect. Then the crack begins to develop, accompanied by a redistribution of the energy of elastic deformations, part of which is converted into the energy of the newly formed surfaces. This part of the energy is the useful energy of the comminution. The rest of the energy is for elastic compression deformations and is dissipated in the form of heat and other types of energy.

A feature of the physics of the process is the wave nature of the deformation and comminution process. In such a process, there is a radial divergence of the motion of particles with the transformation of local compression into tension and movable deformations of the entire volume of the soil element. In addition, the features of soil rheology are used, namely, the dependence of the tensile strength of the soil on the rate of load application (Panchenko, 1999).

In the literature, there are not recognized, physically adequate and scientifically grounded theories that allow to analytically determine the degree of soil comminution during various types of tillage, depending on the parameters of the soil and working bodies. There are only a few attempts to indirectly determine soil comminution. So, V.V. Katsigin theoretically determined the effect of the speed of tillage machines on the intensity of soil comminution, using the hypothesis of I. Pokrovskiy, who believed that the destruction of material occurs at a certain amount of energy absorbed by a unit volume of matter. V.V. Katsigin argues that the comminution of the layer depends not so much on the magnitude of the stress as on the difference in the stress state of neighboring soil elements, that is, on the stress gradient and on the nature of the stress state of the entire soil layer ї.

As it was established above, when the working body moves in the soil, a stressed zone is formed in front of it. The stress gradient in this zone depends on the intensity of absorption of the energy flux, transmitted in the wave process, by the soil. Obviously, the separation of the layer into separate fractions will be the higher, the more energy is absorbed in the shear zone (shearing). Therefore, the smaller the volume of the deformed layer, the greater the effect of soil deformation can be expected.

Real soil is permeated with a large number of different dislocations (defects): cracks, voids, foreign inclusions, etc. As the stress state of the soil increases so the separate fractures occur under the action of the applied load around some defects, for example, cracks as stress concentrators. This occurs primarily in places where the stress is greater than the strength of inter-aggregate bonds. At these moments, avalanche-like destruction of the monolith formation occurs, and it crumbles into separate aggregates.

The intensity of soil comminution with increasing speed first increases sharply, and then this growth practically stops. This is due to the scale factor, that is, the dependence of strength on the size of the destroyed body. The process of coarse comminution, as already mentioned above, can be represented, basically, as the development of existing defects in the structure. With the transition to comminution small fractions, defects become rarer in them, and the conglomerates become more and more durable. The energy consumption for tillage begins to grow rapidly with the approach of working speeds to the speed of propagation of plastic deformations, and the degree of comminution increases insignificantly. Therefore, there are physical limits of working speeds, exceeding which is irrational, for each type of tillage machines.

The classical laws of comminution relate the energy (work) that needs to be spent on the comminution (destruction) process of a material with the dimensions of the finished product.

In order to determine the energy required for the soil comminution, several hypotheses have been developed: the surfaces hypothesis of P. Rittinger indicates the proportionality of the energy of the newly formed surface; the Kirpichev-Kick volume hypothesis – about the proportionality of energy to the volumes or masses of the shredded bodies; the combined hypothesis of P. Rebinder - about the proportionality of the comminution energy to the surfaces formed in the volume of the crushed body (Panchenko, 1999).

All comminution laws can be presented in the form of a generalized form proposed by R. Charles

$$dE = k\frac{dx}{x^n}, \tag{1}$$

where dE - is the energy required to reduce a piece of rock with x size by dx value; k - is the proportionality coefficient; n - is an exponent depending on the size of the parts (constant).

However, the most practical one is the comminution hypothesis, which was proposed by Bond, in the form of the obtained regularity

$$E = k_n\left(\frac{1}{\sqrt{d_{cep}}} - \frac{1}{\sqrt{D_{cep}}}\right), \tag{2}$$

where k_n - is the proportionality coefficient, H·m^{1,5}/kg d_{cep} и D_{cep} - are the average sizes of the initial and crushed product, m; M - is the mass of the material that is crushed, kg.

The Bond equation can be obtained by integrating Charles' generalizing formula (1) with n= 1.5. (2).

The relative linear dimensions d_{cep} и D_{cep} can be defined as geometric or arithmetic mean values.

The objective of this work is to ultimately establish ways to reduce the energy consumption of the work of subsoilers.

Having determined the energy consumption for the soil comminution and comparing it with all the energy that is spent by the subsoiler for loosening the soil mass, we obtain its efficiency, the analysis of which, in turn, will allow us to evaluate the effectiveness of possible ways to improve the technological and economic performance indicators.

The hypothesis proposed by Bond takes into account both the energy for the formation of new surfaces and the energy for deformation of the body. Determining the dimensions of the input and output product, as well as their mass, is not difficult. The biggest problem in using this relation is to establish the value of the proportionality coefficient. For this, the experimental research was carried out.

METHODS AND TECHNIQUES

Experimental research was planned and carried out on a soil canal on the basis of the methods of full factorial experiment, probability theory and mathematical statistics. The soil canal is a rectangular concreted container with dimensions of 9.2×1.8×1.5 m. It is filled with previously loosened and compacted loamy soil with the following fractional content of particles: clayey <0.002 mm - 27%; dustlike 0.002 ... 0.05 mm - 23%; sandy 0.05 ... 2 mm - 45%, gravel> 2 mm - 5%.

A dynamometric trolley with experimental working equipment moves along the rails, which are laid on the upper plane of the vertical concrete walls of the channel. Racks with working bodies are installed on the trolley frame (Figure 2).

Several prototypes of loosening working bodies with different parameters were made to study the effect. The design of the trolley (Figure 2, item 11) made it possible to change the depth of loosening *h* and the final radius of the cross-section of the surface of the loosening working bodies *R*. The height of the tier was changed by moving the working body rack relative to the frame of the trolley, and the radius - by installing replaceable loosening working bodies (Figure 3).

Figure 2. Diagram of a soil channel with a dynamometer trolley (1 – reducer; 2 – gear change box; 3 – electric motor; 4 – creeper gear; 5, 20 – chain drives; 6 - winch of preliminary tension of the cable; 7 – rail track; 8 – soil channel; 9, 12, 15, 16, 17 – guide blocks; 10 – road rollers; 11 – trolley; 13 – electronic dynamometer; 14 – support; 18 – traction cable; 19 – drum)

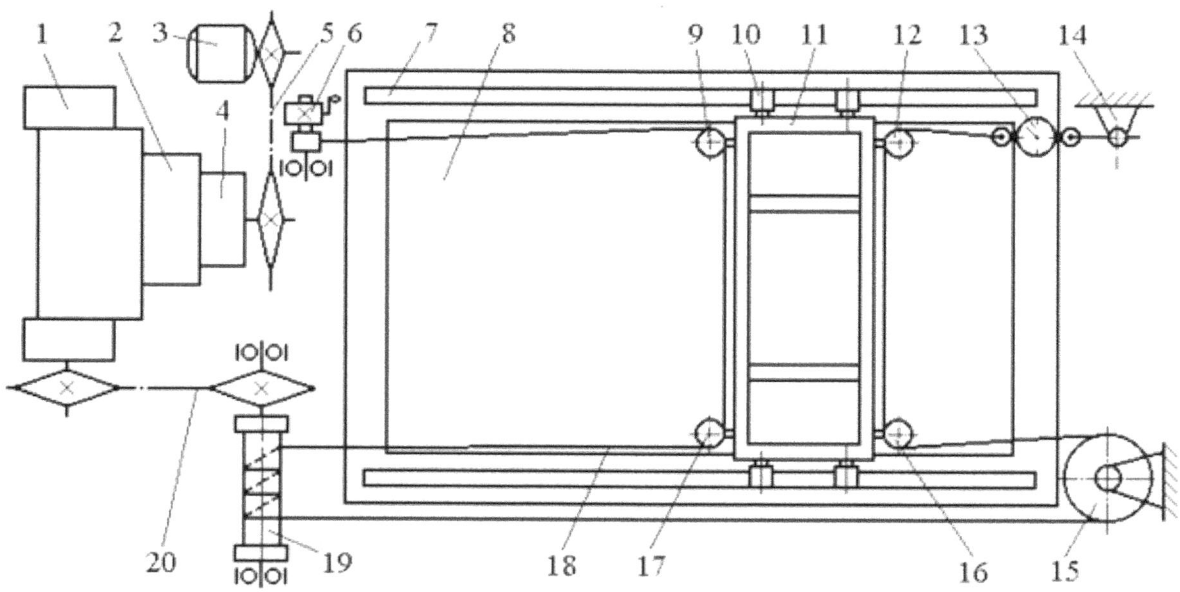

Figure 3. Installation of the research equipment (1 - dynamometer trolley;2 - rack;3 - loosening working body)

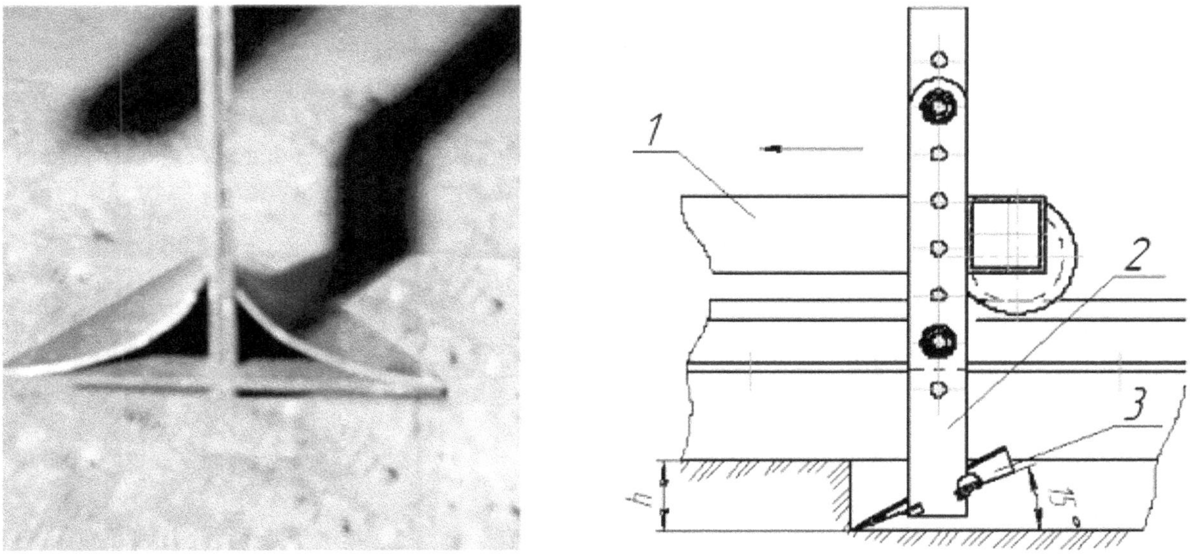

The change in the depth of soil excavation was carried out by the formation of a stepped relief of the soil channel with a height corresponding to the thickness of the soil layer. The length of one step was at least 2 m.

The tractive effort was measured by an electronic dynamometer built into the trolley's hoist system. After passing the trolley, soil samples were taken at each site to determine its moisture content, the fractional structure of the loosened soil was estimated. After obtaining the experimental data, the reliability and adequacy were checked according to the generally accepted method.

In order to establish the significance of the influence of the depth of loosening and the final radius of the cross-section of the working body on the traction resistance, a regression analysis of the data of experimental research was carried out. The regression equations are presented as a two-factor model. The implementation of the program and the results of experiments, according to the planning matrix of the factorial experiment of the 2^2 type, are shown in the Table 1.

Table 1. Traction test result

N°	Factors		Tractive effort, kN				Relative error, ε, %
	h, m	R_k, m	Repeatability			Average	
			1	2	3		
1	0.05	0.15	0.35	0.40	0.43	0.39	7.34
2	0.05	0.25	0.25	0.31	0.30	0.29	8.53
3	0.15	0.15	1.65	1.73	1.86	1.75	4.33
4	0.15	0.25	1.30	1.35	1.45	1.37	4.07
Average value:							6.07

After the transition from code values to natural values, we obtained the following regression model:

$$P_{T(kN)} = 0.218 + 12.167 \bullet h_{(m)} - 2.433 \bullet R_{k(m)} \quad (3)$$

Checking the significance of the coefficients of the regression equation by the Student's test showed that all the coefficients are significant.

Graphical displays of the regression analysis of the response of the dependence of the tractive effort on the depth of excavation and the final transverse radius of the surface according to model (3) are shown in the Figure 4.

Figure 4. The surface of the response of the dependence of the tractive effort on the depth of excavation and the final transverse radius of the surface

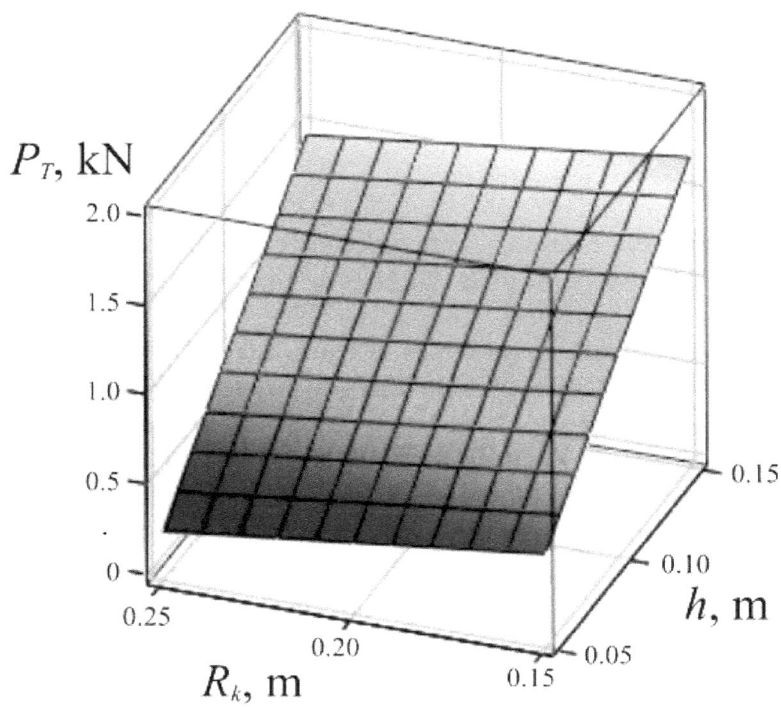

Thus, the obtained results indicate that the tractive effort depends more on the depth of soil excavation than on the final transverse radius of the surface of the working body. At the same time, the presented data of experimental research confirm our preliminary theoretical research (Kravets, Kovan'ko, Lukianchuk, 2015; Lukyanchuk, Ryzhyi, Ihnatiuk, 2017).

As an example of the practical application of the obtained results, we use the obtained regression dependence (1) to select a tractor. Comparative characteristics of three-tier subsoilers with a nominal working width, designed to accommodate 6 loosening working bodies of the same size in the plan of the upper tier, are presented in the Table 2.

Table 2. The comparative characteristics of subsoilers with different loosening profiles

N° p/p	Profile Scheme	Placement of working bodies in a tier	Soil excavation effort, c.u. *	Completeness of loosening of full profile		Relative force of completeness of loosening, c.u. *	
				general	structural	general	structural
1	Rectangular, Fig. 1, a	in-line	18	1.04	1.00	17.3	18.0
2		chess	15			14.4	15.0
3		wedge	15			14.4	15.0
4	Trapezoidal, Fig. 1, b	in-line	12	0.80	0.67	15.0	17.9
5		wedge	10.5			13.1	15.7
6	Triangular, Fig. 1, c	in-line	9	0.63	0.50	14.3	18.0
7		wedge	7.5			11.9	15.0

The note: * c.u. - the effort of soil excavation by one loosening working body, 1 - structural loosening, 2 - partial loosening by a chip

The coefficient of loosening completeness (general and structural) is defined as the ratio of the area of the loosened soil in the frontal plane to the area of the conventional loosening rectangle in the frontal plane. The area of the conventional rectangle is determined by the total width and depth of loosening of the working body. Accordingly, the coefficient of general loosening characterizes the increase in the volume of loosened soil relative to its initial volume, and structural one - the volume of formation of the necessary agronomically valuable structural aggregates in the loosened soil volume relative to the initial volume.

According to the obtained data, the best result, from the point of view of unit costs for the structural moisture-accumulating loosening of the soil, is provided by the option with the wedge-shaped arrangement of the working elements in the tier. Moreover, this result does not depend on the shape of the formed profiles of the loosened soil. With regard to the general loosening of the soil, the formation of a triangular profile is the best option. This form is most expedient to use when loosening heavy loamy and clayey soil.

Depending on the shape of the soil loosening profile, the graphical dependences of the traction resistance on the loosening depth for the three-tier version, calculated using the model (1), will have the following form (Figure 5).

In the general case, for the working equipment of a multi-tiered subsoiler, the coefficient of the total completeness of loosening of the transverse profile of the soil varies within 0.63...1.04, and the coefficient of completeness of the structural moisture-accumulating loosening - within 0.5...1.0. The values of these coefficients decrease with a change in the soil loosening profile from rectangular to triangular one.

Experimental research of the proportionality coefficient is planned and carried out on the basis of the methods of full factorial experiment, probability theory and mathematical statistics.

The tasks that were solved during the experiment: determination of the dependence of the degree of comminution on the type of soil (by the average size of lumps), determination of the comminution energy of the soil sample under research.

To carry out the experiment, we used a soil sample in the form of a cube with a side equal to $D_k = 30$ cm, since this corresponds to the possible size of the loosening layer by a separate working body of the longline subsoiler (Lukyanchuk, Ryzhyi, Ihnatiuk, 2017; Lukyanchuk, Turcheniuk, Prykhodko etc., 2019).

Figure 5. Traction resistance of soil excavation by a three-level subsoiler with a working width of 2 m (1 - rectangular profile; 2 - trapezoidal profile; 3 - triangular profile; dashed line - row arrangement of working bodies in a tier; solid line - wedge placement of working bodies in a tier)

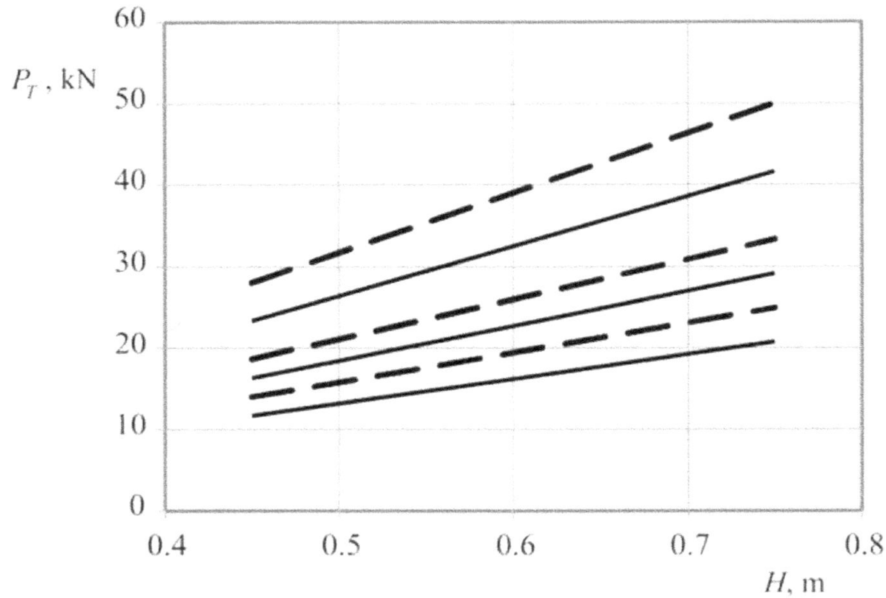

The volume of the selected cube is $V_k = D_k^3 = 0.3^3 = 0.027$ m^3.

The experiment was carried out for sandy and loamy soil types. The factors of the experiment were the height of the fall and the type of soil. The experiment was carried out at three levels.

The characteristics of the soil are shown in the Table 3.

Table 3. Physical and mechanical properties of the researched soil

Soil type (category)	Light sandy soil (I)	Medium sandy soil (II)	Light loamy soil (II)
Relative humidity, %	11.1	13.0	14.3
The number of hits of density, m	3.0	5.0	8.0
Density, t / m3	1.85	2.00	2.15
Plasticity number	2.1	4.4	9.1
Consistency	0.75	0.2	-0.2
Internal friction force, MPa	0.002	0.015	0.06
Internal friction angle, degrees	18	26	25

The comminution process was simulated by providing a soil sample with a certain potential energy. To do this, this sample was raised to a certain height and released. After interaction with the day surface, the soil sample was crushed (Figure 6).

Figure 6. The soil comminution in individual experiments after falling from a height of h = 1.1 m (a - light sandy soil, b - medium sandy soil, c - light loamy soil)

a b c

The required height was determined taking into account the speed at the end of the fall, equal to the speed of movement of the working body ν in the range of 3.1...5.4 m/s according to the free fall formula

$$h = \frac{v^2}{2g} \quad (4)$$

After comminution the soil sample, the obtained fractions were sorted. The following fraction intervals were selected: > 15 cm; 10 ... 15 cm; 5 ... 10 cm; 3 ... 5 cm; 1,5 ... 3 cm; <1,5 cm.

Fractions of > 15 cm; 10 ... 15 cm; 5 ... 10 cm; 3 ... 5 cm were sorted manually, and fractions of 1,5 ... 3 cm;<1.5 cm - using a sieve with apertures of 1,5x1,5 cm, that is, everything that passed through the sieve belonged to the fraction of <1,5 cm, - and what remained on the sieve belonged to the fraction of 1,5... 3 cm.

After separation into fractions, each fraction was weighed for each type of soil and different heights (Table 4).

The average size of the fraction was determined by the equation

$$d_{cp} = \sum_{n}^{j=1} d_j \cdot Q_j / Q_n \quad (5)$$

where d_{cep} – is the average size of the fraction, m; Q_j - mass of lumps in the *j*-th fraction, kg; Q_n is the total mass of the cube of the researched soil, kg.

Knowing the height of the rise of the sample, by means of mathematical transformations, the potential energy, which was expended in comminution the soil sample, was determined.

The degree of comminution *i* was determined by the ratio of the average sizes of fractions before and after comminution

Table 4. Experiment Results

No.	Soil type	Sample mass, kg	*h*, g	> 15 cm	10 ... 15 cm	5 ... 10 cm	3 ... 5 cm	1,5 ... 3 cm	<1,5 cm
				Percentage of fractions, %					
1	Light sandy soil	49	0,5	26	0	11	10	16	37
2	Medium sandy soil	54		36	21	13	9	6	16
3	Light loamy soil	58		69	4	6	5	6	11
4	Light sandy soil	49	1,0	21	5	7	7	9	51
5	Medium sandy soil	54		26	20	10	14	6	24
6	Light loamy soil	58		36	19	18	6	4	16
7	Light sandy soil	49	1,5	10	10	10	12	12	45
8	Medium sandy soil	54		24	10	14	14	18	21
9	Light loamy soil	58		32	17	13	9	10	19

$$i = \frac{D_{cep}}{d_{cep}} \quad (6)$$

The average size before comminution D_{cep} is equal to the side of the cube of the researched soil, that is, $D_{cep} = 0.3$ m.

The proportionality coefficient, according to the hypothesis of soil comminution proposed by Bond, taking into account the expression for the potential energy, was determined as follows

$$k_n = \frac{g \cdot h}{\frac{1}{\sqrt{d_{cep}}} - \frac{1}{\sqrt{D_{cep}}}} \quad (7)$$

where h - is the height of the soil fall, *m*; where h is the height of the soil fall, *m*; d_{cep} and D_{cep} are the average sizes of the original and comminution product, g.

The proportionality coefficient was determined for each experiment and then averaged over the type of soil. Having obtained these data, using the Bond formula, we determined the energy $E_{п}$ for the soil comminution, reduced for convenience to 1 m^3, with the formation of average particle sizes of 10 and 25 mm, as the most desirable in the structure of loosened soil, respectively, arable and lower layers (see the Table 5). The calculation of the d_{cep} value is given in the table 5.

Based on the obtained values, it is possible to calculate the required minimum capacity of technical means for deep loosening and basic tillage. So, with a total working width of 2 m, a working speed of 2 m/s for deep loosening (by 0.75 m) and 4 m / s for the basic tillage (by 0.25 m), it is necessary to have the power of 170-240 kW of technical equipment for sandy soil and 390-440 for loamy soil.

Table 5. Results of calculating comminution indicators

No.	Soil type	Density, t / m3	h, m	d_{cep}, m	i	k_n, H·m1,5/kg	Average value of kn, $H\cdot_m 1,5$/kg	Reduced comminution energy, E_n, kJ	
								10 mm	25 mm
1.			0.5	0.1354	2.22	5.49			
2.	Light sandy soil	1.85	1.0	0.0730	4.11	5.75	5.71	95.1	56.2
3.			1.5	0.0534	5.62	5.88			
4.			0.5	0.1521	1.97	6.64			
5.	Medium sandy soil	2.00	1.0	0.0928	3.23	6.73	6.71	120.8	71.5
6.			1.5	0.0626	4.79	6.78			
7.			0.5	0.1993	1.51	11.84			
8.	Light loamy soil	2.15	1.0	0.1332	2.25	10.73	11.29	218.5	129.2
9.			1.5	0.1023	2.93	11.31			

FUTURE RESEARCH DIRECTIONS

For successful practical implementation of the proposed solution, it is necessary to work out the following tasks:

- improving the methodology for creating technical means of deep loosening of the passive type for differentiated redistribution and maximum accumulation of soil runoff;
- development of models of the physical relationship of the structure of the soil environment with the working elements of adaptive technical means of deep loosening;
- formation of methods for determining the parameters of advanced technical means of deep loosening on energy-saving principles;

CONCLUSION

Taking into account the importance of the influence of the volume of energy costs not only on the cost of production of agricultural products, but also on the environmental component through the technogenic impact on the soil and emissions of exhaust gases during the operation of tractor engines, the correct layout of the working equipment provides potential opportunities for their reduction. From the point of view of unit costs, with different methods of arrangement of the working unit, it is possible to achieve their reduction by 12.7 ... 16.8% relative to the criterion of complete loosening of the soil profile and by 12.3 ... 16.7% - relative to the criterion of structural moisture-accumulating loosening. To achieve this result, it is necessary to choose the option with wedge placement of working elements in the tier, regardless of the shape of the formed profiles of the loosened soil.

Considering the shape of the soil profile, a slightly better option is to form a triangular profile of the loosened soil. It is possible to reduce energy costs by 9.2 ... 17.4% relative to the criterion of complete loosening of the soil profile and by 4.5% - relative to the criterion of structural moisture-accumulating loosening.

In the general case, the most advantageous option in terms of specific indicators will be the option with wedge placement of working elements in the plan of each tier. The traditional truck tractors, which are available in most agricultural enterprises and are widely used in agricultural production, with a traction power of 70 to 300 kW are suitable for aggregation. The established traction resistance varies within a fairly wide range - 11 ... 50 kN. Its efficiency in relation to other possible options for traction will be 12 ... 17% gain in the required tractive effort.

The effectiveness of the Bond's comminution hypothesis was confirmed: with an increase in the energy provided to the soil, the degree of its comminution increases.

The value of the proportionality coefficient of the Bond's comminution hypothesis was experimentally established, that makes it possible to determine the energy saturation of technical means for loosening a certain soil mass.

The required power of technical means for deep loosening of soil is almost 2 times less for sandy soil than for loamy soil. Conclusions can be represented using plain text or numbered using appropriate style.

REFERENCES

Celik, A., & Raper, R. L. (2012). Design and Evaluation of ground-driven rotary subsoilers. *Soil & Tillage Research*, *124*, 203–210. doi:10.1016/j.still.2012.06.010

Gaffarov Kh.R. (2016). Traction resistance of the central working body of the subsoiler. *Technical science. 2*(106), 130-132.

Gaffarov, Kh.R. (2016). What determines the quality and energy performance of machines and implements for deep tillage? *Technical science, 5* (109), 31-32.

Iglesias, A., & Garrote, L. (2015). Adaptation strategies for agricultural water management under climate change in Europe. *Agricultural Water Management*, *155*, 113–124. doi:10.1016/j.agwat.2015.03.014

Kovalenko, P. etc. (2019). Evaluation of climate change in Ukrainian part of Polissia region and ways of adaptation to it. *Journal of Water and Land Development, 41*(IV-VI), 77-82.

Kravets, S. V., Kovan'ko, V. V., & Lukianchuk, O. P. (2015). Scientific bases of creation of earth-tiered cars and underground-moving devices: monograph. NUWEE.

Laziuta, M., Kabulova, B., & Gasanov, C. (2016). Research of soil tillage by advanced design slit ripper. *Agricultural Engineering. Research papers, vol. 45, Nº 2. Policy*, *59*, 152–164.

Li, X., Zhang, D., Zhang, R., & Osman, A. (2012). Performance of an oscillating subsoiler in reducing resistance. Presentation of the American Society of Agricultural and Biological Engineers, St. Joseph, Michigan.

Lukyanchuk, O. P., Ryzhyi, O. P., & Ihnatiuk, R. M. (2017). Design of a tiered operating unit for deep differentiated tillage. [Ukraine.]. *Scientific Bulletin of the National Mining University, Is.*, *4*, 43–48.

Lukyanchuk, O. P., Turcheniuk, V. O., & Prykhodko, N. V. etc. (2019). Necessity and possible approaches to applying deep loosening when cultivating rice. INMATEH - Agricultural Engineering, 57, 199–207.

Mazhayskiy, Y., & Rokochynskiy, A. etc. (2017). Environmental management of Polesie: monograph. Ryazan: Meshcher.f-VNIIIGiM them. A.N. Kostyakova,

Mazhayskiy, Yu., Rokochynskiy, A., Lukianchuk, O., Turcheniuk, V., Volk, P., & Chernikova, O. (2020). Deep loosening as effective adaptive agromeliorative practice on drained mineral soil of european polesie in variable climatic conditions. *Engineering for Rural Development.*, *19*, 28–35. doi:10.22616/ERDev.2020.19.TF009

Odey, S. O., & Manuwa, S. I. (2016). Design steps of narrow tillage tools for draught reduction and increased soil disruption - a review. *Agricultural Engineering International: CIGR Journal*, *18*(1), 91–102.

Odey, S. O., & Manuwa, S. I. (2018). Subsoiler Development Trend in the Alleviation of Soil Compaction for Sustainable Agricultural Production. *International Journal of Engineering Inventions Volume 7*(8).

Odey Simon, O., Ovat Friday, A., & Okon Orok, O. Draughts, (2018). Power Requirements and Soil Disruption of Subsoilers. *International Journal of Emerging Engineering Research and Technology Volume 6*(9), 24-38.

Ogbeche, O. S., Idowu, M. S., & Theophilus, E. (2018). Development and performance evaluation of instrumented subsoilers in breaking soil hard-pan. *Agricultural Engineering International: CIGR Journal*, *20*(3), 85–96.

Panchenko, A. N. (1999). The theory of soil pulverization by tillage implements): monograph. DGAU.

Vetokhin, V.I. (2008). Design of deep rippers taking into account some aspects of soil deformation. *Machinery in agriculture. production, branch mechanical engineering, automation, 20,* 104-109.

Chapter 16
The Role of Drainage Systems in Ensuring the Barrier Functions of the Chornobyl Exclusion Zone

Oleksii Shevchenko
Ukrainian Hydrometeorological Institute, Ukraine

Volodymyr Bublias
Institute of Geological Sciences of the National Academy of Sciences of Ukraine, Ukraine

Mykola Panasiuk
Institute for Safety Problems of Nuclear Power Plants of the National Academy of Sciences of Ukraine, Ukraine

Dmytro Charnyi
The Institute of Environmental Geochemistry of the National Academy of Sciences of Ukraine, Ukraine

ABSTRACT

Drainage systems in the Chornobyl exclusion zone (CHEZ) are considered in the context of this zone's performance of barrier functions to contain the water transport of radionuclides beyond its borders. The principles of assessing the barrier resistance of catchment basins are described, taking into account landscape features and the dynamics of 90Sr removal; the role of drainage systems in the formation of removal of radionuclides at different stages of post-accident works and during periods of different water levels was analyzed, and effective water protection measures were proposed. Natural landscape and geochemical characteristics of catchments, not water protection measures and structures, are responsible for the formation of stable barrier functions. Various man-made facilities mostly only worsen the barrier capacity of the catchment in the long term; for example, the additional artificial drainage network due to the possibility of its regulation helps to reduce the removal of radionuclides in low-water years, but in high-water, it significantly weakens the barrier functions.

DOI: 10.4018/978-1-6684-8248-3.ch016

BACKGROUND

The Influence of the Drainage Network on the Volumes of ^{90}Sr Removal Outside the Chornobyl Exclusion Zone

There are 24 reclamation systems with a total coverage area of 280 km^2 in the Chornobyl Exclusion Zone (ChEZ) and mandatory resettlement zone. The area of the zone is about 2,400 km^2. The exclusion zone has a fairly developed hydrographic network: almost its entire territory lies within the Prypiat river basin. The only exception is the Brahinka river basin which flows into the Kyiv Reservoir from its left bank.

There was released (70–80)·10^{16} Bq of radioactivity was released during the disaster at the Chornobyl NPP on 26 April 1986 (the uncertainty limit for all releases is 50%). The total activity of fission products emission (without radioactive noble gases) was 1.85·10^{18} Bq, which corresponds to approximately 3.5% of the total activity of radionuclides in the reactor at the time of the accident (Baryakhtar et al., 1998). Later, with the help of electronic simulation, slightly higher values of the total activity of radionuclides ejected from the reactor were obtained: 5.3×10^{18} Bq (Steinhauser et al., 2014). According to the estimates of the Scientific and industrial association "Prypiat," there were about 5.55·10^{15} Bq of ^{137}Cs, 4.44·10^{15} Bq of ^{90}Sr, 3.7·10^{13} Bq of ^{239}Pu, and 2.44·10^{13} Bq of ^{241}Am in the soils of the exclusion zone (without taking into account the radioactivity localized in waste burials) in 1998. At the same time, about 63% of radioactive^{90}Sr outside the Chornobyl NPP industrial site and radioactive waste burials was in the area of drainage systems (mainly on the left bank of the Prypiat) (Shevchenko et al., 2011).

After the accident at the Chornobyl NPP, the drainage systems were put out of operation, and most with closed locks. In addition, water protection measures were carried out in the area; they included the construction of numerous cofferdams and dams on the main canals. They led not only to a decrease in the local runoff but also to long-term overwetting of the territory.

Dams and cofferdams constructed during 1986-1992 helped to retain about 2.2 TBq of ^{90}Sr(was determined according to the corresponding modules and flow rates in undisturbed conditions). However, this positive effect was weakened by the following events: an increase in ^{90}Sr and ^{137}Cs concentrations in canals and small rivers due to the flooding of contaminated areas of canal pound constructions with a consequent increase of these processes in the Prypiat. Flooding contributed to faster penetration of radionuclides through the soil profile (Shevchenko et al., 2011). In the areas contaminated mainly by condensation fallouts, ^{90}Sr migrated deeper and even more intensively.

Despite all measures, annually, up to 70% of ^{90}Sr from its total wash-off with Exclusion zone is removed from drainage systems and rivers with partially reclaimed catchments. In specific high-water years (1997-1998) or years with severe floods (1999), the wash-off of ^{90}Sr only from the northern part of the Pripyat reclamation system (about 100 km^2, Figure 1) was up to 20-40% of all removal of ^{90}Sr by the Prypiat from the ChEZ. At the same time, in terms of volume, it is only 0.1-1.3% of the runoff generated within the Zone (Shevchenko et al., 1999).

When the high-water year cycle began (1993-2001), thanks to the increase in precipitation and decrease in evaporation of more than 10%, there was a noticeable increase in the water surface of the Prypiat catchment (flooded and swampy areas) (Figure 1). As a result, the concentration of ^{90}Sr also increased (Shevchenko et al., 1999a).

Figure 1. Water-land change detected to the North from the old left-bank dam based on the multispectral satellite imagery (May 1986 -April 1994): brown color - increase in land surface; blue color - increase in the open water surface

Starting from 1998-2000, the carrying capacity of canal beds and the functioning of regulating hydraulic constructions in the separate drainage systems were restored.

As a result of natural decay and removal, the activity of ^{90}Sr and ^{137}Cs on the catchment area decreased by 2.2-2.3 times over 35 years. Some processes of secondary pollution (wash-off by precipitation from the surface, leaching from bottom sediments and flooded slopes of canals and river banks) lost their importance, while others (removal by groundwater) acquired a dominant role. Over time, climate changes also contributed to shifts in the distribution and course of secondary pollution processes. For example, removal from swamps by spring runoff had almost stopped, and groundwater pollution intensity had decreased due to the reduction of GWT.

Until 2000, increased removal of ^{90}Sr occurred not only during high-water years but also during the years of severe floods and freshets (1991, 1994, 1999), following the relatively high-water years (1990 – 608 mm of precipitation; 1993 – 692 mm; 1998 – 689 mm). Due to the problematic discharge, part of the water from the previous year remained accumulated on the catchment until the next flood, which washed off radionuclides from the unflooded earlier areas. Therefore, the regulated or dammed systems show an increased removal of ^{90}Sr for the next year after the high-water one (Figure 2), after the canals run out of their accumulative capacity. This way regulated watercourses during high-water seasons lost a significant part of dissolved and suspended radioactive substances (through gaps next to gateways and bypass filtration). However, regulated reclamation systems retained and accumulated part of the surface runoff, reducing radionuclides removal during low-water and moderate-water seasons.

Figure 2. Changes in the correlation coefficient (r) between the ^{90}Sr removal and the relative area of reclamation systems within river catchment (1) in comparison with the annual amount of atmospheric precipitation (2) (weather station "Chornobyl")

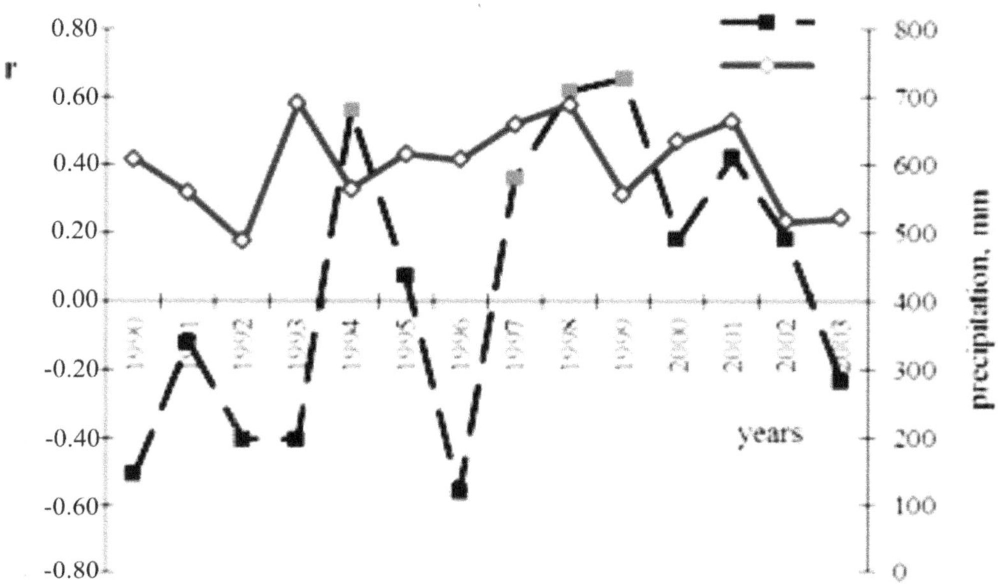

The decrease in ^{90}Sr concentration after the flood ends is faster for watercourses with a smaller area of reclamation systems and lower catchment regulation. In non-covered watercourses the spikes in the volumetric activity of ^{90}Sr (1.5-2.5 times higher than the annual average) are observed mainly only during spring floods or occasionally during winter ice jams. At the same time, in the low-flow canals, an increase in ^{90}Sr concentrations occurs during minor summer floods (1.2-1.5 times increase) and even relatively low-water winter periods.

The catchments of small rivers of the ChEZ are reclamated by 9-90%. Thanks to the extensive network of lateral canals, drainage systems can contribute to the retention of ^{90}Sr even during high-water years (as in 1993) or moderate-water years (1996), following several low-water years. However, different degrees of the systems' regulation make their influence ambiguous even in years with the same water content. Thus, the most reclamated and regulated, the most polluted northern part of the Pripyat reclamation system (flow through the breach in dam #7), made the most significant contribution to ^{90}Sr removal in 1993 - 1795 GBq, while the least reclamated and regulated catchments of the Uzh and Illia rivers made the most considerable contribution in 1996.

The dependence of the ^{90}Sr removal module on the regulation degree of a canal or river is well approximated by power-law graphs (Figure 3).

During the first years of observations (1990-1991), the share of removal (b_n) from non-covered catchments was 10-18 times higher than from covered watersheds; and during 2009-2012, these shares were almost equal: 0.013-0.06% and 0.013-0.055%, respectively. The levelling occurred mainly due to a decrease in b_n in the vast majority of systems, except for the most regulated catchment of the Brahinka River. The most intense decrease of b_n was on the least regulated Uzh River (from 0.234 in 1990-1994 to 0.02 in 2015-2019). This may indicate a positive role of regulation during the first years and a more accelerated "release" of non-covered systems from a significant part of radionuclides "prepared" for migration. On the other hand, the self-purification of surface waters in flowing systems was much faster, while radionuclide concentrations and their removal remained more stable in covered systems.

Figure 3. Dependence of the ^{90}Sr removal modules on the "water protection" regulation degree (dams and cofferdams in 1986) of the ChEZ watercourses for different years: 1 – 1998, 2 – 1999; 3, 4 – corresponding power-law trends

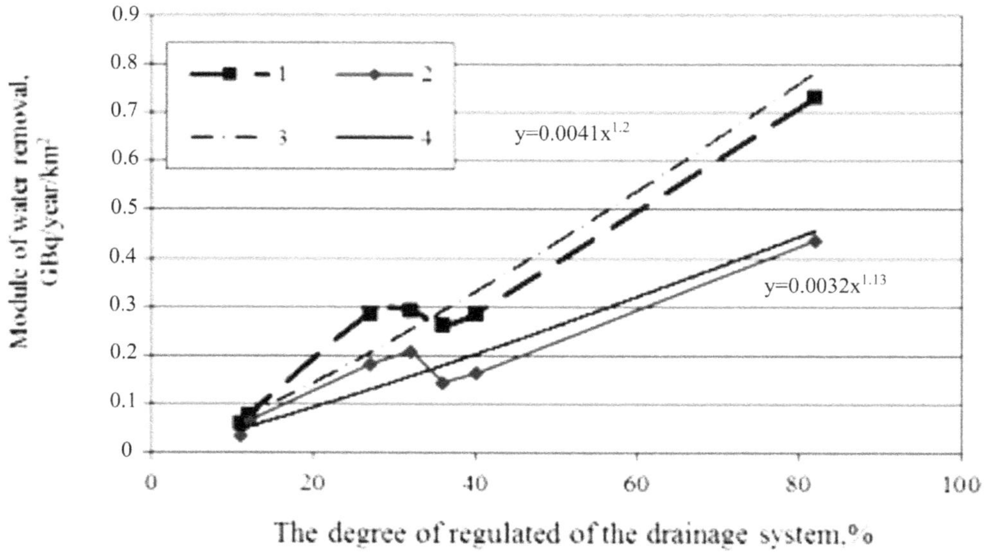

The influence of catchment reclamation (*Lr*) takes into account the linear regression equation (coefficient of determination $R^2 = 0.85$). Using it, we can calculate the removal of ^{90}Sr by watercourse (*W*), having data on the ^{90}Sr activity within the catchment area (Z_{Sr}, GBq) and the number of overwetted and filled with water depressions (*Nwz*):

$$W = -890.867 + 0.002 \cdot Z_{Sr} + 0.618 \cdot Nwz + 11 \cdot Lr \qquad (1)$$

There are five groups of reclamation systems in the Chornobyl exclusion and mandatory resettlement zone, depending on the reasons and forms of manifestation of changes in the water regime and balance:

I. *Permanently heavily waterlogged and flooded nonfunctioning drainage systems.* Waterloggingcovers the whole system or part of it. Waterlogging and flooding cause a breakdown of the barrier functions of catchments, especially in high-water seasons. This group also includes partially flooded former polders.
II. *Seasonally flooded, almostnonfunctioning systems or their sections.* Even having a high level of flooding in high-water seasons, these systems perform barrier functions (excessive removal of radionuclides beyond their borders does not occur).

III. *Self-flowing reclamation systems that are seasonally floodedduring high-water years.* During low-water seasons, over-drying of individual elements of the landscape occurs. It causes theburning of peatlands during droughts. GWT in the area can vary widely: from 0.5 to 4.0 m. The radioecological state is within the expected condition;bursts of removaloccur only in high-water years.

IV. *Locally flooded and almost non-flooded self-flowing unregulated systems* with drainage conditions close to natural, sometimes with active drainage capacity. The supply and discharge ratio is the closest to the natural one, while theremovalof radionuclides is the lowest. GWT is at depths from 0.9 to 3.0 m.

V. *Systems with a restored regulated regime, reconstructed in 1996-2004.* The local backwaters from beaver dams disturb the self-regulating and regulated runoff regime. The removalof radionuclides mostly depends on the density of surface pollution and water content of a year; GWT is at depths of 1.7–3.5 m, sometimes - up to 4.5 m.

When the drainage capacity of the Polisia landscapes is artificially increased, the share of groundwater in the total runoff increases from 30-40 to 60-78%. The washing soil regime under the conditions of the closed drainage leads to a significant increase in the removalof ^{90}Sr and still contributes to the accelerated recovery of contaminated areas.

During the first 3 years after shutting down the drainage systems, there was no significant decrease in the groundwater share in the total runoff because of the low-water content of these years. At the same time, ^{90}Sr existed mainly as non-exchangeable species, and surface water pollution within drainage systems was not too high. ^{90}Sr concentration in groundwater was 70-90% and caused by global fallouts during the years of active nuclear weapons testing. The difference in ^{90}Sr concentrations between surface and groundwater mainly was 4-20 times (Shevchenko et al., 2011), and the share of groundwater in the total removal of ^{90}Sr was relatively high until 1990 - up to 25%. Due to the flooding of large areas, there was accelerated contamination of groundwater. As their share increased in the low-water years (1989-1992), the share of the ^{90}Sr removal also increased to 30-32% (Shevchenko, 2018). When the high-water years begin, the share of the groundwater in the total runoff decreases, and their contribution to the removal of ^{90}Sr decreases to 0.2-2.7%. Since the mass destruction of fuel particles had not yet occurred at that time, the contamination of groundwater was slow, and the concentrations of ^{137}Cs and ^{90}Sr were lower than in surface waters, up to 200 times. After the reconstruction of several drainage systems (1999-2002), the concentration of ^{90}Sr in surface water decreased, and the share of groundwater in the total runoff increased sharply, which caused their share increase in the total removal of ^{90}Sr. The volumetric activity of ^{90}Sr in groundwater at the level of 0.1-0.8 Bq/dm^3 became indicators of the radioactivity level for the drainage systems areas in the exclusion zone. The contribution of groundwater to the removal of ^{90}Sr by canals has increased since 2000 to 36-40% and more, as its concentrations in groundwater and surface water become leveling (a constant decrease in surface water). From 2008 to 2020, the share of groundwater in the total runoff also gradually increased; it was associated with the cycle of low-water years. Moreover, the role of groundwater in the total removal of radionuclides had significantly increased since 2002-2003 due to the following factors: the mass dissociation of fuel particles, the increase in the percentage of ^{90}Sr soluble species, and the intensity of vertical migration of ^{90}Sr.

Over time, the role of other processes of surface water secondary contamination also changes. It is associated with the redistribution and transformation of primary radioactive fallouts. Before the beginning of the mass dissociation of fuel particles and the transition of ^{90}Sr into exchangeable and soluble species, there was an inverse dependence of the share of ^{90}Sr removal against its reserves on the catchment.

The significant influence of the physical and chemical species of ^{90}Sr on its removal formation with wateris also manifested in the dynamics of the removal modules compared to the removal share (Figure 4). Obviously, for the basins where pollution is formed mainly by the condensation (the Uzh river basin), and ^{90}Sr has significantly greater mobility (Kashparov, 2001), there is a clear advantage of removal share over the ^{90}Sr removal module.

Figure 4. Comparison of the average module values (1) and the share of 90Sr removal (2) in the periods: a – dominance of 90Sr in the fuel particles composition (1987-2001); b - dominance of 90Sr migratory species (2002-2011) in the catchment basins of: 1 – the Uzh river; 2 – the Veresnia river; 3 – the Sahan river, 4 – the Brahinka river, 5 – the Main Canal MC-1 of the Prypiat drainage system, 6 – the Illia river. Since 2000, no observations have been made on the Illia and Veresnia rivers

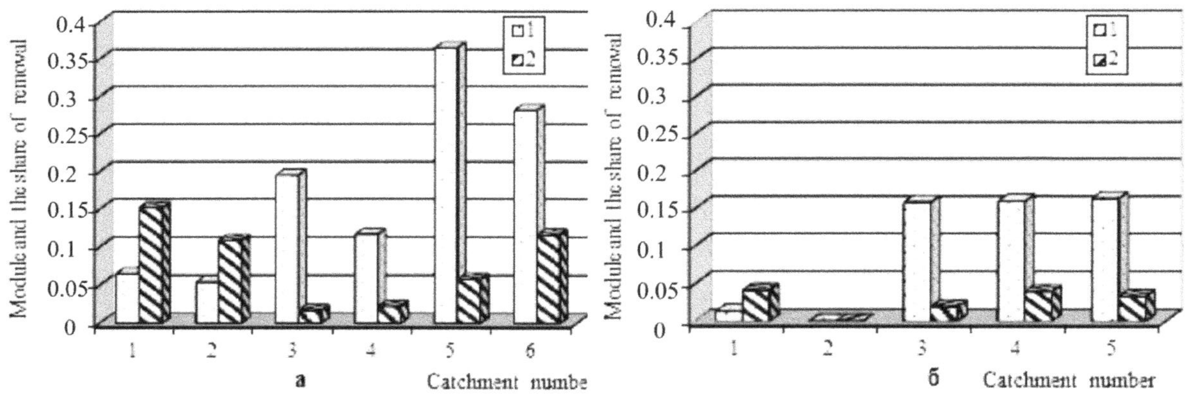

Thus, after the reclamation systems in the ChEZ were shut down, the first 12-13 years were characterized by: a threefold decrease in summer runoff during low-water years compared to the pre-disaster period; a reduction of the total runoff volumes; an increase in the GWT and reduction of its oscillation amplitude, regulating the capacity of the aeration zone, and flow gradients. Difficult runoff conditions led to flooding polluted areas adjacent to canals and rivers and forming eutrophic semi-flowing swamps. In high-water years, these swamps release a significant part of radionuclides reserves that have transformed into a water-soluble species. Although the runoff and removalof radionuclides were reduced in low-water years, the removalmodules from reclamation catchments within regulated canals were higher than from unregulated systems and rivers during high-water years. The accelerated contamination of shallow groundwater and its significantly prolongated drainage has become the major negative radioecological consequences of the runoff regulation.

Conceptual Principles of Handling Drainage Systems in Areas of Radioactive Damage

Differences in the annual volumes of ^{90}Sr removalfor differently drained watersheds are caused not only by the differences in the degrees and forms of radioactive pollution but also by different natural landscapes and the degree of man-made influence. In addition, the volumes of radionuclide removalfrom each basin varychangefrom year to year under the influence of changes in hydrometeorological conditions, moisture level, runoff volumes, the ratio of water balance components, etc. At the same time, the following factors affect the concentration of radionuclides in water: the chemical composition of water, water flow rate, the type of water exchange with groundwater, the nature of redistribution of the radionuclide reserves between soils in the fluctuation zone, bottom sediments, and hydrobionts, etc. All these important factors influence the processes of secondary pollution of surface water by radionuclides.

An artificial drainage network not only changes the landscape, increases drainage, and has a specific effect on the water exchange between surface and groundwater but also provides the possibility of selecting a sufficient, statistically significant number of catchment basins of the main and regulating canals, which allows for correct correlation analysis.

The **main tasks** of the functioning drainage systems of populated areas are regulating groundwater regimes and maintaining drainage rate to minimize the transfer of radionuclides to cultivated plants keeping sufficiently high yields; in the territories of primary recovery - studying and strengthening the processes of self-purification of soils and groundwater, and *in the exclusion zone* - determining the most acceptable scenarios of water exchange and runoff regimes to minimize the removal outside the polluted zone. The treatment of reclamation systems in exclusion zones must be consistent with the basic principles defined by the Concept of operations in the ChEZ and mandatory resettlement zonewithin Ukraine (1998): "any activity in the Zone to improve the radio-ecological situation should be carried out with the maximum use of natural factors and minimal intervention in the natural environment." Thus, since the beginning of the 2000s, drainage systems in the ChEZ have been considered as a barrier to minimize the removal of radionuclides beyond its borders. By default, each natural landscape has a buffer capacity or barrier resistance to a particular pollutant, which can be violated by groundwater regime regulation or other measures.

Barrier resistance is a qualitative indicator that expresses the capacity of a catchment to resist the wash-off of a certain man-made migrant beyond its borders. The degree of barrier resistance varies depending on the predominance of restraining or mobilizing factors. At the physical and chemical processes level, barrier resistance is ensured by the soil properties and composition, its organic parts, and biogeochemical (biological) processes. So, if the natural characteristics of the environment are unchanged, the barrier resistance to a particular substance in quantitative terms is a constant value. The concept of barrier resistance of the catchment is close to the concept of the ecological capacity of the ecosystem, but it refers mainly to its abiotic part. The radiological capacity of an ecosystem is the number of radionuclides that an ecosystem can hold (including due to their circulation) without significant consequences for itself (Kutlahmedov, 2002). A high barrier resistance demonstratesa minimal amount of radionuclides washed off and minimal adverse consequences for the nearby populated areas.

Constant and dynamic factors determine barrier resistance. Constant or relatively constant factors are features of the landscape that change slightly over time, which can affect the migration and removalof radionuclides. These include hydrographic networks, predominant soil types (organic or mineral), mineral and granulometric composition of water-bearing parent rocks, certain types of depressions, forest massifs, and others (Shevchenko, 2013). In order to assess the degree of influence of these factors on the removalof radionuclides, they *are translated into parameters* that can be used in statistical analysis and mathematical models.

Dynamic factors include external hydrometeorological indicators that change by year and within the year: the amount of precipitation, flood origin, and the presence of summer floods; and intra-basin factors: groundwater level, the state of a canal related to its overgrowing, silting, and regulation, the chemical composition of groundwater, flow rate, drainage conditions.

Based on established empirical relations (regression equations) between the content of ^{90}Sr in ground and surface waters or its removal, on the one hand, and constant and dynamic factors, on the other hand, we can perform several successful epignostic reconstructions and predictions of radiological indicators changes (Shevchenko et al., 2016).

Considering the specific behavior of radionuclides on the drained landscapes, there is a set of scientific research and activities within drainage systems and in the adjacent area:

1) The migration of radioactive substances in drained landscapes differs from their migration under undisturbed conditions because of the dense drainage network, subsurface drain, numerous hydrotechnical constructions, and accelerated regulated water exchange.
2) The role of processes and factors of secondary pollution of surface waters changes over time due to the redistribution and transformation of primary radioactive precipitation; following these changes, the course of water protection activities should change as well (Shevchenko, 2005).
3) The quantitative indicator of radionuclides removal by drainage flow is an integral indicator of the barrier stability of the landscape conditions of reclaimed catchments and the technical condition of drainage systems in the territories of radioactive contamination.
4) Analysis of formation conditions and forecast of radionuclide removalshould be performed on a landscape-basin basis.
5) Obtaining scientific results is ensured by three levels of detailing of working operations: detailed - at the balance sites; local - at catchments of individual canals; regional - by summarizing patterns for the entire pollution area.
6) Suppose there are drainage systems on the territory of radioactive contamination. In that case, it is necessary to solve the problem of their adaptation and use it to minimize the removal of radionuclides and recover contaminated areas.

Secondary Radioactive Contamination of Water in Drainage Canals

The processes causing the secondary pollution of surface waters with ^{90}Sr are quite diverse in terms of their level of influence, mechanisms, and dynamics; and the ratio of their contribution changes not only in time but also along the length of the watercourse due to the changeable hydrodynamic regime. ^{90}Sr is distributed in the water bodies more evenly than ^{137}Cs: the relative content of ^{90}Sr in water reaches 50±4%, about 27±5% is accumulated by biota and 23±4% by bottom sediments (for the 13-20th year after contamination).

The processes of secondary radioactive contamination of surface waters are closely related to the processes of mobilization and remobilization of radionuclides from the solid phase. The main depositing components are the *soils of the bank slopes* of the canals and the *bottom sediments*. Solid runoff containing sorbed radionuclides is transferred from the surface of the bank slopes (slope of 20-36°) to the channel bed or bottom sediments. The desorption of radionuclides from bottom sediments occurs more intensively when the water level in the canals decreases against the GWT in the adjacent territory (Shevchenko et al., 2011). The main source of secondary radioactive pollution of the canal's waters is the bank slopes, as they ensure the inflow of radionuclides in three ways: during a flood when the water level rises, during short-term drain of rainfall and snowmelt runoffs along the unflooded upper reaches, and when groundwater seeps above the water level in the canal.

In order to quantify the main processes of water pollution, we conducted: field observations and laboratory experiments with soil monoliths and bottom sediments; calculations of groundwater inflow to canals, water balance and ^{90}Sr balance in canals.

The components of radionuclide inflow that can be determined only by water balance calculations include the inflow of groundwater, surface-slope runoff, amount of precipitation per water surface, and inflow from swamps.

The processes that are not directly related to the water balance but affect the concentration of radionuclides in surface water include (1) desorption and sorption of radionuclides by bottom sediments; (2) absorption by hydrobionts and extraction during necrosis and decay of organisms; (3) leaching (washing) from flooded soils; (4) sorption of radionuclides by the solid phase of the soils of bank slopes during filtration losses from the canal. In the general water-radiation balance equation, such processes are parameterized in the form of coefficients that are determined empirically or by solving inverse problems.

In some cases, when the reduction of water volumes is accompanied by evaporation and underflow of highly concentrated swamp waters, the following processes take the lead: complex formation with organic substances, coagulation with suspended particles, and sorption of radionuclides by the upper layer of bottom sediments.

The *role of various processes and factors of radionuclide removal changes over time* because of the gradual horizontal redistribution under the influence of erosion processes, sinking into the ground, and transformation of the primary forms of fallouts. During the first 13-16 years after the disaster, the secondary contamination of surface waters with ^{90}Sr was determined mainly by the processes associated with the action of water on the environmental components that deposit radionuclides and are in contact with water the longest (soils, bottom sediments), i.e., radiation-balance processes. For the left-bank part of the ChEZ, this is related to the time of complete destruction of fuel particles (13-15 years).

In high-water periods and in the backwater zones, **leaching in the flooded slopes** of canalssignificantly prevails among the secondary processes of radioactive contamination of surface waters. During low-water years in regulated reclamation systems, leaching prevails, too. Nevertheless, inflow with groundwater and leaching from bottom sediments and dead organisms prevails in more flowing canals.

The maximum impact of **wash-off** on the ^{90}Sr secondary contamination of surface water is achieved when the water rises to previously unflooded areas. In the dynamic conditions of the experiment, the maximum release of ^{90}Sr into the solution was observed after 17-21 days (Shevchenko et al., 2011). According to the static experiments and studies on the distribution of ^{90}Sr in the "soil of a flooded slope – solution" system and in the "bottom sediments – solution"system, after 40-42 days and 60-65 days of water exposure on the surface of the solid phase, without changes in the chemical and temperature conditions, there is a decrease and then stabilization of concentrations. The content of ^{90}Sr in water increases with the concentration of low-molecular-weight fulvic acids and with an increase in permanganate oxidizability (PO) (Shevchenko & Dolin, 2002; Shevchenko et al., 2002).

We mathematically described changes in the concentration of radionuclides in surface waters under the conditions of rising water levels and long-term flooding of canal slopes on the example of the MC-1 canal on the left bank of the Prypiatriver. We also tested them on the "Grezlia" system canals. Due to an increase in the roughness coefficient from 0.027 to 0.05 when the canal is overgrown and silted (Shevchenko & Kozytskyi, 2003), the water level in the canal increases by Δh (Figure 5). At a flow rate of 0.61 m^3/s, Δh is equal to 0.17 m.

Figure 5. Transverse profile of the canal and canal bed parameters before and after the water level rise when canal slope m = 2

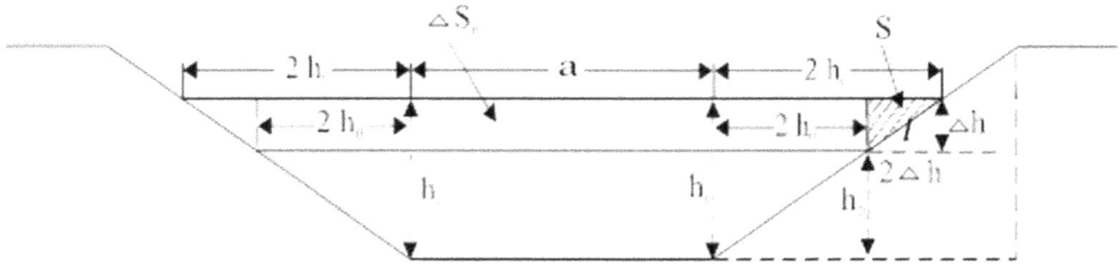

To calculate the concentration of ^{90}Sr in canal water (C_w) when the water level rises, it is necessary to know the leaching coefficient:

$$K_l = \frac{C_{dz} \cdot V_w}{C_s \cdot \rho \cdot V_s \cdot h} \text{(m-1)}, \quad (2)$$

where C_{dz} – the volumetric activity of ^{90}Sr in the water above the contaminated slope, Bq/m^3; V_w – the volume of this water, m^3; C_s – the specific activity of ^{90}Sr in the contact layer of the soil, Bq/kg; ρ – the volumetric mass of this layer, kg/m^3; V_s – the volume of the effective soil contact layer, m^3; h – the water column above the monolith, m.

Let's take the value of K_l, which characterizes an increase of ^{90}Sr in water during 16 days coming from the monolith which has been taken from the slope of the canal on the leftbank of the Prypiatriver and filled with natural water, i.e., 1.25 m^{-1}. From the inverse equation for C_w, it is obvious that the concentration of the radionuclide in water directly depends on the increase in the water level. If the canal slope ratio is $m = 2$, the water will spread over the slope for a distance $c = \sqrt{4 \cdot \Delta h^2 + \Delta h^2} = 2.236 \cdot \Delta h$, the cross-sectional area of the surface water column (Figure 5) will be equal to $S = 2\Delta h \frac{\Delta h}{2} = \Delta h^2$, and the entire additional area will be:

$$\Delta S_n = \left(\left(4h_t + a\right) + \left(4h_o + a\right)\right) \cdot \frac{\Delta h}{2}, \quad (3)$$

where h_t– the depth or water level at time t, which has increased due to the decrease in channel capacity (for MC-1, it is equal to 0.67 m); h_0– the initial depth of water in the canal, which corresponds to the design depth (0.5 m) for the given flow rate;a– thecanal bottom width; $\Delta h/2$ – the average increase of the water column above the flooded slope, given that the cross-section of the canal has a trapezoidal shape (see Fig. 5). We can write $h_t = h_0 + \Delta h$, then

$$\Delta S_n = (4h_0 + 2 \bullet \Delta h + a) \bullet \Delta h \quad (4)$$

Let's calculate the concentration for the sidedwater column or volume of water considering that

$$V_s = 2.236 \bullet \Delta h \bullet H_m \bullet L \tag{5}$$

where H_m – the thickness of the mass exchange layer or the effective contact layer of the soil (mainly0.05-0.1 m); L – the length of the canal (length for calculations is 1 m), $V_w = \Delta h^2 \bullet L$.

Then:

$$C_{dz} = \frac{K_l C_s \rho H_m 2.236 \cdot \Delta h^2}{2\Delta h^2} = 1.118 \cdot K_l C_s \rho H_m \tag{6}$$

where ρ- the specific gravity of the soil, kg/m^3.

For the entire additional water column:

$$C_\Delta = \frac{K_l C_s \cdot \rho H_m 1.118 \cdot \Delta h}{4h_o + 2 \cdot \Delta h + a} \tag{7}$$

When using the experimental data (Table 1), we get $C_\Delta = 0.13$ kBq/m^3, which corresponds to the actual concentrations of ^{90}Sr in the central part of the MC-1 flow. In contrast to the entire additional volume, we get higher concentrations of ^{90}Sr in the sided water column: $C_{dz} = 4.7$ kBq/m^3. This is higher than the maximum permissible concentration (2.0kBq/m^3) by more than 2 times. Due to overgrowth and lack of flow in this layer, there is almost no mixing.

Table 1. Parameter values for calculating the volumetric activity of ^{90}Sr during water level rises in canals

Formula	Parameter							Result
	K_l, m^{-1}	C_s, Bq/kg	ρ, kg/m^3	Δh, m	H_m, m	h_o, m	a, m	C_{dz} or C_Δ, kBq/m^3
(6)	0.94	30	1230	-	0.15	-	-	5.8
(6)	1.25	35	1220	-	0.12	-	-	7.2
(7)	1.25	40	1200	0.17	0.07	0.5	4	0.126

Dispersion, i.e., the movement of radionuclides to the transition zone (between flowing and dead), is subject to molecular diffusion and proceeds rather slowly (assuming that D_m in free solutions is (8…15)$\cdot 10^{-5}$m^2/day (Roshal, 1980)). The dead zone's stagnant mode determines the H_m growth, which leads to an increase in the release of ^{90}Sr into the solution, according to (6) and (7). At the same time, with depth, the specific activity of the soil (C_s) decreases, and the volumetric mass increases (Table 1).

In the transition zone, the diffusion is gradually changed by turbulent mixing, the dispersion between flows of different velocities increases towards the central, most flowing part.

Due to turbulent mixing, especially at high flow velocities and minor overgrowth, radionuclides spread from the zone adjacent to the slope to the entire volume of water flow in the active section. As a result, we get a slightly lower average concentration:

$$C = \frac{V_{rz}C_{rz} + V_c C_c + V_{lz}C_{lz}}{V_{rz} + V_c + V_{lz}}, \qquad (8)$$

Where C – the resulting volumetric activity of the radionuclide after mixing the water volumes of the central zone (V_{cz}) with a concentration (C_c) and of the dead zones of the right (V_{rz}) and left (V_{lz}) banks with concentrations (C_{rz}) and (C_{lz}), respectively.

Therefore, the contamination of watercourses of the ChEZ by radionuclides is subject to their carrying capacity (flowage).

0.18-0.26% of ^{137}Cs and 8.5-38.0% of ^{90}Sr transform from **bottom sediments** under static conditions into a water-soluble form. In the "pore solutions of bottom sediments - lower - upper layer of surface waters of the canal" system in 1998-2001, the directions of ^{90}Sr concentration gradients were negative in 33% of cases (in May-July), mainly in the backwater zones. The greater content of ^{137}Cs than ^{90}Sr in canal water is caused by ^{90}Sr wash-off from bottom sediments. The ratio of ^{137}Cs/^{90}Sr content in all sediment samples taken from the reclamation canals of the left bank of the Prypiat river in 2000 was 20±2 (Shevchenko & Dolin 2002).

Remains of semi-decomposed algae and other dead **hydrobionts** are an influential source of secondary pollution of canal waters: in 10 days, about 0.9% of their initial content of ^{90}Sr in plant remains (per dry mass) is leached; the ratio of radionuclide content in water to its content in plant remains (transition coefficient) will approach 0.01 kg/dm^3.

After the beginning of the reconstruction of the reclamation systems of the ChEZ in 2000-2001, which coincided with the end of the high-water-year cycle, **groundwater**became the dominant source of secondary pollution of surface waters. In areas with subsurface tubular drainage, there are increased concentrations of ^{90}Sr in the underground drainage runoff, even at the lower GWT(Shevchenko, 2018).

Four *types of water exchange between groundwater and canals* determine the redistribution features of man-made radionuclides in the aquatic environment of drained landscapes of the ChEZ. These types of water exchange include: 1) without subsurface material drainage, when the canal works as a filter drain; 2) with active material drainage; 3) under conditions of a constant backwater; 4) underground flows of reverse directions (to and from the canal) in different hydrological phases. The most intense groundwater pollution processes occur in the areas of the 2nd and 3rd types, while surface waters - in the areas of active subsurface drainage (the 2nd type) (Shevchenko, 2002). In drainage waters of subsurface collectors, there are comparable or even higher concentrations of ^{90}Sr than in the surface waters of canals. Here, ^{90}Sr is washed off to a greater extent from the surface layer of the soil and moved to the depth in larger quantities relative to ^{137}Cs than in the areas without drainage.

In sites with the 1st water exchange type, the removal of radionuclides is the least; the barrier functions of the landscape are the best. Here, infiltration supply is the determining factor of ^{90}Sr groundwater contamination unless there is significant lateral inflow from adjacent flooded areas. For areas of alternating water exchange with surface waters (type 4), there is the closest relation between the ^{90}Sr concentration in groundwater and the surface waters of the canal.

During the research period of 1996-2002, the cleanest surface runoff in terms of radiation was formed by its mixing with drainage waters coming from clean areas or with groundwater - without artificial drainage and backwater structures. The share of ^{90}Sr removal by underground runoff from the total removalby canals during the high-water years of 1997-2000 was: for heavily polluted areas – 0.2-2.7% (Shevchenko et al., 2011); for systems or canals in less polluted areas – 1.1-25.0%; in low-water years (2004-2007), this share increased to 35%. In 2010-2020, it increased to 40-45% (Figure 6), and groundwater became one of the leading sources of ^{90}Sr removal to watercourses.

The role of bottom sediments and surface runoff by atmospheric precipitation from slopes in the secondary pollution of canal waters is noticeably decreasing over time (Figure 6). Its maximum was probably in 1986-1992. For 1995-1996, the contribution of these processes to the formation of ^{90}Sr removal by canals on the left bank of the Prypiat River was experimentally calculated: 45% and 2.7%, respectively. Leaching from flooded slopes and runoff from swamps provided 54% and 7.8%, respectively, in 1997-1998.

Figure 6. The main sources of secondary pollution of the surface waters of the canals and changes in their share in the ^{90}Sr removal formation over the years of observations on the left bank systems in the exclusion zone (on the example of the MC-7 canal of the Prypiat system)

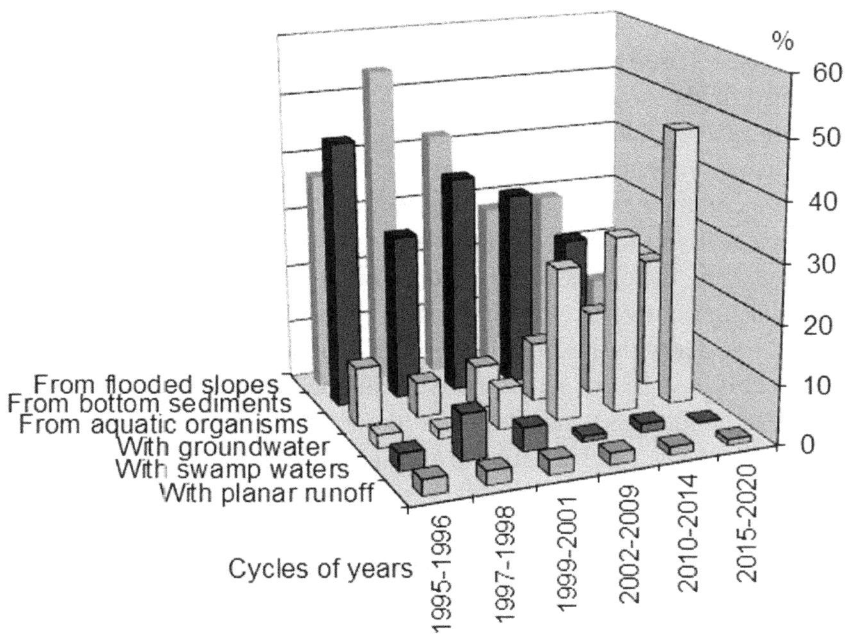

The dependence of ^{90}Sr concentrations in surface waters *on the water level in the canals* is directly proportional. However, its dependence on *flow rate* is ambiguous: if the increase in flow rate is associated with a rise in water level, the dependence is direct, which is typical for main canals; if the water levels change, at the same time, not significant - the dependence is inverted (mainly side canals).

The increase in the share of exchangeable and soluble ^{90}Sr species in the catchment soils is the main reason for the non-temporary but permanent increase in the role of groundwater in the removal of ^{90}Sr to open water receivers.

Factors of Barrier Resistance of Catchments

All factors that determine the course of the above-mentioned processes are divided into those that determine the migration properties of the radionuclide (valence, ionic radius, physical and chemical forms of presence in the environment, etc.), and those that characterize the conditions or environment of its migration.

The hydrogen index (pH of the water extract) of the soils showed a significant degree of influence on the involvement of ^{90}Sr in water migration only in the first years after the disaster. The influence of pH has an inverse relation with the amount of precipitation (r = - 0.58). When the high-water (> 600 mm) period of 1996-2001 started, the influence of this indicator on the ^{90}Sr removal decreased; its influence was also insignificant during the next short high-water cycle, which began in 2004.

When ^{90}Sr starts massive migrating into groundwater (2002-2014) resulting from the large-scale destruction of fuel particles, *the content of* ***organic complexing*** *agents or sorption capacity of rocks* in the aeration zone and water-bearing rocks has become one of the main factors that determine barrier functions.

About 40% (more than 100,000 ha) of the ChEZ territory is occupied by fallows (grassy areas)formed on the site of abandoned fields. At the same time, 39,000 ha of fallow lands are located in areas affected by reclamation systems, and 62,000 ha are under natural moisture conditions.

The ***meadow*** landscape subsystem performs a barrier function between a canal (river) subsystem and other subsystems. It is a common and basic one for modern reclamation systems in the ChEZ, covering about 75% of their area (without forest lands that were not previously a part of the systems' area). On such meadow landscapes, the major share of migratory radionuclides species is retained and circulated in the aeration zone during their vertical infiltration-evaporation movement and absorption by plants, dispersion, and lateral transfer with the filtration flow of groundwater (Shevchenko et al., 2011). Non-draining depressions retain a significant part of mobile radionuclide species. A relatively insignificant part of the total reserves of radionuclides concentrated in meadow catchments, about 0.01-0.14%, enters the canals and is carried out in the water receivers. On average, about 1.8 10^{12} Bq ^{90}Sr is in the plant biomass of meadows and fallows, which is 0.2% of its total reserves in the soils of the ChEZ (Paskevich, 2005).

The inverse dependence of the removalby surfase water^{90}Sr on the relative ***area of the forest*** on the catchment (Figure 7) increases with time from the moment of disasterfallouts. It is explained by the increase in the share of exchangeable and water-soluble radionuclide species that vegetation can absorb. During high-water years, the role of forests is also more significant.

Forest reclamation significantly increases the barrier resistance of catchments against the ^{90}Sr wash-off. According to calculations, an increase in the forest area by 50% would help reduce the removal of^{90}Sr from various catchments by 19-46% compared to the value of 1999.

According to the results of statistical assessments, most of ***the depressions'*** area, especially large and waterlogged ones, retains ^{90}Sr. In small and conductive depressions, which are more numerous, the relative area of fast migration zones is larger (up to 20%), and they transport more ^{90}Sr to groundwater than retain. Active zones make up only 4-6% of their area in large depressions, but here they have a higher transportation level of ^{90}Sr than in small depressions (Bublyas & Shestopalov, 2001). The transit role of conductive depressions increases over time (Figure 8).

In addition to these factors, the composition of the soils of the aeration zone and GWT plays an important role in the performance of barrier functions. However, we did not manage to perform the correct differentiation of catchment areas according to GWT due to the insufficient number of wells.

Figure 7. Chronological changes in the degree of the inverse effect of the relative forest area (F) in the catchment on the removal by water of ^{90}Sr (W) (for 9 catchment basins)

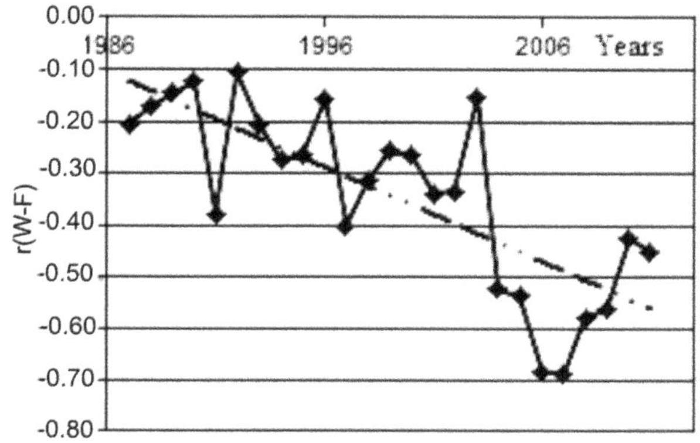

Figure 8. Chronological changes in the dependence of the ^{90}Sr removal on the area of depressions in the catchment, km² (1), their density, units/km² (2), relative area of dry (flowing) depressions, % (3) with polynomial trends of the 3rd degree

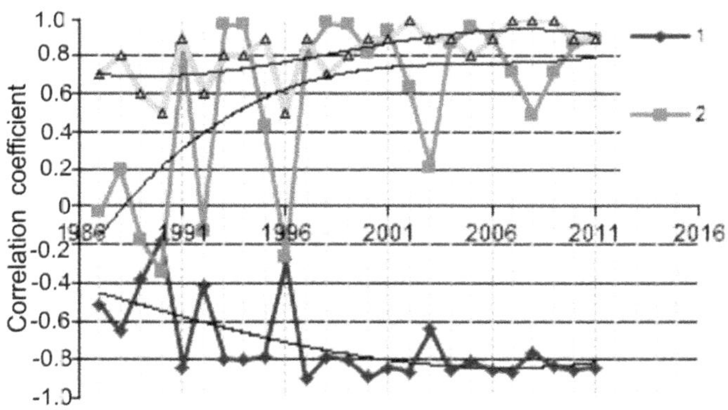

Man-made factors, namely, the degree of reclamation of the catchment area and regulation of the hydrographic network, turned out to be the most influential (according to the results of multiple correlation analyses). Therefore, it is quite possible to adjust the radioecological state of the catchments.

Using regression equations similar to (1), it is possible to select such factor values that keep the concentration and/or removalof radionuclide within the lowest limit. This is a sufficient justification for determining the optimal runoff regimes from drainage systems and preventive water protection measures to minimize the removal of radionuclides. According to the equation's solution that considers the radionuclide reserves in the catchment (Z_{Sr}), a 20% reduction in *regulation* of runoffleads to a 23-29% reduction in removalin a high-water year. A 25% reduction of *drainage conditions* can result in a 3-10% reduction in ^{90}Sr removal. An increase in the share of groundwater flow in the watercourses balance to 35%, according to all regression equations for the scenario of 1999 (at constant flow rate and pH of the solution), would allow for a reduction of ^{90}Sr removalby 9-20%. Increasing the forest area to 50%

of the total catchment area leads to a decrease in ^{90}Sr removalin various canals by 24-44% compared to the value of 1999.

Under conditions of the absence of artificial retaining structures and non-delayed flow passage, the epignostic calculations of removalhelp to simulate an alternative to the actual one multi-year course of ^{90}Sr removalfrom drained catchments after the disaster at the Chornobyl NPP. It would reduce the removal by water flow during the period from 1987 to 2004 by 1.48 TBq. Therefore, the buffer capacity of each natural landscape to a certain pollutant can be not only violated by man-made structures but also strengthened by reclamation measures, forest plantations, and the introduction of sorbent materials.

Barrier Resistance of Catchments in The Exclusion Zone to ^{90}Sr Removal

The ^{90}Sr reserves in the soils of the entire catchment are not a decisive and sufficient parameter in determining the volumes of removal by surface runoff (Shevchenko et al., 2016). The low correlation coefficients between ^{90}Sr removal and its reserves (up to 0.58 for the left bank of the Prypiat river) evidence that.As radiological indicators of barrier resistance for catchment basins it is better to use the given indicators – *removal modules*, which take into account not only the volume of runoff but also the area of the basin. The *share of removal* takes into account the radionuclide reserves in the area of the basin.

The removal module (M_r), which is the ratio of the amount of radionuclide activity removed by a watercourse per year to the catchment area of this watercourse (GBq/year/km^2), can serve as a *quantitative indicator of the barrier resistance of catchments*, both at the local and regional levels:

$$M_r = \frac{W}{F} = \frac{\sum_{i=1}^{n} Q \cdot C \cdot 86.4}{F \cdot 10^3}, \tag{9}$$

where W – the amount of radionuclide removedper year or hydrological period, GBq/year (period), n – the number of days in the calculation period, F – catchment area, km^2; Q – flow rate, m^3/s; C – volumetric activity, kBq/m^3 (Shevchenko & Kireev, 2005).

Removal of ^{90}Sr (GBq) is determined by the dependence:

$$W = 10^{-2} \cdot b_n \cdot Z^o_{Sr} \cdot e^{-0.024t} \tag{10}$$

Where, b_n – the share of removal from the radionuclide reserves on the catchment for n-th year, %; Z^o_{Sr} – reserves of ^{90}Sr on the catchment area of this watercourse in the first year of pollution, t – is the number of years from the beginning of pollution to the n-th year.

The indicator of the depletion of barrier resistance is the module of barrier and mobilization functions (M_{bm}), which is calculated as a difference between the removalmodule and the maximum permissible concentration module (M_{MPC}): $M_{bm'} = M_r - M_{MPC}$

Natural and *disturbed* barrier resistance is evaluated separately. During the assessment of the natural resistance of the catchments in the ChEZ to ^{90}Sr, the following indicators are taken into account: density of depressions, units/km^2; relative distribution area of waterlogged and flooded depressions (% of the area of the basin); dry depressions and depressions on sand; forest; organic mineral soils; soils of organic origin. Forest massifs and organic mineral soils are the most influential natural factors that ensure the barrier resistance of catchments to the ^{90}Sr removal.

The relative distribution of these indicators (landscape features) in the catchments, taking into account the rating (by multiple correlation coefficients), was replaced by points based on the principles of qualimetry. The range of points varies from 18 (area of dry depressions) to +20 (forest area).

The dense drainage network and cooling pond of the Chornobyl NPP significantly reduce the natural barrier resistance of catchments and have a greater influence on waterremovalthan natural factors. This is evidenced by the larger values of positive correlation coefficients. Thus, with the reclamation of 40%, the resistance of the Sahan river catchment decreases 3.7 times and changes from "good" to "satisfactory." With reclamation of 60-70%, the resistance of a separate catchment changes from "average with a positive score" to "low with a negative score"; when the reclamation is greater than 70%, the barrier resistance becomes "unsatisfactory" and "very unsatisfactory and dangerous." It results in a consistently high level of the ^{90}Sr removal. Reclamation at 10-20% level does not significantly affect barrier resistance.

The decrease in resistance is well demonstrated by the removal module and, especially, by the changes over time in the module of barrier and mobilization functions (M_{bm}) to ^{90}Sr. In catchments with higher natural resistance, the rate of decrease in annual ^{90}Sr removal is greater than in catchments with a lower ability to retain the radionuclide. First, the transition to the negative M_{bm} values occurs in those catchments where the sum of points, which determine natural barrier resistance, is greater than 10, and reclamation is less than 20%, *regardless of the degree of radioactive contamination.* The high natural barrier resistance of catchments is significantly reduced after taking into account the drainage network (Figure 9). These man-made factors slow the transition of M_{bm} to negative values. Therefore, the speed of reaching acceptable M_{bm}, when the concentration of ^{90}Sr in the surface runoff is less than the MPC, is inversely proportional to the water network's reclamation and density, making it possible to predict the timing of the ^{90}Sr removalvalues to a safe level.

Modern Problems and Ways to Improve the Situation at Catchments With Drainage Systems in The Exclusion Zone

As stated in the previous sections, reclamation systems have special conditions of secondary radioactive contamination of ground and surface waters related to the nature of water exchange, structural features, and the modern technical state of the systems.

The long-term experience proves that drainage systems are a determining factor influencing the processes of landscape restoration, forest state, fire safety, flooding of burial sites and temporary localization of radioactive waste etc.

If the catchment is heavily polluted and its barrier resistance is very low, it is necessary to consider the possibility of transferring its runoff to adjacent basins with lower pollution density and better barrier resistance. Research in 1997-1998 has proved that the use of reclamation canals to ensure the circulation of polluted water in conditions of good groundwater drainage and the presence of natural bioplateaus ensure their self-purification by 90-96%, depending on the length of the transport network.

The reconstruction of the regulating locks and the canal's extensive network on the left bank of the Prypiat river made it possible to transfer part of the runoff from the waterlogged and highly polluted Northern Basin (runoff to the breach in the dam No. 7) through the MC-1 canal to the mouth of the Brahinka river through less polluted and flooded areas of the ChEZ. The water is noticeably purified when passing a distance of 25-35 km due to natural bioplateaus. In the second year after the clearing and

Figure 9. Zoning of catchment basins in the Chornobyl Exclusion Zone according to the barrier resistance to the ⁹⁰Sr removal, taking into account the drainage network

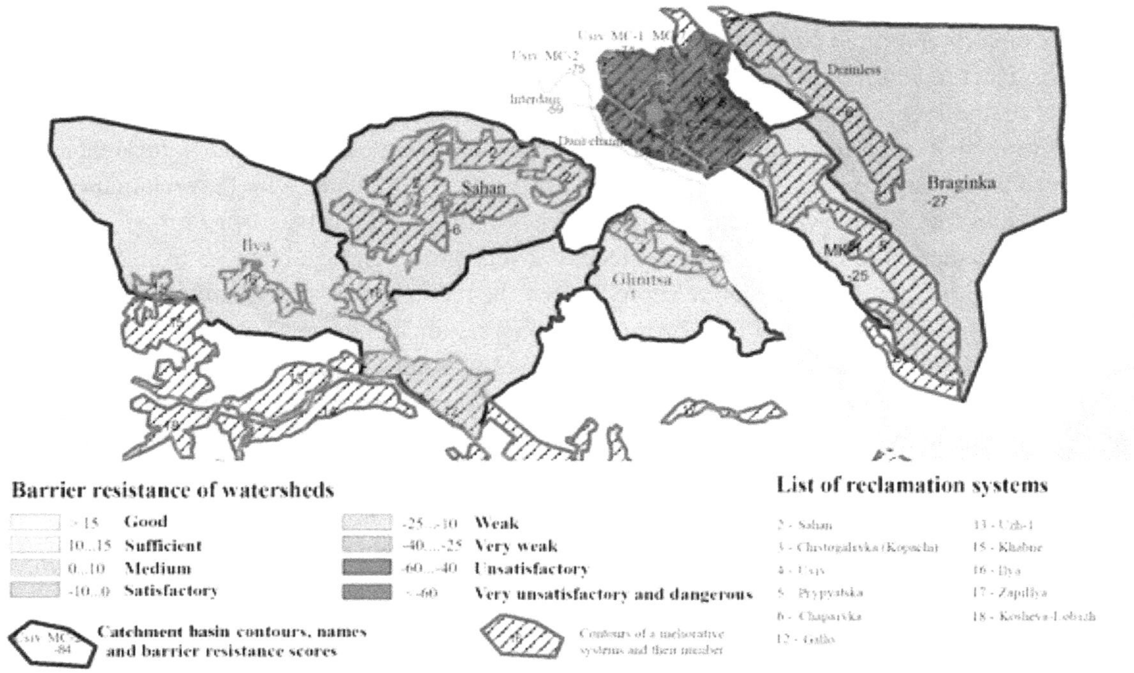

the implementation of the new water distribution system (2002), it was possible to reduce the volume of ^{90}Sr washed off in the MC-1 channel by 57% compared to the similar water content level in 2000. Such a reduction was reached despite the fact that due to the increase in the degree of drainage, volumes of runoff increased from 13.36 million m^3 in 2000 to 16.42 million m^3 in 2002. Transfer of contaminated runoff from a catchment with a high density of ^{90}Sr contamination (1480-3700 kBq/m^2) to a catchment basin with a much lower density of contamination (7.4-18.5 kBq /m^2) turned out to be the most effective regulatory measure on reclamation systems to date.

At the current stage, when determining how to manage the drainage systems, we should consider a step-by-step transfer of the ChEZ to the status of a radioecological reserve. This requires a differentiated approach to the introduction of environmental protection measures in accordance with the maintenance of minimal radionuclide removal outside the zone and taking into account the distribution of drainage systems by functional zones (Kholosha, Sobotovych, 1994).

When choosing a scenario for managing drainage systems, we should take into account the dominant factors and the nature of the radionuclide distribution in the open hydrosystems of the ChEZ (Table 2).

Since 2015 the GWT have been anomalously low, which is explained by a decrease in groundwater infiltration due to global warming and the termination of operation of the cooling reservoir in the Chornobyl NPP.

During 1993-2020, more than 1,800 local and large-scale ***fires*** occurred on the territory of the ChEZ. The forest massifs have a barrier function preventing the spread of radioactive substances. More than half of them covering the ChEZ territory have degraded (wetted, died from pests) or completely destroyed in some places.

Table 2. Recommended measures to strengthen barrier functions and minimize the removal of radionuclides from drainage systems at various stages of the post-accident cycle (on the example of the ChEZ)

Stage	Nature of radionuclides distribution in the surface runoff	Recommended measures
1. 1986 - summer 1987	The predominance of suspended and poorly soluble species. The higher volumetric activities and mobility of ^{137}Cs compared to ^{90}Sr (ratio $^{90}Sr/^{137}Cs$ = 0,81)	Accumulation of runoff in the side network of canals and sedimentation of suspended particles with the help of hydraulic structure. Floating polystyrene foam filters; water flowing through channels at levels below "bank anomalies." Removal and burial of turf from the most polluted bank slopes. Application of slow sand filters.
2. autumn 1987 - 1992 low water years	Mobility advantage of ^{90}Sr over ^{137}Cs (ratio $^{90}Sr/^{137}Cs$ = 2-3.6). The ^{137}Cs removal in the Pripyat decreased by 4 times. There is a high content of colloidal forms. The share of ^{137}Cs in the suspended particles increased to 50%, and soluble species of ^{90}Sr - up to 82%. The removal of radionuclides are mainly during floods. The gradual decrease of volumetric activities of ^{137}Cs and ^{90}Sr in the well-flowing watercourses.	Cleaning of sides and channels in the upper pools of structures on canals. The transfer of polluted runoff through relatively clean areas, and partial accumulation in the network. The use of a dusty fraction of mineral sorbents in watercourses with discharge < 100 l/s; intercanal filtration fields with mineral fillers and sand dams as slow filters; floating filters, bioplateau, static sorption for objects with a slow runoff.
3. 1993 - 2001, 2004 – 2008, 2013, 2016 high water years	Overwetting of the catchment area. Advantage of soluble ^{90}Sr species (up to 96%). A significant decrease in the volumetric activity of ^{137}Cs – below the MPC (co-ratio $^{90}Sr/^{137}Cs$ = 4). The high content of organic suspended particles and dissolved organic substances. Since 1996, there has been no significant decrease in the volumetric activity of ^{90}Sr.	The use of biosorption and the transfer of runoff over long distances through a network of canals. Prevention of waterlogging and flooding. Drainage intensification to increase the inflow of cleaner groundwater into the open network, providing the cascade systems and forced aeration of water. Use of existing pumping stations to prevent flooding of floodplains and water aeration.
4. 2002 - 2003, 2009 - 2020 low water years, except for 2013 and 2016	Advantage of soluble ^{90}Sr species (up to 95%). The groundwater (that contacts with radioactive waste disposal sites, is filtered from the backwater cooling pond, etc.) plays a major role in the ^{90}Sr removal beyond the limits of the ChEZ.	Keeping the runoff within zones with high aeration in fire-hazardous areas; regulation of water levels, isolation of outlets of drainage collectors. Water liming and biological methods of radionuclides sorption. Afforestation of the catchments.

The area of burners with destroyed or damaged wood, dry stands, and cluttered forests reaches about 100 thousand hectares with estimated reserves of 25 million m^3 of dry radioactively contaminated wood. The fire-triggered mechanisms of the increase in the migratory activity of PH contributed to the deterioration of the radiation situation in the exclusion zone. In particular, the combustion products in the places of the fire acquired the characteristics of radioactive waste. Burnt forest litter and ash contain 600-180,000 Bq/kg of ^{90}Sr and 4,100-270,000 Bq/kg of ^{137}Cs. In 2015 and 2022, there were released 130 GBq and 700 GBq of deposited by wood ^{137}Cs, respectively.

Several factors become an obstacle in fire-prevention activities: the imperfection of the regulatory framework, the absence of a protected regime in some areas, as well as a lack of evenly spaced fire protection ponds and a regulated network of canals. Today, the problem of watering peatland areas to prevent large-scale fires and the release of radioactive aerosols into the atmosphere is also critical. Such release of radioactive aerosols can significantly contribute to the collective radiation dose of personnel and the population of the areas adjacent to the ChEZ. In total, peatlands occupy 5,320 ha of drainage systems within the former Chornobyl district - 30.7% of the total area of the systems. The depth of peat distribution in some sites reaches 5.0 m. If the moisture level of lowland peatlands is 69% and upland peatlands - about 72%, these peatlands are considered fire-hazardous (Struts, 1984).

When peatlands were waterlogged, or there was living swamp vegetation on the surface, no fires were recorded. To restore the natural state of peatlands, which indicates an increase or restoration of their capacity, in Soviet times, flooding of previously drained peatlands was used (Maslov & Minaev, 1985).

In 2019-2020, GWL fell below the bottom layer of most peatlands (2.0-2.5 m) - to permeable sands. Due to the low capillary rise in sands (10-17 cm), a break in the capillary moisture supply to peat occurs; as a result, it over-dries.

In the 1980s, the practice of "artificial peat conservation using ultra-deep plowing (1.5- 2.0 m) with plowing the sand on the surface" was used in order to reduce the danger of burning peatlands, wind and water erosion, and improving their water-physical properties. The application of mineral soil of up to 0.2 m achieving an ash content of 0.5 parts of a unit also ensures the prevention of fires on peatlands. In some cases, it may be enough to enrich peat with mineral soil or cover the surface of the peat in narrow strips along the forest, roads, and other fire-hazardous areas. Such strips can be used to divide fire-hazardous peat areas into "sectors" with an area of 100 ha. These measures reduce the probability of fires by 90% (Struts, 1984).

To prevent fires on peatlands in the ChEZ, it is necessary to create a network of observation sites and set tensiometers and humidity sensors to transmit information remotely (via satellite).

One of the most important *modern radioecological problems* is the accumulation of americium-241 in the natural environment and its dosage. Currently, the content of this radionuclide outside the damaged reactor is almost equal to ^{90}Sr and will continue to increase until the middle of this century. At the same time, its radiobiological efficiency significantly exceeds the gamma and beta radiation of dose-forming ^{137}Cs and ^{90}Sr.

CONCLUSION

The effect of natural dynamic and man-made factors determined the stages of changes of the groundwater's role in the general removal of radionuclides from reclaimed and natural landscapes. From 2001 to 2021, the share of underground runoff and ^{90}Sr removal by groundwater under conditions of artificially disturbed regimes increased, and groundwater became one of the leading sources of secondary surface water pollution in the exclusion zone.

As the degree of reclamation increases, the barrier resistance of the catchment decreases. It leads to the increased ^{90}Sr removal. Using the solutions to regression equations, it is possible to strengthen the barrier functions of catchments and achieve a lower removal of radionuclides by regulating individual hydrological and landscape factors.

The general strategy of water protection activities on catchments of the ChEZ with drainage systems should consist of the maximum use of natural barrier elements and functions of the landscape, technical means of the drainage systems (developed network of channels, cross regulators, etc.), and neutralization and localization of the factors that contribute to the mobilization of radionuclides. If the barrier resistance of the landscape catchment system is very low, it is possible to increase the accumulative capacity of these landscapes by applying sorption materials into the soil of micro-depressions, arranging biological barriers, etc.

REFERENCES

Baryakhtar, V., Kuhar, V., Los, I., Poyarkov, V., Kholosha, V., & Shestopalov, V. (Eds.). (1998). *A comprehensive risk assessment of the accident at the Chernobyl NPP*. IAEA.

Shevchenko, A. (2002). Distribution and Migration of Radionuclides in Meliorated Areas. In V. Shestopalov (ed.) *Chernobyl disaster and groundwater*. Lisse/Abingdon/Exton(pa)/Tokyo.

Shevchenko, A.L. (2005). Activities on drying-moistening systems in case of radioactive soil contamination. *Reclamation and water management. 92*, 122-130.

Shevchenko, A. L., Charnyi, D. V., Akinfiev, G. A., & Kireev, S. I. (2016). FactorsGoverningStrontium-90 exportwithSurfaceRunoffintheChernobylNPPRestrictedZone. *Water Resources*, *43*(3), 522–532. doi:10.1134/S0097807816010127

Shevchenko, A.L., & Dolin, V.V. (2002). Natural and man-made prerequisites for the decontamination of the water basin of radiation-contaminated territories. In V. Sobotovych, G.N. Bondarenko (eds.) *Geochemistry of technogenic radionuclides*. Resp.

Shevchenko, A. L., Dolin, V. V., Nasedkin, I. Yu., Kireev, S. I., & Kozitskii, O. N. (2002). Hydrological Changes and their effect on the radiological characteristics in the Chernobyl abandoned Zone. *Water Resources*, *29*(4), 449–464. doi:10.1023/A:1019634609828

Shevchenko, O., Shestopalov, V., Sakhatskiy, O., Nasedkin, I., Gudzenko, V., Akinfiev, G. (1999). Left bank flood plain: pathways to the decision of over moistening and increasing of ^{90}Sr outflow through the hole the dam N°7. *Bulletin of ecological state of Chornobyl Exclusion Zone,* (14), 51-57.

Shevchenko, O. L. (2013). Retrospective analysis of the priority of pollutant mobilization factors.*Bulletin of KNU named after T.* [In Ukraine]. *Shevchenko (Geology)*, *60*, 63–69.

Shevchenko, O.L. (2018). Drainage systems in radioactively contaminated catchments. .M. Gadzala, V.A. Stashuka, A.M. Rokochinsky. (eds.) *Reclamation and development of the Ukrainian Polesie.* Y Kyiv-Rivne.

Shevchenko, O. L., & Kireev, S. I. (2005). The method of integral assessment of the barrier capacity of catchments. [In Ukraine]. *Ecological Restoration*, *11*, 77–86.

Shevchenko,O.L., Kozytskyi,O.M. (2003). Overgrowth and siltation of riverbeds as an initial parameter in the prognostic model of radioactive contamination of reclamation channels. *Scientific works of the Ukrainian Research Hydrometeorological Institute*, *252*, 131–143.

Shevchenko, O.L., Kozytskyi, O.M., Nasedkin, I.Yu., Akinfiev, G.O., Kireev, S.I., Sakhatskyi, O.I., Khodorovsky, A.Ya. (2001). Analysis of efficiency and operation options of the water protection complex of the Left Bank polder system. *Problems of the Chernobyl exclusion zone, 7,* 112–125.

Shevchenko, O.L., Kozytskyi, O.M., Nasedkin, I.Yu., Ryabtseva, G.P., Bublyas, V.M., Ivanushkina, N.I., Osadcha, N.M., & Syzonenko, V.P. (2011). *Patterns of migration of man-made radionuclides on reclamation systems of the Chernobyl Exclusion Zone (based on research results from 1986-2004).* Kherson.

Shevchenko, O.L., Sakhatsky, O.I., Khodorovsky, A.Ya., Azimov, O.T. (1999a). Justification of water protection measures on the left bank of the Pripyat River in the exclusion zone based on the use of multizone space survey materials. *New methods in space geology*. Kyiv: Ukraine.

Steinhauser, G., Brandl, A., & Johnson, T. E. (2014). Comparison of the Chernobyl and Fukushima Nuclear Accidents: A Review of the Environmental Impacts. *The Science of the Total Environment, 470-471*, 470–471. doi:10.1016/j.scitotenv.2013.10.029 PMID:24189103

Struts, V.A. (1984). Fire on peat deposits. Fire prevention measures. *Reclamation Encyclopedic reference book*. Minsk.: Belorus.

Chapter 17
Heat Melioration With the Use of Warm Water Discharge From Energy and Industrial Facilities

Oleg Pinchuk
National University of Water and Environmental Engineering, Ukraine

Ievgenii Gerasimov
https://orcid.org/0000-0001-7194-8251
National University of Water and Environmental Engineering, Ukraine

Ivan Romanyuk
National University of Water and Environmental Engineering, Ukraine

ABSTRACT

Summarized data on an important scientific and applied problem, which is coming out in the development and justification of effective ways, technologies, and technical solutions of using warm waters from industry and energy in agriculture, are given. The use of surface heating with heat exchangers significantly affects the temperature regime of the soil and the surface air layer. Field research has established that heating the soil with sleeves-pipelines with a diameter of 10 cm, in which warm water with a temperature of 20...30°C circulates, increases the temperature of the arable soil layer by 12...16°C and the air layer by 7...12 °C, which provides sufficient conditions for the growth and development of plants in early spring. It has been proven that heating by sleeves-pipelines allows the period of active use of the soil to start earlier in the spring by 50-60 days for the soil protected by shelters, which helps to increase the sums of positive temperatures in the arable layer of the soil during the spring period by 1500...2000°C.

DOI: 10.4018/978-1-6684-8248-3.ch017

BACKGROUND

The problem of getting early and permanent yields of crops, particularly vegetables, and berries of saleable quality is one of the major socio-economic tasks in agriculture of many countries. Climate change and weather conditions, which are becoming unpredictable, exacerbate the urgency of the problem. In the generally accepted tendency of global warming, experts also pay attention to the possibility of local temperature decreases in some areas, prolonged cold periods in spring, and considerable fluctuations in weather conditions (Iglesias & Garrote, 2015; Boichenko, 2005; Piao et al., 2010; Lybbert & Summer, 2015; Rokochynskiy et al., 2019a).

Studying the influence of possible climatic and weather changes on the development and productivity of crops production and developing measures to prevent negative climatic and weather phenomena are becoming relevant in agriculture not only for researchers but also for professional practitioners

The role of climate in improving soil fertility is well-known, scientists have even introduced the concept of "agricultural fertility of the climate" as the ability of the atmosphere and the underlying surface to provide a certain level of soil fertility and crop yield through climatic resources. It is thus emphasized that the soil properties and its natural and effective fertility are manifested only in the presence of a certain amount of moisture in the soil and at a certain thermal regime (Smith, 2000; Rotjanakusol & Laosuwan, 2018; Bita & Gerats, 2013; Rokochynskiy et al., 2019b).

Thermal resources play a significant role in yield formation. Each plant has its own strictly individual thermal-temporal structure of the development cycle from sowing to harvesting. The parameters of this cycle are strictly related to the corresponding level of temperatures in the soil and air. Therefore, purposeful control of plant ontogenesis can be possible by controlling the thermal regime of the microenvironment of its habitat. The thermal regime of the soil and the surface layer of air will be also decisive for the intensification of the processes of growth and early cultivation of crops Fig.1. Taking into account the presence of a significant amount of warm water discharged and its temperature potential (20-35 °C), which is optimal for plant growth, its use for thermal melioration purposes allows for creating better conditions for crop production by expanding the boundaries of the growing season, especially in early spring and improve natural conditions for growing heat-loving crops (Pinchuk, 2012; 2015).

A significant number of methods, methods, and technical decisions of thermal land reclamation have been developed by different scientists in general and with low-potential waste heat in particular. Nowadays, a significant number of agrotechnical and engineering measures, that affect the thermal regime of the soil are known.

The general classification of methods and ways of thermal land reclamation is given by Kurtener D.A. and Chudnovsky A.F. at work (Kurtener & Chudnovsky, 1969; 1979):

1st type – placement of materials with different thermophysical characteristics to the soil surface, which ensure good thermal contact between the soil and materials;
2nd type – changing the temperature parameters of the surrounding environment by air heating, air blowing equipment, infrared emitters;
3rd type – use of shelters made of thin transparent and translucent shells that do not have close thermal contact with the soil surface;
4th type – a purposeful change in the thermophysical characteristics of the soil and the microrelief of its surface;
5th type - the use of heat sources inside and on the soil surface.

The analysis of meteorological observations in the Ukrainian Polesie shows that the danger of low temperatures persists at night and in the early morning hours practically throughout the entire spring period. For instance, the frosts up to -10°C took place daily until March, 25, and the transition of minimum air temperatures to more than -10°C took place on April, 10. During April, nighttime sub-zero temperatures of air persist. Even in April, the air temperature drops to -1 °C (Boichenko, 2005; Rokochynskiy et al., 2019a).

Figure 1. Scheme of the thermal regime formation process of the soil massif

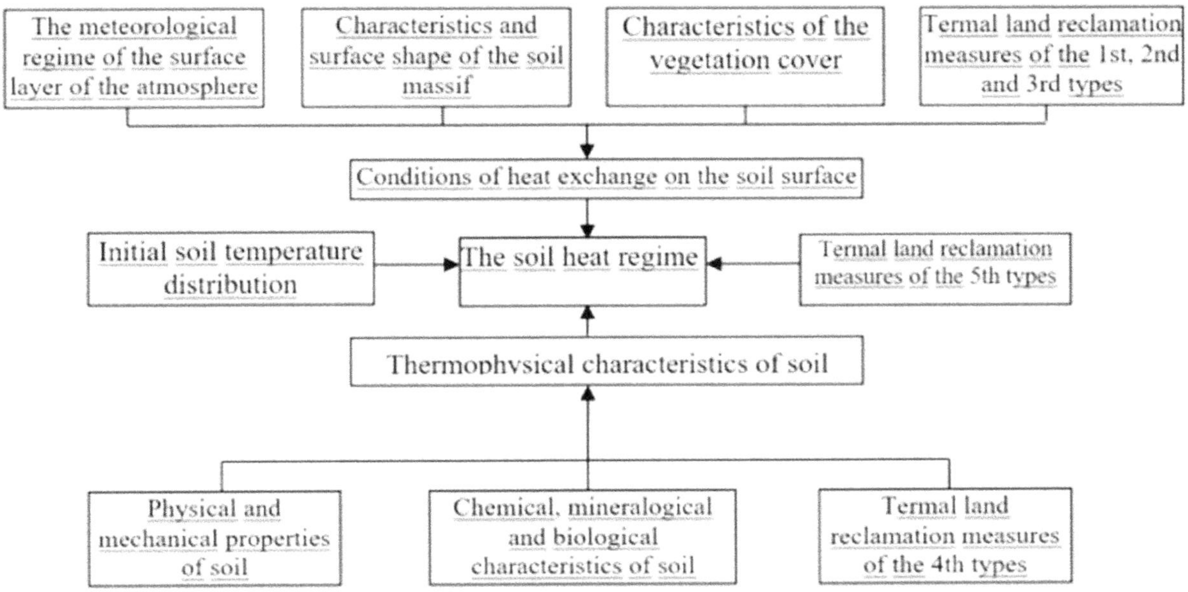

These features of the temperature conditions of the Ukrainian Polesie do not allow the active vegetation to begin in March-April, especially for thermophilic crops, and forced them to postpone the active soil use to May. The task of heat supply for crop production in the early spring periods is traditionally solved by creating greenhouse facilities and by heating the soil. Thus, high-temperature sources of thermal energy are used for heating the soil and air in cultivation facilities, which are obtained by burning high-value fuels, primarily natural gas.

Significant increases in the price of natural fuels and thermal energy put on the agenda the tasks of finding, justifying, and developing alternative ways to regulate the thermal regime of the soil and air, and sources of alternative energy to get early harvests of crops.

One option for solving this problem can be the creation of soil areas heated by special thermal reclamation facilities using thermal wastes from the industry. Hot water (heat-exchange water) from industrial and power-generating facilities with a temperature of 20-40 °C and geothermal water can be used as attractive free sources of thermal energy (Pinchuk, 2012; 2015). Their attractiveness is that such temperatures are optimal for the development of most crops.

In the 80s of the last century in many developed countries of the world, in the USA, France, Germany, the former Soviet Union, and in particular at our National University of Water and Environmental

Engineering, long-term research on the efficient use of the charged hot water from nuclear power plants and thermal power plants for heating the soil by the pipeline systems were conducted (Pinchuk, 2012; Miernicki et al., 2020; Dalampakis et al., 2017; Olszewski, 1978; Margen, 1978; Balligand et al., 1978; Turcheniuk et al., 2022). The pipeline systems showed sufficient efficiency in increasing the productivity of separate crops and enabling earlier harvesting. However, their thermal efficiency about the surface layer of air was low (0.5…1.5 °C), which did not allow them to find wide application as an effective method of heating reclamation, except for some engineering problems (cooling of water in the soil, accumulation of heat in the soil, heating of sports grounds, etc.).

One of the promising and effective methods of thermal reclamation of adjacent areas, in our view, can be the surface heating of soil with warm water with temperature parameters within 25…35 °C.

The working scientific hypothesis is that the maximum thermal effects when using water with such temperatures should be expected by directing the water flows to the area of plants' habitat. The transfer of thermal resources in the amount of 15…20 °C to the environment "soil – air" will theoretically allow shifting the transition through 10ºC to the end of February – beginning of March Figure 2.

Figure 2. Forecast of the efficiency of heat reclamation for the conditions of Ukrainian Polesie: 1 – average ten-day long-term temperatures of the surface air layer (Rivne); 2 – increased air temperatures by 5°C; 3 – increased air temperatures by 10°C

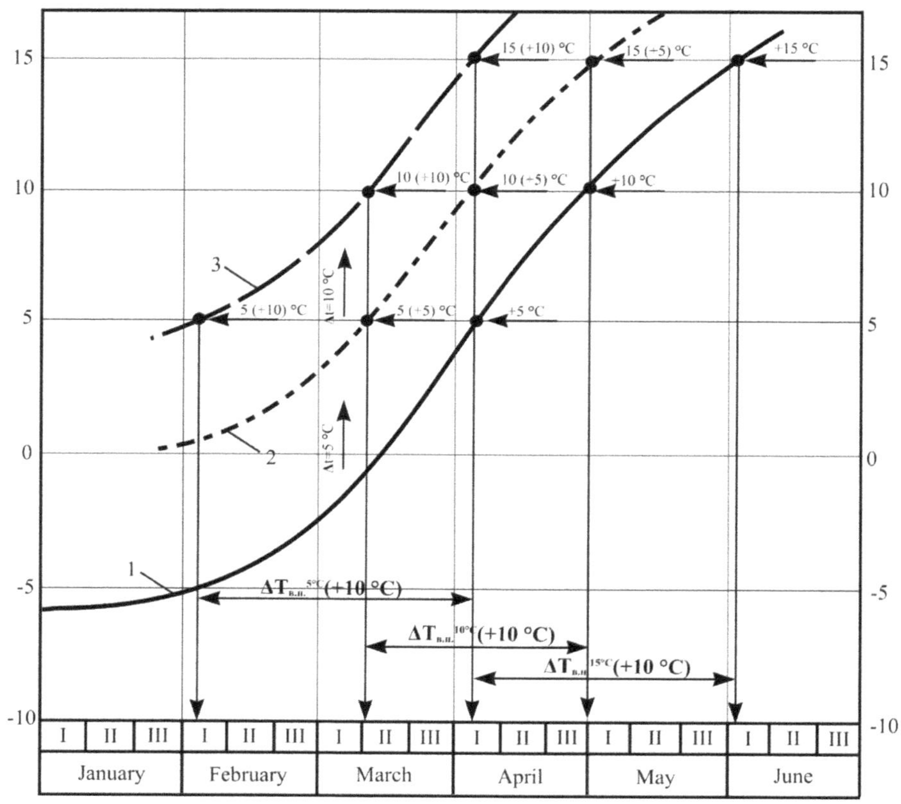

Technologically, this technique can be implemented by directing the water flow in a thin layer on the soil surface, and can be technically executed with the application of special sleeves – heat-exchangers of various designs and geometry, made of thin, long lasting, flexible polymeric materials – films Fig. 3.

The authors proposed a general principle design scheme of the soil surface heating system Fig. 4 with shell-sleeves developed by analogy to the surface irrigation design schemes and drip irrigation systems. According to the small pressures used by the system, it is designed as a low-pressure, gravity system. The intake of warm water from discharge channel 1 is carried out with the help of pumping station 2 and is fed to regulating basin 3. The flow regulator (water outlet) 4 allows the regulation of the water flow to the main pipeline 5, which distributes it between the supply pipelines 6 and the distribution pipelines 7 of the block modules 9 with block sections 8. The network of collecting 10, outlet 11, and discharge pipelines (closed collectors) 12 drains water from block modules 9 with block sections 8 and from flexible shells-sleeves outside the surface soil heating system (SSHS).

Figure 3. A general view of heat-exchanger with plants

METHODS AND TECHNIQUES

The thermal efficiency of soil heating with water-filled flexible sleeves was studied experimentally in a field model experiment performed in the climatic terms of the Ukrainian Polesie on sandy loam and black soils. Strawberry of the "Festivalny" type was used as the main crop-indicator. The soil was heated with specially heated warm water, which circulated by gravity-flow under a small initial pressure (10...20 cm) Figure 5. The temperature of water in the sleeves was automatically maintained within 24…26 °C in March with a subsequent increase in April and May up to 27…30 °C.

The experimental system worked as follows: under the pressure of 0.1 m, through pipeline 2, the heat-carrying medium from the warm water reservoir 1 enters the distributing collector, which distributes water between the sleeves 4. Warm water passing through cover sleeves heats the soil and surface layer of air around the plants, intensifying the processes of their growth and development. The cooled water enters the suction line of the pump unit 6 through collector 5 and then flows through the conveyor tube 7 to the electric heater 8. In electric heater 8, the coolant is heated to the specified temperature and enters the warm water reservoir 1. The flow rate of the heat-carrying medium from reservoir 1 is controlled by valve 12, afterwards, the cycle is repeated. In the heating system, the water temperature was kept within the range of 20-30°C.

To ensure the continuity of measuring temperature parameters in the environment "soil–plant – air-ground interface", an increase in the number of measurements carried out with constant high frequency the researchers applied a specially designed automated data collection system. The authors measured the soil temperature in an automated mode at the air-ground heating area in the center of the greenhouse at depths of 0, 10, 20, and 40 cm, as well as the air temperature in the shelter at heights of 20, 30, and 70 cm. They also measured the air temperature in the shelter at the side of the film at a height of 20 cm from the soil surface and outside of the shelter above the film. The temperature of the soil outside the shelter was measured at a depth of 0 and 20 cm, and the air – at a height of 200 cm. Measurement and logging of temperature parameters were made once a minute, and the data commit process took place on a personal computer at 5-minute intervals.

Figure 4. The schematic diagram of the SSHS with shells-sleeves

1 – warm water channel; 2 – pumping station; 3 – regulating basin; 4 – flow regulator; 5 – main pipeline; 6 – supply pipeline; 7 – distribution pipeline; 8 – block sections made of sleeves; 9 – block modules; 10 – collecting pipeline; 11 – outlet pipeline; 12 – discharge pipeline (channel)

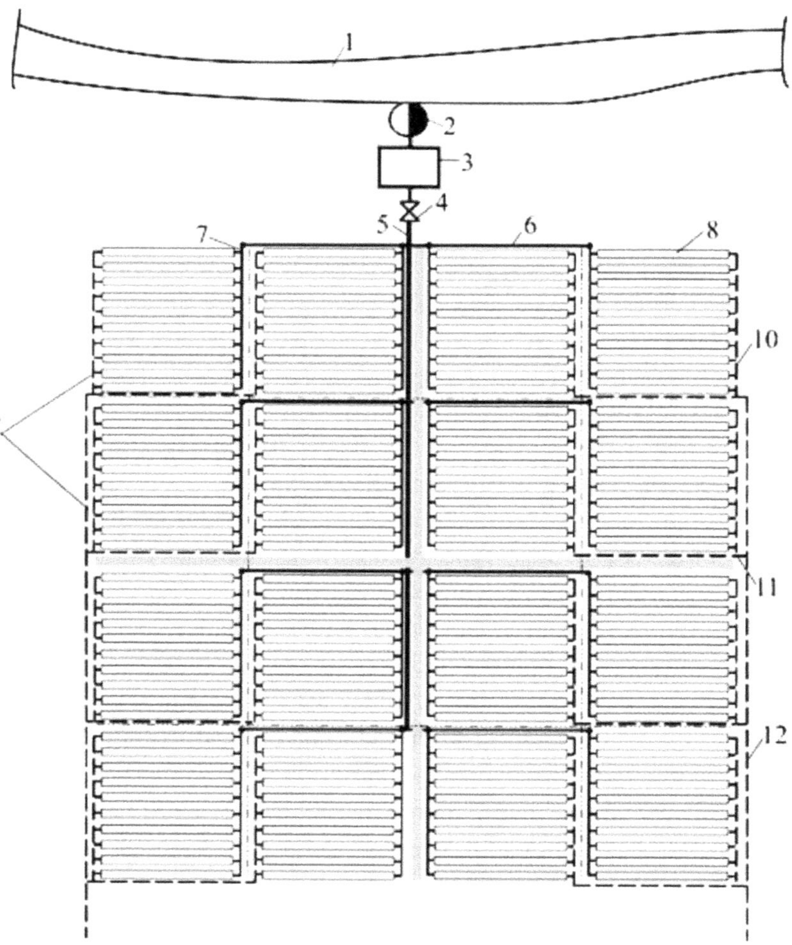

Figure 5. Field Experiment Design
1 – warm water reservoir; 2 – heat supply pipeline; 3 – distributing collector; 4 – cover sleeve; 5 – accumulating collector; 6 – pump unit; 7 – main pipe; 8 – electric heater; 9 – tunnel cover made of polyethylene film (greenhouse); 10 – plants; 11 – temperature sensors; 12 – slide valve

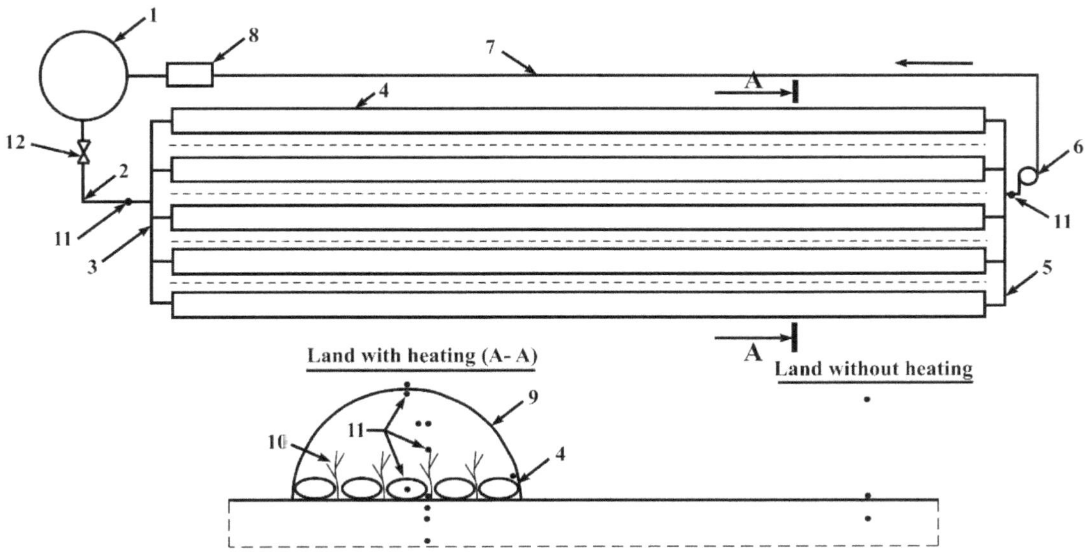

The initial stage of soil heating by heat exchangers was studied by us on March 12, 2019, for two weeks. All parameters were measured twice a day at 7:00 am and 7:00 pm. The basic results of measuring the temperature of soil, water, and air, are shown in Figure 6. An open heated ground in natural conditions was taken as a control option.

The research results show that surface heating is an effective technical method of thermal reclamation of the plant's habitat, which makes it possible to quickly and reliably increase the temperature of the soil and the surface air when using a film cover. On the third day of operation of the heat exchangers, the temperature in the upper 20 cm soil layer rose to 10…14°C. At the same time, due to the heat flow from the surface of heat exchangers upwards, the surface air layer warmed up to 8…11 °C.

On the fourth-fifth day of the heating system operation, favorable temperature conditions were created at the level of 14…16 °C in the upper 20 cm layer of soil and about 10…12 °C in the surface layer of air, which were constantly preserved even on cloudy and frosty days. Therefore, heating with heat exchangers, which is very important, is guaranteed to maintain a positive air temperature at night and protect plants from frost.

The soil temperatures obtained as a result of constant heating (14…16 °C) promote the germination and development of most vegetable and berry crops, even thermophilic ones. In this case, under natural conditions, soil, and air temperature remain very low, varying from - 2 °C to +3 °C. The whole period of observations of the soil in natural conditions remained unprepared for agricultural use due to low temperatures. It should be noted that at this time there was precipitation in the form of snow and cold rain. The snow that fell remained on the surface of the control area, while on the surface of the film cover it melted and turned into water.

Figure 6. The research results of surface air (A) and soil (B) temperature in March 2019

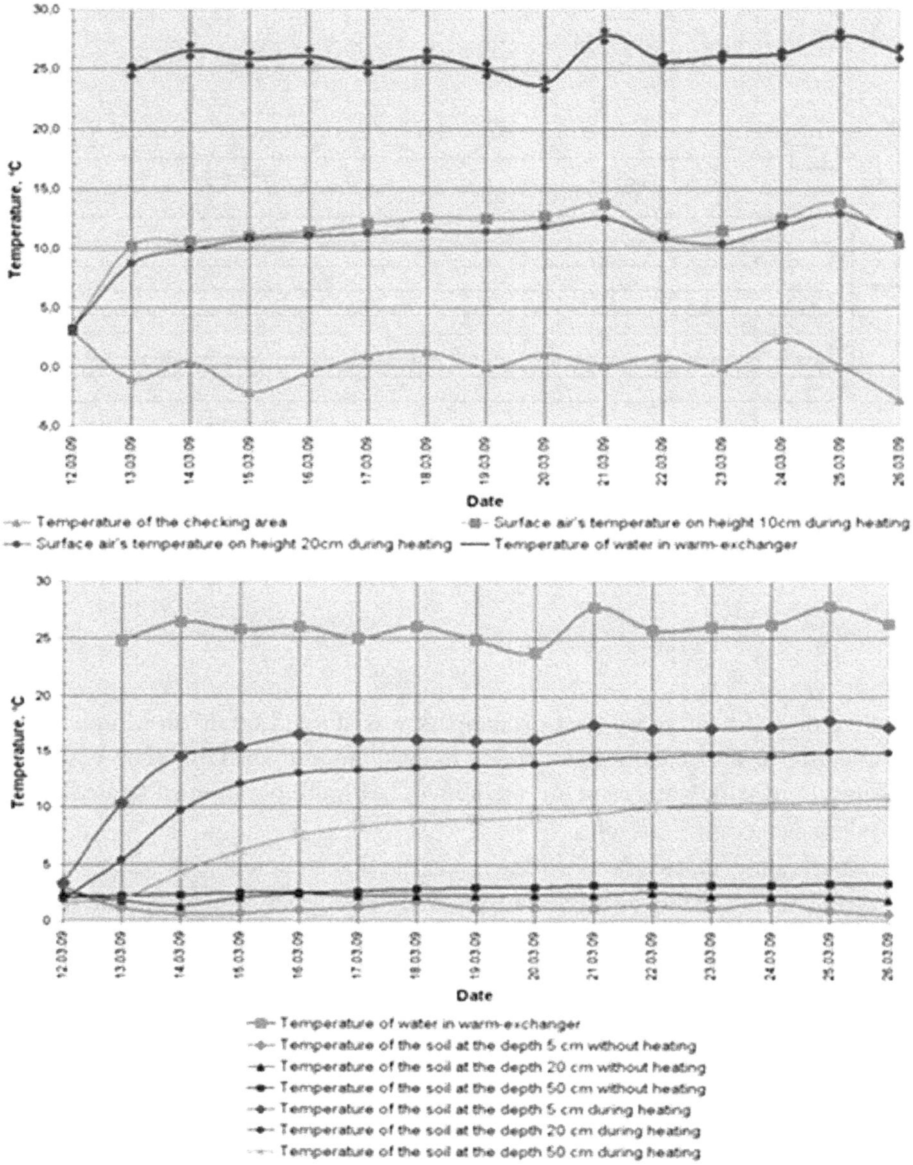

Heat exchangers constantly consumed thermal energy, and consequently, water cooled down. As a result of measurements and calculations of thermal energy, its consumption amounted to 700-800 Wt/h or 200-220 Wt/h per one square meter of soil which was heated, and water in the sleeves cooled down by 0.6...0.9 °C.

Observations of the soil temperature have shown that the effect of heating extends to a considerable depth. Thus, according to the results of measurements in 209, the soil temperature at a depth of 1 m under heat exchangers after two weeks of their operation rose to 6.4 °C, while in the control area, it amounted to 4 °C (table 1).

Soil heating can be carried out at the depth of 2m and more. In this case, an array of heated soil is created under the heat exchanger, accumulating significant reserves of thermal energy. This reserve of thermal energy guarantees the protection of the upper root layer of the soil from penetration of negative temperatures from the lower soil layers.

Table 1. Temperature of the soil in the area that is heated and in the control area in different points of the soil profile, °C (of March 26, 2019 at 7:00am)

Depth, cm	Control area	Heated Soil
0.0	-0.3	15.4-22.8*
5.0	0.6	17.5
10.0	0.8	17.2
15.0	1.6	16.8
20.0	1.9	15.8
30.0	2.7	13.5
40.0	3.0	11.8
50.0	3.3	10.7
60.0	3.4	9.7
80.0	3.6	7.5
100.0	4.0	6.4

*Note: lower value corresponds to the soil temperature at input of the heat exchanger, higher one – to the temperature of the soil above the heat exchanger.

Great importance in assessing the temperature of the heated soil is attached to the peculiarities of its formation during the vegetation period (March-May), as well as during the day. According to the results of our research in 2004-2020, conducted on sandy loam soils of the Ukrainian Polesie, the influence of the surface heating is more noticeable in cold periods (March, April). The constant inflow of heat from the heating system prevents penetration of negative temperatures into the soil and maintains the temperature background in the habitat of plants at the levels close to optimal.

The temperature regime of the soil in the daily cycle, due to the frequency of receiving solar heat, is characterized by a slow change of temperature values from maximum to minimum, which changes with time and depth.

In the course of our research it is found out that the general nature of temperature changes during the day, both in the heating section and in the control one it remained the same. For example, the maximum temperatures of soil at the depth of 5 cm for the heated area are reached at 4:00pm…6:00pm and are within 27…29 °C, while in the control area – at 5:00-6:00pm reaching the temperature 18…19°C. The minimum soil temperatures are observed in the heated area at 7:00-8:00am and are 19…20°C, and in the control area they reach 12…13°C at the same time. The process of heat exchange in the "soil-air" system and the direction of heat flows depend on the thermal gradient conditions in the soil. The heated soil is characterized by the presence of maximum temperatures in the area where heat exchangers are located (surface soil), therefore, the dynamics of merging gradients and their type of movement must be investigated for a considerable soil layer Figure 7.

Figure 7. Profile changes of average monthly soil temperatures in 2019

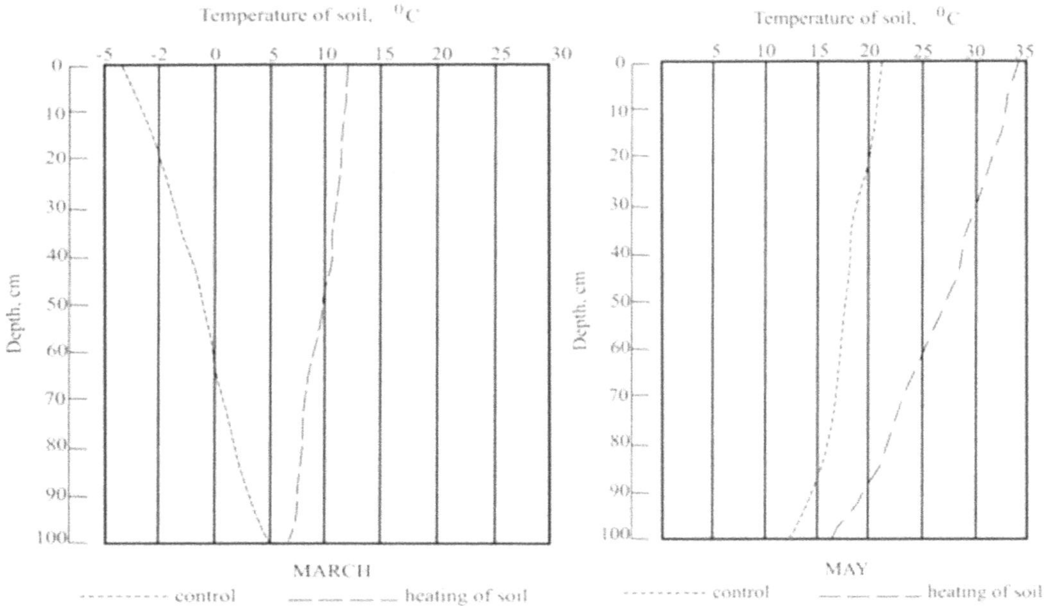

In March, 2019, soil temperature at the depth from 0 to 20 cm in the control area was below zero, while in the heated area it reached 11…13 °C. In the layer from 20 to 80 cm, the temperature of control soil changed from 0.5°C to 3.5 °C, and in the heated area it ranged from 8°C to 11°C.

A completely different picture is observed in May. For example, in the soil layer ranging from 0 to 20 cm in the control area, the temperature is 20…21 °C, and in the heating area – 31…34°C. In the layer from 20 to 80 cm in the control area, the soil temperature is 16…20 °C, and in the heating area – 26…32°C.

Although heating with heat exchangers using warm water cannot create optimal temperatures in the upper soil layer immediately after starting the system, but in late March and early April contributes to its increase by 10°C and more, and in mid-April – by 15°C and more. At the control site, the temperature reaches 10°C in late April, and 15°C in mid-May.

The increase in periods of soil heating and increase in absolute temperatures during these periods result in the increase in the sum of soil temperatures during vegetation period (table 2).

The obtained results demonstrate that from the point of view of positive temperatures accumulation, the option of soil heating with heat exchangers with the use of shelters provides very effective heat reclamation method.

Table 2. The sum of soil temperatures more than 5, 10, 15 and 20 °C at the depth of 20 cm, °C

Temperature, °C	Heating with heat exchangers	Control
5	8125	6348
10	7240	5886
15	6430	4522
20	4055	1925

The difference in sums of soil temperatures at the depth of 20 cm as compared to control area are more than 2000°C.

As our research has shown for the heated soil, the relationship between its temperature (Y) at the depths of 5, 10, 20 cm, air temperature (X) and water temperature in heat exchanger (Z), can be described by the following empiric equations (for values of Z > 20 °C):

$$Y0.05= X+0.14Z–0.48; Rxyz=0.95 \quad (1)$$

$$Y0.1= X+0.15Z–1.76; Rxyz=0.95 \quad (2)$$

$$Y0.2= X+0.12Z–2.40; Rxyz=0.95 \quad (3)$$

When using tunnel-type greenhouses, soil heating with heat exchangers with a water temperature of 25…30 °C helps create all the necessary temperature conditions for shifting the vegetation period and early growth of crops, on average, from one to two months depending on the type of crop.

The moment that determines the beginning and suspension of active vegetation of plants is the date of transition of air and soil temperature through 10 °C. With the onset of the dates of temperature transition through 10°C, the same patterns appear along with the beginning of the active temperatures period. In the area of heating by heat exchangers, the date of temperature transition through 10°C in the soil at the depth of 20cm in 2019 fell on April 1, and in the control area – on April 22, that is 21 days earlier.

Mathematical models of heat transfer and heat exchange form the theoretical basis for substantiating, and designing engineering processes and technical means of heating the soil in a greenhouse. When designing mathematical models of heat transfer and heat exchange in the system "soil-plant-surface air layer" it is necessary to take into account the main factors that determine heat exchange and heat transfer processes (Kurtener & Chudnovsky, 1969; 1979).

As a result of the extensive theoretical studies, the authors developed a mathematical model of the thermal regime of the greenhouse, taking into account the air convection in its volume. This mathematical model is a system of equations, which consists of the thermal balance equation about thermal conductivity and convection components, the Navier-Stokes equation, and the flow continuity equation (Vostrikov et al., 2008; Hud et al., 2019; Turcheniuk et al., 2022).

The theoretical calculations established sufficiently high heat amelioration effects in the greenhouse soil and air under the conditions of surface heating by cover sleeves. A case study simulating the temperature of 0°C and the warm water temperature of 20°C confirmed the creation of the area of guaranteed heating in the air under the cover, and heat penetration of the soil up to a depth of 40 cm.

The research proved that air circulation has a significant influence on temperature distribution, and this impact results in maximum temperature in the center of the greenhouse. The analysis established that the temperature of the soil under the shelter decreases from the middle to its edge by 2-6 °C, depending on the distance to its surface.

It is known that the criterion for assessing the effectiveness of any measures, including heat reclamation is the yield of crops. As a result of the application of surface heating in a root-containing layer, there is a change of a whole complex of processes which, in the end, influences the growth, development, and formation of crops.

Figure 8. Results of computer simulation of temperature distribution in a greenhouse with soil heating system at warm water temperature T_{ww}=20°C and air temperature T_a=0°C

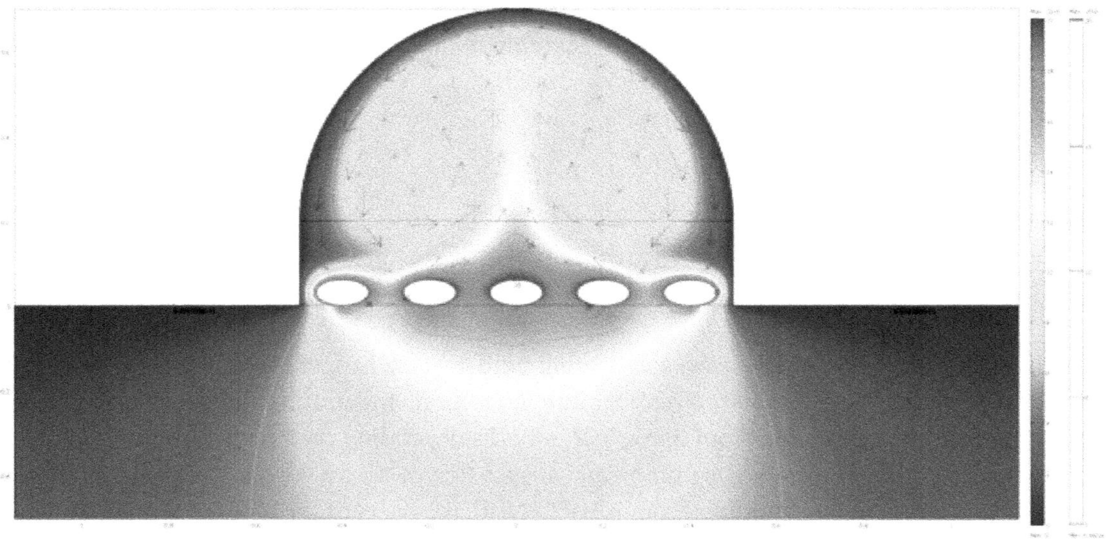

Figure 9. Soil heating pattern in a greenhouse with soil heating system at T_e = 0 °C
T_w = 20 °C: 0 – on the soil surface; 1 – at a depth of 0.1 m; 2 – at depth 0.2 m; 3 – at a depth of 0.3 m; 4 – at a depth of 0.4 m; 5 – at a depth of 0.5 m

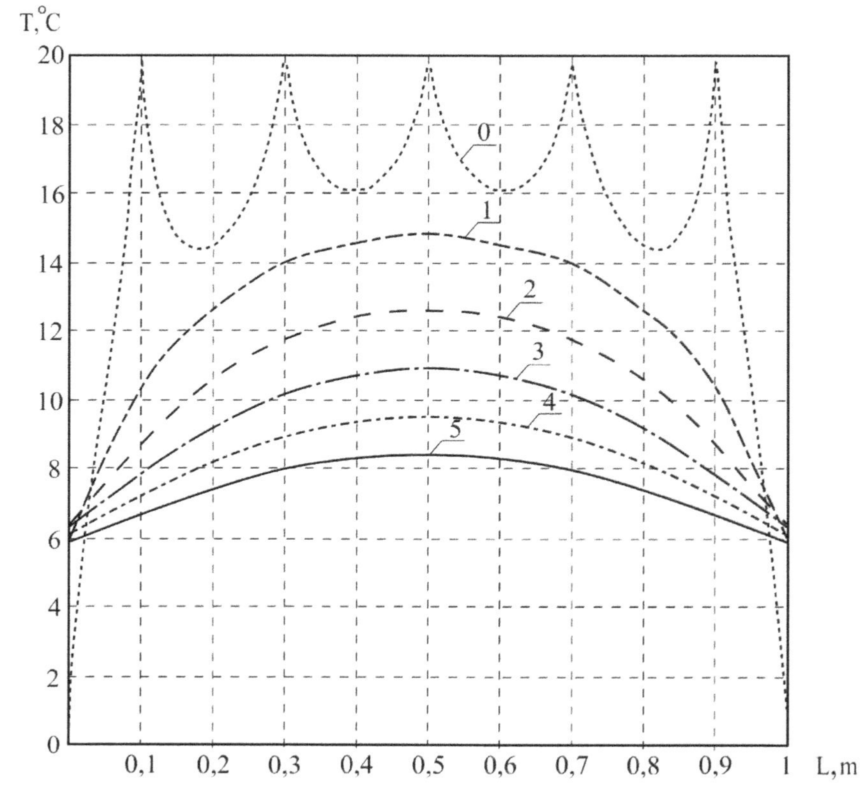

Figure 10. Air heating pattern in a greenhouse with soil heating system at $T_e = 0$ °C
$T_w = 20$ °C: 0 – at a height of sleeves' surface; 1 – at a height of 0.1 m; 2 – at a height of 0.2 m; 3 – at a height of 0.3 m; 4 – at a height of 0.4 m; 5 – at a height of 0.5 m; 6 – at a height of 0.6 m

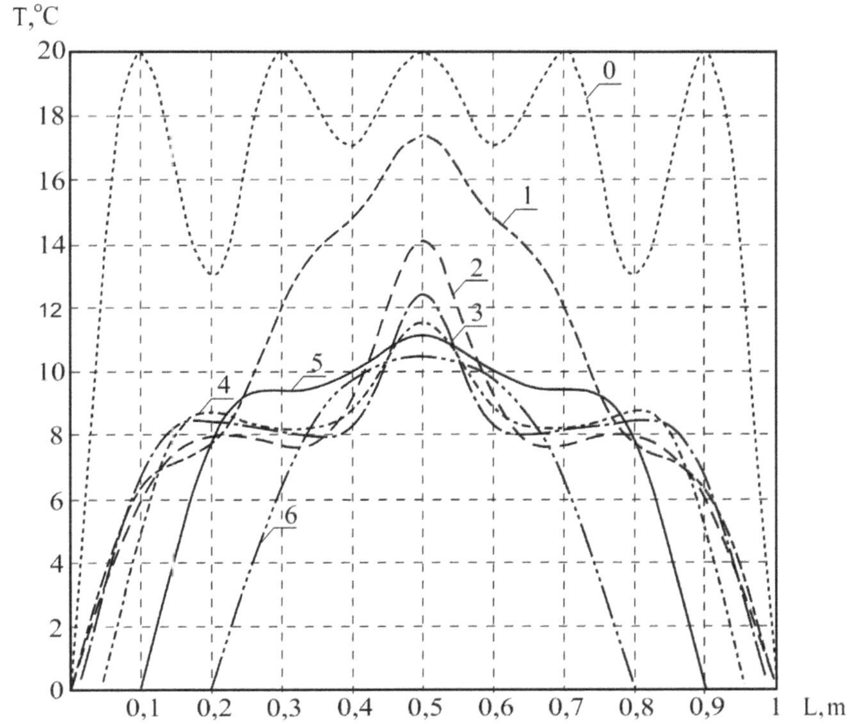

We studied the effect of surface heating on the growth and development of the "Festivalny" strawberry type. The results of observations during March-May 2019 showed that the first leaf on the strawberry appeared on March 19 and at the same time it increased the appearance of the second leaf, and the third one appeared on March 23. Strawberries beginning of stolon (runner) formation in the area of soil heating on April 3 (code 41 according to the BBCH-scale of strawberry), and on April 6, 5 flowers appeared and mass flowering began on April 9 and lasted throughout the second decade of April. Mass bud forming of strawberries was observed on April 21, and the first crop of ripe berries was harvested on April 25.

The dynamics of the strawberry harvest during the study period is presented in Table 3.

FUTURE RESEARCH DIRECTIONS

Thus, surface soil heating systems with water-filled shell sleeves made of strong thin-walled materials are promising from the point of view of thermal improvement effects. Existing scientific research results on the study of such systems are insufficient for conclusions regarding their design, construction, operation, and wide implementation.

Table 3. Dynamics of strawberry crop yield

Date	The mass of berries, g	The average weight of 1 berry, g	crop yield	
			g / m²	g / meter of heat exchanger
25.04.2019	145.41	13.10	22.37	5.59
01.05.2019	475.45	14.28	73.15	18.29
06.05.2019	4754.5	16.06	731.46	182.87
08.05.2019	3670.77	13.97	564.73	141.18
10.05.2019	5306.17	13.04	816.33	204.08
13.05.2019	6716.24	13.86	1033.27	258.32
15.05.2019	3571.61	12.38	549.48	137.37
19.05.2019	2096.42	11.80	322.53	80.63
21.05.2019	1001.22	11.28	154.03	38.51
25.05.2019	645.65	11.63	99.33	24.83
Sum	**28383**	**13.14**	**4366.68**	**1091.67**

Implementation of the surface soil heating systems with water-filled shells-sleeves requires solving several scientific and applied problems, in particular, substantiation and development of rational forms and designs of shells-sleeves that work as heat exchange devices, designs of the hydraulic construction of the whole system itself, which distribute water over the soil surface and are used seasonally, designs of temporary portable transparent shelters, equipment of automation and regulation water distribution systems, regulation of the soil water regime and the surface air layer, determination of rational hydraulic and heat-technical modes of operation of the hydraulic system as a separate object and as a part of the ecological-biological system, verification of design methods, installation and technical operation of surface soil heating systems, as well as methods of growing various crops on thermal meliorative soil.

CONCLUSION

Field research has shown that heating the soil with heat exchangers circulating hot water at the temperature of 25...30°C increases the temperature of the arable layer by 12...16°C in the soil, and by 7...12°C – in the air, which provides sufficient conditions for plant growth and development in early spring.

It is proved that heating with heat exchangers allows the period of active soil use to start earlier in the spring by 40-60 days for sheltered soil, and also increases the number of positive temperatures in the arable soil layer during the spring period by 1500 ... 2000°C. Close linear relationships are established between the temperatures of soil, air, and water in the sleeves.

It is established that the joint action of surface heating and solar radiation during the day leads to a rapid increase in air temperature and the need for ventilation of tunnel greenhouses. Intense ventilation leads to the loss of moisture that evaporates from the plants and drying of the upper 10-cm layer of soil, which causes the need to replenish moisture in the soil during the increase in strawberry yield. It is recommended to carry out two or three watering in late April and early May at the rate of 150-200 m3/ha.

The influence of soil heating with heat exchangers on the growth, development, and yield of strawberries has been studied. Under the conditions of soil heating, the ripening of strawberry berries occurs about 40 days earlier than in the control area without heating, the total yield increases by due to heating by 50% or more when compared to the control area.

Assessment of the introduction of surface heating technology for the soil on the area of 1 hectare as an innovative project based on the methodology for assessing net discounted income gives a net discounted income of 32 thousand dollars when growing strawberries. The payback period for capital investments does not exceed 3 years.

At the same time, several scientific problems need further research, in particular the issues of rational designs of surface heating systems, heat exchangers, methods of soil protection and structures of greenhouses (translucent shelters), water supply and collection regimes, methods of soil water regulation, automation, control and management, operation of heating systems and ventilation systems of shelters, technologies of cultivation of crops, development of methods of thermomechanical and hydraulic calculation, methods of technical operation, etc.

REFERENCES

Balligand, P., Le Gouellec, P., Dumont, M., & Grauby, A. (1978). Experience Gained in France on Heat Recovery from Nuclear Plants for Agriculture and Pisciculture. *Nuclear Technology*, *38*(1), 90–96. doi:10.13182/NT78-A16160

Bita, C., & Gerats, T. (2013, July). Plant tolerance to high temperature in a changing environment: Scientific fundamentals and production of heat stress-tolerant crops. *Frontiers in Plant Science*, *31*. Advance online publication. doi:10.3389/fpls.2013.00273 PMID:23914193

Boichenko, S. (2005) *Hlobalni i rehionalni kolyvannia (zminy) klimatu ta mozhlyvi ekolohichni naslidky vid nykh na terytorii Ukrainy: napivempirychni modeli, stsenarii [Global and Regional Climate Fluctuations (Changes) and Their Possible Ecological Consequences on the Territory of Ukraine: Semiempirical Models, Scenarios]* [Doctoral dissertation, Odessa, Ukraine].

Dalampakis, P., Gelegenis, J., Illias, A., Ladas, A., & Kolios, P. (2017). Technical and economic assessment of geothermal soil heating systems in row covered protected crops: A case study from Greece. *Applied Energy*, *203*, 201–2018. doi:10.1016/j.apenergy.2017.06.029

Hud, V., Pinchuk, O., Martyniuk, P., & Volk, P. (2019). Mathematical modelling of heat transfer in a greenhouse with surface heating system. *Scientific Review Engineering and Environmental Sciences*, *28*(4), 569–583. doi:10.22630/PNIKS.2019.28.4.52

Iglesias, A., & Garrote, L. (2015). Adaptation strategies for agricultural water management under climate change in Europe. *Agricultural Water Management*, *155*(June), 113–124. doi:10.1016/j.agwat.2015.03.014

Kurtener, D., & Chudnovsky, A. (1969). *Raschet i regulirovanie teplovogo rezhima v otkrytom i zashchishchennom grunte* [*Calculation and Control of the Thermal Regime in Open and Covered Soil*]. Hydrometeoizdat.

Kurtener, D., & Chudnovsky, A. (1979). *Agrometeorologicheskie osnovy teplovoi melioratsii* [*The Agrometeorological Foundation of Thermal Melioration of Soil*]. Hydrometeoizdat.

Lybbert, T., & Summer, D. (2015). Agricultural technologies for climate change in developing countries: Policy options for innovation and technology diffusion. *Food Policy*, *37*(1), 114–123. doi:10.1016/j.foodpol.2011.11.001

Margen, P. (1978). Summary - Low-Temperature Nuclear Heat. *Nuclear Technology*, *38*(2), 321–330. doi:10.13182/NT78-A32030

Miernicki, E., Heald, A., Huff, K., Brooks, C., & Margenot, A. (2020). Nuclear waste heat use in agriculture: History and opportunities in the United States. *Journal of Cleaner Production*, *267*, 121918. doi:10.1016/j.jclepro.2020.121918

Olszewski, M. (1978). *Waste heat utilization from electric generating plants*. OSTI.

Piao, S., Ciais, P., Huang, Y., Shen, Z., Peng, S., Li, J., Zhou, L., Liu, H., Ma, Y., Ding, Y., Friedlingstein, P., Liu, C., Tan, K., Yu, Y., Zhang, T., & Fang, J. (2010). The impacts of climate change on water resources and agriculture in China. *Nature*, *467*(7311), 43–51. doi:10.1038/nature09364 PMID:20811450

Pinchuk, O. (2012) *Obhruntuvannia konstruktsii ta parametriv hidrotekhnichnoi systemy poverkhnevoho obihrivu hruntu obolonkamy-rukavamy pry vykorystanni skydnykh teplykh vod [Justification of Design and Parameters of Hydrotechnical Surface Soil Heating System Using Cover Sleeves by Means of Warm Waste Water].* [Dissertation, Rivne]

Pinchuk, O. (2015) Vozmozhnosti teplovoi melioratsii dlia obespecheniia prodovolstvennoi bezopasnosti strany [Possibilities of Heat Amelioration to Ensure the Country's Food Safety]. Prirodoobustroistvo: nauchno-prakticheskii zhurnal [Natural Habitat: Scientific and Practical Journal], 1, 6-11.

Rokochynskiy, A., Volk, P., Frolenkova, N., Prykhodko, N., Gerasimov, I., Pinchuk, O. (2019) Evaluation of climate changes and their accounting for developing the reclamation measures in western Ukraine. *Scientific Review Engineering and Environmental Sciences, 1*(83), 3-13. doi: . doi:10.22630/PNIKS.2019.28.1

Rokochynskiy, A., Volk, P., Pinchuk, O., Turcheniuk, V., Frolenkova, N., Gerasimov, I. (2019) Forecasted estimation of the efficiency of agricultural drainage on drained lands. *Journal of Water and Land Development*, *40* (I-III), 149-153. . doi:10.2478/jwld-2019-0016

Rotjanakusol, T. & Laosuwan, T. (2018) Estimation of land surface temperature using Landsat satellite data: A case study of Mueang Maha Sarakham District. *Scientific Review – Engineering and Environmental Sciences*, *27* (4), 401-409. doi:10.22630/PNIKS.2018.27.4.39

Smith, K. A. (2000). *Soil and Environmental Analysis: Physical Methods, Revised, and Expanded* (2nd ed.). CRC Press., doi:10.1201/9780203908600

Turcheniuk, V., Rokochinskiy, A., Kuzmych, L., Volk, P., & Koptyuk, R. (2022). A Technological System for Using Waste Warm Water from Energy Facilities for Effective Agriculture. *Archives of Hydro-Engineering and Environmental Mechanics*, *69*(1), 13–25. doi:10.2478/heem-2022-0002

Turcheniuk, V., Rokochinskiy, A., Kuzmych, L., Volk, P., Koptyuk, R., Romanyuk, I., & Voropay, G. (2022). The efficiency of waste hot water utilization to improve the temperature conditions for growing plants. *Journal of Water and Land Development*, (54), 1–7. doi:10.24425/jwld.2022.141559

Vostrikov, V., Romaniuk, I., Pinchuk, O., & Vostrikova, N. (2008) Modeliuvannia teplovykh protsesiv v hrunti pry obihrivi vodonapovnenymy rukavamy [Modelling Thermal Processes in Soil While Heating with Water-Filled Sleeves]. Melioratsiia i vodne hospodarstvo: mizhvidomchyi tematychnyi naukovyi zbirnyk [Reclamation and Water Engineering: Interagency Thematic Collection of Research Papers], 96. Kyiv: Ahrarna nauka.

Chapter 18
The Use of Biotechnologies for Treating Underground Waters in North-Western Regions of Ukraine

Oleksandr Kvartenko
National University of Water and Environmental Engineering, Ukraine

ABSTRACT

The chapter presents the results of monitoring studies on the parameters of the quality of underground waters which are the source of water supply for settlements in the northern- western regions of Ukraine. The tendency is revealed concerning their progressing worsening under the impact of the anthropogenic factor by such parameters: the degree of aggressiveness, the content of compounds of iron, manganese, ammonia nitrogen, and in some settlements – of phenols. The analysis was carried out and the low efficiency was pointed out of the operation of traditional technologies on the basis of the method "simplified aeration – filtration". Based on researches conducted a physical model was developed and mechanisms of the comprehensive treatment of underground waters on matrix structures of bio-minerals in inter-pore space of the contact loading of the bioreactor. The technologies were developed, studied and put into practice for the comprehensive water preparation on the basis of combined methods of treatment.

BACKGROUND

The Impact of Climate Change on Several Characteristics of the Sources of Water Supply Systems

The quantity of water available for intake for the water supply system of settlements and industrial enterprises depends on the amount of surface flow, the replenishment of underground water-bearing horizons, their characteristics (the degree of isolation, the depth of bedding, capacity), water quality in water supply source. At present in the world, the issue is not sufficiently studied of the interrelationship

DOI: 10.4018/978-1-6684-8248-3.ch018

between surface flow and underground horizons and the rate of their replenishment (Bates, 2008). As the major part of ground waters has the replenishment on account of the surface flow, then under modern conditions of climate change the regimes of surface flow will render certain influence on several characteristics of underground water-bearing horizons. Not numerous studies of the aftermaths of climate change for ground waters show their strong dependence on a particular locality and a specific climate pattern (Eckhardt, 2003). The work (Bates, 2008) shows the possible consequences of climate change both for sources and for water supply systems. In case of heavy precipitation, there arise unfavorable consequences for the quality of surface and ground waters connected with the pollution of water supply sources. In case of drought risks increase of the deficit of foodstuff and water. It is forecast that the increased water temperature, the growth of precipitation intensity, and more continuous periods of low flow will endanger many forms of water pollution including the contamination by drifts, dissolved organic carbon, increased content of phosphorus, heavy metal ions, phenols, pesticides, salt content, heat pollution (Bates, 2008). These processes in their turn will impact not only ecosystems but also the quantity and quality of groundwater used in water supply systems, increasing expenses on their maintenance.

The Influence of Anthropogenic-Natural Factors on the State of the Quality of Underground Waters in Northern-Western Regions of Ukraine

At present, about 30% of all drinking water in Ukraine is taken from underground sources. For agricultural economic water supply, this figure increases to 90%. However, only 26% of the rural population makes use of services of centralized water supply (NR, 2005). All other ones use for drinking need local sources – mined wells whose water in most cases is in an unsatisfactory condition (NR, 2005). And canalization networks and installations treating waste waters embrace a little more than 8% rural population. Thus, the construction of water supply systems without canalization leads to the pollution of little rivers and underground waters. In particular, only in 2014 did surface objects of Rivne region there was discharge about 111,5 mln.m3 of storm waters with 15,55 thous. t of pollutants (RSER, 2015). Among the main pollutants: are 2544 t chlorides, 2134 t sulfates, 1623 t nitrites, 184 t phosphates, 8,91 t ferrum, 2,6 t SSAS, and 0,996 t oil products (RSER, 2015). The sources of local pollutants in underground waters in Ukraine are filtering accumulators (over 3 thous. pieces), non-regulated dumping grounds of industrial and household wastes, stores of mineral fertilizers, etc. By the data of the State committee of natural resources by 1.01.2004 (NR, 2005) on the territory of Ukraine there were registered 276 stable sources of polluting underground waters.

The progressive pollution of underground water-bearing horizons also influences their insufficient natural protection. Thus, according to (NR, 2005) the protection of the main water-bearing horizon of Ukraine amounts to 39.6%, of protected ones; 24%, conditionally protected; 36.2% of unprotected underground waters. The largest percentage of unprotected water-bearing horizons is in Transcarpathia – 91.2%; in the Ivano-Frankivsk region – 82.2%, Lviv – 68.8%; Zhytomyr – 66.9%. In Volyn and Rivne regions the percentage of protected underground waters is correspondingly 7.8% and 3.4%. Thus, the formation of the qualitative composition of underground waters is influenced both by natural factors (climate, geological structure, hydro-geological conditions) and also an anthropogenic constituent.

The research on the qualitative composition of artesian waters at operating water intakes (NR, 2005; Kvartenko, 2015; Lyuty, 2011) gave the possibility to observe the tendency concerning their progressive worsening by such basic parameters as the degree of aggressiveness, the content of ferrum compounds, manganese, ammonia nitrogen, nitrogen-containing compounds, phosphates, rigidity, common mineralization, microbiological indices.

Thus, the solution to the comprehensive question – the monitoring of underground water-bearing horizons from the point of view of indices of their quality with the simultaneous introduction of present-day power-saving technologies of water treatment is at present-day rather urgent.

The Substantiation of the Necessity of Studying the Qualitative Composition of Underground Water Bearing Horizons Under Designing New and Reconstructing Operating Stations of Water Treatment

Aim and tasks of the work. The aim of the work is the study of the qualitative composition of underground water bearing horizons in some settlements of Volyn and Rivne regions of Ukraine with considering the possibility concerning the introduction of present-day comprehensive technologies for water treatment. To achieve the aim set a number of tasks were solved:

- the monitoring was carried out of indices of water quality in a number of settlements in northern-western regions of Ukraine;
- the impact was analyzed of technogenic natural factors upon the efficiency of the operation of existing technologies;
- a physical model and mechanisms of comprehensive treatment of underground waters from compounds of ferrum and ammonia nitrogen were developed;
- technologies were developed, researched and improved of the comprehensive water treatment on the basis of combined methods of treatment and conditioning.

METHODS OF STUDIES

The work uses a photocolourmetric method for determining the concentrations of ions of ammonium, ferrum, phenols, hydrogen sulphide, a titrometric method for determining hydro carbonate and total alkalinity, the content of Ca^{2+} ions, and also permanganate oxidity by KUBEL method, a potentiometric method for determining valued of pH-Eh, an electronic microscopy with conducting roentgen spectral analysis of matrix structures of biominerals.

Results of Monitoring Studies of Qualitative Composition of Underground Waters in Northern-Western Regions of Ukraine

By results of monitoring research conducted by us of the parameters of the quality of underground waters in 91 settlements in Volyn, Rivne, Zhytomyr, Ternopil, Transcarpathian and Lviv regions of Ukraine there was established the progressive worsening of their quality connected with natural and anthropogenic factors. First of all, it concerns the decreased values of pH and increased degree of the aggressiveness of underground waters in relation to metal and concrete, increased concentrations of compounds of ferrum, ammonia nitrogen, gumin complexes, the seasonal appearance in some water bearing horizons of phenols, ions of heavy metals (Kvartenko, 2011; Lyuty, 2011; Kvartenko, 2015).

In particular, a substantial influence on the qualitative composition of underground waters which are the main source of water supply for settlements situated in northern districts of Volyn, Rivne, Kyiv, Zhytomyr and Chernihiv regions of Ukraine is exercised by landscapes with substantial areas of forests, peat territories, forest lakes and swamps which render conditions for feeding underground waters. The enrichment of underground waters by ferrum takes place as a result of processes of leaching and dissolution of ferrum-like minerals under the action of carbon dioxides and organic compounds contained in weak acid swamps and wood lakes.

Groundwater enrichment with iron occurs as a result of leaching and dissolution of ferrous minerals under the action of carbon dioxide and organic compounds contained in weakly acidic swamps and forest lakes. Forms of iron in natural waters are quite diverse: free ions, hydrocomplexes, and complexes with inorganic and organic ligands (Krainov, 1992). Forms of their existence depend on the geochemical properties of groundwater, namely the pH-Eh values of the aquatic environment. According to (Krainov, 1992; Nikoladze, 1987) we have formed a number of stability of the main iron compounds in natural waters:

$$\begin{aligned} &Fe\Phi K(OH)_2^- > Fe(OH)_3 > Fe(OH)_3^+ > Fe\Phi KOH > Fe(\Phi K)_2^- > FeOH^{2+} \\ &> FeCO_3^+ > Fe(HS)_2 > Fe(OH)_2 > FeCO_3(Fe\Phi K) > FeOH^+ > FeSO_4 > FeCl_3 \end{aligned} \quad (1)$$

Organic iron compounds, hydroxocomplexes and iron (III) hydroxide are apparently the most stable. Humic acids form intrinsically complex compounds with many metals, which, depending on the pH of the medium, become more mobile or form stable, water-insoluble compounds. Soluble organic substances in the form of humates and fulvates of iron give groundwater color and in some cases may be highly colored.

By data of our studies lasting many years on the territory of Volyn region this boundary is from the Svityaz lake between settlements township Ratne – t. Kovel – v. Povorsk – township Lyubeshiv.

In Rivne region the said boundary is from the Bile lake via settlements: t. Volodymyretz –t.Kostopil – v. Bogushi – v. Balashivka. Underground waters of these water bearing horizons may be referred to weak acidic (pH up to 6.0-6.5) with low (<2 mmole/dm^3) or moderate (2-4 mmole/dm^3) alkali reserve which contain dissolved organic substances (PO>5.0 mg O_2/dm^3) in the form of humin (HA) and fulvoacids (FA) and also are characterized by increased (from 3.0 to 30.0 mg/dm^3) content of ferrum compounds.

One of characteristic signs for parameters of quality for underground waters in this region is the increased chromaticity, the content of aggressive carbon (IV) oxider (CO_2^{ag}) and hydrogen sulphide (Kvartenko, 2018). To the south of this boundary are located oxygen less neutral waters with the average value of alkalinity 2.0-4.0 mmol/dm^3 and nonsignificant content of organic substances (PO<5.0 mgO_2/dm^3).

The conducted monitoring studies on the qualitative composition of underground waters concerning the content of Fe^{2+} ions on water intakes to the south of settlements: Novovolynsk – Lokachi –Torchin – Lutsk – Dubno – Rivne – Goshcha – Korets indicated its decreased concentrations to 3 mg/dm^3 (Kvartenko, 2018). By results of monitoring researches of parameters of the quality of underground waters concerning the content ferrum compounds carried out from 2000 to 2018 on 49 water intakes in Rivne and 17 water intakes in Volyn regions it was established that in Rivne region 45% from studied water intakes are based on underground waters with the concentration of Fe(II) up to 3 mg/dm^3, 24.5% with the concentration from 3.0 to 5.0 mg/dm^3, 14.2% – from 5.0 to 10.0 mg/dm^3 and 15% – from 10 to 30 mg/dm^3. The research conducted in Volyn region showed that 52% of underground waters in studied water intakes contained up to 3.0 mg/dm^3, 18% – from 3.0 to 5.0 mg/dm^3, 24% from 5.0 to10 mg/dm^3 and 6% – from 10.0 to 20.0 mg/dm^3 of Fe(II) compounds.

In the content of easily oxidized organic substances 38% of the researched water intakes in Rivne region were based on underground waters with values of permanganate oxidation (PO) from 4.0 to 8.0 mgO_2/dm^3, 12% – on waters with PO from 8.0 to 12.0 mgO_2/dm^3 and 7% – on underground waters with PO > 12) mgO_2/dm^3. In Volyn region 48% of the studied settlements consume water with PO from 4.0 to 8.0 mgO_2/dm^3, 8% – with PO>12.0 mgO_2/dm^3 (Kvartenko, 2018).

The results of monitoring studies carried out on 67 water intakes in Rivne and Volyn regions concerning the content of ammonia nitrogen in underground waters are given in Table 1.

A similar picture was with the pollution of some researched water intakes in phenols. In particular, of 17 studied water intakes suitable for economic drinking water supply there were 29%; polluted (with concentrations of phenols 0.0011-0.2 mg/dm^3) – 41.2%; very polluted (with phenols concentrations 0.21-0.5 mg/dm^3) – 29.4%. By active reaction of pH for underground waters of 12.5% in studied water intakes in Rivne region may be referred to weak acidic ones, 37% – to neutral, 10% – to weakly alkaline, 2% – to alkaline, while in Volyn region – the water supply in most researched settlements was based on neutral waters.

Table 1. The results of monitoring studies concerning the content of ammonia nitrogen in underground waters in Rivne and Volyn regions

Content of ammonia nitrogen, mg/dm^3	**<0.05**	**0.06 – 0.5**	**0.51-1.5**	**>1.5**
Rivne region				
% settlements	16	30	18	36
Volyn region				
% settlements	-	13	31	56

source: own study

An important connecting link in the system Fe – pH – Eh – CO_2 – OP is the hydro-carbonate water alkalinity. With low values of pH<6.0 in underground waters the percentage of the interrelationship between CO_2^{free}/HCO_3^- constitutes (70/30% which characterizes these waters as waters with low hydro-carbonate alkalinity. The application of aeration method in treating aggressive underground waters permits to get water with non-aggressive properties only under values of hydro-carbonate alkalinity >2.0 mmole/dm^3 (Kvartenko, 2016). Neutral waters with hydro-carbonate alkalinity <2.0 mmole/dm^3 as a result of aeration obtain aggressive properties concerning metal and concrete. Thus, to determine possibilities for oxidizing ions of Fe and Mn with the help of aeration method it is necessary to know the value of carbonate rigidity of water. The removal and transfer iron ions into non-dissoluble form is possible to carry out with the help of aeration method only for waters with hydro-carbonate alkalinity>2.0 mmole/dm^3 under the absence of dissolved organic substances.

As a result of monitoring researches conducted on water intakes in Rivne region it was established that in 24.4% of underground waters there is a characteristic low value of hydro-carbonate alkalinity (<2mmole/dm^3), 24.4% were waters with the average alkalinity (2-4 mmole/dm^3), 22.3% – with normal (4-6 mmole/dm^3) and 33.3% with increased (7.0-8.5mmole/dm^3) alkaline reserve. In Volyn region of

20 researched settlements water supply in 16% was based on waters with average, increased alkalinity, 53% normal and 32% with increased alkalinity.

Our monitoring of groundwater aggressiveness degree in over 80 localities in Volyn, Rivne, Zhytomyr, Khmelnytsky, Lviv regions of Ukraine found that water supply in nearly 60% of them is based on using underground waters with different degrees of aggressiveness towards concrete and metal (13% showing ability to carbonate deposits on piped walls, 27% relates to stable water). The greatest degrees of aggressiveness with underground water horizons are northern districts of Volyn and Rivne which are presented in Fig.1.

Joint studies of the aggressiveness of underground waters in this region carried out on the base of the hydro-chemical laboratory under the department of Water Supply, Sewerage and Drilling of NUWEE and the physico-chemical laboratory of Rivne regional center under MPH of Ukraine gave the opportunity to conduct the monitoring if this issue in all districts of the region in the period from 2001 to 2014 Fig. 2.

As is seen from the data presented underground waters by Langelier index should be referred to such aggressiveness categories: from sight corrosion degree (I_L = -0.5) – Berezno Fig. 3. a and Volodymyretz districts (Figure 3b), tendencies to corrosion (I_L = -1.0) – Dubrovytzya Fig. 3. c and Sarny districts (Fig. 3d), very high corrosion degree (I_L > -2.0) in Rokytne district. Moreover, over time there is an increase in the degree of aggressiveness of groundwater Fig.2.

Figure 1. The diagram of groundwater corrosive action

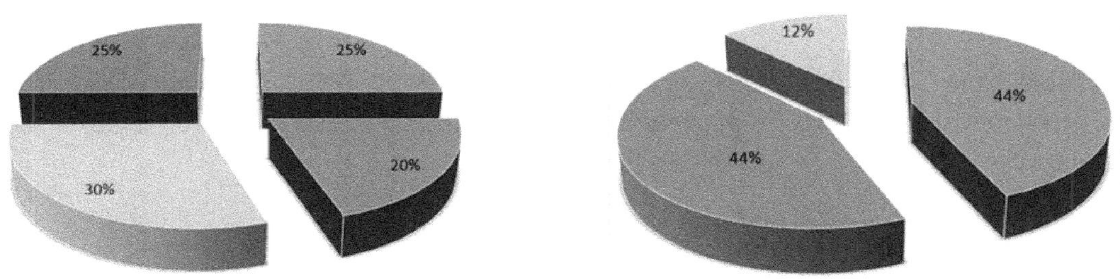

A) North of Rivne region
I – mild corrosion (25%); II – tendency to corrosion (20%); III – very high degree of corrosion (30%); IV – an extremely high degree of corrosion (25%)

B) Volyn region
I – mild corrosion (43.75%); II – tendency to corrosion (43.75%); III – very high degree of corrosion (12.5%); source: own study

Reasons are determined for increased degree of aggressiveness of groundwater in the region: the decrease in acidity, alkalinity and reduced concentration of Ca^{2+} ions. The reused water pollution by corrosion products leads to the deterioration of its quality - namely, increases turbidity, color, appearance of unpleasant tastes and odors, iron hydroxide. The existing iron removal plants commissioned in 60s - 70s of the twentieth century, based on simplified technology or deep aeration followed by filtration - cannot solve this problem. To address the question of initial water purification with further maintenance of its quality during transportation, one must use comprehensive clean groundwater with their subsequent stabilizing treatment.

Figure 2. Changes of stability index in groundwater of northern districts of Rivne region

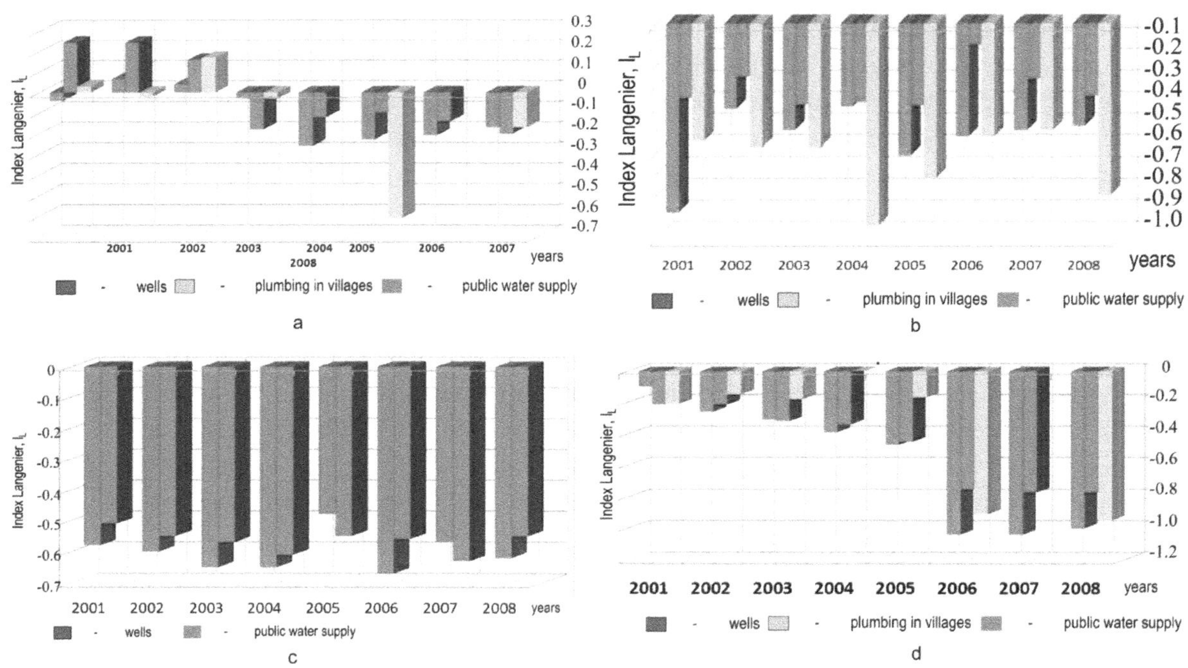

According to the results of monitoring studies (Kvartenko, 2011; 2015; 2018, Grabovska, 2010) of the general geochemical characteristics of groundwater properties of the north-western regions of Ukraine, they can be classified according to a number of indicators:

- **by alkaline reserve**: on waters with low (to 2 mmole/dm^3); moderate (from 2 to 4 mmole/dm^3), average (4-6 mmole/dm^3) and high (> 6.4 mmole/dm^3) hydro-carbonate alkalinity;
- **by value of pH hydrogen index**: weakly acidic (pH to 6.5), neutral (pH 6.5-7.5); weakly alkaline (pH>7.5);
- **by value of permanganate oxidization**: insignificant (<4.0 mg O_2/dm^3), moderate (4.0-8.0 mgO_2/dm^3), average (8.0-12.0 mgO_2/dm^3), high (12.0-20.0 mgO_2/dm^3);
- **by nature and concentrations of ferrum compounds**: in mineral or organic complexes with concentrations: Fe(II) to 3 mg/dm^3; from 3.0 to 5.0 mg/dm^3; from 5.0 to 10.0 mg/dm^3); from 10 to 30 mg/dm^3;
- **by presence of anthropogenic pollutions**: compounds containing nitrogen (NH_4^+, NO_2^-, NO_3^-), phenols, phosphates, SAS, ions of heavy metals;
- **by degree of water aggressiveness** in relation to metal and concrete.

The Analysis of the Efficiency of Traditional Technologies Treating Weak Acidic Underground Waters Containing Iron Organic Complexes

Most existing in Ukraine stations treating underground waters were put into operation by technologies of filtering and simplified aeration which was not designed for conditioning weak acidic waters with low values of hydro-carbonate alkalinity (<2.0 mmole/dm^3). Besides, the method has a number of limitations which by the quality of source water (Fe^{2+}<10 mg/dm^3; H_2S<0.5 mg/dm^3; pH>6.7; permanganate oxidization not more 5 mgO_2/dm$_3$) and does not envisage the comprehensive water treatment from ammonia nitrogen, manganese, humin complexes, phenols. In works (Nikoladze, 1987; Kvartenko, 2019) the deceleration is stressed of autocatalytic processes with the availability in water of humin complexes. The analysis of applying traditional non-reagent technologies (Table 2) for treating weak acidic and neutral underground waters with low alkaline reserve in the presence of dissolved organic compounds showed the insufficient efficiency of their operation and the filtrate aggressiveness. Besides, under modern conditions of increased anthropogenic loading for removing a number of pollutions it is necessary to introduce additionally special methods of water treatment which were not provided by the said technologies. The application of only reagent methods leads to the necessity of using the additional technological equipment which complicates operation, increases the cost of water treated and also does not exclude the possibility of forming chlorium organic compounds, ozonolys. The content of secondary pollutions in drinking water increases with the increased dosage of chlorine, pH, temperature, chromaticity, ozidization, phenol presence. The use of schemes with active catalytic loading is expedient only under the absence in underground waters of dissolved organic compounds, hydrogen sulphide, bacteria.

Table 2. Data of quantitative analysis of groundwater quality and leachate

Water intake	Source water				Filtrate		Source
	pH	Hydro-carbonate alkalinity, mmole/dm^3	Fe, mg/dm^3	PO, mgO_2/dm$_3$	Fe, mg/dm^3		
Technological scheme	**Simplified aeration - Filtration**						
township Bilyy Yar	6.5	3.2	9.55	4.9	1.8		*Kvartenko, 2019*
v. Balashivka (2003)	6.4 – 6.5	1.28 – 1.5	4.0 – 4.8	3.5	1.8		
Technological scheme	**SA → PFI → PFII → WD**						
township Zelenyy Hay	6.4	<2.1	26.9	5.8	19.9	17.8	*Kvartenko, 2019*
township Rokitne	5.0-6.5	0.9 – 1.0	До 30	8.0	15	1.5	

SA – simplified aeration; PFI (PFII) - pressure filters of the I-st and II-st degrees; WD – water disinfection

Guidelines for Intensifying the Operation of Acting Deferrization Stations

One of the possible directions to solve the issue is the introduction of the method of autocatalytic oxidation of iron in the layer of suspended sediment (Orlov, 2008; 2012). Another of possible directions in intensifying the operation of deferrization stations is the application of the biochemical method of oxidation the use of which became possible due to wide spreading of iron bacteria in underground waters in different regions of the world. To the development of the biochemical method of removing from underground waters the compounds of Fe (II) and Mn (II) are devoted works (Safonov, 1984; Czekalla, 1985; Seppanen, 1991; Mouchet, 1992; Hatva, 1998; Bourgine, 1994; Kvartenko, 1997; Tamura, 1999; Søgaard, 2000; Sharma, 2005; Fujikawa, 2010). First research of the method in Ukraine was carried out under the guidance of Professor M.A. Safonov at the department of WD (NUWEE) (Safonov, 1984). By experiments of Mouchet (1992) there were determined optimal parameters of pH – Eh of medium necessary for their development and the limit of their competing with traditional schemes of Fe(II) oxidation by atmospheric oxygen Fig. 3, Fig. 4.

Figure 3. Fields of activity of iron bacteria
(Mouchet, 1992)

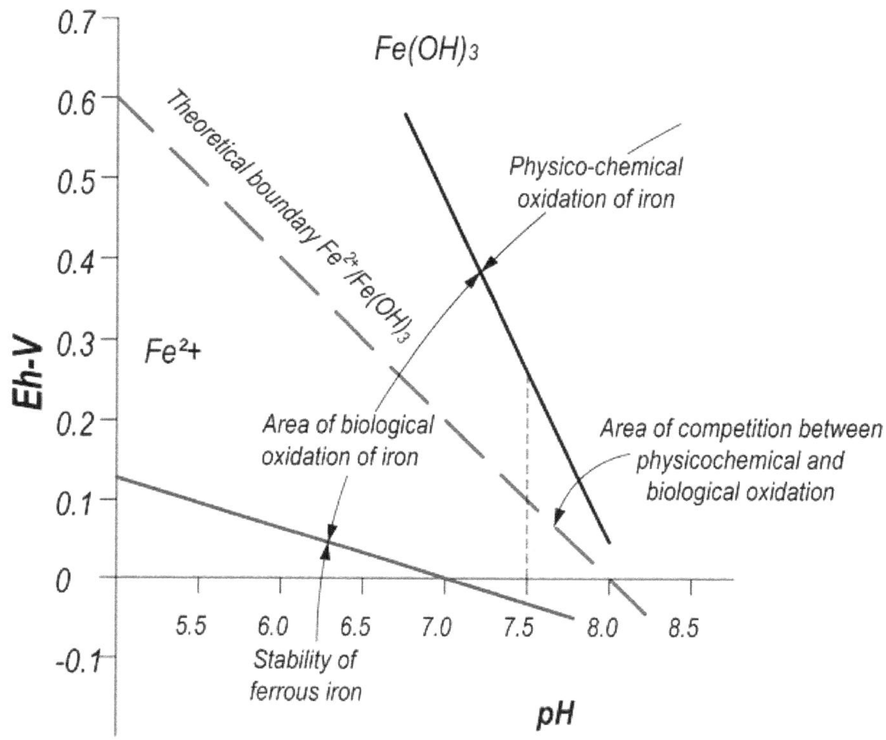

The use of the biochemical method is expedient only for weak acidic and neutral waters with low or normal alkaline medium. In these cases the reconstruction is possible of existing traditional schemes to schemes by the method of biological deferrizing way of introducing such measures: changes of the fraction composition of filtering loading of fine-granular fractions (δ=0.6-1.8 mm) into: gravel δ=10-30 mm (Mouchet,1992); quartz sand of average granularity (δ=2, мм); granite rubble δ= 5.0-10 mm or foam polystyrene δ=3,0-10 (Safonov, 2000); carrying out the correction of values of pH-Eh of the medium (Safonov, 2000; Kvartenko, 2019); the use of preliminary alkalization of source water by the solution of soda ash with deferrization of weak acidic waters with low alkaline reserve (Kvartenko, 2019); the carrying out the correction of the oxygen concentration before bioreactor (Kvartenko, 2019).

The said steps allow intensifying the operation of deferrization stations by way of increasing to 10-20 m/hour speeds of filtration, duration of filter cycles, to decrease expenditures of reagents and volumes of wash water. The increase of the specific mud content in filtering loading under biological deferrization in comparison with traditional non-reagent methods Fig. 4. is explained by matrix structure formed by iron bacteria in the process of life activity Fig. 5. Polymer matrix structures created by bacteria are composed from acidic polysaccharides with carboxyl functional groups which are located in tight spatial connection with ferrum (III) hydroxide. Organic fibres of matrix structures gradually accumulate FeOOH controlling its recrystalization (Kennedy, 2004) and forming *bio*-minerals of nanometric size (Chana Clara, 2009).

Figure 4. Comparison of dirt content of biological and physicochemical methods of deironing
(Mouchet, 1992)

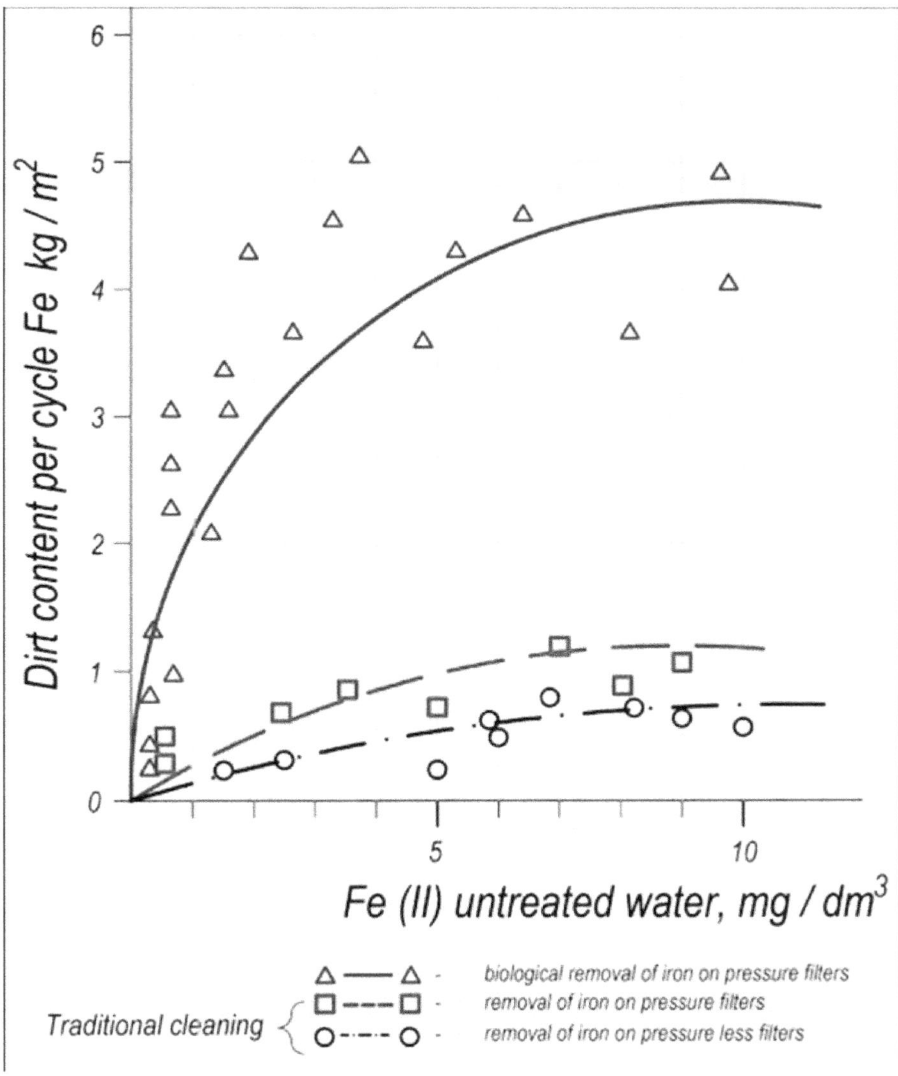

Figure 5. Electronic image of matrix structures from the interporous space of the contact loading of the bioreactor: a – matrix structures are created by iron bacterias Gallionella, Leptothrix SEM 3000 x; b – matrix structures are created by iron bacterias Gallionella SEM 6000 x
Source: own study

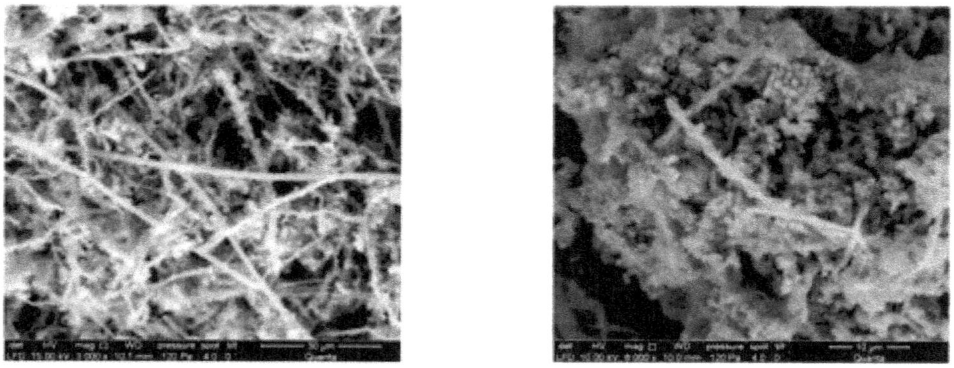

a b

Improved Technology of Comprehensive Biological Treatment of Neutral Aggressive Iron Containing Underground Waters

Technology and equipment. In the period from 200 to 2006 in the system of water supply in t. Berezno for the first time in Ukraine by recommendations of M.G. Zhurba and O.M. Kvartenko there was designed and put into operation the station of the biologic deferrization with the capacity of 2000 cub.m./day Fig. 6. with a bioreactor 2.8 m diameter located in the open air beyond filtration hall, three open polystyrene foam filters of 2.0x2.0x4.3m size equipped with the system of hydroautomatic washout, water disinfection unit.

In 2012 this author made recommendations concerning improved technologies Fig. 6. by way of modernizing the block of aeration in bioreactor, the system of filters washout, and the introduction of the block for stabilizing water processing and also gave recommendations concerning the transfer of system operation into the regime of unevenness of hydraulic loadings.

The improved technology is designed for carrying out the comprehensive treatment and conditioning of aggressive, near-neutral waters with normal alkaline reserve of ammonia nitrogen (to 1,5 mg/dm^3), dissolved organic substances (by PO to 6.0 mg O_2/dm^3), phenols (to 0.03 mg/dm^3).

The contact non-homogeneous loading of the bioreactor was done from foam polystyrene of mark *PE* Industry standard 301-05-92E.The grain size of fractions and the height of its layers was δ =6.0-12 mm, h= 700 mm; δ=3.0-6.0 mm, h=500 mm; δ=2.0-3.0 mm, h= 500 мм.

In the reactor took place the comprehensive water treatment from cations of NH_4^+, Fe^{2+}, dissolved organic substances, phenols. The total time of water presence was from 60 to 20 minutes after which water was coming to filters of *PFF*-1type providing the treatment from ferrum (III) hydroxide suspension and matrix structures of *bio*-minerals.

Figure 6. Technological scheme of treating agressive, near neutral waters: 1- groundwater bore; 2- electric pump; 3 - bioreactor; 4 - contact loading of the bioreactor; 5 - cavitation block; 6-7 - technological ventilation windows and boxes; 8, 18 - sodium hypochlorite dosing unit; 9- water supply pipeline; 10 - air separator; 11- lighting expanded polystyrene filter; 12- washing tank; 13- pocket of filtrate collection; 14- hydrorobot; 15- vacuum break tube; 16- hydro-shutter; 17- canal for collecting washout waters; 19- clean water tank; 20- pump station

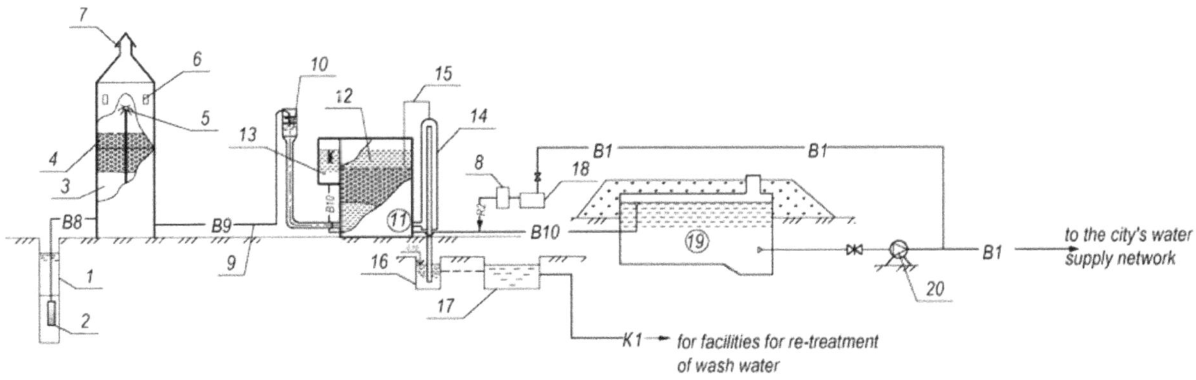

In the capacity of filtering loading we used granular foam polystyrene of mark *PE* or *PE-S* Industry standard 202-05-92 and Industry standard 301-05-92E. The upper distributing layer was adopted of 200 mm thickness with granule diameter of 4-8 mm. The diameter of granules of operating layer was 1.0-2.5 mm and its operating height 1000 mm. The operating rates of filtering were for the bioreactor 7-11 m/hour, for filters 3.5-6.0 m/hour. Before each filter there were arranged air separators of 5.3 m height connected with collectors of lower drainage system of each separate filter. The washout of filters is done in hydro-automatic regime with washout intensity of 14-16 l/sm^2 of 5 minutes duration.

For the disinfection of drinking water the station used electrolysis device "Polum'ya -2". For carrying out the stabilization processing of water were used expenditure tanks with clarifying 1-2% solution $Ca(OH)_2$ and the pump-doser.

Physical Model of Bioreactor and Mechanisms of Comprehensive Treatment of Underground Waters From Ferrum Compounds and Ammonia Nitrogen

The start of the bioreactor into operation condition is possible to be divided into two stages: 1 – the beginning stage of "charging" the connecting loading. During this period through the loading comes water from the bore hole containing not only cations of ferrum, carbon (IV) oxide, phosphorus, dissolved substances, but also microorganisms spread in underground waters. Gradually on granule surface takes place the formation of active catalytic surface from cells and matrix structures. The process takes place as a result of adhesion of bacteria on the cell surface of which has a negative charge to the primary catalytic film of contact loading by the known mechanism (Kvartenko, 2019). This working period is characterized by the passage of the part of Fe^{2+} cations to filtrate. The duration of the "charging" depends on parameters of water quality: values of pH, hydro-carbonate alkalinity, concentrations of cations of Fe^{2+} and bacteria mass per unit of volume of source water. In case of underground waters of t. Berezno this period had three-four days with the filtration rate of 3-5 m/hour and the growth of treatment from 15 to

45%. At the start of reactor operation there took place the transportation together with the water flow of ferrum bacteria, ferrum compounds, carbon (IV) oxide (CO_2) to the upper and middle layers of contact loading. This is explained both by unevenness of loading and by use of the downward water flow in the direction of decreasing the grain size of its granule. In this period due to the significant area of adsorption surface there takes place the formation of starting matrix structure on account of the delay bacteria consortiums which oxidize ferrum compounds and ammonia nitrogen by the following mechanisms:

a) oxidation of Fe^{2+} with the subsequent sedimentation of non-dissolved oxides around the cells of ferrobacteria *Leptothrix* may be the result of the interaction metals ions with products of bacterial metabolism, in particular with H_2O_2. Hydrogen peroxide which emerges in the capacity of the intermediate or final product of oxidation is separated from cells and accumulated in surrounding structures. In neutral or weak acidic medium the oxidizing of cations Fe^{2+} to Fe^{3+} takes place as a result of immediate interaction with H_2O_2:

$$2Fe^{2+} + H_2O_2 \rightarrow 2Fe^{3+} + H_2O + O_2 \quad (2)$$

Besides, the extraction by cells of *Leptothrix* bacteria of hydrogen peroxide into exterior medium contributes to the increased efficiency of destroying iron-organic compounds:

$$R - C(O) - O - Fe - O - C(O) - R + H_2O_2 \rightarrow Fe^{3+} + 2RCOOH \quad (3)$$

b) the fermentative oxidation by ferrobacteria of *Gallionella* type which are able to grow the cell biomass using non-organic carbon which is assimilated from dissolved in water carbon dioxide (Hallbeck, 1991; Hallbeck, 2004; Emerson, 2013).

From each portion of inlet water to contact loading enter new bacteria increasing their total quantity. New bacterial structures start to form their matrixes on previous sedimentations simultaneously with this decrease the admission of necessary for normal growth bacterial consortiums of cations Fe^{2+} and carbon into lower layers transforming them into limiting factors of development. Bacteria which are in these layers pass into anabiosis stage to the stage of active metabolism.

Thus, the bioreactor is a flexible system which may react both at the changing hydraulic loading and at the changing concentration of polluting substances supporting the efficiency of the process of water treatment on the stable level by way of activizing bacteria located in matrix structures in interporous space in anabiosis stage.

The second stage – operation one which characterizes the operation of the reactor between washouts. After the washout in the loading remain bacteria both from the previous cycle and also bacteria entering together with the washout water from bore hole (with total quantity of 10^3-10^4 kl/cm^3). From the flow of outlet water there takes place the adhesion of ferrum bacteria both on the surface of contact loading and on the remaining matrix structures in their interpore space. The fixed in these way bacteria in time begin to form new matrix structures. In achieving on the surface the contact loading and matrix structures biomass concentration which equals or surpasses its limiting value there takes place the breakaway of the part of bacteria with their transfer into water flow. On gradual accumulation of bacteria and matrix structures in layers of contact loading the increase is observed of the intensity of oxidation of ferrum compounds and dissolved organics, of ammonia nitrogen.

The extraction of ammonia nitrogen in bioreactors is possible to be represented by several mechanisms depending on parameters of water quality and qualitative composition of microorganisms.

In neutral and near-neutral underground waters with the presence of chemolito-autotrophic bacteria of *Gallionella* type it is possible to consider the process of sorption of NH_4^+ ions on the surface of matrix structures of *bio*-minerals and also on cells of bacteria themselves due to the availability on their surface of functional groups PO_4^{-3}, COO^-, OH^-.

With time the process takes place of the gradual filling of interporous space by stationary matrix structures under which there will be observed the breakaway of the part of matrix from stationary condition and the subsequent transfer by transient flow to lower layers of contact loading. With the accumulation in interporous space of matrix of matrix structures under concentrations exceeding the calculated value of maximum mud content there takes place the breakaway of matrix from interporous space and its transfer beyond contact loading.

It is to be noted down that the use of great grain foam polystyrene permits to increase the depth of the accumulation of matrix structures to 60-80 cm with subsequent filling of interporous space. With time the processes take place of increased expenditures of head and increased water level under contact loading which leads to keeping in calculated boundaries averaged in time the rate. The washout of contact loading of bioreactor takes place once in three weeks.

RESULTS OF STUDIES

At a water treatment station in t. Berezno in the period of 2016-2018 we carried out comprehensive research concerning the efficiency of the operation of this system Fig. 6. under conditions of uneven hydraulic loadings on water treatment equipment.

Parameters of the quality of initial water: NH_4^+ 0.55-0.65 mg/dm^3, C_5H_5OH 0,025-0.028 mg/ dm^3, Fe^{2+} 1.7-1.9 mg/dm^3, values of permanganate oxidizability 3.6-4.0 mg O_2/dm^3, hydro-carbonate alkalinity 3.96 mmol/dm^3, pH 7.2. Fig. 7. presents the results of a filtration cycle in the period from 12.11.16 to 13.11.16. The consumption of source water for process equipment varied during the day in the range from 13 to 40 m^3/h. Filtration rates were correspondingly: in a bioreactor — from2.5 m/hour to 8 m/ hour; in clarifying filters – from 1.5 m/hour to 3.5 m/hour.

The efficiency of treating water from easily oxidized organic compounds and iron amounted correspondingly from 62 to 79% and from 87 to 98% Fig. 8. It is to be noted down that in an hour after the start of filtration cycle from water were completely removed such complicated for treatment compounds as ammonia nitrogen and phenols Fig. 8. Simultaneously with these values of hydro-carbonate alkalinity were not changed practically and were within 3.84-4.0 mmol/dm^3. Also there were conducted studies concerning relationships of changes in parameters of water quality during filter cycle under increased values of unevenness of specific loadings on bioreactor and filters correspondingly from 5.8 to 10.4 m/hour, and from 3.0 to 5.0 m/hour Fig. 9. Parameters of water quality were: permanganate oxidation 1.35-1.55 mg O_2/dm^3, hydro-carbonate alkalinity 4.0-4.3 mmole/dm^3, pH 7.15-7.2, content of cations Ca^{2+} 84-98 mg/dm^3, Fe^{2+} 2.0-2.2mg/dm^3, total salt content 300-350 mg/dm^3. As is seen from data presented during the studied period there was observed the stable effect of purification at each equipment.

Thus the efficiency of bioreactor operation swayed within 63-65% with filtration rates 6.7-6.5 m/ hour, 57-58% with rates of 7m/hour, decreasing to 48.7% with increased rates to 10.5-11.5 m/hour. Concentration of Fe_{tot} after filters during all the period swayed within 0.02-0.05 mg/dm^3.

Figure 7. Change of filtration rates and efficiency of water treatment depending on hourly water expenditures: 1 — rate in a bioreactor; 2 — rate at filters; 3 — efficiency of deferrization after filters; 4 — efficiency of removing organic substances by PO after filters
Source: own study

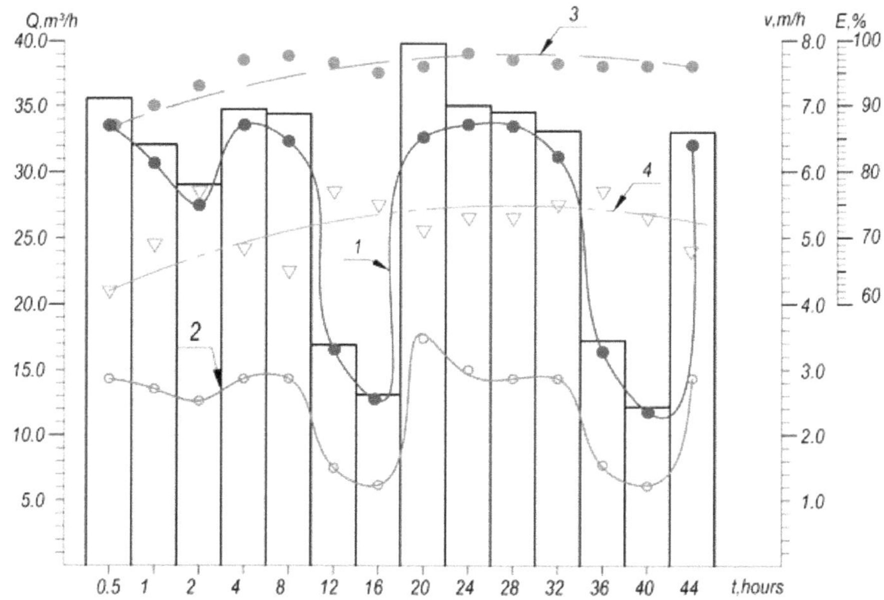

Figure 8. Change of quality parameters of treated water after filters during filter cycle: 1 – pH values; 2 – hydrocarbonate alkalinity; 3 – ferrum compounds; 4 – permanganate oxidization; 5 – ammonia nitrogen; 6 – phenols
Source: own study

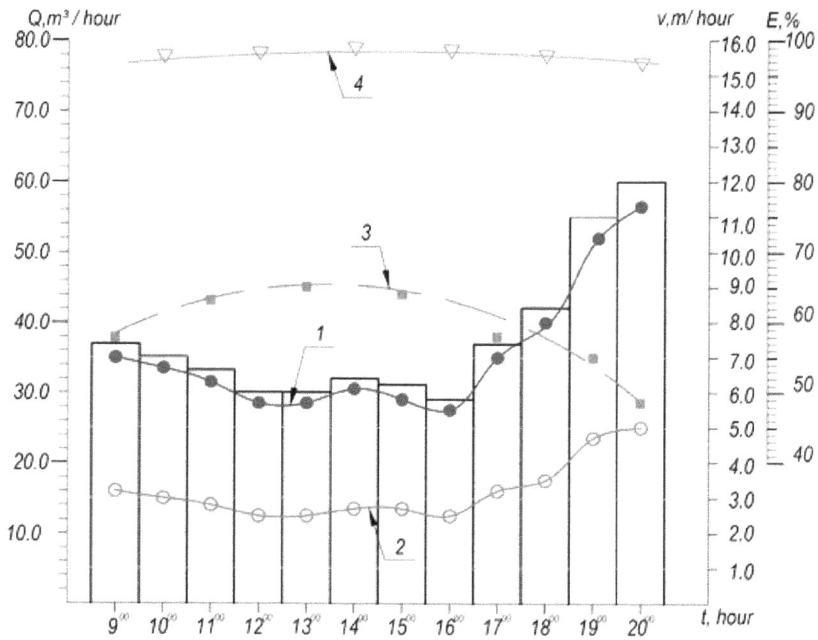

Figure 9. Changes in filtration rates and deironing efficiency during the filter cycle (19.05.15) depending on hourly water consumption: 1 – filtration rate on bioreactor; 2 – filtration rate on filters. The removal efficiency of iron after: 3 – bioreactor; 4 – filters
Source: own study

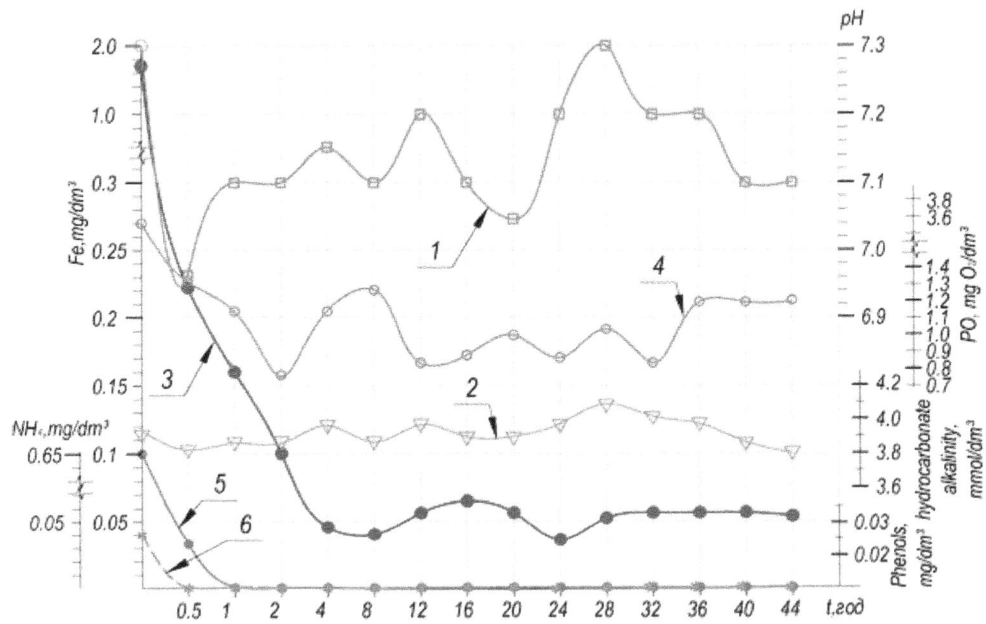

Developed on the basis of combined methods of treatment the technology of the comprehensive water preparation of neutral aggressive iron containing underground waters permits to use equipment under conditions of substantial unevenness of hydraulic loadings with high efficiency (easily oxidized organic compounds and iron amounted correspondingly from 62 to 79% and from 87 to 98%).

The use in essence of two-degree biological treatment permits to increase the duration of filter cycles to 48 hours. The optimal rates of filtering for bioreactor 7-11 m/hour, for filters 3.5-6 m/hour. The washout of filters is possible to be conducted in two regimes: hydro-automatic, with intensity of washout 14-16 l/ s m^2 with duration of 5 minutes, manual – by two degree scheme, with intensity of washout 20-30 l/ s m^2 with duration of each of 2 minutes and five minutes interval.

Technology of Comprehensive Treatment of Aggressive Weak Acidic Iron Containing Underground Waters with Low Alkaline Reserve Which Contain Ammonia Nitrogen and Phenols

Weak (pH 5.0-6.5) aggressive (I_L » -2.0) underground waters with low alkaline reserve (<2.0 mmole/ dm^3), high chromaticity (to 80 degrees platinum cobalt scale) which contain dissolved organic substances (to 8.0 mg O_2/dm^3 in value PO), NH_4^+ (to 4.0 mg/dm^3)·, complex compounds Fe – HA (to 30.0 mg/ dm3) hydrogen sulphide (to 4.0 mg/dm^3), and on some water intakes – phenols (0.01 - 0.12 mg/dm^3) and manganese (to 0.5 mg/dm^3) constitute to 15% of water intakes studied by us in northern-western regions of Ukraine.

The research was done at the operating deferrization station the composition of which included: five pressure filters of the I-st degree H=3.8 m, d=1.5m; two pressure filters of II degree H=3.0 m, d=1,5m; pump of II-lifting (K100-65-200 – 1 piece; KM100-65 – 1 piece); pump of washout water (K100-65-200); compressor of compressed air PM-3142.01 in a set with two receivers, compressor for supplying compressed air for washing out filters. The disinfection of water is done by adding chlorine lime to reservoir of clean water. The first studies concerning parameters of scheme operation by existing technology were conducted 8.02.11 (Table 3) with station productivity of 25 m^3/hour and the rate of filtration 2.5-3.0 m/hour. In the process of treatment there takes place the decrease of values pH, hydro-carbonate alkalinity, and also the content of cations Ca^{2+}, with the simultaneous increase of the degree of filtrate aggressiveness to I_L= -2.85.

Table 3. Monitoring studies results of the efficiency of the existing water treatment equipment of the deironing station

Name of parameters	Fe_{tot}, mg/dm³	Hydro-carbonate alkalinity, mmole/dm³	PO, mg O_2/dm³	pH	NH_4^+, mg/dm³	Ca^{2+}, mg/dm³	I_L
Source water	21	0.9	6.99	6.05	0.8	37.6	-2.52
Filtrate	3.1	0.78	2.83	5.8	0.25	36	-2.85
E,%	85	-	45	-	68.7	-	-

source: own study

As a result of studying the composition of catalytic films on the surface of filtering loading and also the sediment of washout waters there was established the availability of associates and matrix structures of microorganisms of types *Gallionella, Leptothrix, Crenothrix* Fig. 10. Basing on known mechanisms of processes of biological oxidation in improved technology was used the principle of activating biochemical processes on account of introducing into system of additional source of non-organic carbon with the simultaneous increasing values of hydro-carbonate alkalinity and pH to boundaries of their comfort life activity. Together with this the process was speeded of the division of phases on filters of II degree on account of introducing the solution of coagulant and conducting the stabilization preparation of water.

To determine the possibility of increasing the efficiency of scheme operation on account of introducing alkaline reagent at the operating station was carried out an express research. Into the inlet water was added 52 mg/dm³ 0.1 mole-equiv/l of NaOH solution increasing pH value to 7.4 and conducting aeration. After this was added aluminum hydrooxchloride by dose of 20 mg/dm³ which led to decreasing value pH to 6.8-7.0 and creating very fine flocks. After adding 1-3 mg/dm³ 0.1% solution of non-ionogenic flocculant of mark HENGFLOC 70010 there took place the intensive creation of flocks with settling on sand filter. The iron concentration in filtrate was 0.33 mg/dm³. The day calculated expenditure 0.1n of NaOH solution was 25 kg, per average month – 750 kg which for a small station from the point of view of reagent cost was rather much. Therefore in later studies it was decided to use 0.5-1% solution of soda ash.

Parameters of source water were: total iron 22.3 mg/dm³, hydro-carbonate alkalinity 0.78 mmole/dm³, pH 5.8. Results of express research of changes in parameters of water quality depending on doses of soda ash are given in Fig. 11.

Figure 10. Electronic image of matrix structures in the wash water sludge: a – SEM 3000x; b – SEM 6000x
Source: own study

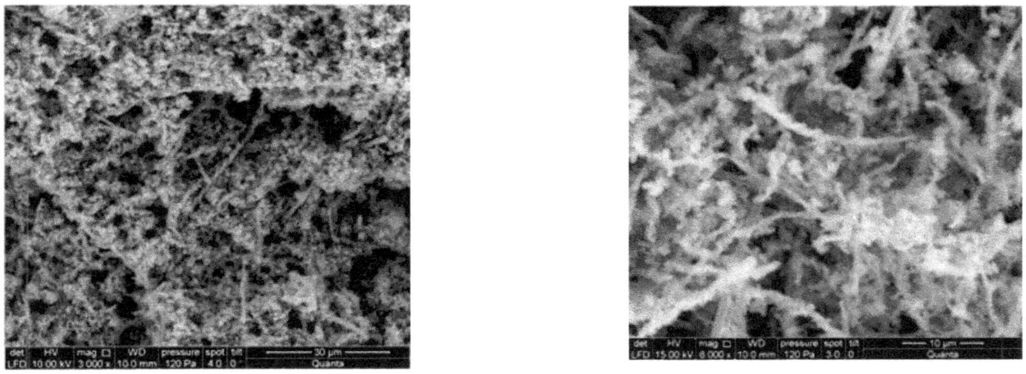

a b

Figure 11. Changing the quality parameters of water that was treated in the suspended layer of matrix structures depending on the dose of 0.5% soda ash solution
Source: own study

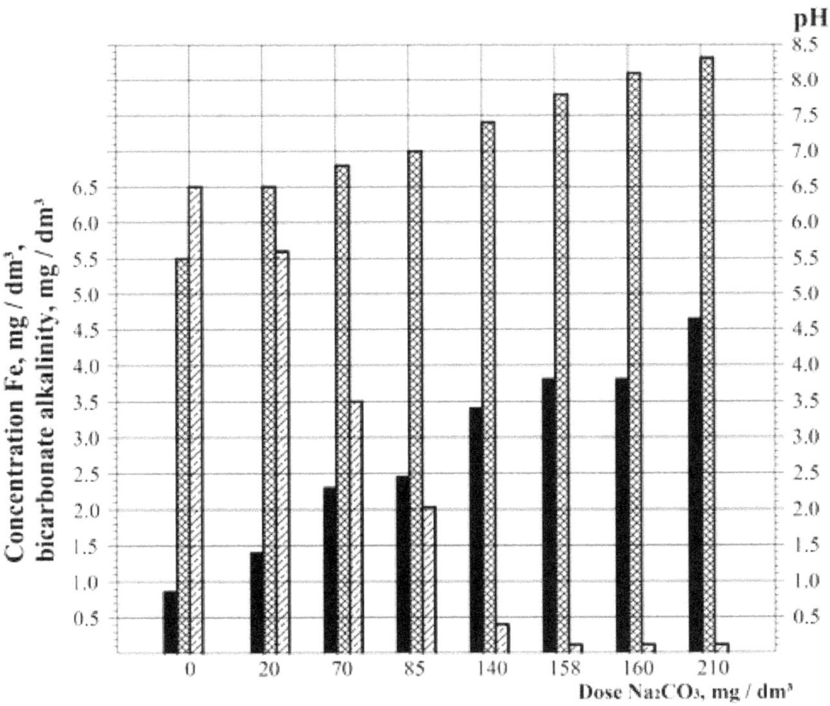

From results obtained it is seen that with increased values of hydro-carbonate alkalinity of water to 2.3-2.4 mmole/dm^3 and pH to 6.7-7.0 under concentrations of 0.5% solution of soda ash correspondingly 70 and 85 mg/dm^3 there took place the increased efficiency of removing cations Fe(II) by 17-24% in comparison with non-reagent scheme Fig. 12. The average expenditure of soda ash in dry matter was 30 kg/day. Simultaneously with this we continued the study with decreased concentration of dissolved oxygen to 4-5mg/dm^3.

Figure 12. Changing water quality parameters during the filter cycle without the use of reagents. Concentrations of iron compounds: 1 – after the 1-st degree filter; 2 – after the filter of the II degree. pH values: 3 – after the 1-st degree filter; 4 – after the filter of the II degree
Source: own study

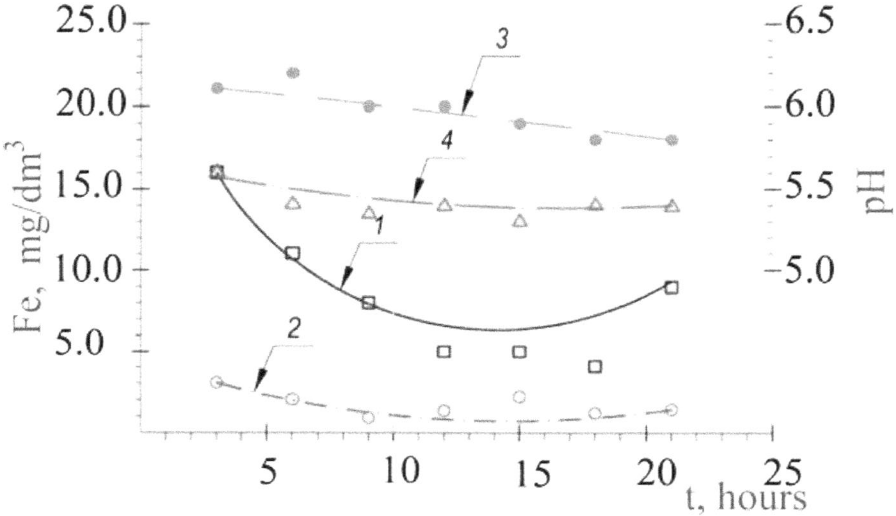

In treating underground weak acidic waters containing bacteria of *Leptothrix* type in contact loading of bioreactor there may be the mechanism of interaction of ammonia nitrogen with hydroxyl radicals •OH formed in the system according to Fenton reaction:

$$Fe^{2+} + H_2O_2 \rightarrow Fe^{3+} + OH^{\cdot} + OH^{-}, \quad (4)$$

$$Fe^{2+} + OH^{\cdot} \rightarrow Fe^{3+} + OH^{-}, \quad (5)$$

$$OH^{\cdot} + H_2O_2 \rightarrow HO_2^{\cdot} + H_2O, \quad (6)$$

$$Fe^{3+} + HO_2^{\cdot} \rightarrow Fe^{2+} + H^{+} + O_2, \quad (7)$$

$$Fe^{2+} + HO_2^{\cdot} \rightarrow Fe^{3+} + HO_2^{-}. \quad (8)$$

The said stages may be implemented in acidic or weak acidic media (pH>2) under following conditions $0.5 < [H_2O_2] / [Fe^{2+}] < 200$ (Luthy, 1964). However, in bioreactors operating in flowing regime with permanent inflow of fresh water containing ions Fe^{2+} and also under the availability of produced by bacteria of *Lepthothrix* type there may be the reaction going mainly by equation (4) with subsequent oxidation of ammonia hydrogen by hydroxyl radicals:

$$\left[NH_4^+ + {}^{\bullet}OH\right] \rightarrow \left[{}^{\bullet}NH_2 + H_2O_2\right] \rightarrow \left[{}^{\bullet}NHOH\right] \rightarrow \left[NO_2^-\right] \rightarrow \left[NO_3^-\right] \qquad (9)$$

As a result of biochemical reactions of oxidation of organic substances and also processes of adsorption of cations NH_4^+ on the surface of cell of ferrobacteria containing negatively charged ions groups (PO_4^{3-}; COO^-; HS^-; OH^-) and tying into non-dissolved complex compounds with polymeric structures of matrixes in contact loading of the bioreactor and filters there took place processes of comprehensive treatment of source water.

In particular, there took place decreased values of permanganate oxidation from 6,56 mg O_2/dm^3 in source water to 4.5-4.0 mg O_2/dm^3 after filters of the first degree and to 2.7-2.3 mg O_2/dm^3 after filters of the second degree Fig. 13 a. Simultaneously with these processes took place of removing ammonia nitrogen from 1.8-3.0 mg/dm^3 in source water to 0.6-1.0mg/dm^3 after filters of the first and to 0.55-0.2 mg/dm^3 after filters of the second degree Fig. 13 b .

Figure 13. Change in permanganate oxidation (a) and ammonium nitrogen concentration (b): 1 – after the 1-st degree filter; 2 – after the filter of the II degree (without reagents)
Source: own study

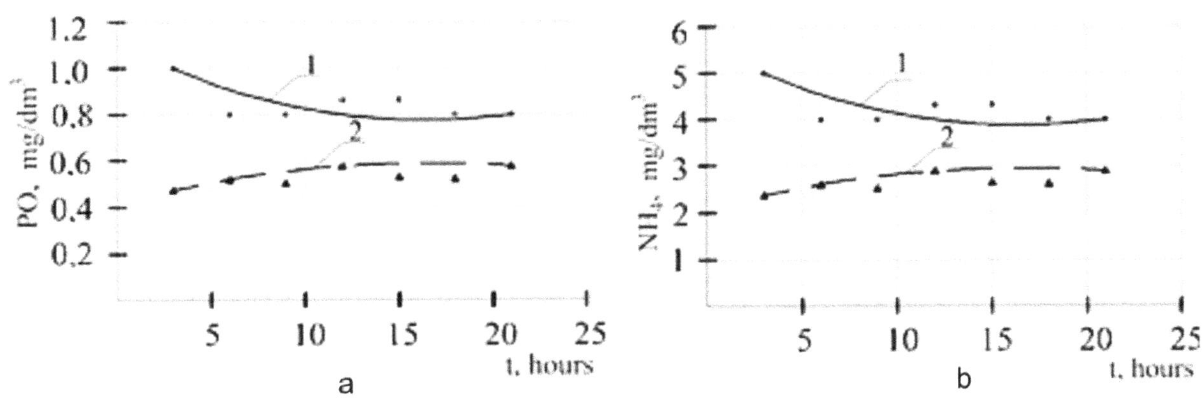

In this way, under concentrations of ammonia nitrogen 1.0-3.0 mg/dm^3 it is recommended to use alkalization after filters of the 2nd degree with the aim of stabilization processing of water. Under the concentrations of ammonia nitrogen to 1.0 mg/dm^3 for intensifying processes of biological treatment it is expedient to carry out the preliminary alkalization of water before filters of 1st degree with doses of soda ash 40-70mg/dm^3 increasing values of pH to 6.5-6.8 and hydro-carbonate alkalinity to 1.8-2.35 mmole/dm^3.

In using doses of soda ash higher 100 mg/dm^3 pH of system increased to 7.5-8.3 Fig. 11. As is known (Mouchet, 1995) under such values of pH there takes place the deceleration of biological processes with the advantage of chemical processes of oxidation which leads to the creation in interporous space of contact loading of amorphic sediment of ferrum(III) hydroxide instead of *bio*-minerals γ-*FeOOH*.

The formed amorphic sediment due to large number of taken water molecules occupies great volume in interpore space which in its turn leads to the quick growth of head expenditures and to decreased duration of filtration cycle (14).

By results of researches carried out it is suggested the technological scheme Fig. 14. defended by the patent of Ukraine and introduced in 2012 into operating project of reconstructing operating station of treating with productivity of 300 m^3/day. The basis of the project is modernized technology grounded on improved biological and physic-chemical methods of treating and conditioning source water.

The technology of treatment included such stages of processing: preliminary processing of water by 0.5% solution of soda ash → aeration → treatment of water in a bioreactor → introduction of coagulant solution → filtering → disinfection → stabilization processing of filtrate. The source water from bore hole before pressure mixing (1) took processing by solution of soda ash. The mixer (1) was equipped with the system of fine bubble aeration. The time of water in the mixer was 4-5 minutes. Air expenditure entering the mixer was planned to be regulated with the help of rotameter 21. This permitted to conduct not only the correction of values pH-Eh, concentrations of dissolved oxygen in a system and also to support their values at the level of optimal for living activity of consortiums of microorganisms.

Figure 14. Technological scheme of groundwater conditioning: 1 – pressure aeration mixer; 2 – pressure filters of the I-st degree (bioreactor); 3 – pressure filters of II degree; 4, 5, 6, 7 – equipment for preparation and dosing of solutions of soda ash, coagulant, flocculant, sodium hypochlorite; 8, 9, 10, 11, 12 – dosing pumps; 13 – hydraulic mixer for preparation of lime solution; 14 – circulating pump; 15 – filter; 16 – supply tank for clarified lime solution; 17 – clean water tank; 18 – pump station; 20 – mixer; 21 – rotameter

Source: own study

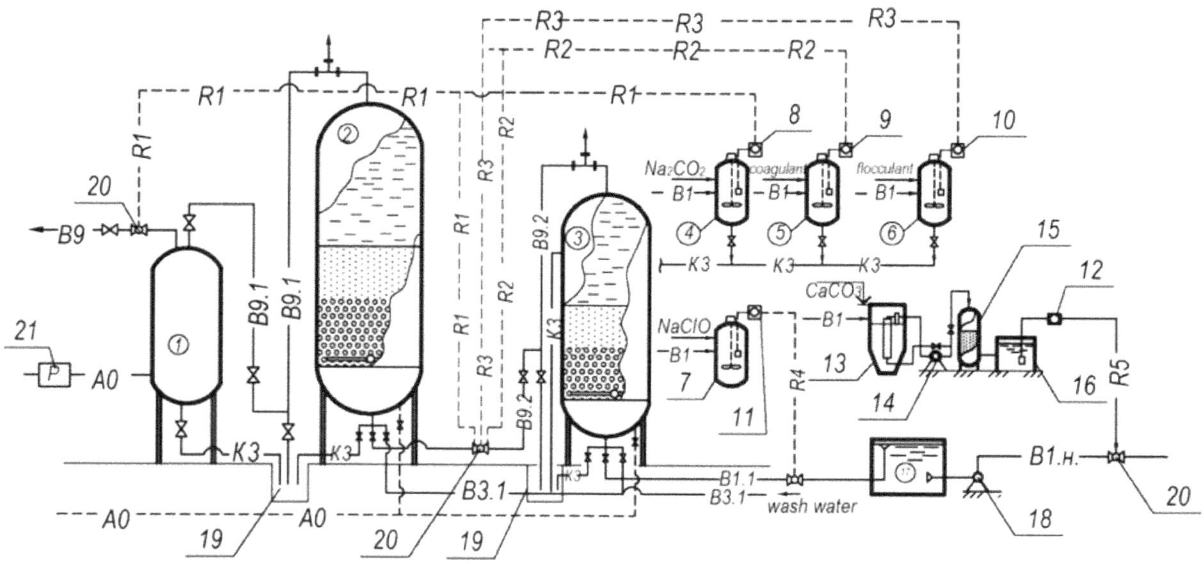

Later on water with changed indices of values pH-Eh, hydro-carbonate alkalinity and also concentrations of dissolved oxygen and non-organic carbon entered the contact loading of the reactor (2). The rate of descending water flow was 2.5-3 m/hour. The contact loading consisted of main filtering layer of large grain fractions δ=2-3 mm; h=1600 mm; and supporting layers δ=3-6 mm; h=200 mm and δ=6-12 mm; h=200 mm.

The time of water contact with matrix structures was 50-55 min. As is known (Mouchet, 1992) matrix structures of *bio*-minerals *(γ-FeOOH, α-FeOOH)* are natural sorbents with high specific surface of contact which permitted by 30% to decrease the expenditures of aluminum hydroxchloride ($Д_к$ 10-15 mg/dm^3) and non-ionogenic flocculent ($Д_{FL}$ 0.5-1.0 mg/dm^3). The introduction of coagulant solution was envisaged under seasonal increases of the content of dissolved organic substances and chromaticity.

The process of coagulation was envisaged to be implemented in over-filter space of filters of II-nd degree 3 (Fig. 14) with subsequent distribution of phases in loading from quartz sand with large grain fractions δ= 0.63-1.2 mm and layer thickness H_f=1400 mm. The designed duration of water contact with sand loading in the project was 20-25 minutes.

To protect water supplying pipes and equipment from corrosion there was envisaged the node of the stabilization processing of water (13-16). The process of disinfection of filtrate took place with the help of the solution of sodium hypochlorite (7) which was planned to be obtained with the help of electrolysis device EP-10-1. The dose of reagent was 1.0 mg/dm^3. The contact of water with chlorine during not less than 30 minutes was provided in the reservoir of clean water (17).

Regeneration of contact and filter loads was planned with the help of air-water flushing with water supply intensity of 18 l/(s.m^2).

The discharge of waste water, according to the project, was planned via the technological pipeline K3 into the existing collection channel, from where it was to be diverted outside the station.

Advantages of Using the Biological Method of Deironing

The application of the biological method of groundwater allows:

- abandon oxidizing columns, cooling towers, contact tanks;
- increase the filtration rate to 20 m /h (depending on the concentration and form of iron in the source water);
- increase the duration of the filrocycle by two or three times;
- to refuse the introduction of strong oxidants at the beginning of the technological scheme for the oxidation of organometallic complexes;
- Refrain from pre-chlorination by introducing disinfectant only before clean water tank;
- abandon powerful energy-consuming devices for saturating water with oxygen;
- reduce operating costs by 50-60%.

FUTURE RESEARCH DIRECTIONS

In the future, it is planned to conduct experiments on the biological extraction of manganese, chromium cations from groundwater and the development of appropriate technologies for the treatment of natural and circulating waters.

CONCLUSION

As a result of conducted monitoring of parameters of quality of underground waters in 91 settlements in northern-western regions of Ukraine: the classification was established of researched waters in values of pH, alkalinity reserve, concentrations of ferrum compounds and dissolved organics; it is proved that standard physic-chemical methods of non-reagent deferrization are unable to carry out the proper treatment of weak acidic, iron containing waters with low (to 2.0 mmole/dm^3) and average (2.0-4.0 mmole/dm^3) alkaline reserve in the presence of humin complexes; the physical model was developed and mechanism of the comprehensive treatment of underground waters on matrix structures of bio-minerals in interpore space of the contact loading of bioreactor; technologies were developed, studied and improved of the comprehensive water preparation on the basis of combined methods of treatment and conditioning: of aggressive (with very high degree of corrosion I_L » -2.0) of weak acidic underground waters with low alkaline reserve (to 1.0 mmole/dm^3), permanganate oxidation (8-12 mg O_2/dm^3), chromaticity (to 70 degrees platinum cobalt scale) which contain compounds of NH_4^+ (to 4 mg/dm^3), Fe-HC complexes (to 30 mg/dm^3), phenols (to 0.03 mg/dm^3), H_2S (to 4 mg/dm^3) with the possible availability of cations Cr^{6+} to 0.5mg/dm^3) by methods: correction of parameters of water quality (hydro-carbonate alkalinity, pH, concentration of dissolved non-organic carbon) by the solution of soda ash – aeration – biological treatment – coagulation – filtering via inert and sorption loading – stabilization processing – disinfection (with productivity of 200 to 2000 m^3/day); neutral aggressive waters (which reveal tendency to corrosion (I_L » -0.5) with the value of permanganate oxidation (to 6.0 mg O_2/dm^3) which contain ferrum compounds (to 5.0 mg/dm^3), NH_4^+ (to 1.0 mg/dm^3), phenols (to 0.03 mg/dm^3) by methods: aeration – biological treatment – filtering – reagent stabilization processing – disinfection (productivity from 1000 to 10000 m^3/day).

REFERENCES

Bates, B. C., Kundzewicz, Z. W., & Palutikof, J. P. (2008). Climate Change and Water Resources. Intergovernmental Panel on Climate Change Technical Paper. IPCC Secretariat.

Chana Clara, S., Fakra Sirine, C., Edwards, D., & Emerson, D. (2009). Iron oxyhydroxide mineralization on microbial extracellular polysaccharides. *Geochimica et Cosmochimica Acta*, *73*(13), 3807–3818. doi:10.1016/j.gca.2009.02.036

Eckhardt, K., & Ulbrich, U. (2003). Potential Impacts of Climate Change on Groundwater Recharge and Streamflow in a Central European Low Mountain Range. *Journal of Hydrology (Amsterdam)*, *284*(1-4), 244–252. doi:10.1016/j.jhydrol.2003.08.005

Emerson, D., Field, E., Chertkov, O., Davenport, K. W., Goodwin, L., Munk, C., Nolan, M., & Woyke, T. (2013). Comparative genomics of freshwater Fe-oxidizing bacteria: implications for physiology, ecology, and systematics. Frontiers in Microbiology / Evolutionary and Genomic Microbiology, 4.

Fujikawa, Y., Sugahara, M., Hamasaki, T., Yoneda, D., Minami, A., Sugimoto, Y., & Iwasaki, H. (2010). Biological filtration using iron bacteria for simultaneous removal of arsenic, iron, manganese and ammonia: Application to waterworks facilities in Japan and developing countries. *J. Human Environ. Studies*, (9), 261–276.

Grabovska, L. L., & Yefremova, O. O. (2010). *Environmental risk assessment in the field of drinking water supply in Ukraine*. Collection of scientific works of the National Academy of the State Border Guard Service of Ukraine. *Technical Sciences. Issue*, (54), 58–62.

Hallbeck, L., & Pedersen, K. (1991). Autotrophic and mixotrophic growth of Gallionella ferruginea. *Journal of General Microbiology*, *137*(11), 2657–2661. doi:10.1099/00221287-137-11-2657

Hallberg, L., Ferris, R., & Grant, F. (2004). Biomineralization by Gallionella. *Geomicrobiology Journal*, *21*(5), 325–330. doi:10.1080/01490450490454001

Kennedy, C. B., Scott, S. D., & Ferris, F. G. (2004). Hydrothermal phase stabilization of 2-line ferrihydrite by bacteria. *Chemical Geology*, *212*(3-4), 269–277. doi:10.1016/j.chemgeo.2004.08.017

Krainov, S. R., & Shvets, V. M. (1992). Hydrogeochemistry. Moscow: Nedra, 463.

Kvartenko, A. N. (2016). Practical stabilization methods of groundwater in North-Western region of Ukraine. *Water Supply and Wastewater Removal: monograph*, 107–120. Lublin University of Technology.

Kvartenko, O. M. (2011). Characteristics of underground waters in the North-Western regions of Ukraine and technological schemes for their conditioning. Problems of water supply, sewerage, and hydraulics. *Scientific and Technical Journal*, (16), 32-40.

Kvartenko, O. M. (2018). System approach to substantiation of new and improvement of existing technologies of air conditioning of multicomponent groundwater. Municipal Economy of Cities. Series: Technical Sciences and Architecture, (140), 98 -103. Kharkiv National University.

Kvartenko, O. M. (2019). Development of scientific principles and improvement of technologies for treating multi-component underground waters. Manuscript for academic degree of doctor of technical sciences in speciality 05.17.21 – technology of water treatment. KPI, Kyiv.

Kvartenko, O.M., & Safonov, R.V. (2015). Analysis of the degree of aggressiveness of groundwater in Rivne region. *Bulletin of National University of Water Environmental Engineering, 1*(69), 58-66.

Luthy, R. G. (1964). New concept for iron bacteria control in water wells. *Water Well Journal.*, *24*, 29–30.

Lyuty, G.G., & Sanina, I.V. (2011). Factors of increasing the density of underground waters in the process of exploitation of water intakes in Ukraine. *Collection of scientific works*, (2), 91-103. Ukrainian state Geological Survey Institute.

Mouchet, P. (1992). From Conventional to Biological Removal of Iron and Manganese in France. *Journal - American Water Works Association*, *84*(4), 158–167. doi:10.1002/j.1551-8833.1992.tb07342.x

Mouchet, P. (1995). Biological Filtration for Iron and Manganese Removal: Some Case Studies. *WQTC 95 (AWWA),* New Orleans LA.

National report on drinking water quality and the state of drinking water supply in Ukraine in 2003. (2005). Rivne.

Nikoladze, G. I. (1987). Improving the quality of groundwater. Moscow: Stroyizdat.

Orlov, V. O. (2008). Deironing of groundwater by simplified aeration and filtration: monograph. Publishing center National University of Water Environmental Engineering.

Orlov, V. O., Martynov, S. Yu., Orlova, A. M., Zoshchuk, V. O., Minaeva, N. L., Kunitsky, S. O., Meddur, M. M., & Trokhimchuk, M. M. (2012). *Purification of natural water on expanded polystyrene filters: monograph/ under the general editorship of V.O. Orlov*. Publishing Center National University of Water Environmental Engineering.

Report on the state of the environment in Rivne region in 2014. (2015). (RSER), Rivne.

Safonov, N. A., Kvartenko, A. N., & Safonov, A. N. (2000). *Self-cleaning water treatment plants (technology, design, calculation): monograph* (S. N. A. Rivne, Ed.). Publishing House Rivne State Technical University.

Safonov, N. A., & Rusak, G. V. (1984). Self-cleaning plant for biological removal of iron from underground waters. In Preparation of water for drinking purposes Conference, Leningrad, LEBI, 162-167. .

Czekalla, C., Mevius, W., & Hanert, H. (1985). Quantitative Removal of Iron and Manganese by Microorganisms in Rapid Sand Filters (In Situ Investigations). [Berlin.]. *Water Supply*, •••, 3.

Seppänen, H. (1991). *Experiences of Biological iron and manganese removal in Finland*. Proc. IWEM ann. Sym. *15* (1), 9-11.

Sharma, S.K., Petrusevski, B., & Schippers, J.C. (2005). Biological iron removal from groundwater. *Water Supply, 54*(4), 239-247.

Søgaard, G. E., Medenwaldt, R., & Abraham-Peskir, J. V. (2000). Conditions and rates of biotic and abiotic iron precipitation in selected Danish freshwater plants and microscopic analysis of precipitate morphology. *Water Research, 34* (10), 2675-2682.

Tamura, T., Tsunai, T., Ishimaru, Y., & Nakata, A. (1999). Iron and manganese removal by iron bacteria in groundwater. *Suido Kyokai Zasshi (J. Japan Water Works Assoc.), 68*, 1–13

Chapter 19
The Importance of Changing Natural and Climatic Conditions in Investment Projects

Nadiia Frolenkova
National University of Water and Environmental Engineering, Ukraine

ABSTRACT

The article focuses on the actual scientific and practical problem of accounting for the influence of meteorological and climatic factors in the technical and economic calculations in the field of environmental management. It has been proven that the introduction of scientifically sound and effective methods of using meteorological and climatic information in economic calculations significantly reduces the loss caused by weather conditions and improves the implementation of an optimal strategy for agricultural production on reclaimed lands. Such calculations are based on economic and statistical modelling of different variants that account for standard hydrometeorological information in the implementation of design, management, and economic decisions. This increases the validity and reliability of calculations, as well as their compliance with the actual operating conditions of environmental and economic facilities. Consequently, this attracts increased interest from both public and private investors.

BACKGROUND

The current development of Ukraine is characterized by significant imbalances in the development of various sectors of the economy. This is mainly due to the scarcity of financial resources, their different investment attractiveness and the presence of negative trends in the sectors that actively use natural resources. At the present time, the situation is exacerbated by the current challenges of the present - energy, food and water crises, climate change at both global and regional levels (Blanc et al. (2017); Dong et al. (2016); Gohar et al. (2016); Kovalenko et al. (2019); Martyn et al. (2019); Martyn et al. (2020); Massey (2012); Openko et al. (2020); Rezaei Zaman et al. (2016); Rokochinskiy (2010).

DOI: 10.4018/978-1-6684-8248-3.ch019

Such changes need special attention and investment in the highest sensitive and the least adaptable industries as most ecosystems are very sensitive to farming practices and resources demand. Therefore, an effective mechanism and toolkit for business planning is needed in such ecosystems to reduce the losses caused by adverse climate change.

That is why special attention should be paid to the development and functioning of the natural-engineering sectors, where the economic growth is limited by the rational use of natural resources and environmental protection. The variability of natural and climatic conditions in investment projects in the field of nature management Land reclamation is one of such sectors (Kovalenko et al. (2019); Rokochynskiy et al. (2020).

The current state of the sector is characterized by a decrease in the efficient use of reclaimed land and its role in agricultural production. It is a consequence of a deep economic crisis that has hit all sectors of the economy. Today, the economic and environmental problems of land reclamation are of particular relevance.

In the last decade, the trends in the sector have been characterized by a complex of unresolved issues, such as underfunding due to the lack of interest from both state and private investors.

These problems in water management and land reclamation can be solved by the restructuring of the system of economic relations, introduction of effective market management methods, and the application of new methodology and methods to economic calculations.

The transformation of economic relations of Ukraine, the emergence of new financing sources for investment projects, and the need for greening modern production require the revision and improvement of traditional approaches, particularly to ecologically and economically optimal design solutions in the areas of environmental management in general and water management and land reclamation in particular (Armeanu et al. (2009); Bierman et al. (2006); Haka (2006); Mohamed, McCowan (2001); Nowak (2005).

This, however, is complicated by the specific functioning of water management and reclamation facilities and the development of projects in the difficult and changing environmental conditions. The above must be taken into consideration when selecting the best option for project implementation (Rokochynskiy (2016); Rokochynskiy et al. (2019).

METHODS AND TECHNIQUES

The most important and distinctive feature of agricultural production, including one on the reclaimed land, is the influence of natural factors. In general, agricultural production in different climatic zones is determined by existing agro-climatic conditions (Rokochynskiy et al. (2010); (2016); Zhukovsky (1981). Agricultural productivity depends not only on biological capabilities of crops, machines and mechanisms, timely and high-quality tillage, fertilizers, crop rotations, etc., but also on a number of meteorological factors and their interaction. Agricultural production is extremely susceptible to weather conditions and changes in hydrometeorological parameters. Therefore, the output of the agricultural activity on the reclaimed land and, accordingly, the effectiveness of investment in land reclamation depend not only on the availability of material resources and their use, but also on the presence or absence of necessary natural and climatic resources.

Transition from the established practice of considering water management and reclamation facilities is needed, since it is not only a technical issue, but complex natural and technical, ecological and economic systems of nature management. Corresponding changes in the methodology, technical and technological strategy of their formation and functioning should follow. This requires direct consideration of volatile natural and climatic conditions. Along with land reclamation factors, they have a decisive influence on the overall natural and reclamation regime of reclaimed land and the corresponding ecological and economic effect.

Therefore, meteorological and climatic factors should be urgently taken into consideration in the technical and economic calculations. The introduction of scientifically sound and effective methods of using meteorological and climatic information in economic calculations will significantly reduce losses caused by weather conditions. In addition, the effect can be boosted by the implementation of an optimal strategy for the organization of agricultural production on reclaimed land.

Thus, according to scientific sources and practical approaches to the substantiation of optimal technical and technological solutions for water management of drained and irrigated lands on ecological and economic grounds, a climate-optimal strategy has been implemented at different intervals in the reclamation projects (Rokochynskiy (2016); Rokochynskiy et al. (2010); Zhukovsky (1981).

Losses due to variable weather conditions and natural phenomena require special attention, since atmospheric processes are continuous in their nature (Rokochynskiy et al. (2010); (2012).

Although at the present stage of development it is possible to avoid some production losses, it is still impossible to stop or change the course of atmospheric processes. It is necessary to take into account these processes, anticipate their adverse impact, apply protection measures, and thus, partially or completely preserve material assets and successfully perform production tasks.

However, in view of the changing nature of climatic conditions, it should be noted that the methodological basis for calculating the economic and environmental effect is still very weak in different production sectors. Results of domestic and foreign studies in the second half of the last century created a theoretical basis for determining the economic effect of different strategies that use predictive and regime climate information (Blanc, Schlenker (2017); Dong et al. (2016); Gohar, Cashma N (2016); Rezaei Zaman et al. (2016).

With regard to water-reclamation measures as a component of agricultural production, in their technological aspect, we need to justify and evaluate different technologies of water management and their influence on reclaimed lands, proper types and designs of reclamation systems and consider their implementation in variable natural and reclamation conditions in the future.

The environmental effect is determined by the projected and practical effects of land reclamation and the economic effect is determined by the agricultural production and the investment in land reclamation.

As practice and experience show, the accuracy and objectivity of relevant calculations directly depend on the accuracy of relevant weather or meteorological forecast used at the appropriate level of decision making.

In fact, such calculations are based on economic and statistical modeling of different options for accounting for standard hydrometeorological information when implementing design, management and economic decisions (Frolenkova at al. (2012); Rokochynskiy at al. (2010).

To build such models it is necessary to obtain the information about:

- laws of distribution of the actual and estimated values of the forecasted element;
- the type and parameters of the user's loss (or profit) function, characterizing losses (or profits) of different values and forecast errors (for alternative forecasts, the so-called payoff matrix).

But to date, such loss (or profit) functions and payoff matrices for the majority of users in various sectors of the economy (including agriculture and water management) have not been determined yet.

General approaches to the building and implementation of the models that optimize business decisions at different levels by using climate data and meteorological forecasts in the creation and operation of complex meteorological and economic systems are set out in the work of Zhukovsky (1981). He introduced the concept of the "meteorological and economic system" and the "climatoogical strategy". When using it, a specific business decision that takes into account data on the climatic conditions of an object has limited efficacy. A climatological strategy can be "climatologically optimal" if the economic decision determined by it ensures the achievement of the extreme of a selected quality criterion.

Climatic or weather conditions in many cases have a decisive influence on the formation of water and natural and reclamation regimes of agricultural lands and crop yields. It is necessary to collect data for a relevant object from a number of previous years. The number of parameters and the selection of specific years depend on the typical long-term interseasonal variability of meteorological conditions for the region.

The application of mathematical methods and technical and economic research into reclamation involves not only various calculations, but such methods and models should also be used as a tool for obtaining theoretical conclusions and practical results.

Once we have necessary information, indicators depending on variable meteorological conditions can be determined by means of generally accepted statistical methods which are the most reasonable when developing reclamation projects.

Given the complexity of such nature management projects, modern BIM-technologies should be used for their automation and optimization. For the effective implementation of such tasks, we have developed a set of optimization, forecasting and simulation models. The models help to assess the following: climatic conditions of the area, meteorological regimes, water regimes, technologies of water regulation on drained lands, and the productivity of drained lands. In Ukraine, the practical application of forecasting and simulation models is regulated by relevant state industry standards (Openko et al. (2017); Rokochynskiy (2016); (2019); Shevchenko et al. (2017); Zhukovsky (1981)

FUTURE RESEARCH DIRECTIONS

The basis for calculating all technological, economic and environmental indicators is a crop yield, which, as noted above, directly depend on changing weather and climatic conditions. While traditional approaches do not take this feature into account, static values of productivity – design, forecast, expected or standard have been used to calculate ecological and economic efficiency of water management projects. Accordingly, economic and environmental results cannot be collated with different meteorological conditions in specific years.

The method we propose enables to avoid such a shortcoming and to bring the estimated forecast values closer to the real ones, while taking into account the effect of actual weather conditions for the operation of an object in a given year.

According to the analysis, the economic result of agricultural production on reclaimed land (E_i), current agricultural (C_a) and reclamation costs (C_M), water costs (C_W), and possible losses (R) are variable and depend on many factors, the main of which are natural and climatic conditions of the production facility.

General approaches to the developing and implementation of the models of economic decision optimization at different levels using climate data and meteorological forecasts when creating and operating complex meteorological and economic systems, are presented in the work of E. Zhukovsky (Zhukovsky, (1981). He introduced the concepts "meteorological and economic system" and "climatological strategy." According to the approach of E. Zhukovsky the selection of a specific economic decision that is made once based on the data of the climatic conditions of production facility functioning. In this case, such a climatological strategy will be "climatologically optimal" if the economic decision, determined by it, will achieve the extremum of the chosen quality criterion.

Due to the seasonal cyclicality of agricultural production on the reclaimed lands, there are different periods of vegetation, in terms of heat and humidity supply, which are grouped in base (typical) yearly groups. The distribution of these groups within the project life cycle is uneven and can be made with the help of a corresponding coefficient in the form of a share (probability of detection) of the corresponding yearly group within the total project implementation period.

In addition, the expected yield of crops and the amount of annual operating costs are influenced by the type and design of the land reclamation system, which determine the technology of water regime control (water regulation) on the reclaimed lands.

For this reason, the effect of agricultural production on reclaimed lands and the operating costs throughout the life cycle of the reclamation facility will depend on three main factors:

1. Weather (meteorological) conditions in the corresponding estimated years, the statistic population - $P=\{p\}, p=1, m.$
2. The share or frequency (probability) of detection of the respective yearly group within the total project life cycle – $\{\alpha p_p$, $p=1, m.$
3. The type and design of the reclamation system (method, water regulation plan), the statistic population - $S=\{s\}, s=1, n.$

Having expressed a variable economic parameter as Y_{ps}, the following utility function can be constructed in a general form

$$Y_{ps}=Y(p,s,\alpha_p),\ p=1,\ m;\ s=1,n. \qquad (1)$$

For better clarity, this function can be expressed as a payoff matrix (Table 1).

Table 1. Payoff matrix view illustration

$\{p\}, p=1, m$	α_p	$\{s\}, s=1, n.$			
		s_1	s_2	...	s_n
p_1	α_{p_1}	Y_{11}	Y_{12}	...	Y_{1n}
p_2	α_{p_2}	Y_{21}	Y_{22}	...	Y_{2n}
...	...	...	...	...	...
p_m	α_{p_m}	Y_{m1}	Y_{m2}	...	Y_{mn}

Source: own study.

Due to the uneven distribution of different by the weather conditions yearly groups throughout the project life cycle, the weighted average but not arithmetic average value by the yearly groups should be used as an annual average variable economic parameter. According to (Kovalenko at al. (2019); Rokochynskiy at al. (2010), the weighted average value by the yearly groups is given by the formula of mathematical expectation, since it is the most important characteristic of a random variable that serves as the center of its probability distribution. This variable shows the most plausible value of the estimated or predicted factor.

Therefore, following the Bayesian approach (Frolenkova at al. (2012); Rokochynskiy at al. (2012); Rokochynskiy at al. (2010), it can be argued that the average (in the statistical sense) values of economic variables for each alternative reclamation project in view of the climatological strategy of managing the reclamation facility over its life cycle, is given by the formula:

$$Y_i = \sum_{j=1}^{m} Y_{ij} \cdot \alpha_{pj}, \; i=1,n. \tag{2}$$

Obviously, determining the values of the coefficients of the share αp_j of the respective estimated yearly group within the project life cycle (probability values corresponding to the given economic parameter by the estimated yearly group) is very important for the future obtained result.

Determination of the indicator αp is due to the need to assess the total heat and water supply of the estimated year by all major determining meteorological factors. These factors form the weather conditions in general (precipitation, temperature, saturation deficit and relative air humidity, etc.). Therefore, such assessment can accurately be carried out for any facility, provided the statistical analysis of the observation years by the relevant meteorological factors.

According to the long-term (50-70 years) retrospective observations at meteorological stations and posts located near (up to 50 km) the investigated facility, the appropriate model of meteorological support for forecasting and optimization calculations (Kovalenko at al. (2019); Rokochynskiy at al. (2010) is based on the investigation of weather factors variability using:

- multivariate statistical analysis using software;
- schematization of meteorological regimes by a comprehensive assessment of meteorological factors and complexes built on their basis;
- identification and formalization of the mechanism of meteorological regimes formation for both long-term and vegetation periods.

Such a model, unlike the available ones, enables to obtain a typical distribution of major meteorological factors (precipitation; average temperature, moisture deficit and relative air humidity); the statistic population is $\{f\}$, $f = \overline{1, n_f}$ for the taken as a basis ten-day period in the typical (estimated) as to the heat and humidity supply vegetation periods, the statistic population is $\{p\}$, $p = \overline{1, n_p}$ (n_p=5): 1 – very wet, p = 10%; 2 – wet, p = 30%; 3 – average, p = 50%; 4 – dry, p = 70%; 5 – very dry, p = 90%.

Such a number of typical models of meteorological regimes is sufficient for the engineering and economic practice of performing forecast and optimization calculations on a long-term basis. The real values of the shares αp, $p = \overline{1, n_p}$ are determined too for the implementation of a climatologically optimal strategy for managing the facility at the design stage.

In the structure of the implementation of the general model, after dividing the statistical sequence into typical groups, the shares of different as to water supply vegetation periods are determined $\{\alpha p_j$, $p=\overline{1,n_p}$ within the analyzed period $\{j\}$, $j=\overline{1,n_j}$

$$\alpha_p = n_{jp}/n_j,\ p=\overline{1,n_p}, \tag{3}$$

where n_{jp} the number of elements of the observation yearly group, the statistic population is $\{j\}$, $j=\overline{1,n_j}$, which were included in the pth yearly group by the water supply, the statistic population is $\{p\}$, $p=\overline{1,n_p}$, of the treated statistical series, the statistic population is n_j.

Otherwise, when long-term observation data is unobtainable the assumptive values of the indicator αp can be assumed based on hydrological simplified calculations or appropriate recommendations.

Knowing the empirical probability of any hydrological or meteorological characteristic, one can calculate its probable frequency over years. The frequency of a hydrological value means a number of years N during which this value occurs on average once.

Probability p (%) and frequency N are related as follows:

$$N = \frac{100}{p} \text{ when } p<50\%; \tag{4}$$

$$N = \frac{100}{100-p} \text{ when } p>50\%. \tag{5}$$

But the values αp calculated using the dependences (4) and (5), firstly, will take the same values for different facilities regardless of their location; secondly, they will not take into account the complex effect of meteorological factors on the environment of meteorological regimes in the estimated years.

Therefore, based on the generalization of the results of our own studies, data from other authors and reference data (Rokochynskiy at al. (2019); Rokochynskiy at al. (2013); Rokochynskiy at al. (2010); Shalay at al. (2004); etc.), the normalized values of the indicator αp by natural zones for drained lands have been determined and shown in Table 2.

Table 2. Recommended normalized values αp for estimated vegetation periods on the drained lands by the natural zones of Ukraine

Natural zone	Probability *p*, %				
	10%	**30%**	**50%**	**70%**	**90%**
Forest steepe	0.10	0.20	0.20	0.30	0.20
Polesie (forest area)	0.15	0.20	0.25	0.25	0.15
Subcarpathia	0.15	0.25	0.30	0.20	0.10
Transcarpathia	0.10	0.20	0.30	0.25	0.15

Source: own study.

These values are well suited for approximate calculations when substantiating design technical and technological solutions for water management on drained lands on a long-term basis based on optimization models, taking into account the climatological strategy for the facility management and estimation of the overall economic efficiency of irrigation implementation. If necessary, the recommended values αp can be adjusted for a specific object, taking into account changes in weather and climatic conditions in the future.

Based on this approach, forecasting and optimization models for selecting technically, economically and environmentally optimal variants of design solutions were developed and implemented when reconstructing existing hydro-reclamation systems in various natural and climatic zones of Ukraine, namely when optimizing agricultural drainage parameters, technology of water management and design of hydro-reclamation systems on drained lands, technological and design solutions in the operation of rice irrigation systems, etc. (Bierman at al. (2006); Nowak. (2005); Haka. (2006); Rokochynskiy at al. (2019); etc.).

Table 3 shows the economic result of agricultural production on reclaimed lands in changing climatic and reclamation conditions when cultivating potatoes. This crop is one of the most profitable in the project crop rotation on drained mineral soils in the Ukrainian forest-steppe zone. Data on yields and technical and economic indicators are necessary for the project to provide forecasting and optimization calculations to substantiate the optimal technology of water regulation in conditions studied.

This assessment was performed as a computer-based experiment using an appropriate set of predictive and simulation models (Rokochynskiy et al. (2010) for different water management technologies (1 – drainage; 2 – preventive locking as periodic watering by partial accumulation of drainage runoff and ponding of ground water table (GWT) in spring period; 3 – locking as regular watering during the growing season by supplying water, rise and ponding of GWT on the system; 4 – sprinkling irrigation of drained lands and estimated as to the heat and water supply vegetation periods ($p = 10\%$ – very wet, $p = 30\%$ – wet, $p = 50\%$ – average wet; $p = 70\%$ – dry, $p = 90\%$ – very dry).

Results in Table 3 clearly demonstrate the necessity and usefulness of taking into account the impact of variable weather and climatic conditions in the economic analysis of agricultural production on reclaimed lands in the projects involving construction, reconstruction and operation of water and reclamation facilities.

Results in Table 3 convincingly show the significant impact of weather and climatic conditions on forecast yields by estimated years and their deviation from weighted average values obtained from Equation (2). This indicates the necessity of taking into account the effect of changing weather and climatic conditions in the economic analysis of agricultural production on reclaimed land in projects involving construction, reconstruction and operation of water management and reclamation facilities.

Table 3. Yields and main technical and economic indicators of potato cultivation in variable natural, agricultural and ameliorative conditions in the Forest-Steppe zone of Ukraine

Estimated heat and water supply for vegetation periods, (%)	Yield, t/ha^{-1}	Agricultural costs	Operational costs	Total running costs	Gross production	Net income
Drainage						
p=10	28.16	497.96	56.48	554.46	750.14	195.68
p=30	32.61	538.52	48.54	587.05	948.20	361.15
p=50	29.77	530.30	47.92	578.22	911.80	333.58
p=70	28.58	507.62	47.92	555.55	797.90	242.36
p=90	23.86	468.97	47.92	516.89	610.16	93.27
Weighted average	28.64	**509.64**	**48.89**	**558.54**	**808.42**	**249.88**
Preventive locking						
p=10	28.16	520.00	89.49	609.49	750.14	140.65
p=30	33.33	563.82	71.51	635.35	964.22	328.87
p=50	31.03	564.20	70.14	634.34	970.42	336.08
p=70	30.89	551.78	73.47	625.25	906.12	280.87
p=90	31.04	551.41	84.28	635.68	905.72	270.04
Weighted average	31.16	**553.43**	**76.18**	**629.60**	**914.91**	**285.31**
Watering locking						
p=10	28.16	531.37	138.64	670.00	750.14	80.14
p=30	33.33	575.19	117.47	692.65	964.22	271.56
p=50	32.03	581.68	115.84	697.53	1000.18	302.65
p=70	31.78	573.84	119.76	693.60	959.72	266.12
p=90	34.41	585.33	132.52	717.84	1013.37	295.53
Weighted average	32.30	**573.72**	**122.95**	**696.67**	**958.48**	**261.80**
Sprinkling irrigation on drained lands						
p=10	28.16	576.32	323.92	900.24	750.14	-150.10
p=30	33.33	620.14	258.48	878.61	964.22	85.60
p=50	36.94	650.83	253.47	904.30	1115.35	211.05
p=70	37.77	661.11	265.58	926.69	1167.19	240.50
p=90	42.18	697.17	304.97	1002.14	1343.60	341.47
Weighted average	36.64	**649.58**	**275.45**	**925.05**	**1109.80**	**184.75**

Source: own study.

CONCLUSION

1. The integration of the long-term variability of natural and climatic conditions in environmental and economic calculations enables to:
   - include them in the economic analysis of agricultural production on reclaimed lands;
   - build and successfully implement forecasting and optimization models to substantiate both optimal technical and technological solutions for water regulation, and the development of

investment projects for the construction and reconstruction of water management and reclamation systems;
   - increase the validity and reliability of the calculations, as well as their compliance with real conditions for the operation of natural facilities. This should attract both public and private investments;
2. Such conditions enable to:
   - maintain a steady development of nature management sectors and their adaptation to global climate change;
   - prevent and reduce losses from uncontrolled environmental factors; and
   - generate additional benefits from more rational economic activity in new environmental conditions

REFERENCES

Armeanu, D., & Lache, L. (2009). The NPV criterion for valuing investments under uncertainty. *Economic computation and economic cybernetics studies and research. Academy of Economic Studies*, (4), 133–143.

Bierman, H. J. R., & Smidt, S. (2006). The capital budgeting decision: Economic analysis of investment projects. 9th ed. Abingdonon-Thames. Routledge.

Blanc, E., & Schlenker, W. (2017, Summer). The use of panel models in assessments of climate impacts on agriculture. *Review of Environmental Economics and Policy*, *11*(2), 258–279. doi:10.1093/reep/rex016

Dong, Z., Pan, Z., Wang, S., An, P., Zhang, J., Zhang, J., Pan, Y., Huang, L., Zhao, H., Han, G., Wu, D., Wang, J., Fan, D., Gao, L., & Pan, X. (2016). Effective crop structure adjustment under climate change. *Ecological Indicators*, *69*(October), 571–577. doi:10.1016/j.ecolind.2016.04.010

Frolenkova, N., Kozhushko, L., & Rokochinskiy, A. (2007). *Ekoloho-ekonomichne otsinyuvannya v upravlinni melioratyvnymy proektamy: Monografiya* [*Ecological and economic assessment in the management of reclamation projects: Monograph*]. NUVGP.

Frolenkova, N., Rokochinskiy, A., Volk, P., Shatkovskyi, A., Prykhodko, N., Tykhenko, R., & Openko, I. (2020). Cost-effectiveness of investments in drip irrigation projects in Ukraine. [IJGE]. *International Journal of Green Economics*, *14*(4), 315–326. doi:10.1504/IJGE.2020.112570

Gohar, A., & Cashman, A. (2016). A methodology to assess the impact of climate variability and change on water resources, food security and economic welfare. Agricultural Systems, 147. doi:. 008. doi:10.1016/j.agsy.2016.05

Haka, S. F. (2006). A review of the literature on capital budgeting and investment appraisal: Past, present, and future musings. *Handbooks of Management Accounting Research*, *2*, 697–728. doi:10.1016/S1751-3243(06)02010-4

Kovalenko, P., Rokochinskiy, A., Jeznach, J., Koptyuk, R., Volk, P., Prykhodk, N., Tykhenko, R. (2019). Evaluation of climate change in Ukrainian part of Polissia region and ways of adaptation to it. *Journal of Water and Land Development*. *41*(IV–VI). . doi:10.2478/jwld-2019-0030

Martyn, A., Openko, I., Ievsiukov, T., Shevchenko, O., & Ripenko, A. (2019). Accuracy of geodetic surveys in cadastral registration of real estate: Value of land as determining factor. *18th International Scientific Conference. Engineering for Rural Development*, Jelgava, Latvia. 10.22616/ERDev2019.18.N236

Martyn, A., Shevchenko, O., Tykhenk, O. R., Openk, O. I., Zhuk, O., & Krasnolutsky, O. (2020). Indirect corporate agricultural land use in Ukraine: Distribution, causes, consequences. *International Journal of Business and Globalisation*, *25*(3), 378–395. doi:10.1504/IJBG.2020.109029

Massey, E. E. (2012). *Experience of the European Union in adaptation to climate change and its application to Ukraine* [online]. Office of the Co-ordinator of OSCE Economic and Environmental Activities. https://www.osce.org/ukraine/104019?download=true

Mohamed, S., & McCowan, A. K. (2001). Modelling project investment decisions under uncertainty using possibility theory. *International Journal of Project Management*, *19*(4), 231–241. doi:10.1016/S0263-7863(99)00077-0

Nowak, M. (2005). Investment projects evaluation by simulation and multiple criteria decision aiding procedure. *Journal of Civil Engineering and Management*, *11*(3), 193–202. doi:10.1080/13923730.2005.9636350

Openko, I., Kostyuchenko, Y. V., Tykhenko, R., Shevchenko, O., Tsvyakh, O., Ievsiukov, T., & Deineha, M. (2020). Mathematical modelling of postindustrial land use value in the big cities in Ukraine. *International Journal of Mathematical. Engineering and Management Sciences.*, *5*(2), 260–271. doi:10.33889/IJMEMS.2020.5.2.021

Openko, I., Shevchenko, O., Zhuk, O., Kryvoviaz, Y., & Tykhenko, R. (2017). Geoinformation modelling of forest shelterbelts effect on pecuniary valuation of adjacent farmlands. [IJGE]. *International Journal of Green Economics*, *11*(2), 139–153. doi:10.1504/IJGE.2017.089015

Rezaei Zaman, M., Morid, S., & Delavar, M. (2016). Evaluating climate adaptation strategies on agricultural production in the Siminehrud catchment and inflow into Lake Urmia, Iran using SWAT within an OECD framework. Agricultural Systems, 147, 98–110. doi:. 2016.06.001. doi:10.1016/j.agsy

Rokochinskiy, A. (2010). *Naukovi ta praktichni aspekti optimizacii vodoregulyuvannya osushuvanikh zemel' na ekologoekonomichnikh zasadakh: Monografiya* [*The scientific and practical aspects optimization of water regulation drained lands on environmental and economic grounds. Monograph*]. NUVGP.

Rokochinskiy A., FrolenkovA N., & Koptiuk R. (2012). Investitsiyna otsinka proektiv optimizatsiї vodoregulyuvannya osushuvanih land of urahuvannyam mainly chinnikiv vplivu [Investment assessment project for optimizing water management of drained lands from the main bureaucrats*]. Tavriys'kyy naukovyy visnyk, 83,* 216–220.

Rokochinskiy A. (2016). Systemna optymizatsiya vodorehulyuvannya yak neobkhidna umova stvorennya ta funktsionuvannya vodohospodars'ko-melioratyvnykh ob″yektiv na ekoloho-ekonomichnykh zasadakh [System optimization of water regulation as a prerequisite for the creation and operation of water management and reclamation facilities on ecological and economic grounds]. *Vodne hospodarstvo Ukrayiny, 104,* 67–71.

Rokochinskiy, A., Bilokon, V., Frolenkova, N., Prykhodko, N., Volk, P., Tykhenko, R., & Openko, I. (2020). Implementation of modern approaches to evaluating the effectiveness of innovation for water treatment in irrigation. Journal of Water and Land Development, 45 (IV–VI), 119–125. DOI doi:10.24425/jwld.2020.133053

Rokochinskiy, A., Jeznach, J., Volk, P., Turcheniuk, V., Frolenkova, N., & Koptiuk, R. (2019). Reclamation projects development improvement technology considering optimization of drained lands water regulation based on BIM. Scientific Review Engineering and Environmental Sciences, 28(3), 193–202. doi:10.22630/PNIKS.2019.28.3.40

Rokochinskiy, A., Stachuk, V., Frolenkova, N., Shalay, S., Koptyuk, R., & Volk, P. (2010). *Tymchasovi rekomendatsiyi z optymizatsiyi vodorehulyuvannya osushuvanykh zemel' u proektakh budivnytstva y rekonstruktsiyi vodohospodars'ko-melioratyvnykh ob˝yektiv* [*Temporary recommendations for optimization of water management of drained lands in projects of construction and reconstruction of water management and reclamation facilities*]. NUVGP.

Rokochinskiy, A., Volk, P., Frolenkova, N., Shalay, S., Koptyuk, R., Zayets, V., & Prykhodko, N. (2013). *Naukovo metodychni rekomendatsiyi do obgruntuvannya optymal'nykh parametriv sil's'kohospodars'koho drenazhu na osushuvanykh zemlyakh za ekonomichnymy ta ekolohichnymy vymohamy* [*Scientific and methodological recommendations for substantiation of optimal parameters of agricultural drainage on drained lands according to economic and environmental requirements*]. NUVGP.

Rokochynskiy, A., Turcheniuk, V., Prykhodko, N., Volk, P., Gerasimov, I., & Koç, C. (2020). Evaluation of climate change in the rice-growing zone of Ukraine and ways of adaptation to the predicted changes. *Agricultural Research*, *9*(4), 631–639. doi:10.100740003-020-00473-4

Shalay, S., Rokochinskiy, A., Stashuk, V., & Bezhuk, V. (2004). *Tymchasovi rekomendatsiyi z obhruntuvannya efektyvnoyi proektnoyi vrozhaynosti na osushuvanykh zemlyakh pry budivnytstvi y rekonstruktsiyi melioratyvnykh system* [*Temporary recommendations for substantiation of effective project yield on drained lands during construction and reconstruction of reclamation systems*]. NUVGP.

Shevchenko, O., Openko, I., Zhuk, O., Kryvoviaz, Y., & Tykhenko, R. (2017). Economic assessment of land degradation and its impact on the value of land resources in Ukraine [online]. *International Journal of Economic Research (IJER), 14*(15), 93–100. https://serialsjournals.com/abstract/34405_ch_11_f_-_ivan_openko.pdf

Zhukovsky, E. (1981). *Meteorologicheskaya informatsiya i ekonomicheskiye resheniya* [*Meteorological information and economic decisions*]. Gidrometeoizdat.

Chapter 20
Peculiarities of Environmental Risk Management in Nature Management Projects

Nadiia Frolenkova
National University of Water and Environmental Engineering, Ukraine

Nataliia Prykhodko
https://orcid.org/0000-0003-1424-2628
National University of Water and Environmental Engineering, Ukraine

ABSTRACT

The selection and justification of the feasibility of any environmental project, including water projects, are always associated with the analysis of future events and uncertainty of the outcome of the project. Therefore, one of the most important tools of project management is a risk management framework. Risks usually arise from the uncertainty and variability of the economic, environmental, political, social environment, as well as the different operation conditions of a particular object. Weather and climate risk, as a subtype of general environmental risk, reflects the degree of loss of potential yields, and consequently the amount of income, as a result of uncontrolled meteorological factors. Reducing such risk provides a decrease in the impact of adverse natural factors on the efficiency of economic activity, and hence on the degree of technological excellence and environmental reliability of the selected alternative design solutions.

BACKGROUND

The existence of global climate change is now recognized by the world community and is beyond doubt. The problem of global warming arose at the end of the last century. Addressing this problem requires the development and implementation of strategies to adapt to global climate change. The implementation of such measures will increase productivity, efficiency and profitability, especially agricultural

DOI: 10.4018/978-1-6684-8248-3.ch020

production as an important component of nature management (IPCC, 2019; Popescu, Dinu et al., 2019; Popescu et al., 2020).

According to the UN Framework Convention on Climate Change, adaptation means adapting natural, social or economic systems to actual or expected climate change and its consequences (Kyoto protocol to the United Nations framework convention on climate change (1998); United Nations framework convention on climate change (1992).

To clarify and make wider use of the concept of "adaptation to climate change", clear definitions of the following categories were considered and given: adaptation potential; propensity to influence; sustainability; risk; sensitivity; vulnerability (Massey, 2012).

Risk category is a combination of the probability (or frequency) of natural disasters and the scale of the consequences of their impact. Risk is a function of the object's propensity to influence and the perception of the consequences of that influence by the community or system.

According to the international policy documents on this issue in the field of climate diplomacy in Ukraine and the EU as a whole, based on real research by IPPC (International Plant Protection Convention) criteria, on climate change in Ukraine until 2100, pursuant to the Decree of the President of Ukraine "On the decision of the National Security and Defense Council of Ukraine" of September 14, 2020 and "On the National Security Strategy of Ukraine" and others, Ukraine, like other member states, has committed itself to allocating funds for adaptation to climate change and preparation for its implementation (Cabinet of Ministers of Ukraine. (2016); IPCC. (2019).

That is why the issue of disclosing the concept of "risk" and its use in the field of nature management, which is most affected by weather and climatic conditions and their changes, is extremely important (Blanc et al., 2017; Gohar et al., 2016; Mohamed et al., 2001; Rezaei et al., 2016).

Ensuring the sustainable development of modern enterprises in the field of nature management is impossible without the introduction of innovative technologies. Innovative technologies reduce or avoid the negative impact of economic activities on the environment and provide more efficient use of natural resources (Ievsiukov et al., 2014; Openko et al., 2015; Openko et al., 2020; Shevchenko et al., 2021).

The selection and justification of the feasibility of any environmental, including water projects, are always associated with the analysis of future events and uncertainty of the outcome of the project. Therefore, one of the most important tools of project management is a risk management framework. Risks usually arise from the uncertainty and variability of the economic, environmental, political, social environment, as well as the different operation conditions of a particular object.

At the present stage, taking into account the specific features of project implementation in the field of nature management, an environmental risk framework has become extremely popular due to the capability to predict the amount of such risk at the stage of project development in most cases will prevent the transformation of a potential environmental risk into a real one.

The urgency of taking environmental risk into account when implementing investment projects is growing due to global climate change, which requires the adaptation of natural and economic sectors to new conditions (Kovalenko et al., 2019; Kryvoviaz et al., 2020; Martyn et al., 2019; Tykhenko et al., 2021).

Numerous aspects of environmental risk are studied in the works of many scientists. However, the issues of environmental impact assessment of projects in the field of nature management and their consideration when economically evaluating project decisions are still insufficiently studied.

METHODS AND TECHNIQUES

In the field of nature management, climatic or weather conditions dramatically affect the formation of technical, economic and environmental indicators (Frolenkova et al., 2021; Kovalenko et al., 2019). Such indicators are the basis for assessing environmental risk.

That is why the following research methods were used: methods of qualitative risk analysis (to identify the main factors and the corresponding risks that affect environmental projects); methods of quantitative risk assessment – statistical analysis and matrix modeling (to assess the probability of achieving project objectives and the degree of risk impact on project results); retrospective statistical observations (to create databases of meteorological conditions in different climatic zones); system approach (to consider of nature management projects as complex ecological and economic systems), methods of economic and mathematical modeling.

When studying climate change and its impact on various aspects of agricultural production on the drained lands of the Forest-Steppe zone of Ukraine, in order to develop appropriate adaptive measures, a set of forecasting and simulation models was used. Such a set includes: a model of the local climate; model of water regime and water control of drained lands; model of development and crop formation of cultivated crops. Such models are mostly implemented on the basis of long-term forecast (Frolenkova et al., 2021; Kovalenko et al., 2019; Kovalenko et al., 2021). In addition, the main methods of general experimental theory were used: analytical, experimental, statistical, mathematical modeling, machine experiment using modern information and computer technologies.

Such research methods are based on the application of systems theory along with a systems approach, systems analysis and modeling. These methods are focused on the widespread use of computers and related software and information support when developing modern approaches to justify technical and technological solutions in the projects in the field of nature management in a changing climate.

FUTURE RESEARCH DIRECTIONS

The basis for quantitative risk assessment is, first of all, the idea of the concept of "risk". In addition, it is a distinction between different types of risk depending on the content and the field of detection.

Currently, there is no unambiguous interpretation of the term "risk". However, most often in the literature, "risk" is considered as the possibility of deviating from the goal for which the chosen alternative is realized or material, environmental, moral and other losses associated with the implementation of the chosen alternative in conditions of uncertainty.

When systematizing risks as to an economic object, it is customary to distinguish between external and internal risks. By the field of detection, the impact of economic, market, political, industrial, financial, environmental and other risks is more often assessed. However, in the process of planning and justifying projects, first of all it is necessary to determine which specific types of risks are a priority and directly affect the effectiveness of their implementation.

The analysis of the implementation of water management projects allows making conclusions. First, the implementation of water management projects has an impact on the environment, thereby generating anthropogenic environmental risk. Secondly, the results of project activities directly depend on changing and difficult to predict meteorological conditions, which leads to natural environmental risk.

Such features of environmental risk have different nature and content. In our opinion, that necessitates the identification of different types of environmental risk, which must be determined by different quantitative methods.

So, depending on the regularity and impact areas, environmental risks of water management project may include:

1. *Weather and climatic risk.* This type of risk arises under the influence of uncontrolled environmental factors that cause uncertainty and cyclicity of meteorological conditions. Weather and climate risk as a type of environmental risk provides extremely important information about the degree of the maximum possible economic result when using different ways of project implementation and allows choosing the best option.
2. *External environmental risk.* External environmental factors that cause this type of risk include: possible natural disasters, accidents, environmental cataclysms; radiation and chemical pollution of soils, water and air; extreme manifestations of natural phenomena (cyclones, hurricanes, downpours, floods, inundations, droughts, frosts, squalls, hail, etc.). Such risk is probabilistic and should be determined on the basis of statistics, long-term studies and forecasts.
3. *Internal environmental risk.* Irrational actions within the project activities can cause some damage to the environment. The most stringent criterion for assessing environmental risks is the assessment of changes in the state of environment: extinction of plant and animal species, changes in the chemical composition of environmental components, and so on. Various economic or other decisions must be made in a way preventing exceeding the limits of harmful effects on the environment.

Considering the planning of projects in the environmental field, including water management projects, it should be noted that one of the most important stages is the development of alternative design solutions, their comparison and selection of the best option for implementation according to certain criteria.

When analyzing the above types of environmental risk, it can be concluded that the weather and climate risk is decisive at the planning stage of project decisions, as meteorological conditions determine the results of economic activities within the project, based on the following arguments:

- agricultural productivity, including on reclaimed lands, is affected not only by biological capabilities of crop varieties, available agronomic tools and mechanisms, timely and high-quality tillage, fertilizer application, crop rotation, etc., but also the results of interaction of a number of meteorological factors;
- randomly compiled weather conditions of each current year and long-term climatic characteristics, specific to a particular area, cause significant fluctuations in crop yields;
- it is the weather conditions in the relevant or estimated year have a decisive influence on the annual effect of the reclamation project and its deviation from the expected scale. Accordingly, the weather conditions will have a great influence on the final integrated effect for the entire period of operation of the facility.

As a result of this analysis, the economic content of weather and climate risk was made, based on the following considerations:

1. The assessment of such risk is carried out at the design stage according to a long-term forecast, which is based on a certain technological basis. Such calculations enable: firstly, to determine the objective potentially possible yield of each crop on a certain object, which is formed in ideal (optimal) conditions as to climate, solar radiation, soil quality, agricultural technology, water regime, etc. Secondly, depending on the weather conditions of a particular estimated year and the technological capabilities of the reclamation system, the projected technological yield is determined for each design solution. The projected technological yield, as a rule, does not reach the potentially possible yield. The difference between them characterizes weather and climatic risk.
2. In general, reclamation measures are carried out in areas with adverse natural conditions. Thus, weather and climate risk has a predominantly negative impact on the economic effect and in this sense it can be classified as "pure" risks.
3. Therefore, in our study, the weather and climatic risk of the water management project is considered as an absolute or relative deviation of the actual effect of the design solution from its potential value due to the mismatch of real weather conditions - optimal. In this context, weather and climate risk can also be considered as environmental losses from the increased potential income as a result of adverse natural factors. After all, the maximum yield is limited by the optimal ratio of environmental components (Frolenkova et al., 2007; Kryvoviaz et al., 2020; Openko et al., 2019).

The economic aspect of the assessment of water management projects on the example of the drainage system reconstruction project is also the aim of our study. Therefore, it is necessary to assess the impact of weather and climate risk on the economic performance of the object. Conditionally, it can be called a production effect (E_i) in the corresponding estimated year. To apply statistical methods to assess the impact of risk on the outcome of production activities on reclaimed land, it is necessary to know exactly how the effect depends on changes in weather conditions.

General approaches to making and implementing the models of business decisions optimization at different levels using climate data and meteorological forecasts when creating and operating complex meteorological and economic systems are outlined in our studies (Kryvoviaz et al., 2020; Martyn et al., 2019).

In view of natural seasonal cyclicity of agricultural production on reclaimed lands, there are different in terms of heat and moisture supply periods of vegetation, which can be grouped into estimated (typical) groups of years. The distribution of these groups of years within the project life cycle is uneven. This distribution can be done using a share coefficient (probability of detection) of the relevant group of years in the total project implementation period (Frolenkova et al., 2021; Kovalenko et al., 2021).

In addition, the type and design of water management system affects the amount of the expected yield and, consequently, the overall economic effect. The type and design of water management system determine the appropriate technology of water regime control (water regulation) on reclaimed lands.

That is why the results of agricultural production on reclaimed lands during the project period of operation of the water facility depend on three main factors:

1. Meteorological conditions in the relevant typical estimated years, an aggregate of sets $P=\{p_j\}$, $j=\overline{1,m}$.
2. Frequency (probability) or detection share of the relevant group of years over the design period of object operation – $\left\{\alpha_{p_j}\right\}$, $j=\overline{1,m}$.

3. Type and design of reclamation system by the appropriate method, water regulation scheme, an aggregate of sets $S=\{s_i\}$, $i=\overline{1,n}$.

Thus, in the general implicit form, the utility function of the water management project can be presented as:

$$E_{ps}=E\left(P,\alpha_{@}S\right),\ j=\overline{1,m},\ i=\overline{1,n}. \tag{1}$$

For greater clarity and simplicity of calculations, this function can be presented in the form of a payment matrix (Kovalenko et al., 2021; Rokochinskiy et al., 2021).

As we have noted, risk can be considered as an absolute (*R*) or relative (*f*) deviation from the potentially possible effect. Then, in general, the elements of the risk matrix are a deviation from the weighted average potential effect ($\bar{E}_?$)

$$R_{ij}=\left(E_{ij}-\bar{E}_?\right),\ i=\overline{1,n},\ j=\overline{1,m}. \tag{2}$$

In economic statistics, the standard deviation has become the absolute variation measure. Therefore, the absolute amount of risk for each of the alternative design solutions is determined by the formula:

$$\overline{R_i}=\sqrt{\sum_{j=1}^{m}(E_{ij}-\overline{E_?})^2\cdot\alpha_{pj}}=\sqrt{\sum_{j=1}^{m}R_{ij}^2\cdot\alpha_{pj}},\ i=\overline{1,n}. \tag{3}$$

When addressing the issues of increasing the economy resilience to the influence of meteorological factors, it is better to consider not the variance of fluctuations, but some complex criterion that takes into account both the variance of results and their average degree. Therefore, a relative degree of weather and climatic risk, as a risk degree per an unit effect in the form of shares or percentages, can be calculated by the formula:

$$f_i=\frac{\overline{R_i}}{\bar{E}_?}\cdot 100,\ i=\overline{1,n}. \tag{4}$$

Thus, in general, the criteria for selecting the best option for the project, including water management project, taking into account the impact of weather and climatic risk on the final result, can be given as:

a) by the criterion of effect maximization

$$E\left(s_o\right)=\max_{\{i\}}\left(\bar{E}_i-\overline{R_i}\right),\ i=\overline{1,n}, \tag{5}$$

б) by the criterion of cost minimization

$$B(s_o) = \min_{\{i\}}(\bar{B}_i + \bar{R}_i),\ i = \overline{1,n}. \quad (6)$$

As an example, the impact of weather and climate risk on the choice of design solution can be considered on the example of the reconstruction of drainage system located in Kiev region (the Forest-Steppe zone of Ukraine) in an area of 360 hectares.

Predictive calculations in simulation are performed on the following multiple variable conditions:

- by the typical for this zone grown crops of the aggregate of sets $\{k\}$, $k = \overline{1, n_k}$ (n_k=6), and the corresponding share of their sown areas f_k: winter grains (with a share f_k=0.2), sugar beets (f_k=0.1), potato (f_k=0.1), root crops (f_k=0.1), perennial grasses (f_k=0.4), corn for silage (f_k=0.1).
- by the typical (estimated) in terms of heat and moisture supply periods of the growing seasons of the aggregate of sets $\{p\}$, $|p = 1, \bar{n}_p$ (n_p=5);

At the stage of formation of design solutions, the most rational of them were selected, which were technologically possible for implementation in natural-agro-ameliorative conditions of the studied object: option 1 – preventive sluicing (PS); option 2 – sprinkling on the background of preventive sluicing (SPS); option 3 – moisturizing sluicing (MS); option 4 – drainage (D); option 5 – sprinkling on the background of drainage (SD).

Based on simulation modeling when the corresponding set of optimization and forecast-simulation models was used, the main technical and economic indicators of the considered options of design solutions were obtained. At the same time, the influence of meteorological conditions on the options of design solutions in different by heat and moisture supply years was taken into account.

Generalized results for the changes in the relative degree of weather and climatic risk for a particular studied object by individual crops and estimated groups of years by heat and moisture supply on the example of moisturizing sluicing are presented in Figure 1. These results showed that the greatest risk of crop loss is inherent for perennial grasses and winter grains in wet and very wet years. In average and dry years, the risk for these crops is reduced by means of humidifying measures.

The expected degrees of weather and climate risk by the options of design solutions in terms of different technology of water regulation of drained lands in the years of different weather conditions are presented in Figure 2.

It is obvious that the use of more advanced technology of water regulation on drained lands can significantly reduce the degree of weather and climatic risk compared to drainage in adverse dry years (40...55%): for preventive sluicing – on average up to 30%, for moisturizing sluicing – up to 20...25% and for sprinkling on the background of drainage – 5…20%.

This convincingly demonstrates the high overall efficiency of double regulation of water regime of drained lands.

These results of the study confirm that the more technologically advanced the design of a reclamation system is, the lower the weather and climatic risk it expected. Such advanced water regulation technology on drained lands provides higher crop yields and reduce the dependence of agricultural production on the effects of unpredictable weather conditions.

Figure 1. Characteristics of the relative degree of weather and climatic risk when applying moisturizing sluicing for the crops of project crop rotation by the years of different weather conditions in the Forest-Steppe zone of Ukraine
Source: Own research

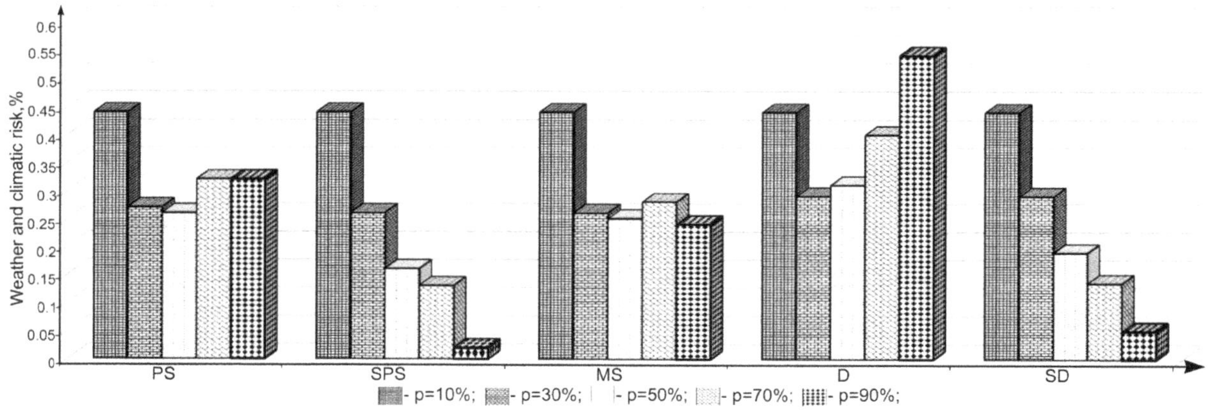

Figure 2. Relative degree of weather and climatic risk by the options of reclamation project in the years of different weather conditions in the Forest-Steppe zone of Ukraine
Source: Own research

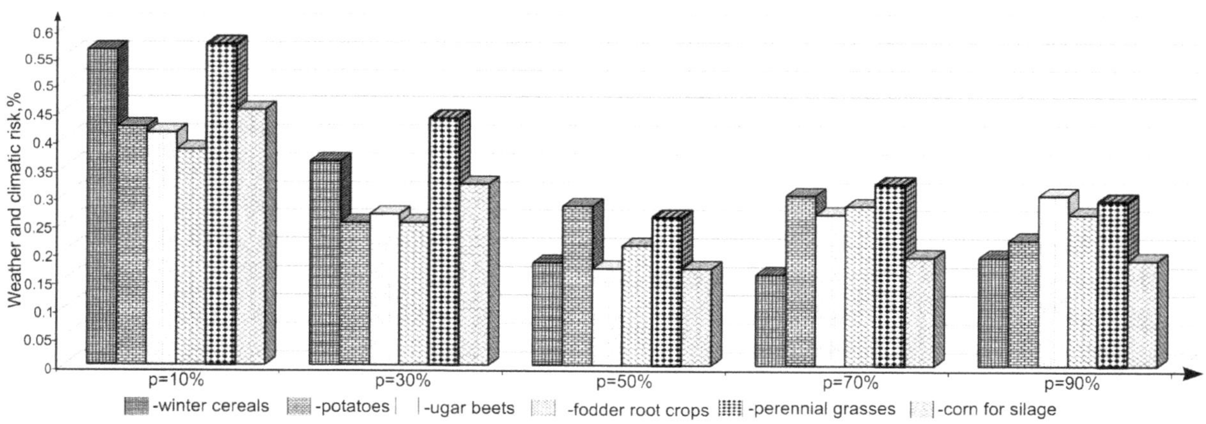

As noted above, the most optimal way to take into account the influence of meteorological factors in economic calculations is to introduce an indicator (absolute or relative) of weather and climate risk in the formula for calculating criteria for comparing alternative design solutions. Traditionally, the criterion for comparing reclamation projects is the minimum given costs (Rokochinskiy et al., 2021).

For example, we calculated the indicators of given costs for design solutions by a traditional approach, which takes into account only the direct production costs, as well as by a modified approach taking into account weather and climatic risk, Table 1.

Table 1. Results of the comparison of the efficiency of water regulation technology by the traditional and modified criteria of cost minimization in the Forest-steppe zone of Ukraine

Nº	By the traditional criterion of given costs		By the criterion of given costs taking into account weather and climatic risk	
	Options of design solutions	**Value**	**Options of design solutions**	**Value**
1.	Drainage	0.867	Moisturizing sluicing	1.321
2.	Preventive sluicing	0.874	Preventive sluicing	1.323
3.	Moisturizing sluicing	0.935	Sprinkling on the background of drainage	1.385
4.	Sprinkling on the background of drainage	1.114	Sprinkling on the background of preventive sluicing	1.408
5.	Sprinkling on the background of preventive sluicing	1.163	Drainage	1.519

Source: Own research

It is obvious that by the traditional approaches, the best option for reclamation project in humid areas is drainage, which is considered to be an obsolete technology for today and provides the least increase in yield. In addition, it is estimated to be characterized by the highest degree of weather and climatic risk that is almost 40%. That is, by the assessment scale it is between medium and high degree. When using the proposed approach, taking into account weather and climatic risk the best options are more technologically advanced options of design solutions that reduce the dependence of yields on meteorological conditions.

Thus, modeling different scenarios of economic activity taking into account weather and climatic risk enables to choose a more advanced technology of water regulation as an optimal model at a particular object in specific natural and agro-ameliorative conditions.

In addition, the forecast assessment of the degree of weather and climate risk can also be effectively used for other purposes – when determining a discount rate in investment calculations; when making decisions on the feasibility of using credit resources to finance projects; when substantiating the degree of agricultural insurance risk, etc.

CONCLUSION

The introduction of scientifically sound and effective methods of using climate information in economic calculations will significantly reduce losses (up to 40%) caused by natural conditions as well as get a greater effect due to the implementation of the optimal strategy for sustainable agricultural production on reclaimed lands.

The introduction of scientifically sound and effective methods of using climate information in economic calculations will significantly reduce losses (up to 40%) caused by natural conditions as well as get a greater effect due to the implementation of the optimal strategy for sustainable agricultural production on reclaimed lands.

Weather and climate risk, as a subtype of general environmental risk, reflects the degree of loss of potential yields, and consequently the amount of income, as a result of uncontrolled meteorological factors. Reducing such risk provides a decrease in the impact of adverse natural factors on the efficiency of economic activity, and hence on the degree of technological excellence and environmental reliability of the selected alternative design solutions.

The use of the proposed author's approaches as to taking into account weather and climatic risk in economic calculations enables:

- to model different scenarios of economic activity and to optimize the number of crops in crop rotation by the years with different climatic conditions to achieve the greatest effect;
- to reduce weather and climatic risk in adverse dry years (on average up to 40...55%): for preventive slicing – up to 30%, for moisturizing sluicing – up to 20...25%, for sprinkling on the background of drainage – up to 5...20% due to introducing the most advanced and relevant to local conditions water regulation technology at a particular object
- planning the measures to reduce the negative impact of weather conditions in the years of the greatest risk.

REFERENCES

Blanc, E., & Schlenker, W. (2017). The use of panel models in assessments of climate impacts on agriculture. *Review of Environmental Economics and Policy*, *11*(2), 258–279. doi:10.1093/reep/rex016

Cabinet of Ministers of Ukraine. (2016). *On approval of the Concept for the implementation of state policy in the field of climate change until 2030*. Cabinet of Ministers of Ukraine. https://www.kmu.gov.ua/ua/npas/249573705

Frolenkova, N., & Rokochinskiy, A. (2007). *Ekoloho-ekonomichne otsinyuvannya v upravlinni melioratyvnymy proektamy*. NUVGP.

Frolenkova, N., Rokochinskiy, A., Volk, P., Shatkovskyi, A., Prykhodko, N., Tykhenko, R., & Openko, I. (2021). Cost-effectiveness of investments in drip irrigation projects in Ukraine. *International Journal of Green Economics*, *14*(4), 315–326. doi:10.1504/IJGE.2020.112570

Gohar, A., & Cashman, A. (2016). A methodology to assess the impact of climate variability and change on water resources, food security and economic welfare. Agricultural Systems, 147, 51–64. doi:. 008 doi:10.1016/j.agsy.2016.05

IPCC. (2019). An IPCC special report on climate change, desertification, land degradation, sustainable land management, food security, and greenhouse gas fluxes in terrestrial ecosystems. In P.R. Shukla, J. Skea, E. Calvo Buendia, V. Masson-Delmotte, H.-O. Pörtner, D. C. Roberts, P. Zhai, R. Slade, S. Connors, R. van Diemen, M. Ferrat, E. Haughey, S. Luz, S. Neogi, M. Pathak, J. Petzold, J. Portugal Pereira, P. Vyas, E. Huntley, K. Kissick, M. Belkacemi, J. Malley, (eds.) *Climate Change and Land*. IPCC. https://www.ipcc.ch/srccl/

Kovalenko, P., Rokochinskiy, A., Jeznach, J., Koptyuk, R., Volk, P., Prykhodko, N., Tykhenko, R. (2019). Evaluation of climate change in Ukrainian part of Polissia region and ways of adaptation to it. *Journal of Water and Land Development, 41* (IV–VI), 77–82. doi:10.2478/jwld-2019-0030

Kovalenko, P., Rokochinskiy, A., Volk, P., Turcheniuk, V., Frolenkova, N., & Tykhenko, R. (2021). Evaluation of ecological and economic efficiency of investment in water management and land reclamation projects. Journal of Water and Land Development, 48 (I–III), 81–87. doi:10.24425/jwld.2021.136149

Kryvoviaz, E., Openko, I., Tykhenko, R., Shevchenko, O., Tykhenko, O., Tsvyakh, O., & Chumachenko, O. (2020). Recovery of losses for inappropriate use of land. *Scientific Papers. Series E. Land Reclamation, Earth Observation & Surveying, Environmental Engineering, 9,* 175–182. http://landreclamationjournal.usamv.ro/pdf/2020/Art2 6.pdf

Kyoto protocol to the united nations framework convention on climate change. (1998). United Nations. https://unfccc.int/resource/docs/convkp/kpeng.pdf

Massey, E. (2012). *Experience of the European Union in adaptation to climate change and its application to Ukraine*. Office of the Coordinator of OSCE Economic and Environmental Activities. https://www.osce.org/ukraine/104019?download=true

Mohamed, S., & McCowan, A. (2001). Modelling project investment decisions under uncertainty using possibility theory. *International Journal of Project Management, 19*(4), 231–241. doi:10.1016/S0263-7863(99)00077-0

Popescu, A., Dinu, T. A., & Stoian, E. (2019). Efficiency of the agricultural land use in the European Union. *Scientific Papers. Series Management, Economic, Engineering in Agriculture and Rural Development, 19*(3), 475–486. http://managementjournal.usamv.ro/pdf/vol.19_3/Art62.pdf

Popescu, A., Dinu, T. A., Stoian, E., & Serban, V. (2020). Variation of the main agricultural crops yield due to drought in Romania and Dobrogea region in the period 2000-2019. *Scientific Papers. Series Management, Economic, Engineering in Agriculture and Rural Development, 20*(4), 397–415. http://managementjournal.usamv.ro/pdf/vol.20_4/volume_20_4_2020.pdf

Rezaei Zaman, M., Morid, S., & Delavar, M. (2016). Evaluating climate adaptation strategies on agricultural production in the Siminehrud catchment and inflow into Lake Urmia, Iran using SWAT within an OECD framework. Agricultural Systems, 147, 98–110. doi:. 2016.06.001 doi:10.1016/j.agsy

Rokochinskiy, A., Frolenkova, N., Turcheniuk, V., Volk, P., Prykhodko, N., Tykhenko, R., & Openko, I. (2021). The variability of natural and climatic conditions in investment projects in the field of nature management. Journal of Water and Land Development, 48 (I–III), 48–54. doi:10.24425/jwld.2021.136145

Shevchenko, O., Openko, I., Tykhenko, R., Tsvyakh, O., Zhuk, O., Kryvoviaz, E., Tykhenko, O., Bavrovska, N., Stepchuk, Ya., Rokochinskiy, A., & Volk, P. (2021). Assessment of economic losses caused by degradation processes of agricultural land use. *Scientific Papers. Series E. Land Reclamation, Earth Observation & Surveying, Environmental Engineering, 10,* 102–109. http://landreclamationjournal.usamv.ro/pdf/2021/Art15.pdf

Tykhenko, R., Tykhenko, O., Openko, I., Shevchenko, O., Bavrovska, N., Zhuk, O., Tsvyakh, O., & Stepchuk, Ya. (2021). The assessment of impact ecological stability of territory on the organization of rational land use of agricultural enterprises. *Scientific Papers. Series Management, Economic, Engineering in Agriculture and Rural Development*, *21*(2), 685–692. http://managementjournal.usamv.ro/pdf/vol.21_2/volume_21_2_2021.pdf

United Nations framework convention on climate change. (1992). United Nations. https://unfccc.int/resource/docs/convkp/conveng.pdf

Chapter 21
Ecological and Economic Efficiency of Investments in Water Management

Nadiia Frolenkova
National University of Water and Environmental Engineering, Ukraine

Leonid Kozhushko
National University of Water and Environmental Engineering, Ukraine

ABSTRACT

The article is devoted to the actual scientific and practical problem of improving methodological and methodical approaches to the evaluation of design solutions in the water management and land reclamation industry based on the ecological and economic principles in conditions of uncertainty. To solve problems, it is necessary to reform organizational-economic relations in the industry, including new sources and forms of financing for water management and land reclamation projects, introduction of new environmentally advanced technologies, and the improvement of the existing ecological and economic evaluation of investments. Based on scientific and methodological recommendations used for evaluating the effectiveness of investment in various spheres of economic activity, the authors developed and implemented an improved methodology for the evaluation of water management and land reclamation projects. The method was tested during the reconstruction of a rice irrigation system in the steppe zone of about 3000 ha in Ukraine.

BACKGROUND

The social and economic changes that have taken place in the country over the last decades are characterized by instability and the lack of a clear, effective economic development strategy in virtually all areas where natural resources are used. In particular, the current water management and land reclamation industry struggles against economic and environmental problems. These problems are primarily caused by improper funding due to the lack of interest on the part of the state and potential private investors.

DOI: 10.4018/978-1-6684-8248-3.ch021

The current stage of the development of the water management sector in Ukraine is characterized by a combination of past negligence and present energy, food and water crises, as well as global climate change. To solve these problems, it is necessary to reform organizational economic relations in the industry, including new sources and forms of financing for water management and land reclamation projects, introduction of new environmentally advanced technologies, and the improvement of the existing mechanism of ecological and economic evaluation of water management projects. This will lead to an increased investment attractiveness of the sector and its development towards greener production. Indeed, the justification of economic feasibility and environmental reliability of such projects is a prerequisite and an incentive for their financing through both public and private investments in modern solutions.

If this problem is not addressed more carefully in other profitable sectors today, ecological and economic evaluation of economic decisions remain unsolved in water management and land reclamation as well. The existing system of economic and environmental evaluation is predominantly static and it is based on methodological approaches used in the past administrative and central planning economy. They were based on the concept of resource utilization cost, which included comparison and selection of economic decisions to minimize the total cost, as well as a static model of calculations without taking into account the life cycle of projects, changes in the cost of resources over time and risk factors. In addition, the existing methodology of economic calculations in the sector does not reflect the real impact of technological (parameters and structures of water management and reclamation systems) and environmental (in particular changing weather and climatic conditions) components on the economic project performance indicators.

METHODS AND TECHNIQUES

In the absence of domestic studies, it is advisable to focus on the foreign experience in and methodological approaches to economically and environmentally optimal business strategies, taking into account modern features of economic relations as well as the financial calculation practice in the water management and reclamation industry in Ukraine.

In the developed countries, project feasibility studies have been based on the methodology developed by United Nations Industrial Development Organization (UNIDO). The methodology was first published in 1978 and became fundamental for making project decisions. The methodology focuses on changes in the value of money over time, evaluation of project effectiveness throughout the its life cycle, investment decisions based mainly on the net present value, and risk evaluation and their impact on the project performance.

The methodology is subject to studies, improvements and adaptation to various industries. Studies have been presented in research papers, including Teichroew et al. (1965), Kwak et al. (2000), Sudong et al. (2000), Mohamed et al. (2001), Nowak (2005), Bierman et al. (2006), Chen (2006), Haka (2006), Armeanu et al. (2009), and Wang et al. (2019).

Currently, the authors focus on the evaluation based on investment efficiency indicators, their comparative analysis in different sectors, evaluation models of alternative investment under uncertainty, and determining of investment analysis criteria, etc.

When analyzing methodological approaches to the evaluation of water management and land reclamation projects regarding their ecological and economic effectiveness, their specific features and differences from investment projects in other sectors should be taken into account. This determines the choice of their evaluation method and calculation of economic indicators. Therefore, we have identified the following factors that are decisive for the improvement of traditional approaches to the evaluation of economic decisions in environmental management:

- dependence of ecological and economic effect and weather and climatic conditions, which is extremely relevant in the context of present and future changes (Rokochinskiy et al. (2017);
- deep relationship between ecological and economic effect and weather and climatic conditions (Martyn et al. (2020); Openko et al. (2020);
- use of natural resources (land, water, etc.) in addition to materials and labor (Martyn et al. (2019); Shevchenko et al. (2017);
- application of specific technology: water management technology on reclaimed land (engineering infrastructure, main reclamation funds, technological processes) and agrotechnics (system of basic agricultural funds, planting, fertilizing methods, etc.).

While taking into account these features in present conditions, land reclamation projects are proposed to be considered not as purely technical but as complex environmental, technical and economic systems.

It is suggested to consider the water management and land reclamation project as an ecological and economic system designed to deliver rational environmental management and optimal and efficient agricultural production on the reclaimed land. This can be achieved by minimizing the impact of adverse environmental conditions on the ecological and economic effects obtained over a certain period.

In the world's practice of economic calculations there are two main approaches to the evaluation of economic investment efficiency of any project, including those in the water sector. According to these approaches, methods of economic investment evaluation can be divided into two groups:

- simple (static) methods;
- discounting methods (dynamic).

The first group became popular in the domestic economic practice. However, they ignore the project life cycle, volatility of cash flows, and risk factor impact on project outcomes. Despite the use of progressive methods of investment analysis in other spheres, domestic design organizations, which develop and implement water-management and land reclamation projects, still use archaic and inefficient methodology for project documentation and feasibility studies. The approach used in substantiating the economic efficiency of water management projects does not take into account changes in the value of money over time, impact of changing climatic conditions on economic parameters, and environmental risks.

The second group includes analytical methods of investment project evaluation that encompass the notion of "time series" and require the application of a special mathematical apparatus and a more thorough preparation of background information. Indicators used are adequate to the main purpose of production under market conditions, and they address revenue and expenditure of a project throughout its implementation period, take into account change in the value of money over time, enable to evaluate risk of a project decision, and provide guidance while selecting a project which best facilitates the growth of the enterprise value.

Usually, indicators used to make investment decisions are the net profit value (NPV), profitability index (PI), discounted payback period (DPP) and internal rate of return (IRR), as well as their modifications. However, conventional methods need to be adapted to domestic economic conditions, especially to the calculation of economic indicators in different sectors.

Generalizing commonly accepted approaches and requirements for the evaluation of investment efficiency of technical projects and in view of characteristics of the land reclamation sector, methodological approaches to the evaluation of water management and land reclamation projects should be based on the following principles:

1. Various options (best design solution chosen from possible alternatives).
2. Change in the value of money over time (water management and land reclamation project dynamics and comprehensive study of effectiveness throughout the life cycle).
3. Specific features of the project implementation environment expressed in the impact of:
   - environmental factors,
   - weather and climate.
4. Multilevel and gradual evaluation of the project (selection of evaluation criteria and indicators according to project cycle stages).
5. Determination of decision making criteria for:
   - projects of national importance which focus less on the economic and more on the environmental and social effectiveness;
   - commercial projects designed to generate desired profits which compensate the uncertainty of their final result.
6. The use of indicators that meet:
   - main goals and objectives of the project,
   - interests of the main parties to the investment process, and
   - project implementation environment.
7. Full return on investment from proceeds generated by the project within the timeframe acceptable to the investor.
8. Where appropriate, qualitative indicators and expert opinions that enable to take into reflect intangible effects (social, environmental).

We have developed these principles based on concept and methodological approaches to evaluate alternative options for water management and land reclamation projects.

Such investment options were tested by the authors when evaluating alternative technical solutions for different climatic zones, in particular, for the irrigated area in the Odessa region in Ukraine (latitude 47°0’0” N; longitude 30°0’0” E).

FUTURE RESEARCH DIRECTIONS

The research helped to develop theoretical and methodological approaches and practical recommendations for the improvement of ecological and economic evaluation in water management and land reclamation projects, set forth by Rokochinskiy et al. (2018).

The main components of the proposed methodology are as follows:

1) in general terms, it is two-tier system for evaluating alternative projects – **preliminary simplified** and **final investment** evaluation based on various criteria and methodologies to enhance decision-making;
2) weather and climatic conditions taken into account in economic calculations, since they have a decisive influence on the formation of economic and environmental effects of water management and land reclamation facilities;
3) quantifies weather and climate risk as a determining factor for comparing alternatives to design decisions (Frolenkova et al. (2015; 2020); Rokochinskiy et al. (2020);
4) quantified ecological effects of water management and land reclamation facilities, which is necessary to compare design solutions.

Thus, a preliminary environmental and economic assessment of alternative project options is advisable at the stage of a concept design before precise information on basic technical and technological indicators and project parameters are defined. Instead of a detailed calculation of revenue and expenditure of each alternative project option, the primary objective to convert them into comparable forms according to selected efficiency criteria to facilitate a final decision in the future.

With that in mind, it is advisable to use simple and generally accepted methods of project effectiveness assessment that are suitable when a preliminary decision is necessary and there is a lack of relevant information. The first criterion is minimal discounted cost adjusted with regard to the level of weather and climate risk.

Thus the general model of preliminary environmental and economic assessment of project options $\{i\}$,under minimal discounted costs *ZP* with regard to **climatologically optimal strategy** (Frolenkova et al. (2007); Kovalenko et al. (2019); Mazhayskiy et al. (ed.) (2017); Rokochinskiy et al. (2017) is as follows:

$$ZP_0 = \min_{\{i\}} \left[\frac{\sum_{j=1}^{m} \left(C_{ij}^{a} + C_{ij}^{o} + A_i + E \cdot K_i + \overline{R_{i_j}} \right) \cdot \alpha_{p_j}}{\sum_{j=1}^{m} V_{ij} \cdot \alpha_{p_j}} \right], \; i = \overline{1,n}. \quad (1)$$

Subject to certain restrictions, the environmental credibility coefficient under i-option of a land reclamation project, which characterizes its environmental acceptability, falls within the following range:

$$0.5 < k_i \leq 1.0, \quad (2)$$

Where: $C_{i_j}^{a}$ – agricultural inputs in growing crops for the *i*-th version of project decisions, totality $\{i\}$, $i = \overline{1,n_i}$, on the totality of estimated years $\{j\}$, $j = \overline{1,n_j}$ within the project time of the facility, USD / ha;

C_j^{o} - respective value of running operational costs of production under the project decision options, USD /ha;

A_i- depreciation costs under the project decision options, USD /ha;

α_{p_j} – coefficient, considering the impact of changeable weather and climate conditions by the set of design years $\{j\}$, $j = \overline{1, n_j}$ within the project period of facilities functioning $\sum_{j=1}^{m} \alpha_{p_j} = 1$

E – normative coefficient of the economic efficiency of capital investment K_i (USD/ha) under the respective project decision options;

V_{V_j} – respective output value under *i*-project option, USD/ha;

R_{i_j} – respective value of weather and climate risk, USD/ ha;

k_i – environmental credibility coefficient under i-option of water and land reclamation project, which is calculated according to the following formula:

$$k_i \sum_{j=1}^{m} k_{ij} \cdot \alpha_{p_j} \tag{3}$$

Environmental reliability can be calculated based on the expert evaluation of physical technological and environmental performance indicators for facilities representative of the ecological and ameliorative status of the territory.

An important component of the proposed approaches is weather and climate risk taken into account in calculations. The risk factor reflects deviation of the actual yield from its potential value in favorable meteorological conditions. Such risk can be identified using statistical analysis of a long-term research data set (Frolenkova et al. (2015).

In the end, weather and climatic conditions as well as the determination of all necessary indicators of technological, economic and ecological efficiency of a particular water management and land reclamation facility under variable natural and ameliorative conditions have been taken into account in proposed complex of interrelated forecast and simulation models. They are implemented on the basis of a computer-aided machine experiment. It allows to develop a long-term forecast and evaluate climatic conditions and meteorological regimes, including water regime, water regulation technologies and productivity of reclaimed lands (Rokochinskiy et al. (2017). Their practical application is regulated by relevant sector standards set by the Ukrainian State Agency of Water Resources (Ukr. Derzhavne ahentstvo vodnykh resursiv Ukrayiny).

Therefore, according to Rokochinskiy et al. (2018), based on the results of the preliminary simplified evaluation at the stage of a concept design, one or more options are selected from the considered set (economically optimal and environmentally acceptable). Then technical designs are developed to determine final values of all project resource indicators and evaluate their investment attractiveness at the final stage.

The final evaluation of selected pre-optimal design solutions is performed according to the UNIDO recommendations based on the indicators listed above and taking into account the impact of meteorological conditions defined according to the following model:

$$\mathrm{PI}_0 = \max_{\{i\}} \sum_{j=1}^{m} \mathrm{PI}_{V_j} \cdot \alpha_{p_j}, \ i = \overline{1, n}, \tag{4}$$

Provided that

$$\begin{cases} \text{NPV}_V \geq 0; \\ \text{IRR}_V \geq d_i; \\ \text{DPP}_V \geq '' . \end{cases} \tag{5}$$

Where: d_i – discount rate under project option; and T – discounted payback period acceptable for the investor.

Each indicator in the general model is calculated while taking into consideration volatile weather and climate conditions and their impact on basic economic parameters.

For instance,

$$\text{PI}_i = \left[\sum\nolimits_{t=0}^{T} \frac{E_{ai}}{(1+d_{i_t})^t} \Big/ \sum\nolimits_{t=0}^{T} \frac{I_{i_t}}{(1+d_{i_t})^t} \right] + 1,\ i = \overline{1,n}; \tag{6}$$

Where: E_{ai} – annual economic benefit under each option of water management and land reclamation project, which is calculated according to the following formula:

$$E_{ai} = \sum\nolimits_{j=1}^{m} \Delta NI_{ij} \cdot \alpha_{p_j} + \Delta A_i - I_i,\ i = \overline{1,n}, \tag{7}$$

Where: I – annual investment, USD/ha; ΔNI – increase in annual anticipated net income as a result of investment project, USD/ha; and ΔA – increase in annual depreciation related to the introduction of new fixed capital, USD/ha.

Other components of the model are calculated in a similar manner.

The proposed methodological approaches were tested while developing optimal parameters for technological and construction decisions under water management and land reclamation projects in different climatic zones (Mazhayskiy et al. (ed.) (2017); Openko et al. (2017); Rokochinskiy et al. (2019); Shevchenko et al. (2017).

As an example, let us consider the evaluation of economic and environmental efficiency of the reconstruction of the Danubian rice irrigation systems in Odessa, Ukraine (about 3000 thous. ha). The goal is to enhance general efficiency of operation based on engineering and land reclamation measures.

According to the analysis and consolidation of research and production data pertaining to the operation of the Danubian rice irrigation system in 1966–2016, a complex regime of technological and technical measures was developed to examine the need and feasibility of the following:

- transfer from the traditional resource-intensive to rational and resource-efficient use of water and energy;
- reuse of drainage and wastewater;
- application of deep loosening using appropriate low energy consumption aggregates;
- periodic leaching of saline soils together with their deep loosening;
- additional equipment added to the existing open drainage network, including subsoil drains;

- construction of additional drainage in addition to existing drainage canals to protect against deformation of the canal bed.

The above-stated measures are IP protected and designed to enhance the overall efficiency of rice irrigation systems in line with current economic and environmental requirements. Parameter substantiation methods were improved.

Thus, the following alternative options have been considered:

- option 1 – is a benchmark for operating conditions in rice irrigation systems during designing, which provides for a rice share in a crop rotation within 70–90% including companion crops (perennial grasses, other grain crops) in largely unfavorable water and salt regimes of irrigated lands; regimes result from progressing waterlogging due to excessive irrigation (up to 25 thous. m3 $\cdot ha^{-1}$) and insufficient drainage conditions;
- option 2 – shows operating conditions in rice irrigation systems when introducing measures aimed at enhancing drainage conditions of rice bays by providing additional equipment in the open drainage network with subsoil collecting drains that maintain the optimal vertical filtration in a rice bay at 8–10 mm·day–1 and provide the optimal rice share of 50–60% in a crop rotation. That shows improved regime and technological aspects of water regulation in rice irrigation systems;
- option 3 – measures under option 2, plus deep loosening of top low permeable soil layers of 0.6 m;
- option 4 – measures under option 2, plus reused drainage water diluted 1:1;
- option 5 – measures under option 2, plus reused drainage water diluted 1:2;
- option 6 – measures under option 3, plus reused drainage water diluted 1:1;
- option 7 – measures under option 3, plus reused drainage water diluted 1:2;
- option 8 – operation of rice irrigation systems in a design mode (option 1) plus deep loosening of soil to the depth of 0.6 m.

Source data for the determining of project decision effectiveness in attracting investment is given in Table 1.

Table 1. Source data for calculating investment decisions under the options of project decisions

Indicators	Value depended on project option							
	1	2	3	4	5	6	7	8
Annual net income	296.9	1285.6	1705.0	1371.8	1367.7	1371.8	1364.7	454.8
Depreciation costs	114.6	127.1	127.1	127.1	127.1	127.1	127.1	114.6
Investments in PD	2291.7	2542.6	2542.6	2542.6	2542.6	2542.6	2542.6	2291.7

1) Measurement unit for all indicators is $USD\cdot ha^{-1}$.
Explanation: options are described above.
Source: own study.

Table 2 contains comparative characteristics and rationale for an optimal project decision based on consolidated economic and environmental effectiveness indices under respective methods.

Table 2. Rationale for the optimal project decision based on the aggregate indicators of economic and environmental efficiency

Indicators	Value depended on project option							
	1	2	3	4	5	6	7	8
Economic efficiency, ZP_i	2.88	0.74	0.63	0.78	0.79	0.75	0.76	2.06
Ecological efficiency and reliability index, k_i	0.41	0.87	0.63	0.75	0.78	0.81	0.83	0.53

Explanation as in the Table 1.
Source: own study.

The data (see Tab. 2) prove that the implementation of the complex of agrotechnical, engineering and land reclamation measures is economically feasible and environmentally acceptable under the existing operating conditions in the Danubian rice irrigation systems.

As regards environmental and economic terms, options 2–7 are acceptable, of which the best is option 3 with the environmental credibility criterion of 0.63.

Generally, results of the alternative project investment valuation are given in Table 3. The results have been used to determine the best possible option.

The calculations (see Tab. 3) prove both general economic and relatively high commercial effectiveness of options 3, 4, and 6.

Table 3. The results of investment valuation of the project decisions to enhance the efficiency of the Danubian rice irrigation systems

Indicators	Value dependent on project option							
	1	2	3	4	5	6	7	8
Investment return index	0.50	1.91	2.47	1,62	2.11	1.30	1.29	1.08
Net current value since the beginning of the project (USD•ha^{-1})	918.0	3160.0	3735.0	3229.0	3194.0	3439.0	3380.0	648.0
Internal rate of return (%)	2.05	3.28	8.19	5.02	9.37	6.66	6.62	5.16
Deterministic payback period (years)	8.65	3.47	2.75	3.28	3.55	3.16	3.19	7.71

Source: own study.

CONCLUSION

1. Unlike traditional ones, the discussed and proposed approaches to assessing water management and land reclamation projects focus on project options with balanced costs and losses due to negative impact of environmental factors. These options support improved technology of water regulation.
2. Since the efficacy of the improved technology for soil water and air regime regulation less depends on unfavorable environmental conditions, the technology provides more stable income. Besides, such technology secures environmental sustainability and rational use of natural resources.

3. Considering volatile and uncertain climatic conditions, the mechanism of a two-tier evaluation of alternative projects while using different methodological approaches enables to enhance economic and environmental feasibility of technical and technological solutions. Consequently, this encourages investment in the industry.
4. Option 3 provides the highest investment return index of 2.47 and the total project cost of 3735 USD·ha^{-1}. The deterministic payback period under this option is 2.75 years, which is quite promising and ensures a rapid return on investment.

REFERENCES

Armeanu, D., & Lache, L. (2009). The NPV criterion for valuing investments under uncertainty. economic computation and economic cybernetics studies and research. Academy of Economic Studies, 4(4), 133–143.

Bierman, H., Jr., & Smidt, S. (2006). The capital budgeting decision: economic analysis of investment projects. 9th ed.: New York. Routledge.

Chen, J. (2006). An analytical theory of project investment: A comparison with real option theory. *International Journal of Managerial Finance.*, *2*(4), 354–363. doi:10.1108/17439130610705535

Frolenkova, N., Kozhushko, L., & Rokochinskiy, A. (2007). *Ekoloho-ekonomichne otsinyuvannya v upravlinni melioratyvnymy proektamy: Monografiya* [*Ecological and economic assessment in the management of reclamation projects: Monograph*]. NUVGP.

Frolenkova, N., & Rokochinskiy, A. (2015). The evaluation of environmental risks in the sphere of water and land reclamation [online]. *Oxford Journal of Scientific Research*, *1*(9), 155–160. https://core.ac.uk/download/pdf/33693269.pdf#page=155

Frolenkova, N., Rokochinskiy, A., Volk, P., Shatkovskyi, A., Prykhodko, N., Tykhenko, R., & Openko, I. (2020). Cost-effectiveness of investments in drip irrigation projects in Ukraine. [IJGE]. *International Journal of Green Economics*, *14*(4), 315–326. doi:10.1504/IJGE.2020.112570

Haka, S. F. (2006). A review of the literature on capital budgeting and investment appraisal: Past, present, and future musings. *Handbooks of Management Accounting Research*, *2*, 697–728. doi:10.1016/S1751-3243(06)02010-4

Kovalenko, P., Rokochinskiy, A., Jeznach, J., Koptyuk, R., Volk, P., Prykhodko, N., Tykhenk, R. (2019). Evaluation of climate change in Ukrainian part of Polissia region and ways of adaptation to it. *Journal of Water and Land Development*, *41* (IV–VI), 77–82. . doi:10.2478/jwld-2019-0030

Kwak, Y. H., & William, I. C. (2000). Calculating project management's return on investment. *Project Management Journal*, *31*(2), 38–47. doi:10.1177/875697280003100205

Martyn, A., Openko, I., Ievsiukov, T., Shevchenko, O., & Ripenk, O. A. (2019). Accuracy of geodetic surveys in cadastral registration of real estate: value of land as determining factor. *Proceedings of the 18th International Scientific Conference on Engineering for Rural Development*, Jelgava, Latvia. 10.22616/ERDev2019.18.N236

Martyn, A., Shevchenko, O., Tykhenko, R., Openko, I., Zhuk, O., & Krasnolutsky, O. (2020). Indirect corporate agricultural land use in Ukraine: Distribution, causes, consequences. *International Journal of Business and Globalisation*, *25*(3), 378–395. doi:10.1504/IJBG.2020.109029

Mazhayskiy, Y., Rokochinskiy, A., Volchek, A., Meshyk, O., & Jeznach, J. (Eds.). (2017). Pryrodoobustroistvo Polesia [Environmental management of Polissya]. Ryazan. VNIIGiM of A. Kostiakov.

Mohamed, S., & Mccowan, A. K. (2001). Modelling project investment decisions under uncertainty using possibility theory. *International Journal of Project Management*, *19*(4), 231–241. doi:10.1016/S0263-7863(99)00077-0

Nowak, M. (2005). Investment projects evaluation by simulation and multiple criteria decision aiding procedure. *Journal of Civil Engineering and Management*, *11*(3), 193–202. doi:10.1080/13923730.2005.9636350

Openko, I., Kostyuchenko, Y., Tykhenk, O. R., Shevchenko, O., Tsvyak, H. O., Ievsiuko, V. T., & Deineha, M. (2020). Mathematical modelling of postindustrial land use value in the big cities in Ukraine. *International Journal of Mathematical. Engineering and Management Sciences.*, *5*(2), 260–271. doi:10.33889/IJMEMS.2020.5.2.021

Openko, I., Shevchenko, O., Zhuk, O., Kryvovia, Z. Y., & Tykhenko, R. (2017). Geoinformation modelling of forest shelterbelts effect on pecuniary valuation of adjacent farmlands. [IJGE]. *International Journal of Green Economics*, *11*(2), 139–153. doi:10.1504/IJGE.2017.089015

Rokochinskiy, A. (2010). *Naukovi ta praktichni aspekti optimizacii̇ vodoregulyuvannya osushuvanikh zemel' na ekologoekonomichnikh zasadakh: Monografiya* [*The scientific and practical aspects optimization of water regulation drained lands on environmental and economic grounds. Monograph*]. NUVGP.

Rokochinskiy, A., Bilokon, W., Frolenkova, N., Prykhodko, N., Volk, P., Tykhenko, R., & Openko, I. (2020). Implementation of modern approaches to evaluating the effectiveness of innovation for water treatment in irrigation. Journal of Water and Land Development, 45 (IV–VI), 119–125. doi:10.24425/jwld.2020.133053

Rokochinskiy, A., Jeznach, J., Volk, P., Turcheniuk, V., Frolenkova, N., & Koptiuk, R. (2018). Reclamation projects development improvement technology considering optimization of drained lands water regulation based on BIM. Scientific Review – Engineering and Environmental Sciences, 28(3), 432–443. doi:. 28.3.40. doi:10.22630/PNIKS.2019

Rokochinskiy, A., Volk, P., Pinchuk, O., Mendus, S., & Koptyuk, R. (2017). Comparative evaluation of various approaches to the foundation of parameters of agricultural drainage. *Journal of Water and Land Development*, *34*(1), 215–220. doi:10.1515/jwld-2017-0056

Rokochinskiy A., Volk P., Pinchuk O., Turcheniuk V., Frolenkova N., Gerasimov IE. (2019). Forecasted estimation of the efficiency of agricultural drainage on drained lands. *Journal of Water and Land Development*, *40* (I–III), 149–153. . doi:10.2478/jwld-2019-0016

Shevchenko, O., Openko, I., Zhuk, O., Kryvoviaz, Y., & Tykhenko, R. (2017). Economic assessment of land degradation and its impact on the value of land resources in Ukraine [online]. *International Journal of Economic Research (IJER), 14*, 93–100. https://serialsjournals.com/abstract/34405_ch_11_f_-_ivan_openko.pdf

Sudong, Y. E., & Tiong, R. L. K. (2000). NPV-at-Risk method in infrastructure project investment evaluation. *Journal of Construction Engineering and Management*, *126*(3), 227–233. doi:10.1061/(ASCE)0733-9364(2000)126:3(227)

Teichroew, D., Robichek, A., & Montalbano, M. (1965). An analysis of criteria for investment and financing decisions under certaint. *Management Science*, *12*(3), 151–179. doi:10.1287/mnsc.12.3.151

Wang, L., Xu, N., Song, Y., Wang, Y., & Song, S. (2019). Research on investment decision of substation project based on life cycle cost. *IOP Conference Series. Earth and Environmental Science*, *242*(2), 022016. doi:10.1088/1755-1315/242/2/022016

Chapter 22
Estimating the Cost of Drained Lands by Using Them in Variable Conditions

Nadiia Frolenkova
National University of Water and Environmental Engineering, Ukraine

Ruslan Tykhenko
National University of Life and Environmental Sciences of Ukraine, Ukraine

Larysa Semenko
National University of Life and Environmental Sciences of Ukraine, Ukraine

Olha Tykhenko
National University of Life and Environmental Sciences of Ukraine, Ukraine

ABSTRACT

The article is devoted to a topical scientific problem in modern conditions – valuation of land in Ukraine. A methodology for determining the market value of reclaimed land based on a differentiated assessment of its productivity through crop yields is proposed, taking into account natural and climatic zones and other conditions of a particular region. The basis of the methodology is the application of long-term forecast and a set of forecast and simulation models, in particular the model of area climatic conditions and the model of water regime and water regulation technologies on reclaimed land. The proposed approaches were tested by the method of large – scale machine experiment using a land plot in the zone of Western Polesie of Ukraine as the example. The obtained results indicate that there is a differentiation in land value, which is a proportional derivative of the yield of cultivated crops depending on the conditions of their cultivation.

DOI: 10.4018/978-1-6684-8248-3.ch022

BACKGROUND

The introduction of the land market in Ukraine determines the relevance of determining the value of the land fund, taking into account all the most important aspects of its use. The imperfection of the existing approaches to the land value causes an active discussion among politicians and scientists, as well as among the owners of land shares, farmers, agricultural holdings, etc. (Martyn et al., 2020).

The value of agricultural land is currently influenced by both objective and subjective factors. Among the objective factors, it is worth noting one of the main indicators that forms a rental income, namely crop yields.

Subjective factors include the desire of landowners to exercise their full rights, namely to alienate their land. This will be possible after the lifting of the moratorium on the sale of agricultural land.

METHODS AND TECHNIQUES

The analysis conducted by various researchers confirms that the forecast price for Ukrainian chernozem is the lowest in Europe (Barvinsky, Tykhenko, 2015; Martyn, Tykhenko, 2015; Novakovsky, 2015).

Thus, according to the calculations of the Ministry of Economy, a hectare of land in Ukraine after the opening of land market can cost about 2,2 thousand. USD according to the common model (sale to residents and non-residents) and 1,5 thousand. USD according to the limited model (sale to residents only). The value of agricultural land in Ukraine is influenced by yields, infrastructure and the owner's desire to sell shares quickly. As the area of the plot increases, the price per hectare almost does not decrease.

For comparison, the estimated price per hectare of land in other countries is: USA – USD7487, Canada – USD5667, the Netherlands – USD81,836, Luxembourg – USD42,700, Italy – USD40,200, Great Britain – USD28,140, Ireland – USD23,884, Denmark – USD20,794, Slovenia – USD20,000. You could buy a hectare of land at USD10,000–15,000 in Greece, Spain, Poland, Sweden, Finland two years ago. In Latvia, Croatia and Slovakia it is substantially cheaper – USD3600. Only the cost of a hectare in Romania (USD2500) is close to the projected price of land in Ukraine (slovoidilo.ua, 2019).

Thus, Ukraine is currently in the transition from regulatory to mass land valuation and the use of the EU INSPIRE Directive on the use of market value for land valuation and taxation in view of Ukrainian legislation will provide a full basis for land taxation.

Agrarian reform has become an integral part of general economic transformations in the country and makes radical changes in relations between producers and the state, organizational and legal forms of management and types of ownership (Martyn et al., 2019, 2020; Tykhenko, 2010b). From the moment when land use in Ukraine became paid, the most important economic function of land – its value as an object of trade, investment and taxation became important (Zakon, 2004).

Due to the development of market relations, the results of land valuation in recent years do not quite correspond to the current state of the economy of agricultural production (Openko et al., 2020). Until recently, the current methods of agricultural land valuation did not take into account the specifics of its direct use (Panas, Malanchuk, 2013; Seryy, 1986; Tykhenko, 2016).

In modern conditions, economic evaluation of agricultural land should be carried out primarily on the basis of differentiated assessment of its productivity through crop yields, taking into account the specifics of climatic zones and other conditions of a particular region (Openko et al., 2017; Shevchenko et al., 2017). At the same time, the current and future changes in weather and climatic conditions determine the increase of the use of agricultural land in almost all regions of Ukraine both in drained and irrigated areas.

In addition, in modern conditions it is necessary to adhere to the major world trends in land valuation, among which the main one is the transition from the regulatory to the market valuation as a basis for land taxation. This will enable to develop a more universal method of land valuation. In the United Kingdom, the process of land valuation is divided into two stages: the first stage is agro-climatic evaluation, and the second one – economic evaluation of the previously identified types of land, which is based on standard net output (Tykhenko, 2010).

In (Adamuscin et al., 2017) analyzed the methodological basis of land ownership valuation in Poland and Slovakia. They note that in the Polish methodology of property valuation a comparative method is the most used while in the Slovak methodology the method of positional differentiation is used.

According to the Standard on Mass Appraisal of Real Property most regions value agricultural land applying a user value. That makes it inappropriate to use a comparative methodological approach for agricultural land, the market value of which exceeds a user value (IAAO, 2017).

In Latvia valuating a monetary (cadastral) valuation of land is carried applying two main models:

- valuation of built-up land;
- valuation of agricultural land.

Cadastral valuation in the Baltic States is based on uniform international valuation principles, which involve the information about the real estate market and the objects registered in the National Information System of Real Estate Cadastre (Barvika, 2020).

FUTURE RESEARCH DIRECTIONS

When concluding civil law agreements on land plots and rights to them, an expert monetary valuation of land plots is used. The current legislation identifies three methodological approaches to land valuation (profit, cost, comparative) which, in turn, can be implemented through the use of certain methods.

The methodical approach of comparing the sale prices of similar land plots (comparative approach), which is based on the principle of substitution, is implemented in practice using the methods of pair-wise comparison and statistical analysis of the market.

According to the comparative (market) approach, the value of a land plot is calculated in view of the comparison results with analogues. Despite the advantages of the comparative approach (the possibility of determining the real market value of the property and ease of use), this method should be used only in developed market conditions. So until now, the agricultural land remains out of a full market circulation. Therefore, there is no base of the main elements of comparison, including the characteristics of such property by its location, physical and functional characteristics, sale conditions and so on.

The cost approach is used only to determine the cost of land improvements and has no independent significance for land valuation.

The profit approach is used in determining the value of plots, and shows the income from their use in future over a certain period. Such methods allow obtaining a valuation of land, based on the expected by the potential buyer income, and are applied only to land plots that bring profit.

The direct capitalization method is based on the assumption of constant and unchanged cash flow from land use (Tykhenko (2010).

Land value is defined as the ratio of net operating or rental income and capitalization rate (KMU, 2018; Ministerstvo APPU, 2017):

$$V_{dc} = \frac{I_o}{R_c} \tag{1}$$

where V_{dc} – land value, determined by direct capitalization (in monetary units); I_o – net operating or rental income (in monetary units); R_c – capitalization rate (in the form of a decimal fraction). Net operating or rental income is the difference between the projected annual income (I_p) and current expenses (C) for the production of outputs, which consist of agricultural and reclamation costs. In turn, the projected annual income by the types of crops depends on the yields of agricultural products and average prices for their sale.

Then:

$$V_{dc} = \frac{I_p - C}{R_c} \tag{2}$$

That is, the distribution of annual rental income having the appropriate capitalization rate converts the amount of expected income to the land value.

To determine the rental income from the land plots used in agricultural production, the methodology proposes to take into account the typical set of crops for the area, which ensures efficient use of the plot, crop rotation and preservation of soil fertility. To determine the normal (typical) yield, the data of long-term observations on the actual soil yields within the valuated land plot or field research data on crop yields on different soils can be used. In practice, farms usually do not have data of long-term studies on yields in the years with different climatic conditions and are not able to predict the expected yields.

The existing yield standardization system used in the projects of construction, reconstruction and operation of reclamation systems, does not fully depict the state and does not take into account the characteristics of reclaimed land in specific conditions of their operation (climate, soils, crop yield, water regulation regimes and technologies etc.). Its standard values cannot be effectively used in the proposed approaches. In this regard, there is an urgent need to create appropriate scientific and methodological approaches that will allow taking into account all these multiple variables, which influence on the project development.

The most important feature and difference of agricultural production from other branches of social production is that it is influenced by natural factors, and the productivity of agriculture, including on reclaimed land. Land profitability is formed not only due to the biological capabilities of crop varieties, the existing set of machines and mechanisms, timely and high-quality tillage, fertilizers, crop rotations, etc., but also due to meteorological factors. Natural phenomena are uncertain and variable and difficult to predict. That is why agricultural production is very sensitive to weather conditions and changes in hydrometeorological factors. The formation of yields of cultivated crops occurs under cyclicality and variability of weather conditions both for multi-year and vegetation periods.

Weather conditions of each year and multi-year climatic characteristics, specific to a particular area, cause significant fluctuations in yields, which depending on weather (climatic) conditions can change 6…9 times.

We have developed theoretical and practical approaches and models for forecasting on a long-term basis the changes in weather and climatic conditions, soil water regime, water regulation technologies, land productivity and, accordingly, crop yields on reclaimed land (Gadzalo et al., 2017; Mazhayskiy at al., 2017). These models are implemented in the process of optimization of technological (water regulation methods) and relevant design solutions for their provision (type, design, parameters of the reclamation system) based on a set of forecasting and optimization models using modern BIM-technologies for construction and reconstruction projects for such facilities (Kovalenko at al., 2019; Kovalenko at al., 2021; Rokochinskiy at al., 2019; Rokochinskiy at al., 2021; Frolenkova at al., 2020). Their practical application is regulated by the relevant sectoral standards of the State Agency of Water Resources of Ukraine.

This enables to move from using the methods of yield standardization to determining the actual possible income depending on the type, design of the reclamation system, as well as the changes in weather conditions. Based on that, the value of land, which is based not on unreasonable regulations, but on more objective forecasts can be calculated (Pavlov et al., 2006).

The provision of a particular crop with external factors is determined primarily by the natural and climatic conditions of the facility location and the vegetation period of the crop.

Therefore, we have proposed an improved classification of yield categories in view of the factors influencing the crop development for reclaimed land in the form of an appropriate structural scheme (Figure 1).

Figure 1. Improved classification of yield categories on reclaimed land; PY – potential yield, CPY – climatically provided yield, MPY – meteorologically possible yield, BPY – yield, provided by the natural soil bonitet, APY – agro-technically provided yield, FY – feasible

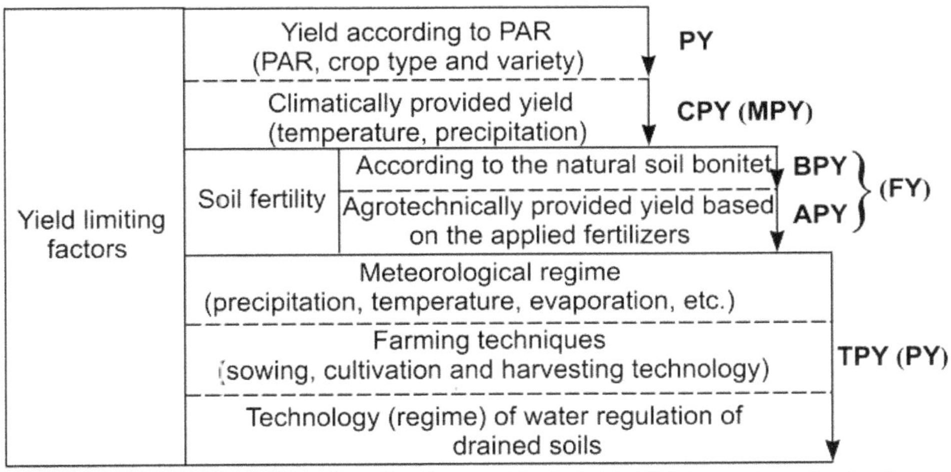

PY - potential yield; CPY - climatically provided yield; MPY - meteorologically possible yield; BPY - yield, provided by the natural soil bonitet;
APY - agro-technically provided yield; FY - feasible yield; TPY - technologically provided yield; PY - production yield.

It is the basis for making a general yield model (estimated analogue of the actual yield in the field conditions).

Based on the generalization of the results of theoretical and experimental research, a model of crop yield on drained land has been developed and proposed in the form of a complex model of the multiplicative type. This model is expressed through the product of climatically provided yield and the functions of the factors impact on the formation of real yield. This impact is estimated by the corresponding multiple coefficients:

$$Y_{k\omega gsp} = Y^F_{\omega kp} \cdot K_1 \cdot K_2 \cdot K_3 \cdot K_4 \cdot K_5 \cdot K_6, \tag{3}$$

where $Y^F_{\omega kp}$ – climatically provided yield of the kth crop over the vegetation period of pth its heat and moisture (depends on solar radiation inflow in a particular place and the degree of its use by the plant according to the efficiency factor of photosynthetic active radiation-PAR); K_1 – influence coefficient of soil bonitet on the yield ($0 \le K_1 \le 1$); K_2 – influence coefficient of applied mineral and organic fertilizers on the yield ($K_2 > 1$, but $0 < K_1 \times K_2 \le 1$); K_3 – influence coefficient of the shifting of crop sowing term from the optimal one ($0 \le K_3 \le 1$) on the kth crop yield under water regulation; K_4 – influent coefficient of current natural and reclamation conditions in the vegetation period (climate and water regulation technologies) on the crop yield ($0 \le K_4 \le 1$); K_5 – influence coefficient of the shifting of crop harvesting term from the optimal one ($0 \le K_5 \le 1$) on the crop yield; K_6 – yield reduction coefficient due to the losses during harvesting and transportation (net yield of grown products) ($0 < K_6 \le 1$).

Due to the natural seasonal cyclicity of agricultural production, according to heat and moisture provision there are different vegetation periods, which can be grouped into estimated (typical) groups of years. The distribution of these groups of years within the life cycle of the project is uneven and can be done using a share coefficient (detection probability) of the corresponding group of years within the total project implementation period (α_p).

This yield model (3) is implemented according to the long-term forecast in combination with the corresponding model of weather and climatic conditions relative to the typical heat and moisture conditions of the estimated years. However, this model does not directly take into account the weather conditions of a particular year and their impact on soil quality and other characteristics.

But according to the structure of construction, this model can take into account the possible risks and reduction of potential soil fertility due to the changes in weather and climatic conditions or using imperfect technology of agricultural production. In addition, the correction factors K_1, K_2, K_3, K_4 in this model in each case of its application are specified in view of the conditions of a real individual object.

Thus, the general yield model is the basis for determining the projected yield on reclaimed land, which can be represented as a weighted average in time and space:

$$\bar{Y}_k = \sum\nolimits_{\omega=1}^{n_\omega} \sum\nolimits_{g=1}^{n_g} \left(\sum\nolimits_{p=1}^{n_p} Y_{k\omega gsp} \cdot \alpha_p\right) \cdot f_\omega \cdot f_g, \tag{4}$$

where $\bar{Y}_k$ – projected yield of the kth crop according to the defined technology of water regulation in the required conditions; $Y_{k\omega gsp}$ – effective (really possible) yield of the kth crop in the corresponding variable conditions of the population: climatic $\{\omega\}$, $\omega = \overline{1, n_\omega}$, soil $\{g\}$, $g = \overline{1, n_g}$, ameliorative (water regulation technologies) $\{s\}, s=\overline{1,n_s}$ conditions in different (calculated) in terms of heat and moisture provision vegetation periods $\{p\}, p=\overline{1,n_p}$; f_ω, f_g – fractional shares of natural-climatic and soil differences within the object, respectively; αp_ known (determined or given) values of repeatability or shares of the population $\{p\}$, $p=\overline{1,n_p}$ of possible meteorological regimes in the estimated vegetation periods within the design period of object operation, $\sum_{p=1}^{n_p}\alpha_p=1$;

In addition to yield, which has a decisive influence on the income and current agricultural costs, a capitalization rate – the coefficient, which converts rent or net operating income into the current value of land, has an important influence on the final value of the land.

Traditionally, any interest rate on the market (R_m) is the sum of risk-free or safe interest rate (*d*) and risk premium (*r*)

$$R_m = d + r \qquad (5)$$

In Ukraine, where the financial market is under development, the basis for the formation of market interest rates for today is refinancing rates and the bank rate of the National Bank of Ukraine.

As for the risks in agriculture, this is a separate big topic that still needs further research. Modern authors propose to include an insurance rate in the risk-free rate in the field of agriculture, which is a value approved by the relevant state authorities in the regions of Ukraine (Bagdonavicius et al., 2004; Barvika et al., 2013; Pavlov et al., 2006) or calculated by insurance companies for each object individually.

A large-scale machine experiment using a computer to determine the forecasted cost of drained land using the average over the past 5 years cost data was performed by us on the example of a land plot of 430 ha, located in Rivne region (Western Polissia zone of Ukraine). Within the plot, the following multiple variable conditions can be specified:

- by the soils $\{g\}, g=\overline{1,n_g}$ (n_g=3), which are characterized by different degree of potential fertility according to bonitet and the share of their distribution f_g within the object: 1 – sod-podzolic silt cohesive-sandy soil (B = 28), f_g=0,1; 2 – soddy slightly podzolic sandy soil (B = 20), f_g=0,3; 3 – peat medium-deep low-ash soil (B = 38), f_g=0,6.
- by typical for the given zone cultivated crops of the population, $\{k\}$, $k=\overline{1,n_k}$ (n_k=3) and the corresponding share of their sown areas f_k: 1 – winter wheat, f_k= 0.3; 2 – potato f_k= 0.2, 3 – perennial grasses f_k= 0.5.
- by the typical (estimated) in terms of heat and moisture provision vegetation periods of the population $\{p\}$, $p=\overline{1,n_p}$, (n_p=5): 1 – very wet, *p*=10%; 2 – wet, *p*=30%; 3 – average, *p*=50%; 4 – dry, *p*=70%; 5 – very dry, *p*=90%.

The generalised results on the differentiation of the forecast values of cultivated crops yields and net rental income as a basis for determining the value of drained land in variable conditions of their use at the object under study are presented in Table 1.

Comparative characteristics in qualitative and quantitative values of the forecast value of drained land under the soil suitability conditions of its use at the object under study, determined using the data presented in Table 1, are presented in Fig. 2 and Fig. 3.

Table 1. Generalized comparative characteristics of forecast values of cultivated crops yields and net rental income on drained land under variable conditions of its use

Soils	Probability by heat and moisture supply	Winter wheat (grain) f_k=0.3		Potato, f_k=0.2		Perennial grasses (hay), f_k=0.5	
		Y_k centner·ha^{-1}	Д$_o$ USD·ha^{-1}	Y_k centner·ha^{-1}	Д$_o$ USD·ha^{-1}	Y_k centner·ha^{-1}	Д$_o$ USD·ha^{-1}
Soil g1, f_g=0.1; B=28	p=10%	20.50	304.39	245.30	2808.3	11.70	50.21
	p=30%	34.90	536.29	305.40	3577.16	18.90	97.00
	p=50%	43.20	676.39	282.40	3509.88	22.10	120.99
	p=70%	42.40	677.47	262.00	3021.08	22.00	127.66
	p=90%	37.10	595.21	222.90	2484.22	17.70	100.21
	weighted average	**37.25**	**583.45**	**266.35**	**3160.8**	**19.17**	**87.80**
Soil g=2. f_g=0.3; B=20	p=10%	12.0	163.46	150.86	1666.49	21.00	105.94
	p=30%	20.40	297.70	156.99	1732.22	34.80	194.74
	p=50%	23.00	341.34	128.77	1407.67	42.70	251.52
	p=70%	24.20	372.98	110.28	1195.03	36.30	225.92
	p=90%	20.20	309.66	101.70	1096.38	29.30	181.28
	weighted average	**20.52**	**304.94**	**129.05**	**1411.6**	**34.94**	**204.56**
Soil g=3. f_g=0.6; B=38	p=10%	35.20	529.74	163.80	1846.08	35.90	209.13
	p=30%	64.10	991.92	195.70	2259.69	50.90	306.01
	p=50%	86.40	1361.58	193.90	2382.94	67.20	422.27
	p=70%	81.80	1324.42	153.90	1934.58	57.80	378.43
	p=90%	71.10	1157.72	141.10	1841.22	45.00	295.48
	weighted average	**71.86**	**1136.95**	**175.10**	**2229.28**	**54.68**	**343.11**

These data convincingly show that there is a differentiation in the value of land, which is a proportional derivative of the differentiation in crops yields depending on the conditions of their cultivation. The variation of the ratio of the difference between the maximum and minimum values to the weighted average value for the studied object is: for the cultivated crops – 393%, and for the given soils – 44,6%.

Thus, within one object alone, the forecast value of land in view of available soils and cultivated crops varies from USD2456•ha^{-1} to USD4005•ha^{-1}, averaging USD3522•ha^{-1}.

It is clear that it is impossible not to take this aspect into account when doing land valuation.

In addition, our previous studies have shown that the influence share of climatic factors on the productivity of reclaimed drained land reaches 80…90%, but the influence share of ameliorative factors depending on water regulation technologies and conditions of heat and moisture provision in the vegetation period is 5…10% when draining and up to 30…40% when irrigating drained land in dry periods (Gadzalo et al., 2017; Mazhayskiy at al., 2017).

Figure 2. Comparative characteristics of the forecast value of drained land under the soil suitability conditions of its use at the object under study; soil 1 – soddy, slightly podzolic, sandy, soil 2 – sod-podzolic silt cohesive-sandy, soil 3 – peat medium-deep low-ash
Source: own study

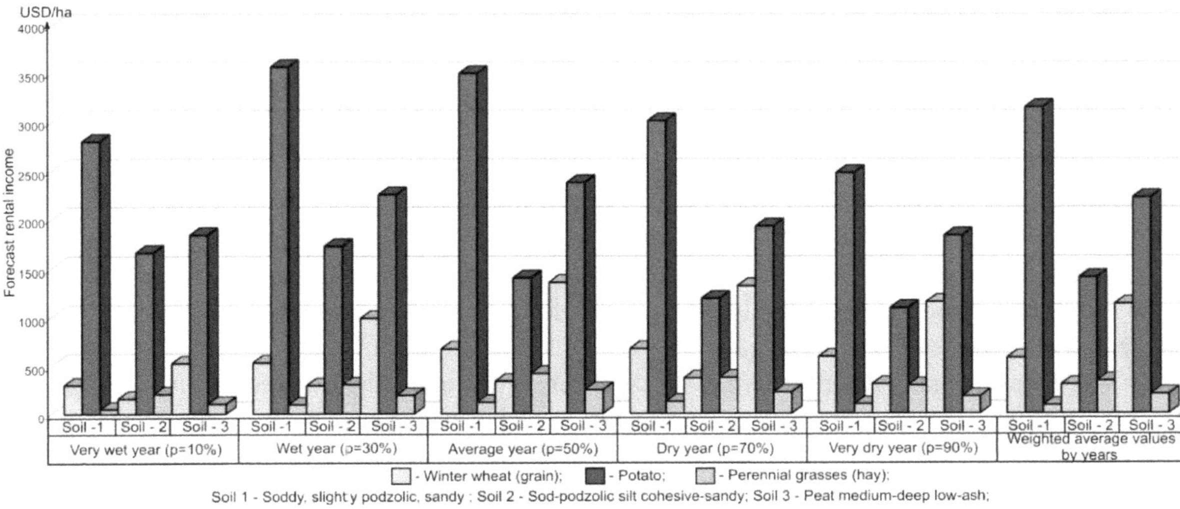

Figure 3. Generalized comparative characteristics of the forecast value of drained land under the soil suitability conditions of its use at the object under study
Source: own study

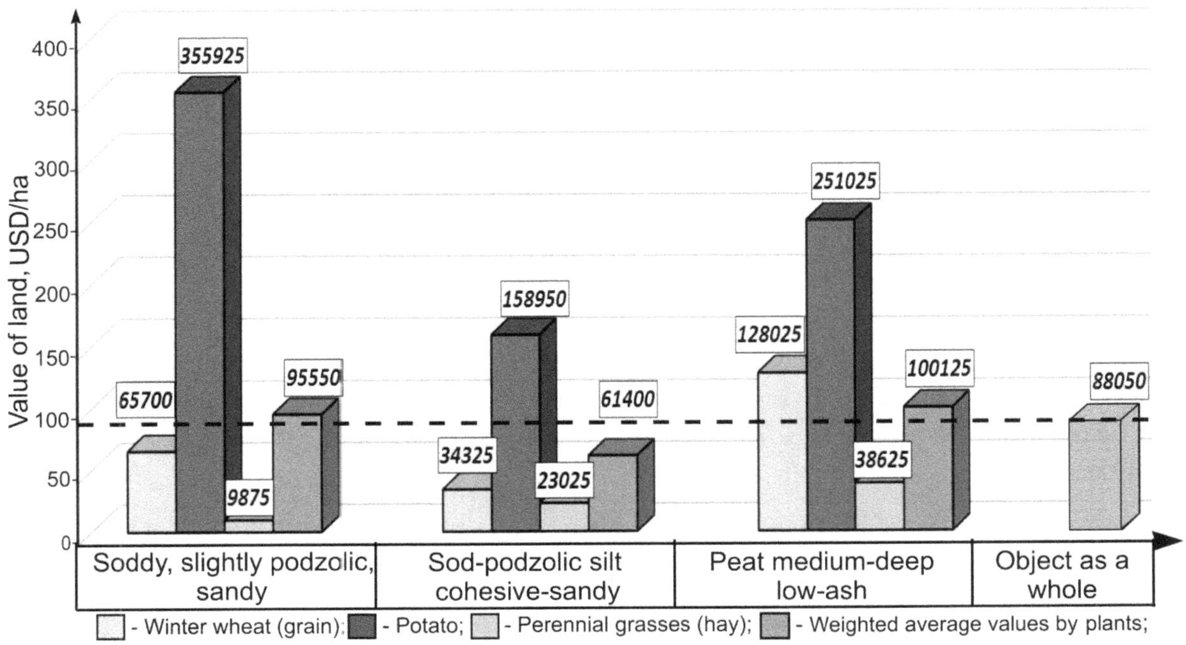

Climate change that we are facing today and is expected to increase in the future, namely an increase in temperature by 5…25% and a decrease in water supply by 15…18% in Polissia region of Ukraine (Rokochinskiy at al., 2019; 2020) will significantly affect the market value of drained land. It can decrease by 10…20% without providing irrigation, and increase by 40…60% and even more when replacing traditional crops with more productive and profitable (vegetables, berries, fruits, etc.) ones and providing regular irrigation during the vegetation period.

CONCLUSION

1. Introduction of a full-fledged land market in Ukraine and existing climate change determine the urgency of improving existing scientific and methodological approaches to determining the value of land, including for agricultural land with regulated water regime, in particular in the Polissia region of Ukraine.
2. It was experimentally determined that the determining factors in the formation of market value of land are their current or perspective productivity, which is determined by multiple variable conditions of its use: soil type, its effective and potential fertility; weather and climatic conditions of the location; type and variety of cultivated crops and agricultural machinery; carrying out the necessary agro-ameliorative and hydro-technical measures to regulate water, salt and other soil regimes, etc.
3. The results obtained on determining the differentiation in the value of drained land in view of the variable conditions of its use convincingly demonstrates the need to revise existing standards in order to increase the objectivity of such a valuation that within one object alone, the forecast value of land in view of available soils and cultivated crops varies from USD2456•ha^{-1} to USD4005•ha^{-1}, averaging 3522 USD•ha^{-1}.

REFERENCES

Adamuscin, A., Palicki, S., & Rącka, I. (2017). Real estate appraisal in central-eastern Europe: Comparative analysis of Poland and Slovakia. *Real Estate Management and Valuation*, *25*(4), 50–62. doi:10.1515/remav-2017-0029

Bagdonavicius, A., & Ramanauskas, R. (2004). Introducing a computerised market value-based mass appraisal system for real property taxation in Lithuania. *FIG Working Week*, Athens, Greece. https://www.fig.net/resources/proceedings/fig_proceedings/athens/papers/ts27/TS27_2_Bagdonavicius_Ramanauskas.pdf

Barvika, S. (2020). The property tax in Latvia: system structure and current challenges [online]. *Working Paper WP20SB1*. Cambridge. Lincoln Institute of Land Policy. https://www.lincolninst.edu/sites/default/files/pubfiles/barvika_wp20sb1_0.pdf

Barvika, S., Rausis, A., & Berzina, I. (2013). Opportunities for the development of the Latvian property tax administration system through improvements in the property registration system and the implementation of European Union requirements for geospatial information [online]. *Proceedings REAL CORP 2013: Planning times,* Rome, Italy. https://programm.corp.at/cdrom2013/papers2013/CORP 2013_201.pdf

Barvinsky, A., & Tykhenko, R. (2015). *Otsinka i prohnoz yakosti zemel: Monografiya* [*Assessment and forecast of land quality: Monograph*]. Medinform.

Frolenkova, N., Rokochinskiy, A., Volk, P., Shatkovsky, A., Prykhodko, N., Tykhenko, R., & Openk, O. (2020). Cost-effectiveness of investments in drip irrigation projects in Ukraine. *International Journal of Green Economics*, *14*(4), 315–326. doi:10.1504/IJGE.2020.112570

Gadzalo, Y., Stashuk, V., & Rokochinskiy, A. (2017). Melioratsiya ta oblashtuvannya Ukrayinskoho Polesie: Monografiya [Land reclamation and arrangement of the Ukrainian Polesie. Monograph]. Oldi-Plus.

IAAO. (2017). *Standard on mass appraisal of real property.* International Association of Assessing Officers. https://www.iaao.org/media/standards/StandardOnMassAppraisal.pdf

KMU. (2018). Postanova vid 7 lyutoho 2018 r. Pro provedennya zahalnonatsionalnoyi (vseukrayinskoyi) normatyvnoyi hroshovoyi otsinky zemel silskohospodarskoho pryznachennya ta vnesennya zmin do deyakykh postanov Kabinetu Ministriv Ukrayiny. No. 105 [Resolution of February 7, 2018 On conducting a nationwide (all-Ukrainian) normative monetary valuation of agricultural lands and making changes to some resolutions of the Cabinet of Ministers of Ukraine. No. 105]. Kabinet Ministriv Ukrayiny. [Access 30.04.2020]. Available at: https://zakon.rada.gov.ua/laws/show/105-2018-п#Text

Kovalenko, P., Rokochinskiy, A., Jeznach, J., Koptyuk, R., Volk, P., Prykhodko, N., & Tykhenko, R. (2019). Evaluation of climate change in Ukrainian part of Polissia region and ways of adaptation to it. *Journal of Water and Land Development*, *41*(1), 77–82. doi:10.2478/jwld-2019-0030

Kovalenko, P., Rokochinskiy, A., Volk, P., Turcheniuk, V., Frolenkova, N., & Tykhenko, R. (2021). Evaluation of ecological and economic efficiency of investment in water management and land reclamation projects. *Journal of Water and Land Development*, *48*, 81–87. doi:10.24425/jwld.2021.136149

Martyn, A., Openko, I., Ievsiuko, V., Shevchenko, O., & Ripenko, A. (2019). Accuracy of geodetic surveys in cadastral registration of real estate: value of land as determining factor. *Proceedings of the 18th International Scientific Conference on Engineering for Rural Development*, Jelgava, Latvia. https://www.tf.llu.lv/conference/proceedings2019/Papers/N236.pdf

Martyn, A., Shevchenko, O., Tykhenko, R., Openko, I., Zhuk, O., & Krasnolutsky, O. (2020). Indirect corporate agricultural land use in Ukraine: Distribution, causes, consequences. *International Journal of Business and Globalisation*, *25*(3), 378–395. doi:10.1504/IJBG.2020.109029

Martyn, A., & Tykhenko, O. (2015). *Formuvannya kadastrovo-reyestratsiynoyi systemy v Ukrayini: Monografiya [Formation of cadastral registration system in Ukraine: Monograph*]. Medinform.

Mazhayskiy, Y., Rokochinskiy, A., Volchek, A., Meshyk, O., & Jeznach, J. (2017). Pryrodoobustroistvo Polesia. [Environmental management of Polesie]. VNIIGiM of A. Kostiakov.

Ministerstvo, A. P. P. U. (2017). *Nakaz vid 23 travnya 2017 r. Pro zatverdzhennya Poryadku normatyvnoyi hroshovoyi otsinky zemel silskohospodarskoho pryznachennya. No. 262 [Order dated May 23, 2017 On approval of the procedure for normative monetary valuation of agricultural lands. No. 262]* Ministerstvo ahrarnoyi polityky ta prodovolstva Ukrayiny. https://zakon.rada.gov.ua/ laws/show/z0679-17#Text

Novakovsky, L. (2015). Problemy metodychnoho zabezpechennya normatyvnoyi hroshovoyi otsinky zemel silskohospodarskoho pryznachennya [Problems of methodological support of normative monetary valuation of agricultural lands]. *Visnyk ahrarnoyi nauky, 12,* 11–16.

Openko, I., Kostyuchenko, Y., Tykhenko, R., Shevchenko, O., Tsvyak, H., Ievsiuko, V., & Deineha, M. (2020). Mathematical modelling of postindustrial land use value in the big cities in Ukraine. International *Journal of Mathematical. Engineering and Management Sciences*, *5*(2), 260–271. doi:10.33889/IJMEMS.2020.5.2.021

Openko, I., Shevchenko, O., Zhuk, O., Kryvoviaz, Y., Tykhenko, R. (2017). Geoinformation modelling of forest shelterbelts effect on pecuniary valuation of adjacent farmlands. *International Journal of Green Economics (IJGE), 11*(2), 139–153. doi:10.1504/IJGE.2017.089015

Panas, R., Malanchuk, M. (2013). Osoblyvosti bonituvannya tekhnohennykh gruntiv [Features of assessment of technogenic soils]. *Heodeziya, kartohrafiya i aerofotoznimannya, 77,* 74–80.

Pavlov, V., Fesina, Yu., Zaremba, V., & Mazuryk, S. (2006). Determinatsiya sil's'kohospodars'kykh zemel' na rynku nerukhomosti: Monografiya [Determination of agricultural land in the real estate market. Monograph]. Rivne: NUVGP.

Rokochinskiy, A., Bilokon, W., Frolenkova, N., Prykhodko, N., Volk, P., Tykhenko, R., & Openko, I. (2020). Implementation of modern approaches to evaluating the effectiveness of innovation for water treatment in irrigation. *Journal of Water and Land Development*, *45*, 119–125. doi:10.24425/jwld.2020.133053

Rokochinskiy, A., Frolenkova, N., Turcheniuk, V., Volk, P., Prykhodko, N., Tykhenko, R., & Openko, I. (2021). The variability of natural and climatic conditions in investment projects in the field of nature management. *Journal of Water and Land Development*, *48*, 48–54. doi:10.24425/jwld.2021.136145

Rokochinskiy, A., Jeznach, J., Volk, P., Turcheniu, K. V., Frolenkov, A. N., & Koptiuk, R. (2019). Reclamation projects development improvement technology considering optimization of drained lands water regulation based on BIM. Scientific Review Engineering and Environmental Sciences. Vol. 28. *Iss.*, *3*(85), 432–443. doi:10.22630/PNIKS.2019.28.3.40

Rokochinskiy, A., Volk, P., Pinchuk, O., Turcheniu, K V., Frolenkova, N., Gerasimov, E. 2019. Forecasted estimation of the efficiency of agricultural drainage on drained lands. *Journal of Water and Land Development, 40*(I–III), 149–153. doi:. doi:10.2478/jwld2019-0016

Seryy, A. (1986). *Bonitirovka pochv: Metodychni rekomendatsiyi* [*Assessment of soil: Methodical recommendations*]. USHA.

Shevchenko, O., Openko, I., Zhuk, O., Kryvoviaz, Y., & Tykhenko, R. (2017). Economic assessment of land degradation and its impact on the value of land resources in Ukraine [online]. *International Journal of Economic Research (IJER), 14*(15), 93–100. https://serialsjournals.com/abstract/34405_ch_11_f_-_ivan_openko.pdf

Slovoidilo.ua. (2019). Rynok zemli: vartist hektara v krayinakh Yevropy ta Pivnichnoyi Ameryky [Land market: the cost per hectare in Europe and North America]. *Slovoidilo.* https://www.slovoidilo.ua/2019/11/13/infografika/ekonomika/rynok-zemli-vartist-hektara-krayinax-yevropy-tapivnichnoyi-ameryky

Tykhenko, O. (2010). Porivnyalnyy analiz metodiv bonituvannya gruntiv Ukrayiny [Comparative analysis of soil grading methods in Ukraine]. *Visnyk Kharivskoho natsionalnoho ahrarnoho universytetu im. V. Dokuchayeva*, *6*, 226–230.

Tykhenko, O. (2016). Ratsionalne vykorystannya silskohospodarskykh zemel: problemy ta perspektyvy yikh vyrishennya [Rational use of agricultural land: problems and prospects for their solution]. *Zemleustrij, kadastr i monitorynh zemel, 1–2,* 76–81. http://journals.nubip.edu.ua/index.php/Zemleustriy/article/viewFile/7204/6989

Tykhenko, R. (2010). Ekoloho-ekonomichna efektyvnist zemleustroyu v umovakh transformatsiyi zemelnykh vidnosyn v Ukrayini: Monografiya [Ecological and economic efficiency of land management in the conditions of transformation of land relations in Ukraine: Monograph]. Kyyiv. Anva-print.

Zakon Ukrayiny Pro otsinku zemel [Law of Ukraine on land valuation]. (2004). Vidomosti Verkhovnoyi Rady Ukrayiny. https://zakon.rada.gov.ua/laws/show/1378-15

Chapter 23
Mathematical Modeling of Economic Losses Caused by Forest Fire in Ukraine

Ivan Openko
https://orcid.org/0000-0003-2810-0778
National University of Life and Environmental Sciences of Ukraine, Ukraine

Ruslan Tykhenko
National University of Life and Environmental Sciences of Ukraine, Ukraine

Oleksandr Shevchenko
National University of Life and Environmental Sciences of Ukraine, Ukraine

Oleg Tsvyakh
National University of Life and Environmental Sciences of Ukraine, Ukraine

Yanina Stepchuk
National University of Life and Environmental Sciences of Ukraine, Ukraine

ABSTRACT

In this chapter, the authors analysed the current state of economic and environmental damage caused by forest fires in Ukraine. Uncontrolled spread of fire not only causes significant economic damage, but also has dire consequences for the ecology. The burning of large areas of forest leads to a radical change in the ecosystems of the affected region, which can eventually lead to unpredictable results. Thus, the negative indicators of forestry are: quantity of forest fires, units; area of forest lands covered by fires; losses caused by forest fires; burned and damaged the forest. It should be noted that the indicators of the consequences of forest fires do not mathematically correlate with the loss of wood cover in Ukraine, which are determined by the data remote sensing of the earth.

DOI: 10.4018/978-1-6684-8248-3.ch023

BACKGROUND

Solution any nature management problems, including forest lands, requires in-depth and comprehensive study. At the same time, any managerial, organizational decisions should be based on certain scientific forecasts on the consequences of nature management in the project, development of measures to prevent the impact of negative factors on the environment and improvement of approaches to economic planning taking into account the environmental factor.

Indeed, effective management of forest resources and prevention of the negative consequences of forest fires require an integrated approach based on scientific knowledge and careful planning. Effective fire prevention and management strategies, including early warning systems, improved firefighting technologies, and better land use planning, must be developed and implemented to protect the environment and minimize economic losses associated with wildfires (Openko et al., 2022).

At the same time, it is important to take a broader look at the management of forest resources, taking into account not only the economic benefits of forest use, but also the ecological and social value of these resources. This requires an integrated approach to forest management that takes into account the needs of local communities, the protection of biodiversity and other ecological values, and the long-term sustainability of forest ecosystems.

To achieve these goals, it is necessary to improve scientific knowledge and research in forest ecology and forest management, as well as to develop more effective partnerships between public institutions, the private sector and local communities.

Also, for the protection of forest lands requires a powerful material and technical base, taking into account modern advances in technology, in particular, Global Navigation Satellite System – technologies that allow you to record forests (areas) with high accuracy; large-scale introduction of technologies of waste-free (low-waste) production; increasing the volume of production of synthetic (artificial) substitutes for natural materials (Cai, Liu, Yang, and Di, 2012; Gensiruk, 1979; Vorob′ev, Mukhamedshyn, & Devyatkyn 1984).

The issue of studying the losses due to forest fires is the subject of a publication H. Cai, Zl. Liu, G. Yang, X. Di. (Cai, Liu, Yang, and Di, 2012). Scientists say that forest fire loss statistics is becoming a crucial tool in decision-making and in helping to reduce loss to life and property due to fires around the world. Estimating forest fire direct loss is important pieces of data when assessing fire at local and national levels. So far, the fire direct loss assessment work abroad has been greatly developed. And current forest fire direct loss assessment only stays in simple statistics and lacks the deep data intellectualized and analysis. Based on the current situation status of forest fire direct loss assessment, they divided the direct loss of forest fire into standing trees resource loss, commodity wood loss, fixed assets loss, current assets loss, forest by-product loss, agriculture and animal husbandry product loss, fire fighting expenses, personnel casualty loss, resident's property loss and made a quantitative and qualitative assessment of forest fire loss through the data of self-made forest fire case.

O. V. Akinola, J. Adegoke (Akinola & Adegoke, 2019) emphasize that forest fire is one of the major disasters that distresses the terrestrial environment and causes economic disruptions for people and communities in areas prone to forest fire. Information on forest fire risk zones is therefore essential for effective and sound decision-making in forest management. Forest fire risk assessment is a critical part and the most important step in forest management because it enables us to know where the risk is higher in order to minimize threats to life, property and natural resources. This study used a hazard assessment model to assess forest fire risk in Missouri based on several measurable environmental parameters influencing forest fire risk vulnerability. Using the four ecological zones in Missouri as the basis of analysis, three forest risk zones were identified. These were high forest fire risk zones, moderate forest fire risk zone and low forest fire risk zone. Strategies for the mitigation of the hazard of forest fire in the state were also recommended.

In a scientific article by M. Zhong, W. Fan, T. Liu, P. Li (Zhong, Fan, Liu, Li, 2003) considered statistical data on forest fires in China are analyzed in order to reveal new features regarding the forest fire safety situation as the economy of China is growing. The causes of forest fires in China are also discussed. The classifications of forest fire hazard used to improve forest fire prevention and management in China are reviewed, and the current status of the fire researches in China is introduced.

However, the destruction of large forests areas that regulate climatic conditions and water regime of huge basins, as noted H. I. Vorobiov, K. D. Mukhamedshyn, L. M. Deviatkin (Zibtsev & Borsuk, 2012) Tedim, F., Xanthopoulos, G., Leone, V. (Tedim, Xanthopoulos, Leone, 2015), Ubysz, B., Szczygiel, R. (Ubysz, Szczygiel, 2002) violates the ecological balance established for millions of years, which in turn leads to constant devastating floods, mudslides, water erosion processes, dust storms, droughts or development of wetlands depending on the specific climatic conditions and the nature of woody vegetation.

At the same time, the problem of forest fires, the numerous environmental and economic losses they cause, in recent years has attracted special attention in the context of growing influence global processes such as declining global forest resources, numerous biodiversity losses, global climate change and land use change. All this is due to the complexity and ambiguity of the impact of forest fires on forest lands, the environment and safe living conditions for the population in settlements (united territorial communities), which are located near forests (Voronin, Senkevich & Bugaev, 1975; Sirin, Maslov, Makarov, Gulbe and Hans, 2021).

METHODS AND TECHNIQUES

In order to assess losses to forestry as a result of the harmful effects of forest fires, it is necessary to take into account not only direct losses caused by fire-damaged plantations, but also collateral losses that may occur. To fully understand the impact of wildfires, a comprehensive assessment that takes into account all direct and indirect effects is necessary. Such an assessment requires a multidisciplinary approach involving experts from various fields, including forestry, ecology, economics and social sciences.

There are various methods and approaches for determining the economic damage caused to forest resources from fires, while they are grouped according to the type of damage caused. Thus, a timber damage assessment involves an assessment of the damage caused to timber resources in the affected area and an estimate of the value of the timber lost. In addition to estimating the value of lost timber, estimating timber damage may also include estimating "salvage" timber, timber that can be sold after a fire. Such sales can generate income and offset some of the economic losses resulting from the fire. However, clearance felling can also have negative environmental effects, such as soil compaction and damage to remaining trees, which should be taken into account when assessing economic damages. In addition, timber damage assessment can also consider the impact of fire on future timber production. For example, if a fire has damaged or destroyed a seed source, it may take years or even decades for the forest to regenerate and produce wood again. This loss of future timber production can have a significant economic impact on the forest industry and local communities that rely on timber as a source of income.

Therefore, this method can provide valuable information on the economic impact of wildfires on the forest industry, but it is important to consider the long-term ecological and economic consequences of fire.

Ecosystem service assessment involves assessing the damage to ecosystem services provided by the forest, such as carbon sequestration and water regulation, as well as supporting services such as soil formation and nutrient cycling. When forests are affected by fire, the ecosystem's biodiversity can be severely affected, potentially leading to the extinction of species and the loss of important ecological functions. This method can provide information on the economic impact of fire on the environment and develop more effective strategies for forest fire prevention and mitigation, as well as restoration of ecosystem services in burned areas.

A recreation and tourism impact assessment involves assessing the indirect economic effects of wildfires and determining the damage to recreational and tourism resources in the affected area, such as campsites, hiking trails, and the destruction of scenic views. This method can provide information on the economic impact of fire on local economies that are heavily dependent on tourism and natural resources.

One of the important methods is the evaluation of the processes of extinguishing fires and subsequent recovery. This method involves estimating the costs of extinguishing a fire, such as the cost of firefighting equipment and the wages of special personnel, as well as the cost of restoring damaged infrastructure. Restoration costs may include the costs of planting new trees and other vegetation, monitoring the restoration process, and implementing measures to prevent erosion and soil degradation. This method can provide information on the economic impact of fire on state budgets and the costs of preventing and responding to future fires.

Depending on the purpose of assessing economic losses, other methods can be used, such as the analysis of statistical data on the production of forest products, assessment of the loss of income for forest users, assessment of the cost of lost biodiversity, etc. In addition, the assessment of economic damages may include not only direct costs of restoring forest resources, but also indirect costs of reducing the quality of air, water and soil, as well as the impact on human health and the economy of the region as a whole. It is important to consider all possible aspects to ensure a complete and accurate assessment of economic losses from wildfires.

A necessary component of the calculation of the estimate of losses from forest fires, in addition to fire damage to plantations, should be the estimate of the losses of already harvested wood as a result of the fire. Destroyed timber, harvested logs and other forest products not only cause direct economic losses to the forestry industry, but also reduce the overall supply of timber in the region, which can lead to higher prices for wood products and reduced production capacity of the forest industry. Loss of harvested timber from fire can also have long-term effects on forest management practices, as fire can destroy critical infrastructure such as roads, bridges, and log depots, as well as buildings and structures. The cost of restoring or repairing damaged structures can be significant, and in some cases it may be more cost-effective to demolish and rebuild rather than repair.

Firefighting costs include personnel, equipment and materials and are usually substantial. These costs are usually borne by the government or other organizations responsible for managing and protecting forests, and can strain budgets and resources, while diverting resources from other important priorities. This can have a long-term economic impact on the affected area, as investment in these areas may be reduced.

The protective and sanitary-hygienic properties of forest plantations decrease sharply as a result of the fire, and the ecological stability of the territory decreases. Forest fires can cause soil erosion, which leads to nutrient loss and reduced soil fertility not only under the forest, but also in the surrounding areas. Forest fires can weaken or destroy trees, making them more vulnerable to pests and diseases. damage the protective tree canopy, making the forest more vulnerable to future fires. There are also negative impacts on biodiversity, such as losses of ecosystem services: pollination, nutrient cycling, and water regulation.

Not only wildfires, but also firefighting can have a negative impact on the environment, such as water pollution from firefighting chemicals, soil erosion from heavy equipment, and damage to wildlife habitat.

The indirect use of forests includes various types of economic activities, such as tourism, recreation, hunting, collection of non-timber forest products. When a forest is damaged or destroyed by fire, it can have serious consequences for the secondary use of the forest, which in turn can have a negative economic impact on the affected area. Forests provide a variety of non-timber forest products such as mushrooms, berries and medicinal plants. When a forest is affected by fire, these products can be lost, resulting in reduced profits for harvesters and related businesses. Also, limited or destroyed opportunities for hiking, camping and other outdoor activities such as picnicking or bird watching can lead to a decline in tourism. Wildlife populations may decline due to wildfires, leading to reduced hunting opportunities and related businesses.

Indirect but significant losses are investors' doubts about the profitability of investments in the forest industry. The potential for loss of income due to fires may reduce the attractiveness of investing in the forestry sector, leading to less investment. Damaged roads, bridges, and other infrastructure used for forest management and timber harvesting result in significant repair and replacement costs and reduce the return on investment in the forestry industry. Forest fires can take years or even decades to fully recover, and uncertainty about how long it will take for a forest to fully recover can make it difficult to estimate future returns, leading to reduced investment.

In order to economically assess the loss of natural resources, scientists I. V. Voronin, A. A. Senkevych, V. A. Buhaiev developed a method for determining the total losses to forestry as a result of forest fires, which foresee the following indicators (State Statistics Service of Ukraine, 2018):

- damages from fire damage to plantations;
- the cost of work to collect damaged trees;
- losses of harvested wood due to fire;
- the cost of burned or damaged buildings and structures;
- losses of crops, hay and more;
- fire extinguishing costs;
- losses from deterioration of protective, sanitary and hygienic properties of forest plantations as a result of fire;
- losses from side use, secondary use of the forest (collection of juice, fruits, berries, mushrooms, etc.);
- losses of the forest industry in the amount of investments for the destroyed timber stock.

Damage from damaging of plantations (V) is determined by formula (1):

$$V = B \times K - T \tag{1}$$

where B – the cost of burned plantings at the age suitable for felling; T– the cost of wood damaged by forest fire; K – coefficient that takes into account economic conditions of reforestation works after the fire.

At the same time, mathematical modelling allows to determine the total scope of economic losses from forest fires, based on dependent factors, determinants, which are pre-determined on the basis of correlation.

In order to quickly assess the economic losses caused by forest fires, we proposed a regression model, the adequacy of which is confirmed by Fisher's criterion – 107.659 and the coefficient of determination $R^2 = 0.923$ (formula 2) (Figure 1) (Strand, 2015, Shevchenko et al., 2021).

$$Y = -4171.26 + 1.12 \times X_1 + 0.13 \times X_2 \quad (2)$$

where Y – losses caused by forest fires, thousand UAH; X_1 – area of forest lands covered by fires, ha; X_2 – the quantity of forest that burned and was damaged on the stump, m^3.

Figure 1. Image of the adequacy of the mathematical model developed on the basis of regression for determining the damage caused by forest fires, thousand UAH
Source: own calculations.

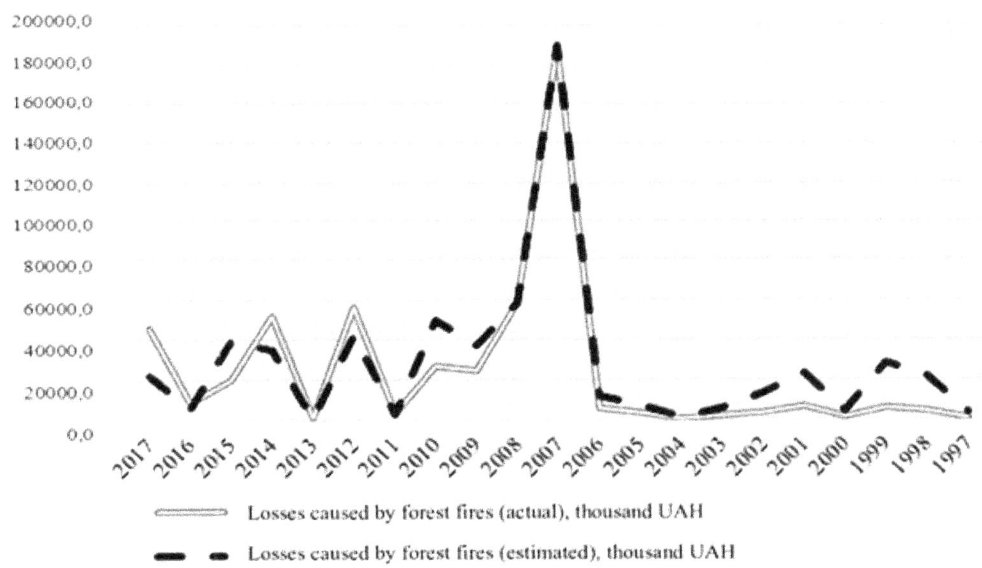

RESULTS

Uncontrolled spread of fire not only causes significant economic damage, but also has dire consequences for the ecology. Loss of biodiversity brings the greatest consequences. Forest fires can destroy the habitats of plants and animals that rely on these habitats for survival. After a forest fire, the composition of plant and animal communities can change, with some species becoming more dominant and others becoming less common or disappearing altogether In addition, some species of plants and animals may not be able to survive a fire or may take a long time to fully or partially recover, leading to further ecological imbalance and cascading effects leading to reduced ecosystem function. Even if a forest fire does not destroy a habitat completely, it can fragment it, making it difficult for species to move between different parts of their range. This can isolate populations and reduce genetic diversity, further contributing to biodiversity loss (Rokochinskiy et al., 2021).

The heat from a wildfire can cause soil to become dry and brittle, making it more susceptible to erosion. Soil erosion can lead to destruction nutrient- rich upper layer in the soil, which leads to a decrease in fertility and productivity not only of forestry, but also of territorially adjacent agriculture and other types of land use. The ability of forests to absorb carbon also deteriorates, as carbon is stored in soil organic matter. Fires even affect the water cycle, disrupting it, reducing the amount of water absorbed by vegetation and soil. This can lead to drought, a decrease in water quality. Eroded soil can then wash into nearby waterways, causing water pollution and increasing the risk of flooding, as it can clog waterways and reduce their water-carrying capacity.

Forest fires can release large amounts of smoke and other pollutants into the air, leading to respiratory and other health problems in both humans and animals. They contribute to the formation of smog and acid rain, which can have a negative impact on the environment. Forest fires are the main source of carbon emissions, as the accumulated carbon is released into the atmosphere during the burning of trees. At the same time, a large volume of other greenhouse gases is released into the atmosphere, which contributes to global warming. According to some estimates, wildfires cause approximately 3-4% of global greenhouse gas emissions each year. This can contribute to climate change and further environmental problems.

Wildfires can have a significant impact on forest ecosystems, reducing the carbon sequestration capacity of forests and increasing the risk of future wildfires. This can lead to a feedback loop where increased wildfire activity leads to additional carbon emissions, further warming the planet and increasing the risk of future wildfires.

All of these factors can have significant environmental consequences, leading to long-term ecological imbalances and potentially affecting the economic and social well-being of local communities.

The burning of large areas of forest leads to a radical change in the ecosystems of the affected region, which can eventually lead to unpredictable results. Thus, the negative indicators of forestry are (Table 1) (Martyn et al., 2019; Hall et al., 2021):

- quantity of forest fires, units;
- area of forest lands covered by fires, ha;
- losses caused by forest fires, thousand UAH;
- burned and damaged the forest, m^3 (Table 1).

It should be noted that the indicators of the consequences of forest fires do not mathematically correlate with the loss of wood cover in Ukraine, which are determined by the data remote sensing of the earth (Forest Global Watch) (Table 2). Methodology for obtaining data on forest cover losses with the help remote sensing of the earth Global Forest Watch described by a team of scientists Hansen M. C., Potapov P. V., Moore R., Hancher M. et al. (Global Forest Watch, 2018).

At the same time, a data set Global Forest Watch on the loss of wood cover concerted with a number of institutions, including the University of Maryland, Google, USGS i NASA, and this resource uses satellite imagery Landsat to display the loss of yearly tree cover with a resolution of 30 × 30 meters (Global Forest Watch, 2018; Shevchenko et al., 2021) (Table 2).

Table 1. Indicators of the consequences of forest fires in Ukraine

Year	Consequences of forest fires			
	Quantity of forest fires, units	**Area of forest lands covered by fires, ha**	**Burned and damaged the forest, m³**	**Losses caused by forest fires, thousand UAH**
2017	3131	5939.0	149775.0	45877.6
2016	1249	1249.0	32559.0	8619.2
2015	3813	14691.0	170686.0	20164.5
2014	2003	13778.0	144694.0	51701.8
2013	1113	418.0	2496.0	1376.2
2012	2163	3479.0	289291.0	56062.7
2011	2526	1049.0	11804.0	3215.9
2010	3240	3668.0	343840.0	26728.4
2009	7036	6315.0	223764.0	24686.4
2008	4042	5529.0	395257.0	58750.3
2007	6100	13787.0	1304271.0	188412.2
2006	3842	4287.0	53119.0	5917.6
2005	4223	2325.0	32101.0	3535.0
2004	1876	595.0	1944.0	428.7
2003	4527	2817.0	19720.0	1817.5
2002	6383	4983.0	59206.0	3378.9
2001	3205	3772.0	139604.0	6204.3

Source: own calculations according to the State Statistics Service of Ukraine (Hansen et al., 2013).

Table 2. Correlation of consequences of forest fires and loss of wood cover in Ukraine

R^2	Loss of wood cover in Ukraine according to data remote sensing of the earth (Forest Global Watch), thousand ha	Indicators of the consequences of forest fires
Correlation coefficients	-0.325	Quantity of forest fires, units
	-0.002	Area of forest lands covered by fires, ha
	0.162	Burned and damaged the forest, m³
	0.259	Losses caused by forest fires, thousand UAH

According to the results of analytical calculations, the losses caused by forest fires, which during 2001-2017 amounted to UAH 506.877 million, mainly depend on the area of forest land covered by fires (R = 0.632) and the quantity of burned and damaged forest on the stump (R = 0.956).

The diagram (Figure 2) shows that 2007 was marked by the largest environmental "catastrophe" associated with forest fires, in particular, in the Kherson region. As a result, a forest fire was extinguished on an area of about 4 thousand hectares.

Given the current dynamics and cyclical nature of the negative consequences of forest fires, we predicted the area of forest land covered by forest fires, the quantity of burned (damaged) forest on the stump, and the damage caused by forest fires by 2030.

Figure 2. Diagram of the dynamics of damage caused by forest fires, thousand UAH
Source: own calculations according to the State Statistics Service of Ukraine.

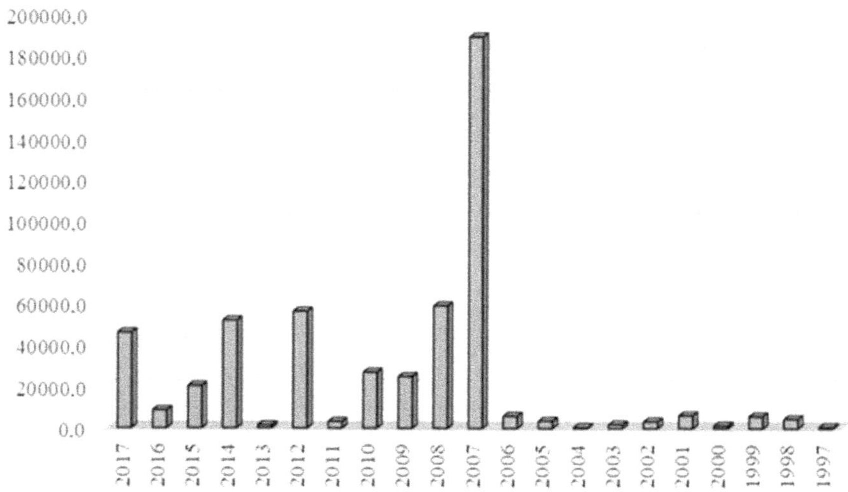

According to the calculations we have found for the period 2019-2030 in Ukraine, the total projected area of forest land covered by fires will be 49.991 thousand hectares, the total quantity of burned and damaged forest on the stump – 4.086 million m^3, losses caused by forest fires, – UAH 906.211 million or UAH 1.817 billion (taking into account the consumer price index for 2020).

CONCLUSION

Therefore, to reduce the amount of environmental and economic losses from the effects of forest fires in the future it is necessary to implement effective management, organizational measures to protect forest lands from such negative phenomena. First, one of the important measures is investment in forest fire prevention. Investments in firebreaks, controlled burning and forest thinning can reduce the risk of wildfires and minimize their impact if they do occur. Good forest management practices can improve the health and resilience of forest ecosystems by making them more resilient to fires and better able to recover from their effects. Improved monitoring and early warning systems can help detect and respond to wildfires more quickly, reducing their spread and minimizing their impact.

By implementing these measures, it is possible to reduce the amount of environmental and economic damage from the consequences of forest fires in the future. This can help protect both the natural environment and the forest industry, ensuring the long-term sustainability of these valuable resources.

A forest fire monitoring system has been developed, in particular the European Forest Fire Information System (EFFIS), which supports forest fire protection services in the EU and provides services to the European Commission and the European Parliament. updated and reliable information on fires in Europe, allows the relevant services to respond quickly to protect forest lands from such negative phenomena. The system allows to determine the area of burned lands from forest fires and to compare these data with the indicators of previous years (average).

According to the results of the study, the forecast indicators of economic losses from forest fires were calculated, taking into account the existing dynamics of these negative phenomena. The proposed mathematical model will allow to calculate the economic losses from forest fires in order to compensate for them.

REFERENCES

Akinola O. V. & Adegoke J. (2019). Assessment of forest fire vulnerability zones in Missouri, United States of America. *International Journal of Sustainable Development & World Ecology, 26*(3), 251-257. doi:10.1080/13504509.2018.1551815

Cai, H.-y., Liu, Z.-l., Yang, G., & Di, X.-y. (2012). Study on forest fire direct loss assessment. *World Automation Congress 2012*, (pp. 1-4).

Gai, C., Weng, W., & Yuan, H. (2011). GIS-Based Forest Fire Risk Assessment and Mapping. *Fourth International Joint Conference on Computational Sciences and Optimization,* (pp. 1240-1244). IEEE. 10.1109/CSO.2011.140

Global Forest Watch. (2018). Tree Cover Loss in [selected area name]. *Global Forest Watch.* www.globalforestwatch.org

Hall, J. V., Zibtsev, S. V., Giglio, L., Skakun, S., Myroniuk, V., Zhuravel, O., Goldammer, J. G., & Kussul, N. (2021). Environmentaland political implications of underestimated croplandburning in Ukraine. *Environmental Research Letters, 16*(6), 064019. doi:10.1088/1748-9326/abfc04 PMID:34316296

Hansen, M. C., Potapov, P. V., Moore, R., Hancher, M., Turubanova, S. A., Tyukavina, A., Thau, D., Stehman, S. V., Goetz, S. J., Loveland, T. R., Kommareddy, A., Egorov, A., Chini, L., Justice, C. O., & Townshend, J. R. G. (2013). High-Resolution Global Maps of 21st-Century Forest Cover Change. *Science, 342*(6160), 850–853. doi:10.1126cience.1244693 PMID:24233722

Kryvoviaz, E., Openko, I., Tykhenko, R., Shevchenko, O., Tykhenko, O., Tsvyakh, O., & Chumachenko, O. (2020). *Recovery of losses for inappropriate use of land.* Scientific Papers. Series E. Land Reclamation, Earth Observation & Surveying, Environmental Engineering. http://landreclamationjournal.usamv.ro/pdf/2020/Art26.pdf

Martyn, A., Openko, I., Ievsiukov, T., Shevchenko, O., & Ripenko, A. (2019). Accuracy of geodetic surveys in cadastral registration of real estate: value of land as determining factor. *18th International Scientific Conference. Engineering for Rural Development,* Jelgava, Latvia.

Martyn, A., Shevchenko, O., Tykhenko, R., Openko, I., Zhuk, O., & Krasnolutsky, O. (2020). Indirect corporate agricultural land use in Ukraine: Distribution, causes, consequences. *International Journal of Business and Globalisation, 25*(3), 378–395. doi:10.1504/IJBG.2020.109029

Openko, I., Kostyuchenko, Y. V., Tykhenko, R., Shevchenko, O., Tsvyakh, O., Ievsiukov, T., & Deineha, M. (2020). Mathematical Modelling of Postindustrial Land Use Value in the Big Cities in Ukraine. International Journal of Mathematical. *Engineering and Management Sciences.*, *5*(2), 260–271. doi:10.33889/IJMEMS.2020.5.2.021

Openko, I., Stepchuk, Ya., Tykhenko, R., Shevchenko, O., & Tsvyakh, O. (2022). Modeling of indicators of economic efficiency of sectoral land use. *Proceedings of the 2022 International Conference. Economic science for Rural Development,* Jelgava, Latvia.

Openko, I., Tykhenko, R., Shevchenko, O., Tsvyakh, O., Stepchuk, Ya., Rokochinskiy, A., & Volk, P. (2022) Mathematical modeling of economic losses caused by forest fire in Ukraine. *21st International Scientific Conference Engineering for Rural Development,* Jelgava, Latvia. 10.22616/ERDev.2022.21.TF004

Rokochinskiy, A., Bilokon, V., Frolenkova, N., Prykhodko, N., Volk, P., Tykhenko, R., & Openko, I. (2020). Implementation of modern approaches to evaluating the effectiveness of innovation for water treatment in irrigation. Journal of Water and Land Development, 45, (IV–VI) 119–125. doi:10.24425/jwld.2020.133053

Rokochinskiy, A., Volk, P., Frolenkova, N., Tykhenko, O., Shalai, S., Tykhenko, R., & Openko, I. (2021). Differentiation in the value of drained land in view of variable conditions of its use. Journal of Water and Land Development, 51 (X–XII), 172–178. doi:10.24425/jwld.2021

Shevchenko, O., Openko, I., Tykhenko, R., Tsvyakh, O., Zhuk, O., Kryvoviaz, E., Tykhenko, O., Bavrovska, N., Stepchuk, Ya., Rokochinskiy, A., & Volk, P. (2021). Assessment of economic losses caused by degradation processes of agricultural land use. Scientific Papers. Series E. Land Reclamation, Earth Observation & Surveying. *Environmental Engineering.*, *X*, 102–109.

Sirin, A., Maslov, A., Makarov, D., Gulbe, Y., & Joosten, H. (2021). Assessing Wood and Soil Carbon Losses from a Forest-Peat Fire in the Boreo-Nemoral Zone. *Forests*, *12*(7), 880. doi:10.3390/f12070880

State Statistics Service of Ukraine. (2018). *Main indicators of forestry activity (1990 – 2017).* UK Stats. http://www.ukrstat.gov.ua/operativ/operativ2014/sg/lis/lis_u/lgd2016_u.htm

Strand, J. (2015). Modeling the Marginal Value of Rainforest Losses: A Dynamic Value Function Approach. *Policy Research Working Paper;No. 7445.* World Bank, Washington, DC.

Tedim, F., Xanthopoulos, G., & Leone, V. (2015). ForestFires in Europe: Facts and Challenges. Wildfire Hazards, Risks and Disasters, 77–99.

Tykhenko R., Tykhenko O., Openko I., Shevchenko O., Bavrovska N., Zhuk O., Tsvyakh O., Stepchuk Ya. (2021). The assessment of impact ecological stability of territory on the organization of rational land use of agricultural enterprises. *Scientific Papers Series "Management, Economic Engineering in Agriculture and Rural Development, 21*(2), 685–692.

Ubysz, B., & Szczygiel, R. (2002). Fire Situation in Poland. *International Forest Fire News*, *27*, 38–64.

Voronin, I. V., Senkevich, A. A. & Bugaev, V. A. (1975). *Ekonomicheskaya effektivnost' v lesokhozyaystvennom i lesomeliorativnom proizvodstvakh,* Moskow: Lesnaia.

Zhong, M., Fan, W., Liu, T., & Li, P. (2003). Statistical analysis on current status of China forest fire safety. *Fire Safety Journal*, *38*(3), 257–269. doi:10.1016/S0379-7112(02)00079-6

Zibtsev, S. V., & Borsuk, O. A. (2012). Protection of lisiv vid pozhezh u sviti that in ukraine – [wiki of the 21st century and prospect of development]. *Lisove i sadovo-parkove hospodarstvo, 1*. http://journals.nubip.edu.ua/index.php/ Lis/article/download/9630/8613

Chapter 24
Ecological Condition of River Basins of Western Polesie of Ukraine

Tetiana Basiuk
National University of Water and Environmental Engineering, Ukraine

Igor Gopchak
National University of Water and Environmental Engineering, Ukraine

Artem Yatsyk
National University of Water and Environmental Engineering, Ukraine

ABSTRACT

This section presents the results of the assessment of the ecological state of the river basins of the Western Polesie of Ukraine according to the level of anthropogenic load. The study was carried out based on systematic observations within the basins of small rivers, using the system model of the hierarchical structure «Basin of a small river». This model classifies the basins of small rivers according to the degree of anthropogenic loads exerted on them and makes it possible to monitor the state of the river basin according to various indicators within individual subsystems (radioactive contamination of the territory; land usage; river runoff usage; water quality). In order to improve and restore the already disturbed natural states of small river basins, a number of water protection measures have been proposed. The water management and ecological zoning of the basins of small rivers in the territory of the Western Polesie of Ukraine was carried out taking into account the quantitative and qualitative ecological and managerial indicators of the states of rivers and river basins.

DOI: 10.4018/978-1-6684-8248-3.ch024

BACKGROUND

Today, more and more attention is being paid to the issue of the ecological state of rivers and their catchment areas, because they are their main key elements. The growing influence of society on the environment, which leads to a significant deterioration of their ecological conditions, requires comprehensive research to analyze and evaluate anthropogenic changes in the basins of small rivers of Ukraine.

An integral indicator of the state of the environment of the catchment area is the state of the geosystem as a whole, the hydrological regime of the territory, water content, and water quality in the river. The final link in the pollution of this ecosystem is the water body itself, in which changes in the physical and chemical parameters of the environment occur, the introduction of an excessive amounts of biogenic, toxic, and radioactive substances, the massive growth of toxin-producing species of algae, the reduction of the reproductive function of aquatic biota, the inclusion of toxic substances into the biotic cycle, changes in the hydrochemical composition, physical and biological properties of water, contamination of drinking water sources occurs (Snizhko, 2001; Yatsyk, 2004; Processes..., 2016; Gopchak et al., 2019).

The anthropocentric approach to the assessment of the state of the aquatic environment for the water resources protection purposes has led to the disturbance of the ecological balance in river basins and the emergence of a number of water resource management and ecological problems, in particular: along with the degradation of small rivers, the biodiversity of the river ichthyofauna and fish productivity of the aquatic environment of the native fish species have decreased, water quality has deteriorated as well as it's recreational characteristics, there is intensive aging and silting of the river network. Particularly significant anthropogenic changes began to manifest themselves in recent decades. Currently, most of the river networks of Ukraine has been transformed both in terms of channels and basins, which require the introduction of a system of measures to improve their condition.

Anthropogenic load is an indicator of the magnitude of the permanent generalized impact of human activity on geosystems, which cause certain changes in their structural and functional organization (Tolstousov, 2007). The theoretical basis for determining the anthropogenic load on river basins is based upon the principles of modern landscape ecology and constructive geography, first described in the works of V. Vyshnevsky (Vyshnevsky, 2003).

The anthropogenic load on river basins should be considered from two perspectives: the transformation of the surface of the catchment basin: deforestation, plowing of the lands in the basin, land reclamation, irrational application of fertilizers or careless storage of pesticides, excessive grazing of livestock in river floodplains, settlement of the basin and transformation of the rivers themselves: regulation of the river bed, water intake and drainage, discharge of polluted water into the river, etc. (Yatsyk, 2004; Methodology..., 2012). The current ecological state of catchments is an indicator of anthropogenic load, primarily, on water and land resources and reflects their irrational use. This leads to the deterioration of the ecological condition in the river basin, in particular the small rivers of the Western Polesie of Ukraine.

Therefore, studies of anthropogenic transformations in small river basins to understand the correlation between the changes in those basins and the intensity of anthropogenic loads and are gaining particular relevance. The assessment of the anthropogenic load on the river basin is very important, primarily for the formation of environmental protection activities in the river catchment and understanding of what indicators have the greatest impact on its ecological condition. The study of anthropogenic influence, the degree of their manifestation and directions of further changes in river basins is an important element in the development of a further strategy for the restoration, preservation and management of these landscape complexes.

The theoretical, methodological and methodical aspects of the study of anthropogenic influence on the ecosystems of river basins have been described in the scientific works of scientists-geographers, hydrochemists, hydromeliorators and ecologists P. Vyshnevsky, O. Kirylyuk, S. Levkivsky, I. Myskovtsia, V. Morokova, O. Rybalova, T. Solovey, Z. Tymchenko, V. Khilchevskoho, O. Tsvetova, T. Yasenchuk, A. Yatsyk and others. The basin approach is the basis of research into the current ecological state of catchment areas, various aspects of anthropogenic impact on these geocomplexes, and the rational use of water resources (Kyryliuk, 2007; Stashuk et al., 2014; Methodology..., 2015). These scientific publications reflect the findings of applied research based on the above-mentioned methodological approach.

The purpose of this research is to assess the ecological state of the river basins of the Western Polesie of Ukraine according to the level of anthropogenic load.

Small rivers of the territory of the Western Polesie of Ukraine were chosen as the object of the study.

The determining factor in choosing the object of research was the availability of initial data for performing calculations and materials from monitoring observations, as well as the fact that according to natural, physical-geographical and socio-economic conditions, the data of the basins is reflective of the general state for the region of Western Polesie and provides the possibility of performing calculations and carrying out an assessment of their ecological condition.

METHODS AND TECHNIQUES

The assessment of the ecological state of the river basins of small rivers of the Western Polesie a of Ukraine was carried out using the criteria of anthropogenic load in accordance with the «Methodology for calculating the anthropogenic load and classification of the ecological state of the basins of small rivers of Ukraine» (Methodology…, 2007).

The logical-mathematical model of the hierarchical structure built according to the ecosystem principle (Figure 1) is intended for the classification of anthropogenic conditions in river basins and consists of an analysis of four subsystems (radioactive contamination of the territory; land use; use of river runoff; water quality) and the river basin as a whole.

This makes it possible to assess not only the general condition of the river basin, but also to form an idea of how changes in individual indicators of subsystems affect the condition of the entire basin system, which is very important for the formation of environmental protection activities in the basins of specific rivers. Using the logic-mathematical model «Basin of a small river», it is possible to gradually approximate for each river those load values that will not lead to the loss of the self-cleaning capacity of its ecosystems.

Each subsystem is characterized by a set of criteria and indicators, by comparison of which the state of the river basin is classified in relation to each indicator, and according to their assessments, the entire subsystem is classified. At the top level of the hierarchy is the «Coordinating Decision-Making Algorithm», where the level of anthropogenic load on the river basin is calculated based on the estimates of the lower level and the overall ecological condition of the river basin is assessed. As a result, the anthropogenic state of the river basin is assessed quantitatively and qualitatively, that is, each quantitative assessment has a qualitative characteristic and vice versa. Such a structure of the calculation model not only provides an assessment of the overall ecological state of the system, but also provides an opportunity to predict how a change in individual indicators of subsystems will affect the state of the entire system, its economic potential, therefore, to direct efforts to improve the indicator that most affects the state of the entire system. An important feature of the proposed system model is the ability to increase the number of indicators in subsystems and even introduce new subsystems during its improvement. In addition, it is positive that the evaluation of the states of the system and subsystems in this model is carried out in parallel in two directions - quantitative and qualitative: the qualitative state of the indicators of the subsystems is evaluated, and a quantitative measure is determined based on the set of states of individual subsystems, and on the basis of quantitative measures of individual subsystems the quantitative measure of the entire system is determined (Methodology... 2007; Yatsyk et al., 2015; 2018; 2019).

Figure 1. Structural diagram of the system logical-mathematical model and classification (assessment) of the state of the small river basin

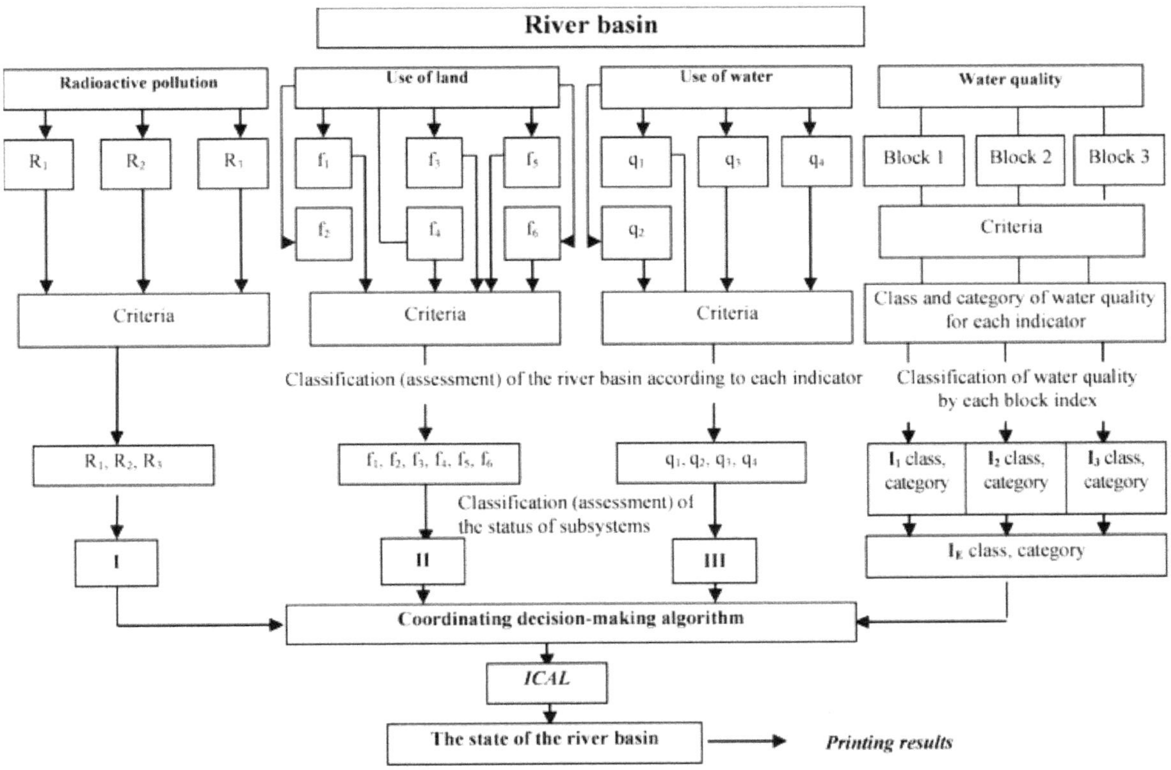

The initial data for calculating the anthropogenic load and assessing the ecological state of the river basin are:

1) Subsystem «Radioactive contamination of the territory» (C_i in Ki/km^2), which is evaluated by indicators:
   - cesium-137, (R_1) – (Cs_{137});
   - strontium-90, (R_2) – (Sr_{90});
   - plutonium-239, 240 (R_3) – ($Pu_{239\ i\ 240}$).

2) Subsystem «Land use», which is evaluated according to indicators of anthropogenic impact on land resources (f_i):
    - forest cover (f_1) – the ratio of the total area of forests, forest strips and tree-shrub vegetation to the total area of the river basin, %;
    - primordial state (f_2) – the ratio of the area of lands in their natural state (swamps, water areas, forests of natural and artificial origin, protective water protection plantations, protected areas, as well as the area of pastures, hayfields, deposits) to the total area of the river basin, %;
    - agricultural development of the territory (f_3) – the ratio of the area of all agricultural land to the total area of the basin, %;
    - arability (f_4) – the ratio of the area of arable land, including homesteads, gardens, orchards to the total area of the basin, %;
    - urbanization (f_5) – the ratio of the land area of settlements, industrial and transport enterprises to the total area of the basin, %;
    - erosion (f_6) – soil washing, t/ha per year.

The system of classification of the state of land use in river basins includes two types of classifications: classification of the level of land use in the river basin in relation to each indicator (*fi*); classification of the state of land use in the river basin according to the level of joint influence of all the indicated indicators (*fi*) on the state of the entire subsystem.

3) The «Usage of river runoff» subsystem in river basins is evaluated by indicators:
    - the actual use of river water (q_1);
    - irreversible water consumption (q_2);
    - discharge of water into the river network (q_3);
    - discharge of polluted sewage into the river (q_4).

Each value is calculated separately using data such as: volume of water intake from the river network (W_I); volume of river runoff losses due to the withdrawal of groundwater that is hydraulically connected to the river network (W_W); the factual volume of river runoff (W_F); volume of water discharge into the river network (W_D); the volume of discharge into the river network of polluted wastewater (W_{PW}).

The assessment of the state of use of river flow in the river basin is carried out, as in the previous subsystem, on the basis of two classifications: classification of the level of use of river flow by each indicator q_i and classification of the general state of the use of river runoff according to the level of impact of all the specified indicators, which characterize the impact of anthropogenic load on the state of the entire subsystem «Usage of river runoff".

4) The «Water quality» subsystem was evaluated in accordance with the «Methodology for ecological assessment of the quality of surface water according to the corresponding categories» (Methodology…, 1998) and «Methodology for establishing and using ecological standards for the quality of surface water of land and estuaries of Ukraine» (Methodology..., 2001) according to indicators, which were grouped into three blocks:
    - block of water salinity indicators (I_1);
    - block of trophosaprobiological (environmental and sanitary) indicators (I_2);
    - block of specific toxic substances (I_3).

On the basis of the obtained results for each block of indicators, an integral ecological index of water quality was determined (I_E), which is the arithmetic mean of the previous three indices.

Among all four subsystems, «Water quality» is the most significant - its contribution to the overall state of the river basin is 50%, while the contribution of the «Land use» subsystem is 30%, and the «River flow use» subsystem is 20%.

According to the methodology (Methodology..., 2007), the value of the primary indicators of the land and water resources use subsystem was transformed into points and a qualitative characteristic was given to each of them. Subsequently, a comprehensive indicator was calculated and the class of subsystem use was determined.

As a result, the level of combined influence of four subsystems on the state of the river basin is determined by the values of the induction coefficient of anthropogenic load (*ICAL*). The initial data for its calculation are the current values of the numerical measure of their state assessment determined for each subsystem. According to the value of *ICAL*, as a numerical measure of the qualitative state of the entire system, the state is determined: «good», «insignificant changes», «satisfactory», «bad», «very bad», «catastrophic». One or another assessment (class) of the state of the basin reflects the degree of anthropogenic load and the reaction of the ecosystem to this load.

With such a structure of the model, it is possible to assess not only the general state of the river basin, but also to create an idea of how changes in individual indicators of subsystems affect the state of the entire system of the basin as a whole. The application of this model extends to small and some medium-sized rivers. The general requirements and common criteria laid down in it are the basis for implementing water management and ecological zoning and clarifying the trends of changes in the ecological state of large river basins (Yatsyk et al., 2015).

The research was carried out using general scientific and special research methods. The following methods were used: theoretical; scientific generalization; system and complex analysis; geographically-hydrological; geosystem-hydroecological; water management-hydrological (for landscape - hydrological analysis and generalization of the results of ecological - hydrological observations); analytical (to assess the environmental condition); mathematically - statistical (for processing observations); GIS technologies (for building maps of small river basins).

The initial data for the research were statistical and cartographic data on the ecological state and the use of land and water resources in the basins of small rivers of the Western Polesie of Ukraine. Open sources of information of the Ministry of Ecology and Natural Resources of Ukraine, monitoring data of the State Agency of Water Resources of Ukraine, materials of reconnaissance studies of the Ukrainian Research Institute of Water Management Problems, data of the State Land Cadastre of Ukraine and the Institute of Land Management of the National Academy of Agrarian Sciences of Ukraine, intra-farm land management projects, materials of soil survey of lands and rivers, technical documentation on the establishment of water protection zones and coastal strips of rivers and reservoirs, regional schemes for anti-erosion measures, archival materials, river passports. The initial information was grouped, systematized and formed into relevant databases according to the direction of research.

The river network of the Western Polesie of Ukraine in the study area is quite well developed. This is explained by the sufficient amount of precipitation, geological structure, relief and temperature regime. The hydrographic network is quite dense, although its distribution is uneven. Most rivers have wide valleys with often swampy floodplains. The flow of rivers is slow (Palamaruk, Zakorchevna, 2001).

For this study, 24 river basins located within the territory of the Western Polesie of Ukraine were selected.

The rivers that flow in the study area belong to the Dnipro basin and are tributaries (right) of the Pripyat River. The hydrographic network of Western Polesie consists of 1506 rivers with a total length of 11.12 thousand km, of which 1494 are small rivers with a total length of 9.19 thousand km. For the rivers of the Western Polesie of Ukraine and their basins, a special effect is the draining of swamps. Small rivers, characterized by an insignificant current speed (0.1 – 0.2 m/s) turned out to be especially sensitive to such an influence (Yatsyk et al., 2018; Gopchak, 2021).

The river basins of the Western Polesie of Ukraine are under a considerable anthropogenic load. All processes that take place in the river basin and that contribute to the transformation of the rivers themselves, directly affect the river, its morphometric and hydrochemical characteristics, which in turn can act as an indicator of the anthropogenic load on the river basins.

When calculating the anthropogenic load in the river basins of the Western Polesie of Ukraine, the assessment of the state via the **«Radioactive contamination of the territory»** subsystem was not taken into account, since no radioactive elements were detected during the study. Therefore, it can be assumed that the ecological state of the river basin is assessed as «satisfactory» with a quantitative measure of 0. In the future, this subsystem will not affect the calculation of the induction coefficient of anthropogenic load.

«Land use» subsystem. In the basins of the vast majority of the studied rivers, land resources are used in an unsatisfactorily manner.

Among the indicators of the use of land resources, the indicator of forest cover of the basin occupies a special place. The forest is the main factor in the formation of the landscape and ecosystems, plays a major climate-forming role. The afforestation of the territory also has a positive effect on the water level of the river. In particular, the forest directly affects the amount of absorbed moisture and is therefore a factor that reduces surface runoff, and also creates more favorable conditions for the absorption of water by the soil, improves the conditions of groundwater nutrition, i.e. transfers surface runoff underground. The greater the forest cover of the basin, the smaller probability of occurrence and development of erosion processes. The lower the value of the coefficient, the less it corresponds with the natural (original) forest cover and the higher the impact of human economic activity (Khimko et al, 2003; Danylchenko, 2019).

Under the influence of the forest, the amount of precipitation increases by 3-8% or more, the surface runoff decreases sharply (by 2-10 times), the maximum losses decrease, and the share of underground runoff in the river runoff increases (Budz, 2000). The soil-forming and soil-protecting functions of forests do not require special explanation.

The forest cover of the study area fluctuates significantly (Figure 2). According to calculations, out of the 24 basins of small rivers of the Western Polesie, only 35% have forests occupying 33 or more percent of the territory, and in the rest, forest cover is mainly down to 27%. The largest forested areas are noted within the separate basins of the small rivers Veselukha (50.5%), Kormyn (49.3%), Lva (48.0%), Stubla (46.4%). The smallest indicators of forest cover are in the basins of the rivers Polkva (5.1%) and Chornohuzka (7%), Lypa (10.4%).

It is important to note that for small rivers of the Western Polesie of Ukraine, the optimal forest cover of watersheds is more than 50%, while the total forest cover for the zone of mixed forests is up to 40%. The average forest cover for the territory of Ukraine is 15.9%. For comparison: in Germany, 29.6% of the territories are occupied by forests, in France – 27.8%, in the USA – 24.3%. And this is an average indicator for all natural zones that are on the territory of these countries (Gopchak, 2021). With a much smaller ecological need for forests in Ukraine in general, and in Western Polesie in particular, the agricultural development of the country's territory exceeds 72%, and the plowed area reaches more

Figure 2. Forest cover of river basins of small rivers of the Western Polesie of Ukraine, %

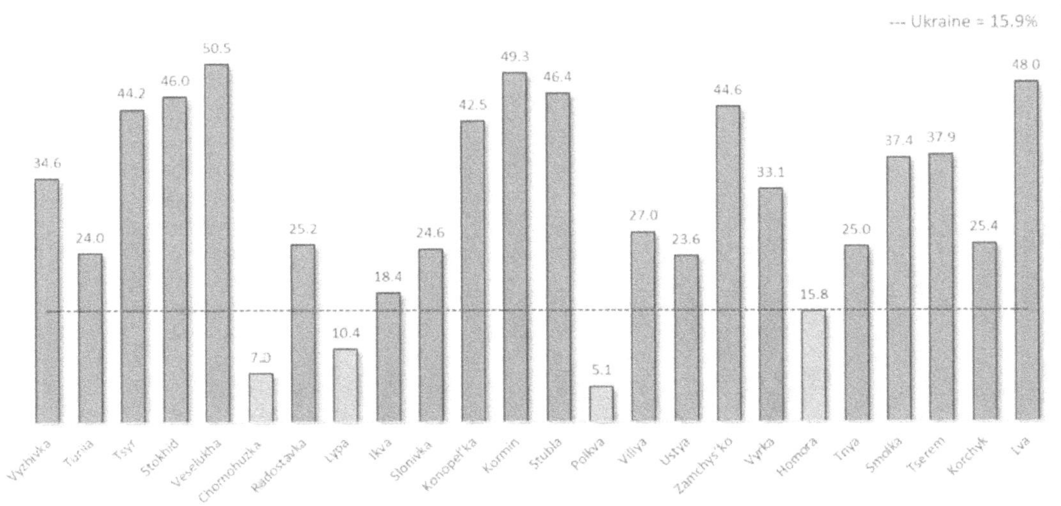

than 57%; the share of arable land in the total area of agricultural land is almost 80%, while meadows and pastures occupy only 12.9% of these lands (Yatsyk, 2004).

The share of the catchment area used by agriculture (agricultural development) varies from 31.6% (Stubla river) to 85.2% (Polkva river) with the prevailing values from 52 to 69%.

The plowability of watersheds also varies significantly - from 17.8% (Veselukha river) to 70.1% (Chornohuzka river) with predominant values of 40-60%. Particularly significant, are limits of fluctuations of such an indicator as as erosion - soil erosion, which fluctuates between less than 2 t/ha per year (Tsyr river) and 33.2 t/ha per year (Lypa river). Fig. 3 shows a map chart of indicators of erosion and plowing for small river basins on the territory of the Western Polesie of Ukraine.

Table 1 shows the state of the main indicators of land resources in the studied basins of small rivers. The obtained results show that 45.8% of rivers are rated as «below normal» and «unsatisfactory» in terms of forest cover; according to the degree of natural state of such rivers – 54.2%, according to agricultural development – 42.1%, according to plowability – 54.2%, according to urbanization – 20.9% and according to erosion – 54.2%. Therefore, the analysis of the state of most indicators of the use of land resources in the river basins of the region shows that almost half of them (except for urbanization) are unsatisfactory and far from ecologically acceptable.

The general assessment of the anthropogenic load and the state of use of land resources according to all indicators of this subsystem is given in the Table 2.

As evidenced by the research results (Table 1, Table 2), the state of land resource use in river basins is far from ecologically safe. Only the basins of the Veselukha and Cyr rivers have a «good» ecological status, which is 8% of the total number of studied small rivers. In five rivers – the Stohid, Kormyn, Stubla, Viliya, Lva rivers (21% of the total), the condition of their basins is assessed as «close to normal», 33% – the condition is «satisfactory». However, nine (38%) small river basins in the Western Polesie of Ukraine have an «unsatisfactory» and «extremely unsatisfactory» condition. The state of the Chornohuzka, Lypa, Slonivka, Polkva, Homora river basins was assessed as «extremely unsatisfactory».

Figure 3. Erodibility and plowing of small river basins in the Western Polesie of Ukraine

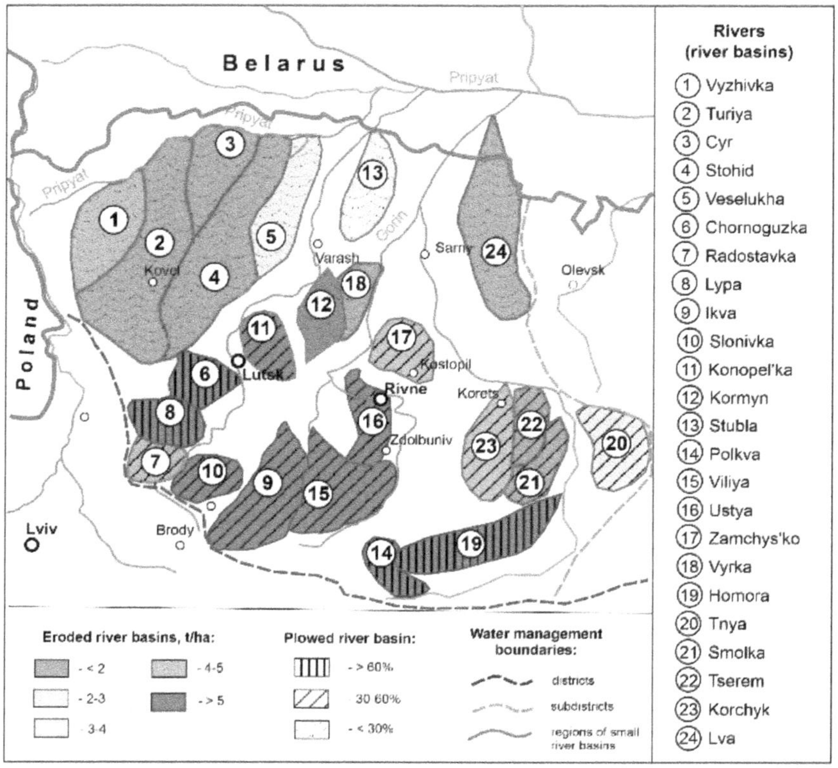

Table 1. The state of the main indicators of land resources in the basins of small rivers of the Western Polesie of Ukraine

Condition assessment	Indicator											
	Forest Cover		The degree of natural state		Agricultural mastery		Plowability		Urbanization		Erodibility - soil washing away	
	q'ty	%	q'ty	%	q'ty	%	q'ty	%	q'ty	%	q'ty	%
Good	4	16.7	1	4.2	9	37.5	7	29.2	10	41.7	8	33.3
Improved	7	29.2	5	20.8	5	20.8	4	16.7	6	25.0	2	8.3
Normal	2	8.3	5	20.8	0	0	0	0	3	12.5	1	4.2
Below the norm	5	20.8	4	16.7	2	8.3	3	12.5	4	16.7	3	12.5
Unsatisfactory	6	25.0	9	37.5	8	33.3	10	41.7	1	4.2	10	41.7
In total	24	100	24	100	24	100	24	100	24	100	24	100

Table 2. Classification (assessment) of the state of use of land resources of small river basins in the Western Polesie of Ukraine

Nº	River	Condition assessment		Nº	River	Condition assessment	
		ICAL	state			ICAL	state
1	Vyzhivka	1.1	satisfactory	13	Stubla	1.9	close to the norm
2	Turiya	1.0	satisfactory	14	vPolkva	-3.6	extremely unsatisfactory
3	Cyr	3.1	good	15	Viliya	1.4	close to the norm
4	Stohid	2.0	close to the norm	16	Ustya	-0.7	satisfactory
5	Veselukha	3.7	good	17	Zamchys'ko	0.6	satisfactory
6	Chornohuzka	-3.5	extremely unsatisfactory	18	Vyrka	0.8	satisfactory
7	Radostavka	-2.5	unsatisfactory	19	Homora	-3.1	extremely unsatisfactory
8	Lypa	-3.2	extremely unsatisfactory	20	Tnya	-3.0	unsatisfactory
9	Ikva	-1.4	unsatisfactory	21	Smolka	-0.6	satisfactory
10	Slonivka	-3.2	extremely unsatisfactory	22	Tserem	-0.4	satisfactory
11	Konopel'ka	0.0	satisfactory	23	Korhyk	-2.3	unsatisfactory
12	Kormyn	1.5	close to the norm	24	Lva	1.7	close to the norm

Thus, it can be concluded that the state of use of land resources in the basins of small rivers of the Western Polesie of Ukraine as a whole is far from ecologically safe. Almost half of the considered basins in terms of the state of land resource use are assessed as «unsatisfactory» and «extremely unsatisfactory» and only two river basins have a «good» state. The given conclusions are the basis for determining the appropriate measures to improve the state of land resource use of small rivers of the Western Polesie of Ukraine.

Subsystem **«Use of river flow»**. The conditions and state of the use of river water resources are closely related to the level and characteristics of economic activity. Direct withdrawal of water from rivers and underground aquifers has a significant impact on this state. The characteristics of the main indicators of their use are determined by the fact that in almost all rivers accepted for research, water intake is carried out directly from riverbeds and from underground aquifers, which have a hydraulic connection with the river and are drained by it. In some river basins (Cyr, Veselukha, Radostavka, Slonivka, Tserem), the entire water intake is the underground component of the river runoff, that is, water intake is carried out only from underground horizons. For half of the small rivers selected for the study, water intake from underground horizons is greater than or equal to withdrawal from the stream.

Almost everywhere, small rivers are the source of agricultural water supply – in certain rivers, up to 70% of water resources involved in economic use are spent on these needs. 10-25% of the total volume of river water used is used for household needs (Gopchak, 2021).

An important indicator of the state of use of water resources is the amount of irreversible water consumption, which quantitatively characterizes the degree of depletion of the river. In most small rivers, the decrease in river flow in average years due to water use does not exceed 10%; in low-water years, these values increase, reaching 25-40% in very low-water years. However, in the rivers Slonivka and Tserem, the discharge exceeds the intake. In general, the state of the water resources of the small rivers of the Western Polesie is particularly impacted by the total volume of water consumed by the river when taking water directly from the watercourse and underground horizons, as well as by the values of direct water consumption of the river runoff, inflows into the river network of sewage waters.

According to the results of the assessment (classifications) of the state of use of water resources of small rivers of the Western Polesie of Ukraine (Table 3), taking into account the joint influence of all indicators (qi) it was established that the state of the basins was assessed as «good» in 63%; in 21% – «satisfactory»; in 4% – bad; in 12% of the considered rivers, the value of the actual use of the river flow reached catastrophic limits (Chornohuzka, Ustya, Zamchys'ko rivers).

Table 3. Classification (assessment) of the state of use of water resources of small rivers of the Western Polesie of Ukraine

N°	River	Condition assessment		N°	River	Condition assessment	
		ICAL	state			ICAL	state
1	Vyzhivka	3.0	good	13	Stubla	3.0	good
2	Turiya	3.0	good	14	Polkva	2.2	satisfactory
3	Cyr	3.0	good	15	Viliya	3.0	good
4	Stohid	3.0	good	16	Ustya	-3.8	disastrous
5	Veselukha	3.0	good	17	Zamchys'ko	-2.8	disastrous
6	Chornohuzka	0.4	disastrous	18	Vyrka	3.0	good
7	Radostavka	3.0	good	19	Homora	0.8	bad
8	Lypa	1.4	satisfactory	20	Tnya	1.8	satisfactory
9	Ikva	2.2	satisfactory	21	Smolka	3.0	good
10	Slonivka	3.0	good	22	Tserem	2.2	satisfactory
11	Konopel'ka	3.0	good	23	Korhyk	3.0	good
12	Kormyn	3.0	good	24	Lva	3.0	good

The situation when no more than 10% of the flow from the river is used is considered to be the international norm of water use. If water use exceeds 20%, then the water body is among those unable to ensure the socio-economic development of the region in which it is located, which is observed in this case. In addition, the ecological situation worsens due to the fact that small rivers react quickly to the influx of sewage, even when their volume in relation to the actual volume of runoff is classified as satisfactory, because in some cases they are unable to dilute polluted sewage to an ecologically safe level.

The subsystem **«Water quality»** is intended for ecological assessment of the quality of surface water and classification of the state of the river basin according to the level of anthropogenic water pollution.

It should be noted that the chemical composition of river waters is quite diverse even within one natural zone. This is due to the physical and geographical features of their formation, which is valid both now and when economic activity was still insignificant and did not affect the quality of water. Currently, the variety of local physical and geographical factors has been joined by an even greater variety of factors of people's economic activity. A change in water quality can be the result of chemical, physical, bacterial and radioactive pollution (Snizhko, 2001).

To characterize the level of chemical pollution of the rivers of the Western Polesie of Ukraine, the main attention was paid to the following indicators: dissolved oxygen, which determines the self-cleaning capacity of the river; bichromatic oxidizability and biochemical oxygen consumption for 5 days, which characterize the presence and state of organisms, substances; compounds of ammonium, nitrite and nitrate nitrogen; total phosphorus, phenols and petroleum products. These indicators are used to control and assess the quality of surface waters and characterize a wide range of chemical and toxicological properties of natural waters and quickly respond to their changes and best reflect the dynamics of water quality.

According to the results of the calculations, the following was established:

1) Water pollution of each small river has its own characteristics related to the water content of the watercourse, the intensity of self-purification and the specifics of the economic use of the basin.
2) The regime of dissolved oxygen largely determines the chemical and biological state of the water body. In the summer months, the content of dissolved oxygen exceeding 10 mg/dm^3 has: Smolka (10.8 mg/dm^3), Viliya (11.4 mg/dm^3). Chornohuzka (10.2 mg/dm^3), Lypa (10.8 mg/dm3), Slonivka (10.2 mg/dm^3), Kormyn (10.5 mg/dm^3). The lowest content of dissolved oxygen in water (4-6 mg/dm^3) was recorded in the Hnylop'yat' and Huiva rivers. The poor state of the oxygen regime is caused by the discharge of polluted sewage, as well as due to the insufficient self-cleaning capacity of these rivers.
3) The most common pollutants in river basins are nitrates, ammonium nitrogen, petroleum products and phenols. Their high concentration in water indicates a violation of water quality standards. The waters of the Ikva, Lypa, Slonivka, Chornoguzka, and Gujva rivers are polluted with nitrates (from 0.6 to 80 MPC), and the waters of Hnylop'yat' - with organic substances (the values of dichromate oxidizability and biochemical oxygen consumption in 5 days are from 1 to 4 MPC). Downstream (from the source to the mouth) of the Ikva, Hnylop'yat', Huiva, Korhyk, Viliya rivers, there is a tendency towards increased water pollution, and in the Kormyn, Lypa, Veselukha rivers, some stabilization and reduction of pollution is observed downstream. The water of the Ikva River is especially noted for its high content of phenols and petroleum products (phenols 10-50 MPC);
4) under the influence of the constant discharges of polluted industrial and municipal wastewater, there is an increase in concentrations of polluting biogenic, organic, toxic substances that violate water quality standards in the Ikva and Hnylop'yat' rivers above the MPC. The water of some rivers is constantly polluted with nitrates, the concentration of which reaches: in the Lypa River – 0.283 mg/dm^3 (142 MPC), the Slonivka River – 0.083 mg/dm^3 (41 MPC), the Chornohuzka River – 0.115 mg/dm^3 (57 MPC).

Therefore, the quality of water in the rivers of Western Polesie is determined by the interaction of natural and anthropogenic factors. Pollution by organic substances is mostly from natural origins. Violation of the oxygen regime is mainly associated with anthropogenic pollution.

The step-by-step application of the «Basin of the Small River» model within the «Water Quality» subsystem, as well as the results of research and bacterial contamination data, made it possible to make a general assessment (classification) of water quality (Table 4). The rivers Radostavka, Ikva, Stubla, Vyrka have water of satisfactory purity (III quality class). Rivers the Slonivka, Lypa, Chornohuzka, Ustya have very polluted water (VI quality class).

Table 4. Classification (assessment) of water quality of small rivers of the Western Polesie of Ukraine

N° з/п	River	Water quality		N° з/п	River	Water quality	
		ICAL	state			ICAL	state
1	Vyzhivka	-1.0	polluted IV	13	Stubla	0.0	satisfactory cleanliness III
2	Turiya	-1.0	polluted IV	14	Polkva	-1.0	polluted IV
3	Cyr	-1.0	polluted IV	15	Viliya	-1.0	polluted IV
4	Stohid	-1.0	polluted IV	16	Ustya	-4.0	very polluted VI
5	Veselukha	-1.0	polluted IV	17	Zamchys'ko	-3.0	dirty V
6	Chornohuzka	-4.0	very polluted VI	18	Vyrka	0.0	satisfactory cleanliness III
7	Radostavka	0.0	satisfactory cleanliness III	19	Homora	-3.0	dirty V
8	Lypa	-4.0	very polluted VI	20	Tnya	-1.0	polluted IV
9	Ikva	0.0	satisfactory cleanliness III	21	Smolka	-1.0	polluted IV
10	Slonivka	-4.0	very polluted VI	22	Tserem	-3.0	dirty V
11	Konopel'ka	-1.0	polluted IV	23	Korhyk	-1.0	polluted IV
12	Kormyn	-1.0	polluted IV	24	Lva	-1.0	polluted IV

The calculation of the anthropogenic load and the assessment of its impact on the ecological systems of the rivers of the Western Polesie of Ukraine was carried out based on the results of the classification (assessment) of the state of the main natural systems (subsystems) - land and water resources, water quality according to chemical, toxicological, bacteriological and radiation pollution (Table 5).

Table 5. Assessment of anthropogenic load and ecological status of small river basins in the Western Polesie of Ukraine

N°	River	General condition of the basin		N°	River	General condition of the basin	
		ICAL	state			ICAL	state
1	Vyzhivka	0.43	satisfactory	13	Stubla	1.17	insignificant changes
2	Turiya	0.40	satisfactory	14	Polkva	-1.14	bad
3	Cyr	1.03	insignificant changes	15	Viliya	0.52	insignificant changes
4	Stohid	0.70	insignificant changes	16	Ustya	-2.97	very bad
5	Veselukha	1.21	insignificant changes	17	Zamchys'ko	-1.88	bad
6	Radostavka	-0.15	satisfactory	18	Vyrka	0.84	insignificant changes
7	Slonivka	-2.36	very bad	19	Homora	-2.27	very bad
8	Lypa	-2.68	very bad	20	Tnya	-1.04	bad
9	Ikva	0.02	satisfactory	21	Smolka	-0.08	satisfactory
10	Chornohuzka	-2.97	very bad	22	Tserem	-1.18	bad
11	Konopel'ka	0.10	satisfactory	23	Korhyk	-0.59	bad
12	Kormyn	0.55	insignificant changes	24	Lva	0.61	insignificant changes

According to the results of the conducted research (Table 5), there is not a single small river that is in a «good» ecological condition. In 33% of the studied rivers «changes are insignificant», 25% have a «satisfactory» condition, in 42% the condition is «poor» (the rivers Polkva, Zamchys'ko, Tnya, Tserem, Korhyk) and «very bad» (the rivers Slonivka, Lypa, Chornohuzka, Ustya, Homora). Overall, the general condition of the basins of small rivers of Western Polesie a (which implies the medium rivers as well) is generally "unsatisfactory".

So, despite all the complexity of the issue of normalizing the anthropogenic load on river basins, using the logic-mathematical model «Basin of a small river», it is possible to gradually establish for each river the values of the load that will not lead to the loss of the self-cleaning capacity of its ecosystems.

The tense ecological situation that has developed in recent decades in the subsystems of small rivers under the influence of anthropogenic load has led to losses or a sharp decrease in the self-cleaning ability of rivers. Calculations showed that in all river basins it is necessary to increase the forest cover and reduce the area of agricultural land and the rate of plowing. This will not restore the natural state of those territories and rivers, but it will help to reestablish lost functional connections, as well as contribute to the reproduction of, if not natural, then natural-anthropogenic environment of acceptable quality at the water intakes. The analysis of the state of river waters, taking into account the economic development of the regions and the impact of economic activity on them, indicates the possibility of reducing the anthropogenic load on the basins of small rivers of the Western Polesie. The conducted research and the obtained results of the performed calculations prove the possibility of improving the water quality of its rivers to the III class through certain changes in water management.

Fig. 4 gives a detailed diagram of the current ecological state of rivers. It not only allows you to visually show the achieved changes in the natural states of the studied rivers under the influence of anthropogenic load, but also makes it possible to identify the basins (rivers) with the greatest violations from the point of view of ecology, which require priority investments for their improvement. The given map chart is simultaneously a scheme of water management and ecological zoning of the territory of the Western Polesie of Ukraine.

To improve the ecological condition in the basins of small rivers of the Western Polesie of Ukraine, as well as for the purpose of further rational use, prevention and elimination of surface water pollution of rivers, a number of water protection measures were proposed (Figure 5). The purpose of pollution control measures is to improve the water environment of water bodies with moderate or low status, and to prevent the deterioration of the water environment in areas with good and very good water status. First, it is necessary to make measures for improvement, and secondly, to develop a set of protection measures.

The development of a set of water protection measures was primarily based on the basin management principle (EU Water Framework Directive 2000/60/EC), with basic principles being: comprehensiveness (coverage of all problematic situations and areas), interconnectedness (complexity of measures leads to their greater flexibility), sequentiality (allows to put the complex into operation block by block, increasing the efficiency of its work as a whole) and complementarity (consists in the expediency of developing special additional measures, which allows to ensure the formation of sufficiently sustainable territories in terms of ecology). During their development, consideration was made, that all the measures carried out must first of all be ecologically oriented, have a comprehensive nature, include organizational and economic, agroforestry and technical components, which will allow organizing monitoring of changes in the indicators of the ecological state of the territory, establishing their optimal values and ensuring optimal, safe for humans and the environment conditions for life and economic activity. The need to: establish water protection zones; monitoring the quality of surface and underground waters; preventing

Figure 4. Zoning map of the territory of the Western Polesie of Ukraine according to the anthropogenic load and the ecological state of the basins of small rivers

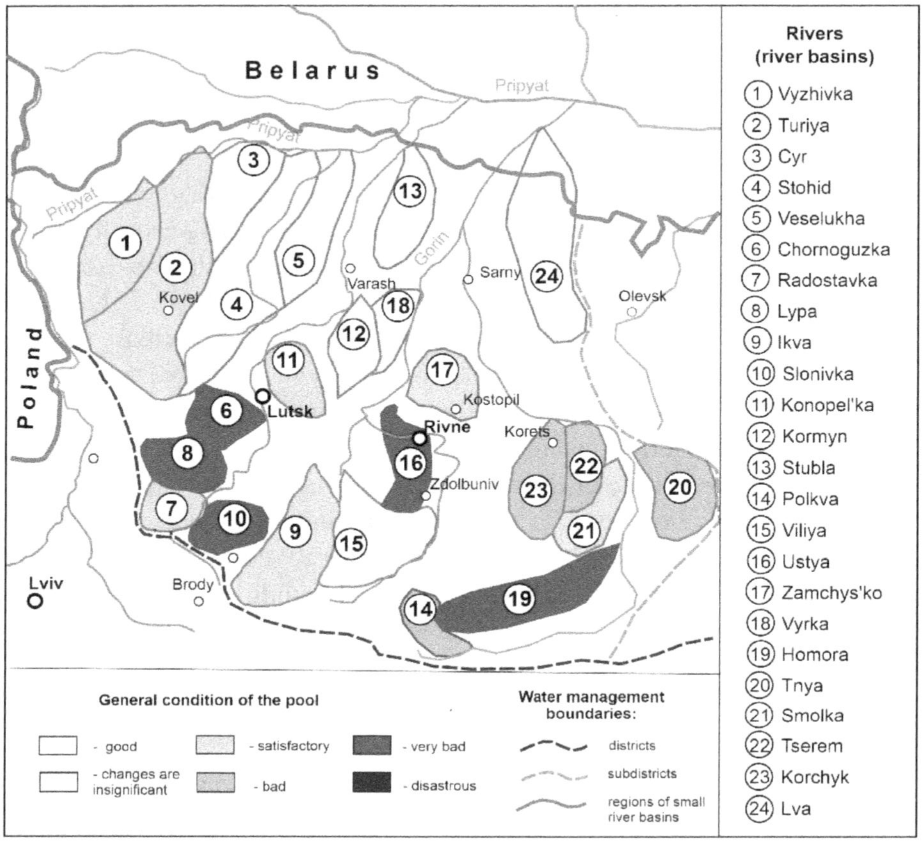

the degradation of agricultural land, etc. is foreseen. In the future, it is necessary to pay special attention to the use of land within the river basins and the pollution of their surface waters.

FUTURE RESEARCH DIRECTIONS

The results of the research can be used to improve the mechanism of management, preservation and revival of small rivers, namely, in the development of recommendations, technological and technical solutions for their protection from quantitative and qualitative depletion, prevention of harmful effects of water, etc. They contain methods of quantitative and qualitative assessment of the impact of individual indicators of anthropogenic factors, as well as characteristic pollutants that directly affect the ecological state of the river and its basin. This will make it possible to carry out its comprehensive assessment and determine the most affected areas of the river basins in terms of the level of anthropogenic load, which will contribute to the concentration of efforts of all managerial assets on the planning, development and implementation of water protection measures to improve their water management quality and ecological

Figure 5. Algorithm for carrying out environmental protection measures within the basins of small rivers of the Western Polesie of Ukraine

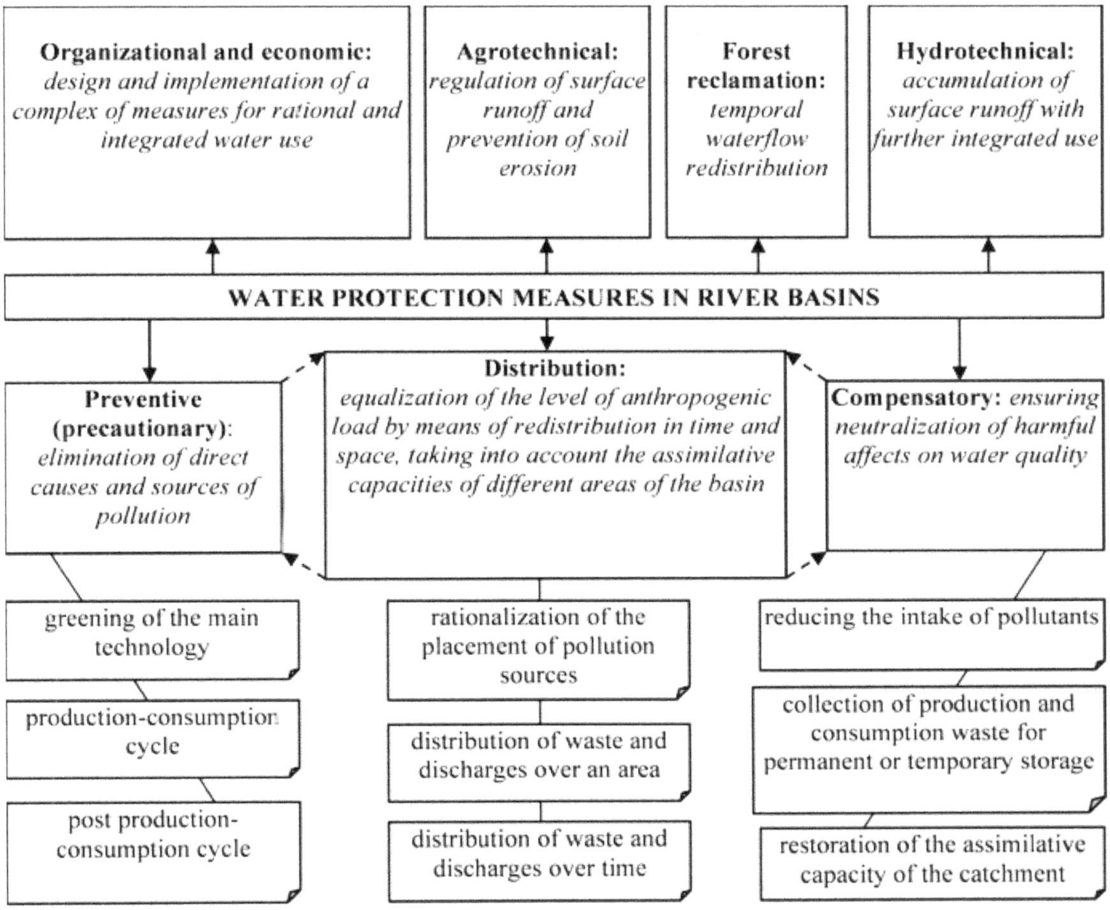

status, as well as optimization of decision-making in the water resources management of the basins of small rivers of the Western Polesie of Ukraine.

CONCLUSION

According to the results of the research, it was established that the current state of water resources within the territory of the Western Polesie of Ukraine is rather complicated.

The analysis of the state of the majority of indicators of the use of land resources in river basins shows that most of them (except for urbanization) are unsatisfactory and are far from ecologically acceptable. The actual use of river runoff reached catastrophic limits only in the basins of the Ustya and Hnylop'yat' rivers (24-25%), and in the basins of the Chornohuzka, Zamchys'ko, Homora rivers – bad (11-16%). The rivers of Chornohuzka, Ustya, Zamchys'ko and Hnylop'yat' are in a catastrophic condition. The water quality assessment of small rivers of the Western Polesie of Ukraine showed that there are no rivers with «very clean» and «clean» water among the small rivers studied. The rivers Slonivka, Lypa, Chornohuzka, Ustya belong to the VI quality class and are characterized as «very dirty». The integral assessment of the state of the basins of small rivers of the Western Polesie of Ukraine is characterized by the following indicators: 33% of the studied rivers have a rating of experiencing «insignificant changes», 25% have a rating of «satisfactory», 21% have a rating of «bad», 21% have a rating of «very bad», with no rivers being in a «catastrophic» state. In order to improve the ecological condition of all river basins, it is essential to increase the forested area and a decrease the area of agricultural land and the rate of plowing in the basins.

A comprehensive assessment of the ecological state of river basins made it possible to determine the most "dangerous" areas in terms of the level of anthropogenic load, which require priority environmental protection measures to preserve or restore the natural state of the ecosystem and ensure ecologically safe living conditions for the population. It is expedient to use water management and ecological zoning as a methodical basis for regional differentiation of water usage fees depending on the quality of water resources in a given area. This will contribute to the concentration of efforts of all managerial assets on planning, development and implementation of water protection measures to improve the water management and ecological conditions, as well as decision-making in water resources management of small rivers of the Western Polesie of Ukraine. All this outlines the perspectives for further research of small rivers of the Western Polesie of Ukraine, which should be focused on issues of detailed assessment of the ecological state of their basins.

REFERENCES

Budz, O. (2000). *Rozrakhunok maksymalnykh vytrat vesnianoi poveni pravoberezhnykh prytok r. Prypiati: dys. kand. tekhn. n.*: hidravlika ta inzhenerna hidrolohiia [*Calculation of the maximum costs of the spring flood of the right-bank tributaries of the Pripyat River*] [Dissertation, Rivne, Ukraina].

EU Water Framework Directive 2000/60/EC. (2006). Kyiv, Ukraina.

Gopchak, I. (2021). *Naukovi zasady zberezhennia ta vidrodzhennia malykh richok Zakhidnoho Polesie Ukrainy*: avtoref. dys. [Scientific principles of preservation and revival of small rivers of the Western Polesie of Ukraine. [Thesis, Kyiv, Ukraina].

Danylchenko, O. (2019). *Richkovi baseiny Sumskoi oblasti: heoekolohichnyi analiz*: monohrafiia [*River basins of the Sumy region: geoecological analysis*: monograph]. Sumy, Ukraina. 271

Gopchak, I., Basiuk, T., Bialyk, I., Pinchuk, O., Gerasimov, I. (2019). Dynamics of changes in surface water quality indicators of the Western Bug River basin within Ukraine using GIS technologies. *Journal of Water and Land Development, 42* (VII–IX), 67–75. . doi:10.2478/jwld-2019-0046

Kyryliuk, O. (2007). Istoriia stanovlennia baseinovoho pidkhodu u heohrafii ta ekolohichnomu rusloznavstvi [The history of the formation of the basin approach in geography and ecological river studies]. Nauk. vypysky Vinnytsk. derzh. ped. un-tu im. Mykhaila Kotsiubynskoho. Seriia: Heohrafiia. Vyp., *14*, 40–47.

Jatsyk, A., Pasheniuk, I., Gopchak, I., & Basiuk, T. (2019). Zoning of the territory of western polissya of Ukraine for anthrogenic loading and ecological status of small river basins. *Bichuk*.

Gopchak, I. (2018). Retrospective Analysis of the dynamics of changes in the surface water quality in the Western Bug river. *Land Reclamation and water Management.*

Yatsyk, A., Yatsyk, I., Hopchak, I., & Basiuk, T. (2019). Assessment of state of aquatic ecosystems of volynsk region as to the level of anthropogenic load. *UDC.*

NJD Science. (2020). Ecological state of the Oskil River in the Kharkiv Region. *Norwegian Journal of Development of the International Science, 43*(3).

Palamarchuk, M., Zakorchevna, B. (2001). *Vodnyi fond Ukrainy*: dovidkovyi posibnyk [*Water fund of Ukraine*: refe rence guide]. Kyiv, Ukraina. 392 s.

Osadchyy, V. (2016). Processes determining surface water chemistry [Processes determining surface water chemistry]. *Volodymyr Osadchyy*. Springer.

Snizhko, S. (2001). *Otsinka ta prohnozuvannia yakosti pryrodnykh vod [Evaluation and forecasting of the quality of natural waters].* Kyiv.

Stashuk, V., Mokin, V., Hrebin, V., Chunarov, O. (2014). *Naukovi zasady ratsionalnoho vykorystannia vodnykh resursiv Ukrainy za baseinovym pryntsypom*: *Monohrafiia* [*Scientific principles of rational use of water resources of Ukraine according to the basin principle*: *Monograph*]. Kherson, Ukraina.

Tolstousov, A. (2007). *Ekolohichna entsyklopediia [Ecological encyclopedia].* Kyiv, Ukraina.

Khimko, R., Merezhko, O., & Babko, R. (2003). *Mali richky – doslidzhennia, okhorona, vidnovlennia* [Small rivers - research, protection, restoration]. Ukraina.

Vyshnevskyi, V. (2003). *Antropohennyi vplyv na richky Ukrainy: avtoref. dys. na zdobuttia nauk. stupenia doktora heohr. nauk: 11.00.11 [Anthropogenic influence on the rivers of Ukraine: author's abstract.* [Thesis, Geogr. Sciences].

Yatsyk, A. (2004). *Vodohospodarska ekolohiia [Water management ecology].* Kyiv.

Yatsyk, A., Hopchak, I., & Basiuk, T. (2018). Otsiniuvannia antropohennoho vplyvu na zemelni resursy v mezhakh baseiniv malykh richok Zakhidnoho Polesie Ukrainy [Assessment of anthropogenic impact on land resources within the basins of small rivers of the Western Polesie of Ukraine]. *Vyp., 108*, 66–71.

Yatsyk, A., Hopchak, I., Basiuk, T. (2019). Normuvannia antropohennoho navantazhennia na baseiny malykh richok Ukrainy, yak osnova yikh zberezhennia ta vidrodzhennia [Normalization of the anthropogenic load on the basins of small rivers of Ukraine, as a basis for their preservation and revival]. *Hidrolohiia, hidrokhimiia i hidroekolohiia, 3*(54), 93-95.

Yatsyk, A., Hopchak, I., Pasheniuk, I., & Basiuk, T. (2015). Naukovi zasady normuvannia antropohennoho navantazhennia richkovykh baseiniv. *«ETEVK-2015»: zbirka dopovidei Mizhnarodnoho Konhresu.* [Scientific principles of normalization of anthropogenic load of river basins. *«ETEVK-2015»: a collection of reports of the International Congress,* Kyiv

Chapter 25
Features of the Hydrological Regime of Shatsk Water Bodies of the Volyn Region

Mykhailo Yatsiuk
Institute of Water Problems and Land Reclamation f the National Academy of Agrarian Sciences, Ukraine

Olena Sydorenko
Institute of Water Problems and Land Reclamation f the National Academy of Agrarian Sciences, Ukraine

Olena Tsvetova
Institute of Water Problems and Land Reclamation f the National Academy of Agrarian Sciences, Ukraine

Olga Turaieva
https://orcid.org/0000-0002-3367-404X
Institute of Water Problems and Land Reclamation f the National Academy of Agrarian Sciences, Ukraine

ABSTRACT

This is an analysis of the state of natural waters of the Shatsk lake in the conditions of climate change. Analyzed are hydrological regime of water bodies and hydrogeological state of areas of formation and feeding. It is noted that there is a slight increase in the levels of ground and pressure aquifers, and, accordingly, surface aquifers, which is mainly due to atmospheric factors. It is noted that the chemical composition of the water of the lakes of the Shatsk group has remained stable enough for decades and the formation of its composition is determined not only by the peculiarity of the water supply of the lakes, but also by the level of anthropogenic load on the water area. It was established that in order to improve the condition and water supply of the Shatsk lake, it is necessary to develop and implement measures to prevent and minimize the possible impact of harmful effects of water, develop integrated approaches and measures to optimize water use, improve water quality, conserve and restore water resources, ensure sustainable use, and development of natural reserve fund objects.

DOI: 10.4018/978-1-6684-8248-3.ch025

BACKGROUND

Preservation of unique natural complexes and water bodies for the future encourages the relevance of the study of nature conservation areas – as reserves of untouched and slightly disturbed nature. At the same time, the key role belongs to the study of the current state, features of functioning, and the degree of anthropogenic transformation of water bodies of the Shatsk lakes area, as well as the development of principles of integrated water management of the studied territory.

The first network of observation wells for studying the regime and balance of ground and pressure water was created after the construction of the Kopayiv drainage system by the Lviv Hydrogeological and Reclamation Expedition in 1969. The network consisted of three observation wells (between Svityaz Lake and Pripyat River valley; Svityaz Lake – Pulemetske Lake and Svityaz Lake – Krymno Lake), a water-balance site on the northern shore of Svityaz Lake, and gauging stations on canals and lakes, where routine observations were carried out (Formation of the regime of natural waters of the Shatsk lakes area in modern conditions, 2004).

To protect rare natural complexes in the area of Shatsk Lakes, Shatsk National Nature Park (32.500 hectares) was established in 1983. The nature of the lake complex, located on the Main European watershed, the territory of the park has no analogs in Ukraine.

Shatsk National Natural Park (SNNP) is included in the international biosphere reserve «Western Polesie», as well as in the European Bug Eco Corridor, which unites the park-reserve territories of Ukraine, Poland, and Belarus with an area of about 120 thousand hectares. Therefore, environmental monitoring in the territory of the Shatsk NNP and the valley of the Prypyat River (not only in its upper reaches) becomes especially relevant. This enables control of the changes in the components of the natural environment and makes timely management decisions. At the same time, it becomes possible to determine the patterns of changes in natural complexes under various anthropogenic loads, assess their consequences, and determine measures to neutralize negative processes.

The most important component of the Shatsk NNP is water resources, which include surface (lakes, rivers, artificial reservoirs) and underground (soil and pressure) waters. Their main users in the park are agriculture, the communal economy, and the population of the district, which directly takes water from lakes, rivers, and aquifers, using it for economic needs (Khomik N.V., 2013).

The natural waters of the Shatsk Lake District are a unique object of conservation that requires constant scientific attention to assess and forecast their level regimes and hydrochemical characteristics, the changes of which depend both on natural factors and anthropogenic activity in the basin of the

Western Bug River, to which the Shatsk Lake District belongs to. Its territory is located in the watershed area of the Western Bug and Pripyat rivers, which practically coincides with the boundaries of the Shatsk National Nature Park (SNNP). Within these boundaries, there are 23 lakes with a total area of 6338.9 hectares (13% of the area of the Shatsk NNP), among which Svityaz Lake is the deepest in Ukraine (58,8 m) and the largest in the area among the Shatsk Lakes.

The main water arteries of the Volyn Polesie are the Western Bug and Prypyat Rivers, as well as the tributaries of the Western Bug River – Kopaivka and Ryta. But only the Kopaivka (2.0 km) and Ryta (about 1,0 km) Rivers are included in the territory of the Shatsk National Nature Park.

The Prypyat River, the right tributary of the Dnieper River, belongs to the Black Sea basin, and the Western Bug River is the right tributary of the Vistula River and belongs to the Baltic Sea basin. The line of the Main European watershed between these rivers, due to the flatness of the territory, passes through the elevations along the chalk ridge. From the watershed line, the sloping plain extends further north beyond the group of Shatsk Lakes to the Mukhovets River. Svityaz Lake is separated from the Prypyat Valley by a low watershed, located lower on the slope, and has a watermark 4 m lower than the water level of the Prypyat River. The underground watershed exceeds the river level by 1.5 m on average, and the lake level by 5 m.

The Kopaivka and Ryta Rivers are the right tributaries of the Western Bug River. The Kopaivka River originates from Luka Lake (it receives water through artificial channels from Svityaz and Pulemetske Lakes) and in the territory of Shatsk Park, it is almost channelized. The Ryta River flows in the extreme west of the park territory on the border with Belarus. It originates from Krymno Lake. The Ryta River is the right tributary of the Maloryta River. Kopaivka River is a river in the northwestern part of the park, and currently, it is completely channelized. It has a total length of 57 km, a catchment area of 264 km^2, and a slope of 0,6 m/km. It has a length of 25.6 km within the Shatsk district. To the north of Hrypsk village, it crosses the border with Belarus and flows into the Western Bug River at the point of 376 km from its mouth north of Domachevo village of the Brest region of Belarus.

The wetlands of the Shatsk NNP are included in the List of Wetlands of Ukraine, which, according to the Ramsar Convention on Wetlands, are of international importance, mainly as a habitat for waterfowl. In total, they occupy almost 9% of the territory of the park, of which swamps make up about 2.7%. In total, there are approximately 20 swamps in the territory of the park.

The purpose of the research is to analyze the state of the natural waters of the Shatsk Lake area in the current climatic conditions.

To achieve this goal, scientists conducted monitoring studies of natural waters in the territory of the Shatsk district, analyzed the hydrological, hydrogeological, and chemical regime of water bodies, established the impact of climatic and anthropogenic factors on the water resources of the Shatsk lakes area. Analytical and field research methods were applied. The methodology of the works included a reconnaissance survey of the existing monitoring network, control measurements of groundwater levels, hydrological measurements, and analytical calculations of water balances.

The physical and chemical elements of natural water quality were analyzed in the laboratory according to the normative and methodological documentation in force in Ukraine. Sampling was carried out quarterly using a monitoring network of observation wells and gauging stations.

ORGANIZATION AND CONDUCT OF OBSERVATIONS (MONITORING)

Because of the increasing anthropogenic load on the natural environment, it is clear that the uncontrolled use of natural resources leads to negative consequences, so there is a need of obtaining systematic objective information about the state of the natural environment.

Since 1972, a network of observation wells has been operating in the territory of the Shatsk NPP to study the natural and affected by reclamation measures water regimes. Before 1990 all observation wells were concentrated in the southern part of the Kopaiv drainage system and did not cover the entire currently studied territory with stationary observations.

In 1990, a new monitoring network (33 wells) was formed in the area of influence of reclamation systems and protected areas of different statuses for comprehensive monitoring. Hydrogeological and ameliorative zoning was used to establish this network. To date, 18 observation wells are operating in the Shatsk NNP (Natural Resources Management in Polesie, 2017).

Three observation wells are equipped in the protected areas of the park between the Pisochne and Moshno lakes. The observations allow us to determine the degree of changes in natural complexes under the influence of economic activity in the protected area.

In the buffer zone, 9 wells located in the latitudinal direction from Pisochne Lake to Pulemetske Lake were restored. The depth of wells for the pressure water table is 40-60 m, and for groundwater - 5 m. The wells are located in the zone of influence of the Kopayiv drainage system, which allows for assessing the impact of drainage reclamation on the protected lands of the park adjacent to the system. To control the state of the ground and surface waters in the inner zones of the park, one observation well for groundwater and one gauging station for large lakes were organized. Observations on the table regime and chemical composition of groundwater (in Quaternary sediments) and pressure (in Upper Cretaceous sediments) aquifers are carried out through a network of observation wells. Observations of lake levels are carried out by benchmarks tied to the above wells, as well as by previously identified wells (Yatsiuk M.V. et al, 2022a).

In the watershed area of Svityaz Lake – Pripyat River, 12 observation wells were previously equipped. The distance between them is 500 m. This number is sufficient for all types of control - hydrochemical, level, and water balance.

Observations on underground and groundwater tables in the territory of the Shatsk NNP are carried out using the observation wells of Pisochne and Moshno Lakes (N° 4y, 4н, 5y, 6y) as well as Lutsymer and Ploske Lakes (N° 12y, 13y, 14y, 15y). Observations on underground tables are carried out using the wells near Moshno (N° 4y, 4н), Svityaz (N° 12y, 13н), and Chorne Velyke Lakes (N° 2y, 2н). Some wells provide the observations by groundwater table, while others by the pressure water table.

The carried-out observations are the basis for constant monitoring of the state of natural waters in the territory of Shatsk NNP and the impact on them and the reclamation systems bordering the park (Natural Resources Management in Polesie, 2017).

RESULTS OF RESEARCH

According to morphometric indicators, most lakes are small and shallow. The distribution of lakes by their functional zones is given in Table 1. The surface area of the lakes ranges from 2622.0 to 1.9 hectares, the maximum depth is from 58.4 to 1.8 m, and their water volume is from 180.8 to 0.1 million m^3. Eight lakes (Svityaz, Pulemetske, Lyutsymer, Luky, Ostrivyanske, Pisochne, Krymno, Peremut) have an area of the water mirror is more than 100 hectares, and the maximum depths exceed 10 m in four of them (Svityaz, Pulemetske, Pisochne, Lyutsimer). The largest lake in the park territory and the deepest one in Ukraine is Svityaz Lake.

Small own surface water collection is typical for the lakes. The bank line in large and deep lakes is clearly defined, and in very small ones (Navrattya, Kruhle, Dovhe, etc.) it is indistinct and merges with the surrounding swamps. The banks of small lakes are very peaty, the bottoms are muddy. They overgrow intensively, turning into swamps.

The main sources of replenishing the water capacity of the lakes are precipitation, pressure feeding, and inflow from adjacent territories. The discharge components of the water balance are water evaporation from lake water surfaces and water infiltration into the soil horizons located below. For deep lakes located in chalk deposits, the main source of feeding is headwater and precipitation, and for shallow ones – precipitation, surface, and soil runoff. All this determines the intensity of water exchange in the lakes and their hydrochemical and hydrothermal regimes. The results of long-term studies on the water balance of the lakes of Svityaz, Pulemetske, Lyutsymer, Luky, Ostrivyanske, Pisochne, Krymno, and Chorne are given in Table 2.

Table 1. Distribution of the lakes of the Shatsk NNP by functional zones

Nº	Lake name	Total area, ha	Area, ha by functional zones			
			Conservation area, ha	Area of regulated recreation, ha	Area of fixed recreation, ha	Economic area, ha
Lakes directly subordinated to the park						
1.	Svityaz	2622.0	92.0	2530.0		
2.	Pulemetske	1568.0		1568.0		
3.	Luky	673.2		673.2		
4.	Lyutsymer	430.0		430.0		
5.	Ostrivyanske	255.0		255.0		
6.	Krymno	147.0		147.0		
7.	Peremut	142.0		142.0		
8.	Chorne Velyke	83.0		83.0		
9.	Moshne	36.0	36.0			
10.	Chorne Male	31.0	31.0			
11.	Dovhe	19.0	19.0			
12.	Plotychchya	11.0				11.0
13.	Kruhle	9.0	9.0			
14.	Klymivske	29.0	29.0			
In total		**6055.2**	**216.0**	**5828.2**		**11.0**
Lakes of other land users						
15.	Pisochne	187.0			187.0	
16.	Somynets	43.0				43.0
17.	Karasynets	15.0				15.0
18.	Ozertse	13.7				13.7
19.	Lynovets	9.0				9.0
20.	Oleshno	5.9				5.9
21.	Rytets	4.4				4.4
22.	Zvedynka	3.8				3.8
23.	Navrattya	1.9				1.9
In total		**283.7**			**187.0**	**96.7**
In total in SNNP		**6338.9**	**216.0**	**5828.2**	**187.0**	**107.7**

Table 2. Water exchange of the lakes, millions m^3/year

Water exchange factors	In total	Lakes							
		Luky	Lyutsy-mer	Chorne	Svi-tyaz	Ostri-vyanske	Pulemet-ske	Krym-no	Pisoch-ne
Water replenishing									
Precipitation	33.0	3.7	2.3	0.4	14.8	1.3	8.8	0.7	1.0
Surface influx	12.7	1.3	1.9		2.7		1.4	4.2	0.2
Underground influx	10.7	1.3	1.4	0.2	5.6	0.3	1.7		0.2
In total	**56.4**	**6.3**	**6.6**	**0.6**	**23.1**	**1.6**	**11.9**	**4.9**	**1.4**
Water discharge									
Evaporation	37.3	4.1	2.6	0.5	16.6	1.6	9.9	0.9	1.1
Surface outflow	10.7	1.0	4.2	0.1	1.2	0.1	—	4.1	—
Underground outflow	9.7	1.2	0.3	0.2	5.7		2.1	0.2	—
In total	**57.7**	**6.3**	**7.1**	**0.8**	**23.5**	**1.7**	**12.0**	**5.2**	**1.1**
Lake volume, million m^3		**13.0**	**19.0**	**2.3**	**180.0**	**5.9**	**72.0**	**5.7**	**5.5**

The most stable water regime is inherent to the largest lakes of the park Svityaz and Pulemetske, as well as to the lake of Ostrivyanske, which is connected by a hydrographic network only with Pulemetske Lake. The stabilizing factors of the water regime of the first two lakes are a significant volume of their water mass, a large area, as well as a close connection with the transit underground flow, which is the main regulator of the water system of the region. The surface inflow to these lakes is not very large, and their water balance is formed mainly due to precipitation, evaporation from the water surface, and water exchange with the pressure aquifer of Upper Cretaceous sediments.

Lyutsymer and Krymno Lakes being a part of the transit water system, which naturally connects the upper reaches of the Prypyat and Ryta Rivers, are characterized by the greatest dynamics of water regime and intensity of water exchange. The main source of feeding for these lakes is the surface inflow from catchment areas. The underground inflow, which is formed by local infiltration and soil runoff, plays an essential role in the formation of the water regime of Lyutsymer Lake. The water regime and balance of Krymno Lake are completely dependent on the surface inflow from the side of Lyutsymer Lake and the adjacent swamp massif as well as on the outflow to the Ryta River, the regime of which entirely depends on its artificial regulation.

The water balance of Svityaz Lake during the observation period until 2015, had a rather stable character with certain deviations from the average indicators of both individual components of the balance and its resultant, which influenced the level of the lake mirror. The amplitudes of fluctuations of the latter were in the range of 0.2-0.4 m, having long-term maximums up to 0.5-0.6 m. The main replenishing sources of the water balance were precipitation (50-80%), and pressure feeding (20-50%) with the subordinate value of the surface inflow and groundwater inflow. The main discharge sources were evaporation (up to 90%), replenishment of soil flow (10-15%), and surface leaks (5-10%). The last component of the water balance was not unambiguous due to various attempts to artificially regulate the leakage. The stabilizing factor in maintaining the water balance was pressure water, the inflow of which always rose when increasing water discharge and decreasing the lake level (Yatsiuk M.V. et al, 2021).

The period from 2013 to the present, when a steady trend of lowering the water level of Svityaz Lake was observed, was anomalous in the formation of the water balance of the lake. Against this background, from 2015 to 2018, the level stabilized slightly lower, and from 2018 to the end of 2019, it fell to a multi-year minimum, below the absolute mark of 163 m. This period is associated with an increase in annual air temperatures by an average of 2°C, evaporation up to 924 mm, and a decrease in the levels of groundwater and pressure water by 1.0-1.2 m. Much greater than the decrease in the lake level the drop in the groundwater table caused an increase in the discharge from it to feed the soil flow, which reached 575 mm in the total water balance (2018), or 45% of the total discharge, which has not been observed during the entire observation period. The share of the pressure feeding of the lake increased, which, however, could not fully compensate for the increase in discharge for evaporation and replenishment of groundwater.

The water regime and balance of Luky Lake are formed by surface inflow and outflow, underground water exchange, and meteorological factors. Surface inflow to the lake occurs from the side of Svityaz Lake, and the outflow – into the reclamation network by filtering through the soil. The water exchange with the lower aquifers is balanced.

The water balance of Pisochne Lake is formed mainly by precipitation and evaporation from the water's surface. The catchment of the lake occupies a small area, therefore, depending on the formation of moisture reserves in the forest catchment, water replenishment may occur due to surface inflow during the period of snow melting.

Chorne Lake has a specific water regime. It practically does not have its catchment, but there is a constant flow to Lyutsymer Lake. The lake is characterized by intensive water exchange, the components of which are precipitation, evaporation, water exchange with the lower-pressure aquifer, and surface outflow.

Fluctuations in the water level of the park's lakes are determined by the ratio of water balance components, primarily precipitation, which plays an essential role in the replenishing part of the water balance of the territory. So, the main feature of the hydrological regime of the park's lakes is a relatively stable water level, which fluctuates by the seasons of the year. Maximum water levels are observed in spring and autumn time, and minimum water levels – in summer and winter time. The maximum rise in the water level is observed in the period from the end of April to the beginning of May after the melting of snow and ice, and the minimum is observed in the period from the end of August to the beginning of September.

An important feature of the water balance of the lakes of the Shatsk NNP is their interconnecting by flow redistribution channels, which were created over a long period and which currently ensure the uncontrolled flow of a significant amount of water both within the territory and beyond it (Romashchenko M.I. et al, 2020).

The specificity of the location of the natural park is that its water sources are practically not connected with the surface flow of the water arteries limiting it – Pripyat and Western Bug Rivers, and water resources are formed both due to precipitation and the water of marl-chalk deposits, the feeding area of which is outside the territory of the park.

The nature of the geological structure and diversity of the lithological composition of water-bearing rocks caused a great variety of conditions of groundwater formation in different stratigraphic complexes, which, in general, forms a complex hydraulic system.

Quaternary sediments are watered everywhere and they are the first aquifer complex from the surface. Sandy water-bearing rocks are interlayered with moraine loams and form a series of aquifers of complex configuration, which are hydraulically interconnected and drained by surface watercourses.

The underlying Quaternary fissured marl and chalk rocks contain pressure waters, the feeding area of which is located within the main watersheds of the Volyn-Podilsky Upland. On the territory of Volyn Polesie, the discharge of this aquifer occurs in floodplains, marshes and lake depressions (Dyatel O.O. et al, 2019).

The specified aquifers make up the zone of active water exchange, feed water bodies, and form surface runoff from the studied territory. Groundwater is fed by precipitation and the inflow of pressure water. For this district, the area of groundwater supply coincides with the area of its distribution. Water discharge occurs due to partial drainage through a network of natural and artificial watercourses, replenishing the lakes, as well as evapotranspiration (Rokochinskiy, A. et al, 2020).

By the data of long-term observations, well-expressed seasonal fluctuations in the groundwater table (GT) caused by weather factors were recorded. Long-term cycles of fluctuations of the GT were estimated, the amplitude of which in dry years is 1.0-1.4 m, in average and wet years – 0.6-0.8 m. Fluctuations of the piezometric level of pressure waters of the Upper Cretaceous sediments relative to the GT were recorded with a small (for 2 weeks) shift of maximums and minimums. That is, seasonal fluctuations of pressure and groundwater levels occur under the influence of the same factors.

It was specified that the aquifers are hydraulically interconnected, and there is an intensive water exchange between them. The ratio of levels and their nature allow us to determine the area of water exchange and its temporal and spatial trends.

Recently, there has been an increase in anthropogenic load on the natural environment, so it has become obvious that the uncontrolled use of natural resources leads to negative consequences, which, in turn, determines the need to obtain systematic objective information about the state of the natural environment. This can be achieved through the organization of ecological monitoring, which is the system of complex observations of natural environment components (water, soil, plants, etc.) at the level of geographical systems (geosystems), assessment, and forecast of their conditions. It is necessary to provide this information to interested national management bodies (Yatsiuk M.V. et al, 2021).

Climate change is the most significant factor of negative impact on the water supply of the Shatsk Lake area. The modern climate of the Polesie of Ukraine is characterized by uneven warming, pronounced in winter and summer months. Over the past 120 years, the average annual air temperature in Polesie has risen by 1.8 °C (Svityaz weather station). Based on the observation results at the Svityaz meteorological station for the observation period of 1985-2022, it was established that the most precipitation fell in 2010 (863.0 mm), and the least in 1986 (471.2 mm). During the above observation period, the highest average annual air temperature was observed in 2019 (+10.4°C), and the lowest one was in 1987 (+6.0°C). The dynamics of air temperature and precipitation during 2021-2022 differed from long-term data (Figure 1-2). The average annual air temperature rose by 1.1°C in 2021, and by 3.3°C in 2022 compared to the norm for the same period. Average monthly air temperatures (by the Svityaz weather station) in 2022 are higher than in 2021.

Figure 1. Dynamics of multi-year average air temperature and precipitation for 1951-2022
(Svityazh weather station)

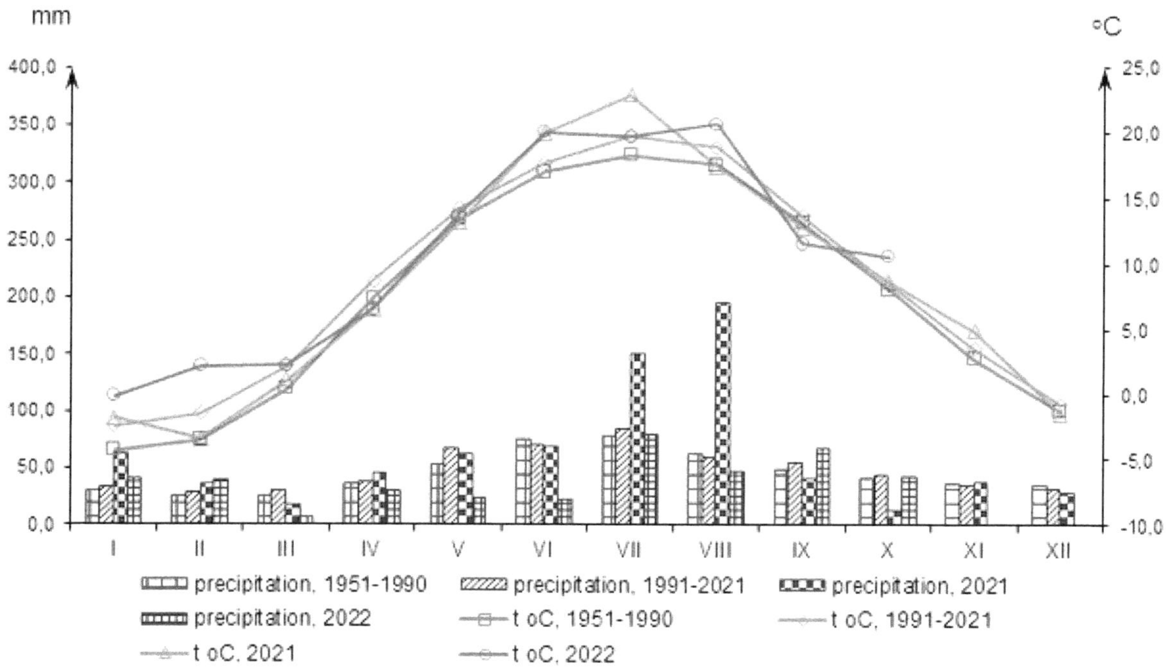

Figure 2. Changes in the amount of precipitation in 1991-2022 relative to the climatic norm (1951-1990)
(Svityaz weather station)

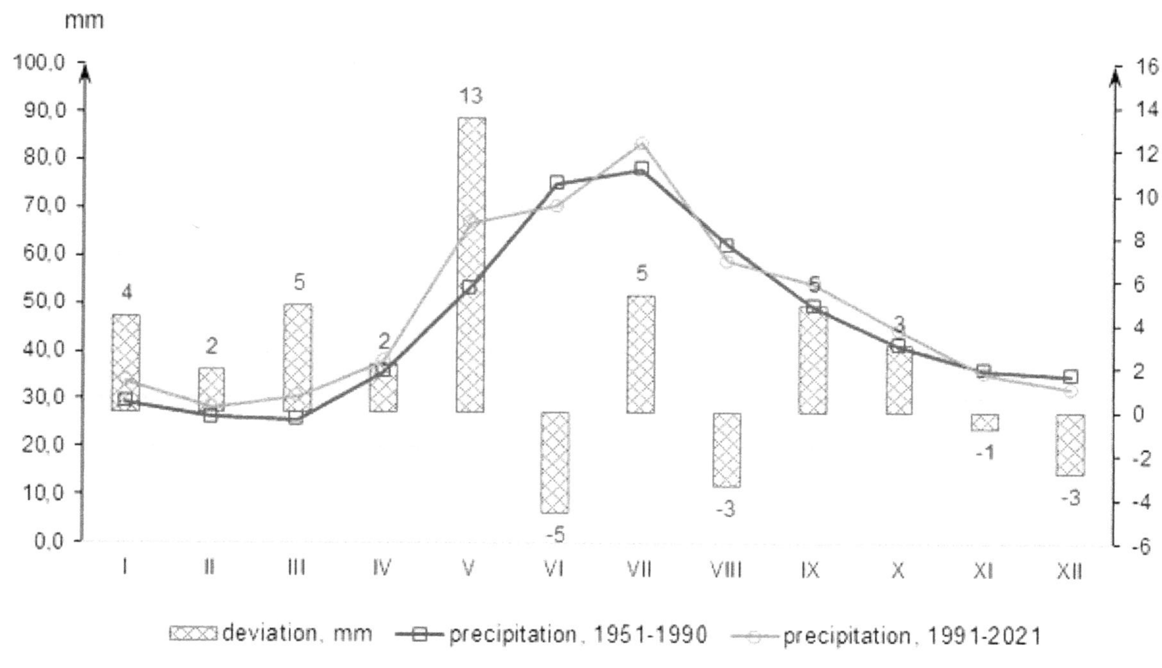

The hydrothermal conditions of the territory are characterized by the separate indicators of soil and air temperature and moisture regimes, as well as a complex indicator – the Selyaninov hydrothermal coefficient (HTC), the calculation results of which show that over the last decade, 2018 and 2019 were the most critical (HTC was equal to 1.01 and 0.95, respectively), which indicates climate aridification. But in 2021, this indicator increased and amounted to 1.8, which is explained by the abnormal amount of precipitation in May, July, and August, that is, according to Selyaninov's classification, this area is classified as a zone of excessive moisture, which is typical for the Polesie zone. Therefore, it is possible to conclude the unevenness of precipitation (Yatsiuk M.V. et al, 2022b).

A hydrological feature of the territory of the Shatsk National Nature Park is the close hydraulic and dynamic interconnection of aquifers, which do not have a clearly defined waterproof layer. This regularity indicates that the violation of at least one component of the water regime leads to the deterioration of the state of all components of the water ecosystem. Changes in the hydrological situation can affect the redistribution of water masses, which, in turn, can lead to changes in the natural state of wetland complexes, which, by the Ramsar Convention, have been of international importance since 1995, as part of the «Western Polesie» International Biosphere Reserve.

Groundwater is fed by precipitation through the infiltration process and, to a lesser extent, by the inflow of surface water during floods. In some parts of the lake area, where there is no siltation zone in the lakes, groundwater is fed by the flow of pressure waters of the Upper Cretaceous aquifer (Yatsiuk M.V. et al, 2022c).

Analyzing the hydrological and hydrogeological state of the territory that developed during 2022, it should be noted that there is a slight increase, compared to 2019, in the tables of groundwater and pressure aquifers, and, accordingly, of surface aquifers. This is mainly explained by atmospheric factors (Figure 3-4).

Figure 3. Graphs of fluctuations of groundwater table of Quaternary sediments in wells 13y, 14y and 15y

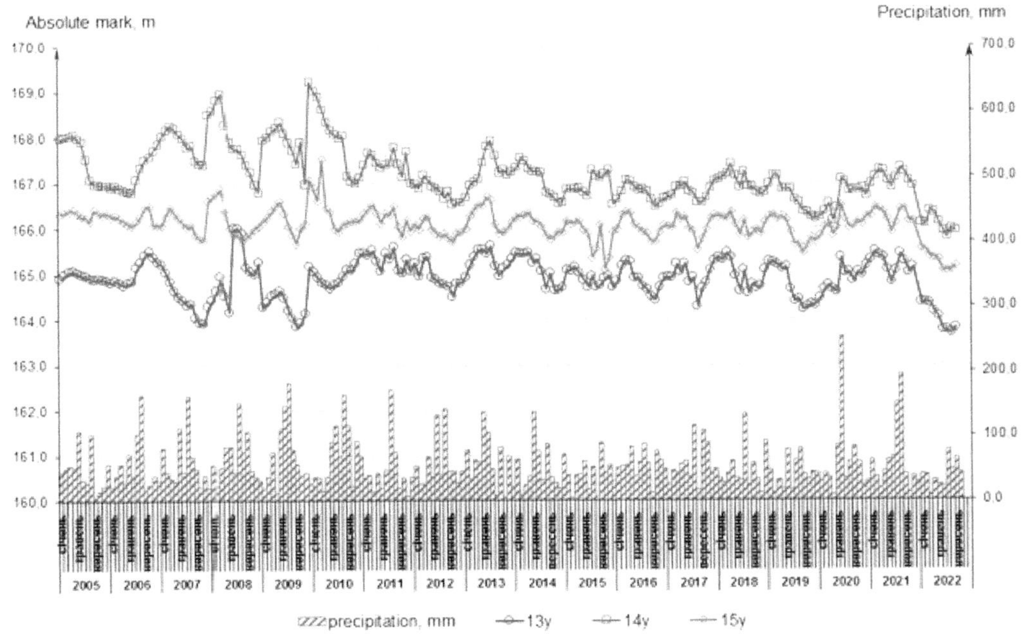

The average level of Svityaz Lake in 2020 was 163.05 m above sea level, the maximum was 163.19 m, and the minimum was 162.95 m. In 2021, the average water level in the lake was 163.26 m, the maximum was 163.32 m, and the minimum was 163.13 m. For 9 months of 2022, the average level was 163.37 m, the maximum was 163.47 m, and the minimum was 163.28 m (Figure 5-6). Just for comparison: the highest level of Svityaz Lake was observed in 1999 – 163.76 m, and the lowest ones were observed in 1973 (162.91 m) and 1974 (162.96 m) after drainage reclamations were carried out, but later the level stabilized and fluctuations occurred by the seasons of the year and cyclically «dry-wet» years, depending on meteorological conditions. A very low level in the lake was also observed in October 2019 – 162.94 m.

Figure 4. Graphs of fluctuations of groundwater table of Upper Cretaceous sediments in wells 2н, 4н and 19н

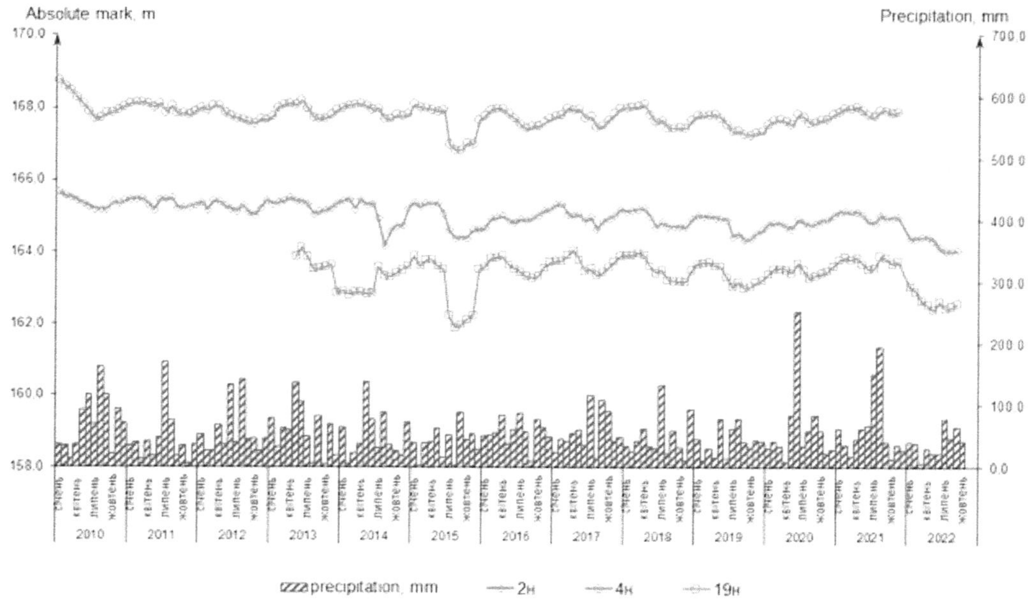

Starting from 2014-2015, there is a decrease in the water level in the lake, especially the long-term minimum one, which cannot be explained only by changes in climatic conditions (Figure 5). It is obvious that the hydrological regime of the lake is also affected by man-made factors (Yatsiuk M.V. et al, 2022d).

One of the most important characteristics of natural waters is their chemical composition, the change of which over time allows us to conclude the influence of various factors on natural waters as a whole.

The chemical composition of the water in the lakes of the Shatsk lake area has remained fairly stable for ten years, despite the anthropogenic load on the water area.

In all lakes of the Shatsk lake area, hydrocarbons and calcium dominate in chemical composition. The content of sulfate and chlorine ions in the water is insignificant. Sometimes there are slight excesses of nitrite ions and chlorine in the water of Svityaz Lake, which is caused by the ingress of organic sewage from settlements. At the same time, all compounds are within the limits of permissible norms.

Figure 5. Dynamics of the water levels in Svityaz Lake by the data of long-term observations, 2005-2022

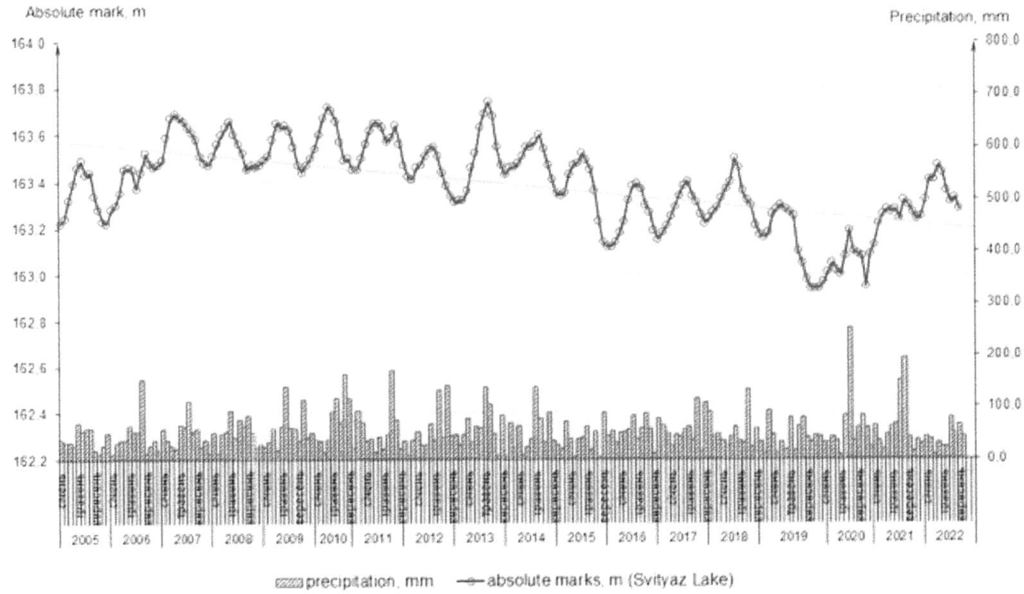

The lakes of the Shatsk NNP belong to the low mineralized lakes with low general water mineralization - 110.5-392 mg/dm^3. The pH ranges from 6.9 to 7.4, which indicates the general neutrality of natural waters. The content of mineral forms of nitrogen in the lake waters is insignificant. During long-term observations, it was recorded an increased iron content in both surface and underground waters. The iron content in the lakes is different and ranges from 0,9 mg/dm^3 in Krymno Lake, up to 0.1 mg/dm^3 in Pisochne Lake (Tsvetova H. et al, 2011).

The chemical composition of the water of the lakes is in a certain way consistent with their genesis. Although the lake of ancient origin, Moshno, has the same chemical composition as other lakes, it is characterized by significantly less mineralization than the lakes of karst or glacial origin. Karst lakes, which are fed by pressure waters, have the same chemical properties as the latest, the predominant components of which are $HCO3^-$ and Ca^{2+}. However, the mineralization of their water is lower than that of pressure water, which is mainly due to the influence of precipitation.

The changes in the chemical composition are determined, to a large extent, by the properties of the soils through which precipitation is filtered. The self-purification process for groundwater is insignificant and pollutants can remain in the aquifer for a long time, moving over long distances. Control of the chemical composition of water is quite important because it is very difficult to eliminate groundwater pollution; it is easier to prevent it, especially where there is accumulation or uncontrolled use of mineral fertilizers and toxic chemicals, as well as disorderly storage of waste from farms or a lack of a sewage network in the settlement. As a result of the specified pollution, soil and underground waters become unfit for consumption.

The negative impact of climate change on the water supply of the Shatsk lakes area also increases the inefficient functioning of the existing reclamation systems, the main reasons for which are structural imperfections (the impossibility of implementing the dual regulation function) and unsatisfactory technical condition due to the low level of operation and wear-and-tear.

Figure 6. Dynamics of average monthly water levels in Svityaz Lake, 1995-2022

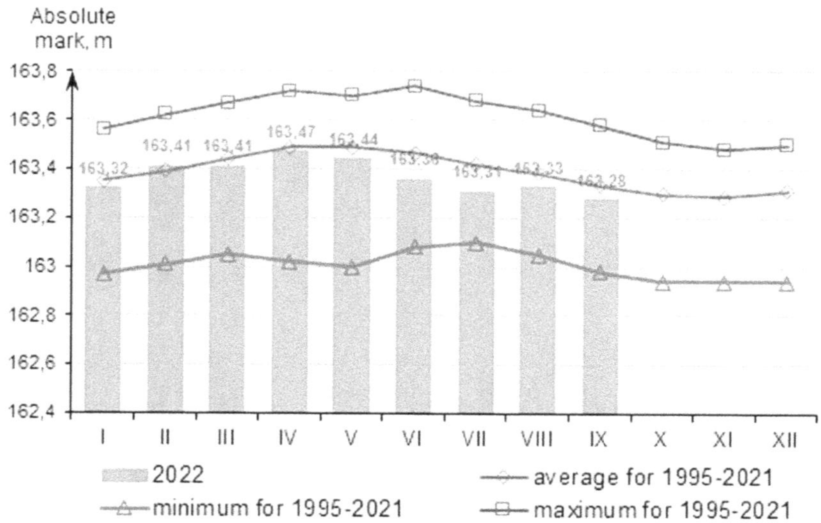

The technological focus of reclamation systems mainly on drainage in the conditions of a significant decrease in the water supply is a very important factor in aggravating the processes of lake shallowing, which requires urgent measures to modernize and reconstruct the existing reclamation systems to change their function from drainage to water regulation through accumulation, redistribution, and minimization of drainage.

THE DIRECTION OF FUTURE RESEARCH

Climate change is a potential threat to the recreational and health opportunities of the region and a catalyst for reducing the water availability of the Shatsk Lakes in general. The urgent scientific task is the development of the State Program for the conservation of the Shatsk Lakes, aimed at optimizing water use, preventing and minimizing the possible impact of harmful effects of water and droughts, conservation and restoration of water resources, improving water quality, ensuring sustainable use and development of nature reserve fund objects, to implement a system of integrated water resources management based on the basin principle, to increase environmental sustainability and balanced development of the Shatsk Lakes area, to preserve aquatic ecosystems as unique components of the nature.

The implementation of the State Program of Ukraine for the Preservation of the Shatsk Lakes will significantly increase the efficiency of water resources management and contribute to the sustainability of the natural reserve fund on the territory of the Shatsk Lakes.

CONCLUSION

1. For today, the most significant factor influencing the natural waters of the Shatsk Lake area is climate change, which is confirmed by the change in the value of the hydrothermal coefficient, which characterizes the balance of water inflow and evaporation, the value of which for the territory of Shatsk Lake area varies between 0.95-1.01. This indicates the existence of a stable process of aridification of the climate, as a result of which the predominance of water inflow, which was typical until 2018, changed to the predominance of evaporation. Anthropogenic influence and climatic change have caused an increase in evaporation from the surface, and the torrential and uneven nature of precipitation does not contribute to its accumulation in the catchment area, which leads to an increasing shortage of water resources, especially in the summertime.
2. The analysis of observation data of hydrological, hydrogeological, and hydrochemical regimes of groundwater and pressure water in the network of existing wells showed that water table fluctuations are synchronous and reflect the dependence on climatic conditions. It was found that the current amplitude of water table fluctuations (2022) in the observation wells does not exceed critical values, and there are no significant and sharp drops in the tables, although a trend towards a decrease in groundwater and pressure water tables was recorded for some wells, which does not exclude anthropogenic influence. The hydro-ecological situation in the territory of Shatsk NNP is rather stable.
3. The chemical composition of the lake water in the Shatsk lake area has remained fairly stable for ten years, and its composition reflects not only the features of the water supply of the lakes but also the anthropogenic load on the water area.
4. To improve the water supply of the territory of the Shatsk lake area, it is necessary to develop and implement comprehensive measures to optimize water use, increase the efficiency of the reclamation systems, prevent and minimize the possible impact of harmful effects of water and droughts, preserve and restore water resources, improve water quality, increase environmental sustainability and balanced development of the territory of Shatsk lake area, preserve water ecosystems as unique components of the natural environment.

REFERENCES

Dyatel, O. O., Sydorenko, O. O., Tsvetova, O. V., & Turaieva, O. V. (2019). Preservation of the natural geo-ecosystem of the cross-border territory of Volyn Polissya. *«Water for All»: dedicated to World Water Day: International. scientific-practical Conference,* (pp. 127-128).

Formation of the regime of natural waters of the Shatsk lakes area in modern conditions. (2004). *Collective monograph (edited by M.I. Romashchenko, Y.Y. Bakhmachuk). Kyiv: Agricultural science.*

Jatsiuk, M., Sydorenko, O., & Nechai, O. (2022b). Water accumulation capacity of forest zone (Polissia) of Ukraine under global warming conditions. 5th International Scientific and Technical conference «Innovative development of resource-saving technologies and sustainable use of natural resources». (pp. 31-33). Universitas Publishing Petroşani.

Jatsiuk, M., Tsvetova, O., & Turaieva, O. (2022a). Sustainability of natural waters within the area of potential impact of the «Khotyslavsky» quarry. 5th International Scientific and Technical conference «Innovative development of resource-saving technologies and sustainable use of natural resources», (pp. 76-78). Universitas Publishing Petroşani. .

Khomik, N. V. (2013). Water resources of Shatsk National Nature Park: current state, protection, management. Kyiv: Agricultural science.

Natural Resources Management in Polesie. (2017). Monograph in 4 books (edited by Y.A. Mazhaiskiy, A.N. Rokochinskiy, A.A. Volchek, O.P. Meshik, E. Eznach). Ryazan: A.N. Kostyakov Meshchersky branch of VNIIGiM.

Rokochinskiy, A., Korobiichuk, I., Kuzmych, L., Volk, P., & Kuzmych, A. (2020). *The System Optimization of Technical, Technological and Construction Parameters of Polder Systems. Automation 2020: Towards Industry of the Future. Advances in Intelligent Systems and Computing* (Vol. 1140). Springer. doi:10.1007/978-3-030-40971-5_8

Romashchenko, M. I., Yatsiuk, M. V., Sydorenko, O. O., Nechay, O. M., Voropay, G. V., Nasedkin, I. Y., Tsvetova, O. V., & Saydak, R. V. (2020). Reasons of Shatsk lakes shallowing and ways of water balance regulation. Kyiv. *Bulletin of Agrarian Science*, (8), 5–13.

Tsvetova, H., Riabtseva, G., & Turaeva, O. (2011). Monitoring of the chemical composition of several lakes of the Shatsk group in the Western Bug river basin exposed to recreational pressures. In K. Dygus (ed.) The Water environment. Problems of evaluation and protection: monograph. Warsawa.

Yatsiuk, M. V., Sydorenko, O. O., Nechay, O. M., Tsvetova, O. V., & Turaieva, O. V. (2022c). Impact of climate change on water availability in Ukrainian Polissya. *International. scientific-practical Conference.* Kyiv. https://doi.org/ doi:10.31073/mivg2022

Yatsiuk, M. V., Sydorenko, O. O., Tsvetova, O. V., & Turaieva, O. V. (2021). Monitoring and features of hydrogeological and hydrological conditions of shatsky lakes in modern climate conditions. *Sciences of Europe*, (67), 51–59.

Yatsiuk, M. V., Sydorenko, O. O., Tsvetova, O. V., Turaieva, O. V., & Nechay, O. M. (2022d). The current state of natural waters in the zone of possible impact of the Khotyslavsky career. *International. scientific-practical Conference,* Kyiv. https://doi.org/ doi:10.31073/mivg2022

Chapter 26
Integrated Management of Water Resources in the Polesie Region

Vasyl Stashuk
National University of Water and Environmental Engineering, Ukraine

Nadiia Frolenkova
National University of Water and Environmental Engineering, Ukraine

Leonid Kozhushko
National University of Water and Environmental Engineering, Ukraine

ABSTRACT

The chapter underlines the necessity and suggests the pre-conditions of the introduction of the integrated management of water resources in Ukraine, and in Polesie region in particular. The authors state that the current state of the water resource management, water protection and water resource restoration of Ukraine are classified as a complicated system of branch, department and local functions and structures of the public administration which have a branch and administrative-territorial direction, an imperfect and unbalanced mechanism of water use, water protection and restoration of water resources. In view of this, prior trends in the implementation of the principles of stable water use were worked out, in particular, the introduction of the principles of the integrated management of water resources. A strategic aim was formulated, namely, to supply a basin balance and development of water management, water protection and restoration of water resources, to ensure stable water supply for the population and the branches of the national economy.

DOI: 10.4018/978-1-6684-8248-3.ch026

BACKGROUND

The problems of the natural resource rational use, the harmonization of the relations between man and nature, and natural environment protection are among the most urgent problems of today, affecting every inhabitant of the planet, since the future of all mankind depends on the solution of these problems. In Europe, Ukraine is one of the countries with few local water resources, and these resources are distributed very irregularly. In Ukraine, the efficiency of water resource use is extremely low, and the water capacity of the gross domestic product, on the contrary, is too high.

Today the strategic goal of ecologically safe water resource usage is to ensure the priority of the environmental protection functions over the economic ones, the rational use of surface and underground waters, and the wide introduction of water-saving technologies in all sectors of the economy. The implementation of this goal should be carried out based on the development and observance of the water legislature, which is an integral part of the environmental legislation and is directed towards the regulation of the relations between the subjects of the economic activity to strengthen the responsibility of legal and physical entities to preserve and maintain water resources and water facilities in a proper condition, as well as to ensure the interest of the enterprises, institutions, organizations, and citizens in the conservation of water resources, the implementation of water-saving technologies, the development, and production of low water-consuming technological equipment (Pavlov, Stashuk, 2013)

The strategic goal of the state's water management policy is sustainable water usage with constant economic development.

The priority directions for the implementation of the principles of sustainable water use are as follows:

- providing the population and economic sectors of Ukraine with high-quality drinking water in the required amount and appropriate quality;
- restoration and preservation of a sustainable ecological state in river basins;
- restoration and protection of water bodies, reproduction of water resources;
- justification of new groundwater intakes and expansion of their usage for drinking water supply;
- artificial replenishment of underground water reserves;
- prevention and elimination of the consequences of harmful effects of floods, flooding, and water erosion;
- protection and rational use of water resources and restoration of small rivers;
- increasing the management level of reservoir operation modes;
- exercising safety control over the water management facilities and systems;
- reduction of production water consumption, water losses, improvement of drinking water quality;
- increasing the structure reliability for the engineering protection of the territories against the harmful effects of water;
- the creation of an economic mechanism for water usage based on the market conditions of the stimulation of the rational usage of water resources and the sufficient financing of water management and environmental protection activities;
- improvement of the system of environmental regulation and standardization in the field of water resource use;
- formation of a rational regional structure of the economy and a water management complex based on the water factor.

As it is not possible to quickly solve water-ecological problems, public policy should be based on the principles of defining strategic goals, on which all water-ecological activities are oriented.

The national priorities for environmental protection and the rational use of natural resources include:

- environmental safety guarantee of the nuclear facilities, the protection of the population and the environment from radiation, the prevention of the harmful effects of the accident consequences at Chornobyl NPP;
- improvement of the ecological condition of the Dnipro river basin and the quality of drinking water;
- stabilization and improvement of the ecological condition in the cities and industrial centers of the Donetsk-Dnipro region;
- prevention of the pollution of the Black and Azov seas and improvement of their ecological condition;
- structural reconstruction of the production potential of the economy, the greening of the technologies in the industry, power engineering, construction, agriculture, and transport.

The existing organizational structure of water use management is not effective enough due to the division of competence between many departments and the absence of clear coordination and coherence of actions, the imperfection of the economic mechanism founded on budget financing. In Ukraine, the tendency to concentrate the functions of the environmental and economic management of water in the hands of the state is still traditionally preserved, which is typical for a planned economy.

In the absence of a single state body of basin management, which is responsible for the state of water bodies in the basin, the right of such management is granted by law to a large number of the state bodies, for which it is not the main duty. As a result, parallelism and duplication are observed in the implementation of several management functions. The departmental interests of some state administration bodies come into conflict with this process, preventing a clear legislative division of the functions of environmental and economic management.

The water management and reclamation complex of Ukraine, powerful in terms of its volume and extremely complex in terms of its content and structure, was created mainly 60-70 years ago. Today, these state material funds require significant funds for their reconstruction and capital repair. To successfully solve these tasks, it is necessary to introduce into production modern economically effective achievements of science and technology and, first of all, the principles of the integrated management of water resources, that is, a comprehensive solution of water management and environmental problems based on the rational use of water resources of Ukraine and the creation of the conditions for the transition to a sustainable and effective functioning of the entire water management complex of our state.

Public management in the field of water usage and protection and water resource reproduction should be carried out according to the basin principle based on state, interstate and regional programs (Water Code, 2000). The definition and scientific substantiation of the strategic goals and directions aimed at increasing the ecological sustainability and balanced development of water management, providing the population with high-quality drinking water, minimizing damages and social stress due to the harmful water effects, preserving water ecosystems – this is far from a complete list of the priority tasks, the solution of which is envisaged by complex state programs.

The transition of Ukraine to the position of sustainable development must be preceded by national economic reform and the development of an appropriate environmental and economic policy. The water management complex and the water resource management system also need to be reformed and transferred to the basin principle, the creation of institutions for the collective management of water use, the introduction of new economic leverages, management decentralization, etc. (Yatsyk, 2004).

The introduction of the basin principle management and the necessity to take into account modern European demands require the coordination of the water management areas with the basin and sub-basin boundaries, which will be allocated in the territory of the country, according to the EU demands.

The entire complex of the listed problems and issues regarding the management and effective use of water resources is related directly to the Polesie region of Ukraine, which is characterized by a powerful and developed hydrographic river network, is of trans-border significance, because it borders on the hydrographic network of the corresponding regions of Belarus, Poland, and Russian Polesie.

Besides, this problem becomes highly urgent and complex during military operations as a result of Russian aggression, especially in the post-war period. This causes the extreme urgency of solving the issue of the water resource integrated management by a basin principle, taking into account these characteristic features.

METHODS AND MODELS

The main water artery is the Horyn River (its length within the region is 386km) with its tributaries: on the right – Sluch with Stav, Bober with Serehivka, also Zamchysko; on the left – Melnytsia, Vyrka. Among other important rivers, one should mention Styr, Lva, Stvyha, in the marginal north-west – Prypiat. The largest lakes of the territory are Nobel (499 ha, Zarichnenskyi district), Bile (453 ha, Volodymyretskyi district), Luka, Velyke Pochaivske, Richytske, Ozertse. The main direction of the rivers is from the south to the north which is due to the general lowering of the surface in this direction. Within Polesie area, the rivers have small slopes (0.3 – 0.6 m/km), wide, waterlogged floods, there are many floodplains, lakes in the valleys. In Polesie the average river net density is 0.3 – 0.43 km/km^2. The main source of river feeding is the melted snow waters (55 – 60%). In view of hydrology, Polissia is situated within three artesian basins of the underground waters: Volyno-Podilskyi, Prypiatskyi and Ukrainskyi basin of cracked waters. The resources of the underground water in the region were estimated as 345 th. m^3/day, including: Volyno-Podilskyi artesian basin – 265th. m^3/day, Prypiatskyi artesian basin – 12. th. m^3/day, Ukrainskyi basin of cracked waters – 68.8 th. m^3/day.

The introduction of the integrated management mechanism is aimed at the organization of the rational use of water resources of the country. The task of the rational (from Latin ratio-reason), i.e. reasonable use of water resources consists in satisfying the needs of the consumers along with ensuring the requirements of their protection and restoration.

The preconditions for the principles of the water resource management were founded in the scientific works of the leading scientists in the sphere of water economy and land reclamation. The rational use of water depends not only on the improvement of the methods of its extraction, it depends more on the improvement of the whole process of its purification, supply, transportation, distribution and consumption.

For instance, at the beginning of the 60ties the issue of the rational water resource use was urgent in land reclamation, when the irrigated lands increased and the irrevocable water consumption and drainage began. At the beginning of the 60ties of XX century, a scientific-technical trend appeared in the world which was aimed at the development of automatic regulators of the level and water consumption in the irrigated canals. The following scientists took an active part in it: (Stashuk & Kozhushko,2018).

When designing and constructing large irrigation systems with long main canals, the management of the water distribution in the extensive network of canals became an urgent issue.

A scientific trend in the watering management developed intensively under the leadership of V. Ostapchyk. A serious work to create an automated system of a dispatching control over water consumption started (Current state, the main problems of water amelioration, 2001).

Worked out a new management technology of water resources in the irrigated arable farming which was based on the controlling principles over water distribution systems as the objects of stochastic nature. At present the main attention is paid to the management of inter-system and inter-business water distribution in the conditions of a severe deficit of electric power, irrigation water and water consumption decrease.

In the middle of the 80ties of XX century, the scientific optimization principles of water use for the area of excessive and unstable moistening were initiated by Professor M. Lazarchuk and his colleagues (Lazarchuk et al., 1991). They worked out the optimization methods and models and the predicted assessment on a long-term basis of the efficiency of water regulation of drained lands in variable natural-agro-ameliorative conditions (Hadzalo et al., 2018).

Specialists from all the countries work at the problem how to avoid water losses and to find the places of its sources on time. Among the countries of NIS, a serious work is carried out at the research institute of communal water supply and water purification named after (Honcharov, Dupliak,1992; Charnyi, Horuzhyi,1996; Stashuk et al., 2014).

Thus, the management of water resources is a complicated technological process which involves the organization of the proper measures so that they can ensure the optimal use of the available resources (water, energy, human) to reach a set goal, namely, to supply all the consumers with the required amount and quality of water with the estimated reliability. So, management is a complex of the well-coordinated measures aimed at reaching a definite goal. It is based on the methodology and principles of the science of management which uses the achievements of the theory of information, cybernetics, economy and responds to the changes of the main concepts of society's political life (Stashuk, 2006)

Certain scientific methods can be used to solve complicated (problem) situations; their connection with the stages of decision making is shown in Fig. 1.

The approaches to the consideration of the water economy problems were changed drastically in this country; presently their solution is based on ecological and economical techniques. The management of water resources is in priority in a public policy and it is considered to be one of the most important factors for the stable development of society. Nowadays these problems are dealt with the participation of the public, taking into consideration social factors, educational and informational measures directed towards the conservation and rational use of water resources. The water resource management has to be oriented on a concrete individual.

The most important conceptual principles of the water management system are to be:

- legislative introduction of the principle of decentralization;
- considering water, taken from the water object, as "a commodity";
- implementation of the principle "he who pollutes water has to pay".

Figure 1. Connection of the management process with scientific methods of the research

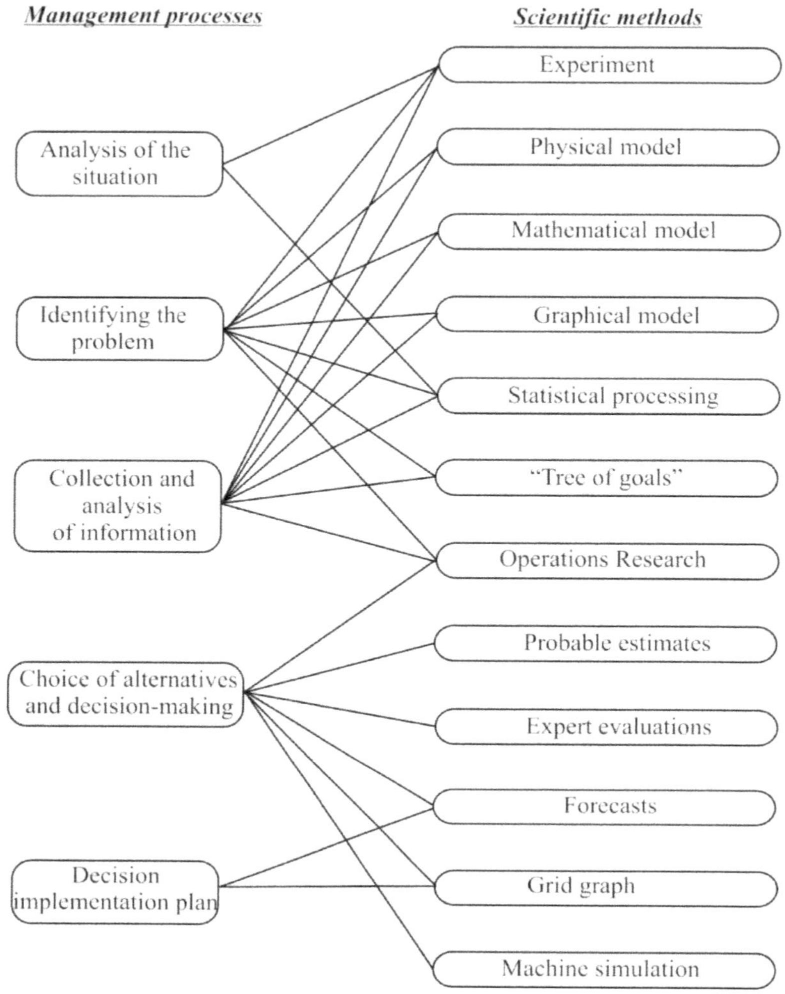

A river basin is a system of watercourses of different orders, limited with watersheds, which fall into one water receiver. Usually it is a sea but also it can be a basin/water situated in the middle of the mainland (e.g., lake or swamp). An underground basin is an underground isolated water-permeable soil layer which keeps water reserves. Managing water resources, a basin is classified as an independent hydrological unit. Various terms are used in the countries such as "basin", "catchment", "catchment area", but a term "basin" is used in most cases.

The active use of water resources requires the necessity to establish hydrological, social, economical and ecological interconnections in the basins. This mutual dependence demands some integrated approaches to the development and management of water resources. Taking into consideration a multi-vector nature of the water resource management, an integrated approach to the water resource management is introduced on a national and basin level in the majority of the countries which requires the improvement of both institutional structures and executive practice.

The water resource management starts with the planning of the modes of water-economic systems (WES), the basis of which is the analysis of water-economic balances. A water-economic system is a complex of water resource sources, water consumers and managerial means of water resources, connected hydraulically and organizationally, so that they can be used rationally, restored, preserved and protected from contamination. A water-economic system consists of numerous elements meant to meet economic, social-ecologic, aesthetic and other needs of society in water of proper quality and amount.

In addition to the needs of the production sphere and society in water of proper amount, mode and quality, WES is to maintain an ecological balance in the territory of a basin and to protect the environment from a harmful water effect.

Hence, WES can belong to a class of large (complicated) systems, as it comprises water resources as a basis, contains several artificial engineering facilities and has extended connections with the natural, social and economic environment. Functionally WES represents a unity of sub-systems of water-supply and water-consumption Fig.2.

Figure 2. Functional and sectoral composition of the water management system

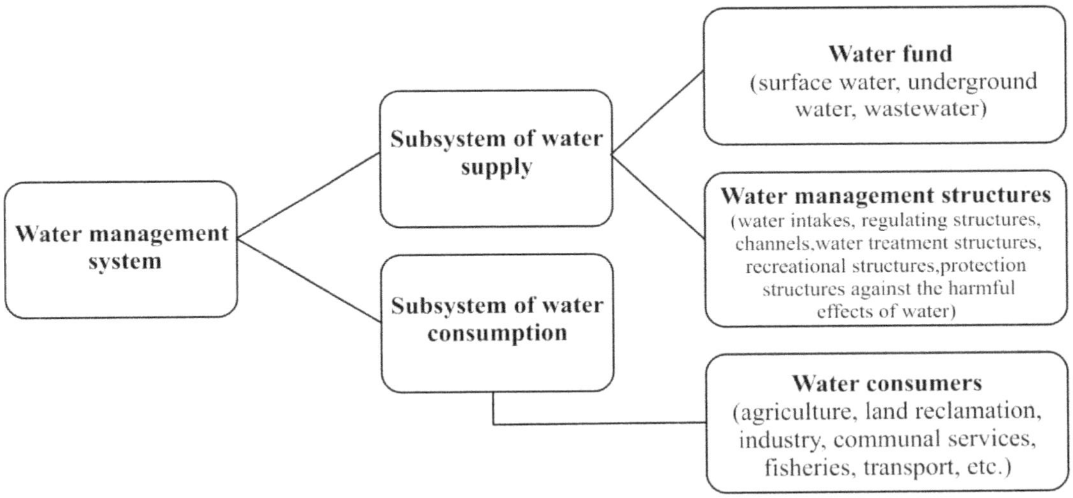

A sub-system of water supply includes a water fund and water-economic facilities. A water fund of the country consists of superficial, underground and waste water. Rivers, lakes, ponds, water reservoirs, seas, etc. belong to the superficial water, i.e. the one which is above the land surface. The underground water is located in pores, cracks of mountainous rocks at the depth of dozens or hundreds of meters from the land surface. Waste (sewage) water is formed in the residential settlements, in the territory of the industrial and agricultural enterprises; after it has been used by the consumers this water has wastes and contamination. Also, melt and storm water belongs to waste water. In many branches of the industry this water can be used repeatedly (e.g., in power engineering) in the system of reverse and repeated-consecutive water supply which results in a significant decrease of the share of irreversible losses in the total scope of used water.

The reason for a crisis situation in the water management is first of all the extensive development of the country's economy which has lasted for many years. To stabilize and improve the ecological state of water economy and to develop the directions how to come out of the crisis situation, it is expedient to identify the reasons which have led to it. First of all, they are as follows:

- high resource- and energy-consuming old technologies which exceed similar indicators of the industrialized countries by 2...3 times;
- a high level of the concentration of industrial objects/premises;
- the lack or insufficient capacity of waterworks and poor purification technologies;
- the lack of legal and economic mechanisms to stimulate the development of ecologically safe/ friendly technologies and nature-protective systems.
- The state of crisis in water-ecological and water-economic conditions requires the development of an ecological approach to all economic measures which are taken, the improvement of the management system of water objects, water resources.
- The ecological approach envisages:
- a complex assessment of the effect on water resources from the side of all elements of a social-ecologic-economic system of a basin;
- the coordination of the administrative-territorial-economic interests, the interests of different categories of water users with the requirements to preserve water resources of the whole basin;
- the determination of an allowable anthropogenic load on the basin and the development of complex-target-programs of a step-by-step achievement of a self-reproductive water use;
- the identification of man as a central element of a basin eco-system.

The implementation of a public policy of a stable development of water economy is based on the achievement of strategic goals, which include the restoration and protection of water objects, the enhancement of the technical and sanitary reliability of water-economic systems, the rational water use. A list of priority measures includes: to implement the measures concerning the improvement of an ecological state of ameliorative systems and to adjust the ecological and resource condition of water objects to the requirements of the main standards which would guarantee the life safety for the population and the balance between the levels of the anthropogenic effect on water objects and their renewable capacity.

The major tasks are the following:

- to process the state standards as to the identification of resistance limits of water objects to the anthropogenic and techno-genic loads, to determine the parameters of the balance between the use and reproduction of a water-resource potential and, on this basis, to introduce the limits of water intake and the discharge of pollutants into water objects;
- to work out and to apply the analytical methods of the risk assessment of a negative effect of the factors of water-economic activity on people's health and life expectancy, to develop automated evaluation systems of water resource and ecological situations, to predict a harmful impact on water objects.

The problems of water economy, which are of a complex nature and their solution is not possible without a government support and the coordination of inter-branch connections, are solved through the development and implementation of state target programs. These are the problems which are typical for the whole territory of Ukraine – the supply of the population with quality drinking water, the elimination of the harmful water effect, the restoration of the water quality in water objects and other regional water-economic ecological problems. The implementation of these state target programs is very important for the ecological safety, the transition to a stable development, the creation of the conditions to enhance the development efficiency of water-consuming branches of the economy.

The priority measures aimed at the improvement of the existing situation are the following (Khvesyk, Smolenskyi, 2000):

- in communal economy – the reconstruction, repair and modernization of the equipment of water supply, sewage networks and facilities; the substantiation and introduction of the rational rates of water supply, updated technologies of water preparation and purification of waste water, new techniques of water accounting, the improvement of the control over water consumption, the increase of the improvement level of public services of the settlements, the adjustment of the drinking water standards to the world ones;
- - in industry – the largest water-consuming industry (electric power engineering) must provide a proper level of the rational use of water resources through the modernization and reconstruction of the acting HES (hydro-electric station), the improvement of the water supply system of HES (heat-electric central/station), the introduction of new energy-saving technologies, the putting of new capacities into operation and the restoration and development of small hydro-power-energy; in metallurgy, due to a well-balanced production, oriented on the manufacture of the competitive output with the mandatory observance of the ecological requirements, the introduction of new technologies, protective measures, it is envisaged to decrease water-consumption and the scope of the waste water pollution; in a chemical and oil-processing industry, due to the use of capacity reserves of reverse systems (4.0 km3), the technological renovation, reconstruction and improvement of the production, it is possible to decrease significantly general water needs and the scope of the waste water pollution; in a machine-building and metal-working industry the planned structural changes will make it possible to release the capacities of reverse water supply, and the expected needs of the industry in fresh water will not exceed 300...330 mln m3/year;
- in agriculture – one of the most water-consuming industries (land reclamation systems), the formation processes of the agrarian sector of the national economy, land relations, land and water resource assessment to the agrarian capital turnover, the establishment of new land ownership forms will facilitate some positive changes in the reconstruction of an agrarian-ameliorative complex.

It is envisaged to reconstruct and improve the acting ameliorative systems, to adjust irrigation to the optimal correlation with other land reclamation measures, to improve soil properties and soil fertility, to apply updated water-saving technologies and irrigation modes and to prevent land flooding.

A serious attention has to be paid to the problem of water supply for the rural areas. The main task is to provide the centralized water supply along with the drainage systems for at least 50% of the villages, particularly in the areas where people consume delivered water.

FUTURE RESEARCH DIRECTIONS

The Major Directions of the Improvement of the Water Use Management by a Basin Principle

The management of the subjects of water use by a basin principle is directed towards the sustainable water supply of the national economy industries and the population through the introduction of the promising technological standards of water use, the prevention of a harmful water effect. To better implement a basin principle, it is necessary to create managerial structures which, within water intake basins, will deal with the functions of planning, coordination and control over water resource use, and the powers will be delegated among the national governmental bodies and local authorities. Being developed, these principles are to be approved at a legal level; also it is expedient to create a complex basin geo-information system with a bank of cadastre information about a water fund, water resources and the means of their regulation as well as the structure of land and ameliorated land.

A planned basis of the basin principle management is a target program of the usage, protection and restoration of water resources in a water intake basin based on the anthropogenic load rationing on them, the requirements to water quality to preserve the ability of self-renovation and water-regulating functions of the rivers.

When the national ecological legislature becomes closer to the EU norms and directives, it will facilitate the implementation of the legal acts which are to ensure the Integrated management of water resources.

The improvement of the organization structure and functional scheme of the introduction of the water resource management by a basin principle. A basin principle is considered to be one of the efficient methodological techniques to solve global ecological problems of water use, water protection and restoration of water resources. A strategic goal of the water resource management by a basin principle consists in providing a basin balance of the water-economic development, water protection and restoration of water resources based on the coordination of legal principles and managerial actions of the subjects of water use by a basin principle, directed towards stable water supply for the population and the branches of the national economy, the introduction of promising technological standards of the use of water resources, the prevention of harmful water effects.

One of the directions of the development of a basin management is the creation and guarantee of the functioning of this mechanism which will ensure a direct connection between the charge for water use and funding of priority water protection measures within a basin. And, as a foreign experience proves, the best results of the water management are reached where there exists a simple, fair and understandable/clear for all, and first of all for water users, system of charging for water intake and irrevocable consumption and discharge of standardized substances with reverse water.

Here comes the task to build a renewed system of the water resource management of Ukraine in compliance with current views and principles taking into account the latest achievements of science and engineering, and with its help to get the system out of crisis and to make it serve for the social- economic reconstruction of a new democratic country which has taken the path of independence.

As it is not possible to quickly solve all water problems, the public policy should be based on the principles of defining strategic goals, which all water-economic activity is oriented on, which all current actions and the ones planned for near future are matched to, and also the goals of a transitive period. The strategic goals are predetermined by the interests of society; their achievement will make it possible to solve water-economic problems, to eliminate water contradictions and to remove water restraints while doing the social-economic tasks.

Scientific principles, concept and plan of the introduction of a basin system in Ukraine. A river basin is a natural unit of ecological zoning, as a rule its boundaries do not correspond to the administrative borders which creates problems for ecological management. Usually a basin of a large and medium river, which has the area of water intake equal to 2 th. km^2 and more (it is over 50 th. km^2 for a large river), covers several administrative regions.

In the latest decades a scientific justification of a basin system management with the attempt to introduce it has been carried out in many countries. Keeping in mind that practice is the criterion of the choice, the attention has to be paid, first of all, to the results of the introduction of a basin approach to the management while developing a concept of this very approach in Ukraine. As the studies of the foreign experience prove, the example of all-round justification of the introduction of a basin system management can be the system implemented in France; the results of its functioning have been generally acknowledge, and the number of the advocates of the idea to introduce this system in different countries is growing with every coming year.

According to the document "Main directions of the public policy of Ukraine in the sphere of the environment protection, the use of natural resources and the provision of ecological safety" adopted by the Verkhovna Rada (Supreme Council) of Ukraine, the long-term goals of the policy of the rational use and restoration of water resources and ecosystems are the following:

1. To reduce the anthropogenic load on water objects.
2. To reach ecologically safe use of water objects and water resources to satisfy economic needs of society.
3. To ensure ecologically stable functioning of a water object as an element of natural environment with the preservation of the property of water ecosystems to restore water quality.
4. To create an efficient management structure and the mechanisms of economic regulation of the protection and use of water resources.

For a stage-by-stage implementation of the above-mentioned goals, it is necessary to take a complex of measures following the priority directions, namely:

1. The protection of surface and underground waters from contamination.
2. The ecologically safe use of water resources.
3. The rebirth and maintenance of a favorable hydrological condition of the rivers and controlling measures over water harmful effect.
4. The improvement of the management system of the protection and use of water resources.
5. The reduction of the radio-active contamination effect.

Creation of the International basin councils. The necessity to consider the interests of water use in river basins, a serious aggravation of a water contamination problem in international rivers and the degradation of their ecosystems – all this requires to create or strengthen an international-legal and institutional base which will be able to oppose a water-ecological threat. Presently a well-organized cooperation exists in many large international rover basins. It is exercised through joint bodies (commissions, committees) which function in the whole basin, the agreements signed by two or more countries. About

150 two- and multi-sided agreements on the protection and use of trans-border watercourses function in Europe and North America alone. In Europe the experience of cooperation through the conclusions of the agreements and the creation of the international bodies was accumulated in the basins of such rivers as Danube, Rein, Elba, Oder, Maas, Moselle, Saar; on the other continents – in the basins of the rivers Mekong, Gang, Nile, St. Lawrence and others.

A positive fact is the awareness of the international community that the regulation of nature use, protection and restoration of natural-resource potential can be done within the whole river system, that is, in accordance with a basin principle.

The Basin Council is created to establish an efficient mechanism of the coordination of the efforts of the borderline states as to the sanitation of river basins, the drawing and signing of the inter-government agreements and normative acts, the efficient information exchange and a wide participation of the parties involved. The creation of the Basin Councils is to be founded on the existing, well-developed legislative base and practice of the cooperation between the countries of river basins in the sphere of the environmental protection, the joint use and protection of trans-border watercourses. The countries, Ukraine borders on, have accumulated a significant positive experience in the introduction and improvement of a basin principle management of water resources and their protection. The implementation of these tasks is closely connected with the solution of the issue of a complex evaluation of water use; they are a required precondition for the formation of the natural basis of society and water use.

A key issue of the pre-condition of the development of a basin management system is the creation of such administration body for the whole river basin which will include the representatives of every segment of society who deal with water problems (both those who use water and those who are responsible for water supply). The purpose of setting up such representative body is on a local level to help solve any conflict of interests which may occur between industrial and farm producers, fishermen, officials, members of non-profit organization s and, quite natural, local population. On a local level the decisions are made and put into action by a basin body which is to guarantee the maintenance of required water quality and the availability of the formulated requirements for this.

It is recommended to introduce the Basin management system in the following stages:

First stage (preparation). The working out of the changes and amendments which are necessary for the introduction and development of the basin management system for the main water basins of Ukraine, to the Water and Land Codes, Laws on the environment protection, self-governing and other legislative and legal acts and the coordination among them. The creation of the Concept and the Program of the development of the basin management systems and the Plan of their introduction.

Second stage (organizational-experimental). The creation and provision of the activity of the Main basin coordinator in charge of water resource use and restoration, water protection and water object sanitation in the composition of the Councils in ecological issues (environment-investment management, audit and entrepreneurship). The creation and provision of the activity of the Basin administrations of water resources and their structural and functional sub-divisions. The development and approval of the basin and regional programs of water resource use and restoration, water protection and water object sanitation and their introduction.

Third stage (introduction). The creation and provision of the activity of the basin management structures of water resource use and restoration, water protection and water object sanitation of Ukraine. The development and introduction of the state, basin (including inter-state ones) and regional (oblast) programs of water use and protection. The creation and provision of the performance of the State and basin water-environment-investment funds.

The main tasks of the basin management bodies must be the following:

1. To plan water quality on the river, that is, to define the goal of the functioning and development of a water-economic complex of the basin, the means to reach and observe it, to make a plan of coordinated measures. It requires to model water object and the actions which have an impact on its condition, to predict changes of this condition with its potential regulation. Also it is necessary to have the assessment of a current river state, the requirements to water quality, basin rationing of the water state with the definition of the characteristics of "mouth-outlet" for the rivers which cross the state borders as well as the administrative boundaries of the regions.
2. To organize the management which will unite the administrative-territorial structure of the state management of waters, the basin consideration of water protection tasks, beginning with the establishment of kinds and places of water use and ending with the organization of an economic mechanism of water use in the conditions of market relations. It requires the creation of the organization structure of a basin management and its organizational provision in the form of proper rules, instructions and bylaws.
3. To exercise the regulation of water relations: legal administrative, economic and technical. First of all, it requires the creation of the necessary normative provision in the form of system normative-legal and other bylaws, which would take into consideration, on the one side, concrete hydrological and hydro-chemical conditions, characteristic of a given river basin, and on the other side, environmental-economic and social interests of the population of the regions which belong to this basin. According to the adopted standards, the permits have to be issued for special water use; other measures of legal regulations have to be taken along with the economic, administrative-managerial and technical regulation: the collection of fees for water use, fines for the violation of water legislature and compensation for the losses caused by water contamination, limits in reverse water discharge (waste, drainage) up to the closure of the enterprise, the development and implementation of the investment water protection projects, etc.
4. To exercise control in the process of managing water quality. It implies basin monitoring of water state and intake from water objects of fresh water and discharge of reverse water, ecological expertise of production and other facilities, which are designed, technical means, technologies and materials which influence the water state, and also the operational control over the implementation of the managerial decisions and environmental inspection.
5. To carry out the accounting, i.e., summing up the performance of the management for a certain period of time. An efficient information system has to be created which would allow getting operational systematized data to correct current plans and work out the management system of water quality for the next period.

Among the major principles of the basin system introduction, one can mention the following:

1. The phasing in the creation and improvement of the system.
2. The urgent legislative provision of the creation of the basin management system (the development of the main directions of the national water policy, the addition of the corresponding changes to the Water code and other bylaws of Ukraine and their approval by the Verkhovna Rada of Ukraine).
3. The follow-up normative-methodological support of a new management system (bylaws of the Cabinet of Ministers of Ukraine, ministries and other central bodies of the executive power with

their stage-by-stage introduction into action as the creation and development of the basin management takes place).

4. The reliance on the existing staff of the bodies of the state administration when the personnel issue of the basin system is solved.
5. The maximal use of the positive foreign experience while creating the organization support of the management system (the formation of a functional, technical, organizational structure of the system, the working out of the instructions and regulations for the operational personnel concerning the work of the system).
6. The development of the integrated basin management system of the environmental data and information. The creation of the monitoring system of a sustainable basin development.
7. The use of all existing means to promote the introduction of the management system: legal, administrative, economic; maximal objective information for the population about the state of things/affairs, the formation of the public support for the introduction of the system; the application of coercive measures, if necessary.

The Experience of the Introduction of the Integrated Management of Water Resources in Polesie Zone of Ukraine With the Consideration of International Approaches

In Ukraine a great deal of work was done as to the introduction of a basin principle of the water resource management. In addition to the above-mentioned, corresponding changes have been made in the Water Code of Ukraine, a complex of state programs, concepts and methodologies has been approved, those which envisage data monitoring and the introduction of the water resource management taking into account a basin principle [National program of the environmental sanitation of the Dnipro basin]. However, a separation of theoretical principles from practical recommendations takes place. For instance, there is no domestic methodology of the calculation of the water-economic balance, instead only the Soviet one exists. It is necessary to create and improve other aspects of the support of decision making as to the optimal water resource management as well.

In Ukraine almost all large river basins are trans-border ones – the basins of the rivers Dnipro, Siverskyi Donets, Dnister, Zahidnyi Buh, Tysa, Danube. According to the principles, presented in several Conventions which Ukraine joined, the International basin commissions are established to manage trans-border (international) river basins. Nowadays dozens of such commissions function in the world. The International Commission on the Protection of the Danube River (ICPDR) is the most authoritative among them and Ukraine is its member. The international cooperation in this sphere is coordinated by the international body – Global Water Partnership (GWP) and its 11 technical committees (in particular, in Bratislava – for the countries of East Europe). This will make it possible to receive methodological, technical and other assistance in the introduction of the integrated river basin management, to use the international experience as to the assessment of the condition of water ecosystems.

The main task of the inter-state basin commissions is to have the methodologically coordinated (between the countries of a basin) analysis and assessment of the water resource condition and the cooperation concerning the development of a well-agreed Plan of the river basin management, as well as the implementation of the programs of the measures, aimed at reaching the goals in a river basin – the appropriate water quality and the sustainable environmental condition of water ecosystems.

The important stage in the introduction of a basin principle of the water resource management is the development of the plans of the river basin management. The plans of the river basin management must include the boundaries of the basins and sub-basins, put/charted on the map, which belong to this district of a river basin. For this purpose, it is necessary to carry out a new hydro-graphic zoning of the territory of Ukraine. Hydro-graphic zoning is the division of the territory of Ukraine into hydro-graphic units which is done to make the plans of the river basin management. Hydro-graphic units are the districts of river basins and sub-basins within their boundaries. The districts of river basins are the main unit of the management in the sphere of water object use and protection which consist of the basins of definite rivers and underground water objects connected with them.

The European countries used systematically a basin approach in their research and the water resource management before, but river basins became a central concept of the water resource management only when the Water Framework Directive came into effect. The territory of a river basin can be the basin of one large river or, in another case it can include several basins of small rivers. On the other side, the territory of a river basin can be situated within the boundaries of one country, or within two or more countries. In many cases, the territory of a river basin is shared among several countries (between the countries-members of the EU, or between the countries-members of the EU and those which are not). The majority of the large European river basins are within the boundaries of several countries. Accordingly, the tarns-border cooperation is necessary to analyze a river basin and to manage it.

The management of the implementation of the Directive is distributed among regional and national levels. Most of the countries have competent bodies both at a national level and the level of the river basin territory. The international commissions in charge of river basins play an important role in the coordination of the WFD implementation in the countries-members of the EU. The most active commissions in river basins of Europe are the following: the International commission for the protection of the river Oder from contamination (ICPOAP), the International commission for the protection of the rive Elba (IKSE), the International commission for the protection of the river Danube (ICPDR), the International commission for the protection of the rive Rein (ICPR), the International commission for Shelda (ISC). Among the above-mentioned commissions, the International commission for the protection of the river Danube has some countries-members which are not in the composition of the EU. This commission is the example of the organized executive structure in the international territory of a river basin which goes beyond the boundaries of the EU. In general the determination of the territories of river basins in the countries-members of the EU is done on the basis of hydrographic boundaries.

The necessity to consider the interests of water use in the river basins, the aggravation of the problem of water contamination in the international rivers and the degradation of their ecosystems result in the need to create or strengthen an international-legal and institutional base capable of ensuring the counteraction/resistance to water-ecological threat. Nowadays the organizationally established cooperation exists in many international river basins. It is exercised through joint bodies (commissions, committees) which function in the whole basin, or the agreements signed by two or more countries.

A present-day task is to create a structure and a functional scheme of the main chain of the basin administration, the delegation of powers and tasks of managerial structures vertically: the National Water Council – a basin Council (commission) – a basin water administration – regional administrations of a water management.

To solve the problem of the Integrated water resource management, inter-department and basin councils have been established in Ukraine, the work is being carried out in the framework of the international ecological projects.

The basin councils perform their activity in team with the Basin administrations of water resources, Regional offices of water resources, Ukraine's Ministry of ecology and natural resources, State water agency, local authorities of the executive power, territorial bodies of the central executive authorities and local self-governing authorities which implement their obligations/powers within the sub-basins in accordance with a new hydrographic zoning of the territory of Ukraine, other organizations involved, institutions, businesses, public associations Fig. 3.

Figure 3. System of basin councils of Ukraine by the data presented by the State water agency at the end of 2021

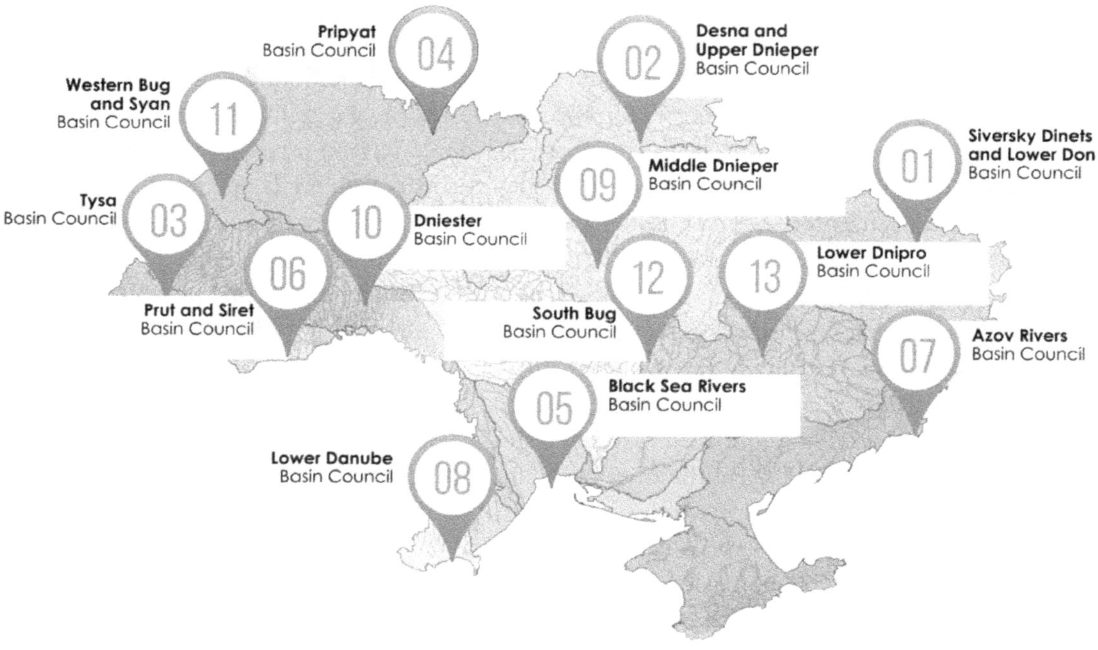

The composition of the basin councils includes the representatives of the parties involved: central and local bodies of the executive power, local self-governing authorities, Basin administrations of water resources, Regional offices of water resources, water users, enterprises/businesses, institutions, organizations, public associations, which function within the boundaries of a certain sub-basin or their activity is connected with a certain river basin.

The Basin council is an advisory body within the territory of a river basin, established at the central body of the executive power, which implements the public policy in the sphere of a water-economic development, with the aim of ensuring the rational water use and protection and water resource restoration, and the integrated management of them. The Basin council is created by the central body of the executive power which implements the public policy in the sphere of a water-economic development with the aim of making proposals and support of the coordinated interests of the businesses, institutions and organizations which deal with water use and protection and water resource restoration within a basin.

Based on the provisions, bylaws which were used to create preconditions of the reformation of the organizational structure of water-economic organization by a basin management principle, as a current approach to the water resource management, and by the corresponding orders/resolutions of the State water agency in the district of the Dnipro river, the Basin administration of water resources was established: it includes the management of the Prypiat river within Zhytomyr, Volyn and Rivne regions in a proper interaction with four (4) other regions, which are covered by the basin of the Prypiat river, namely: Kyiv, Lviv, Ternopil and Khmelnytsk regions Fig. 4.

Figure 4. Territorial scheme and hydrographic network of the Basin council of the WRBM of the Prypiat river

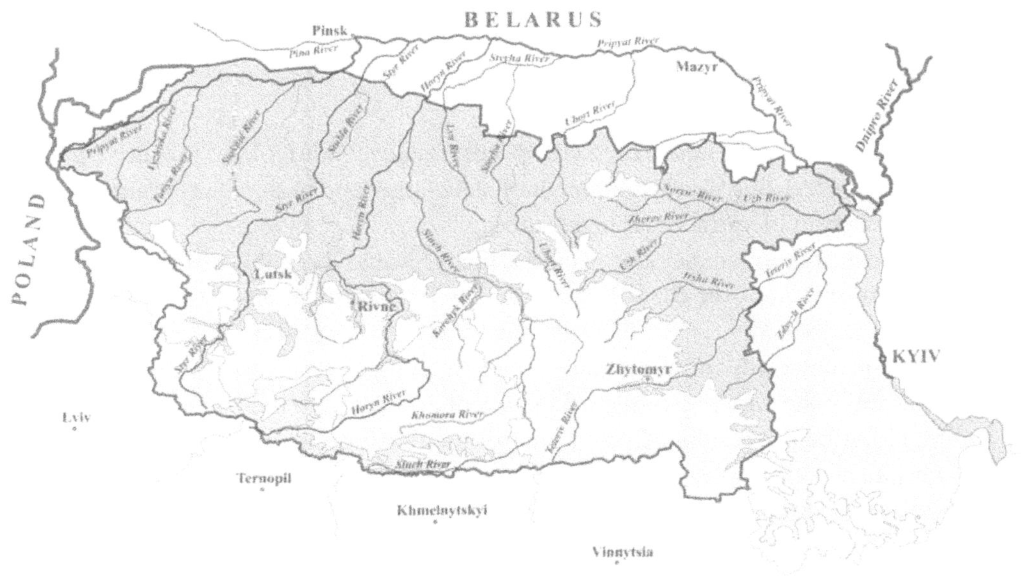

There are 4.429 watercourses with a total length of 20.075 km flowing into the Prypiat River basin, of which 4.010 are rivers less than 10 km long, with a length of 11.304 km - 56% of the total length; 50 water reservoirs with a total volume of 230.19 million m^3, 2130 ponds with a volume of 265.9 million m^3.

In the Polissia zone the basin management of water resources belongs to the management sphere of the central executive body, which implements the state policy in the water management and hydrotechnical land reclamation field, management, surface water resource usage and reproduction of the State Water Resources Agency of Ukraine.

The main tasks of the basin management are to ensure the water resource state management to achieve "good" ecological state of surface and underground water bodies, which are defined by the approved Regulation on the water resource basin management. The Prypiat River Water Resources Basin Management is designated as the organization responsible for forming the composition of the Prypiat River Basin Council and preparing the meetings of the representatives of the parties involved.

As an element of the river basin integrated management, in order to coordinate issues which are related to the rational and comprehensive water resource usage in the Prypiat River basin, the establishing of the optimal operating modes of multi-purpose reservoirs, the Interdepartmental Commission for Coordination of Operating Modes and Water Resources Management in the Prypiat River Basin has been established, which is an advisory body at the Basin Administration of the Prypiat River Water Resource.

The main tasks of WRBM are the following:

1) to ensure the implementation of the state policy in the management field, to use and reproduce the surface water of water resources within the relevant river sub-basins;
2) to direct and coordinate the organization activities which belong to the sphere of the administration of the State Water Agency in charge of management, usage and reproduction of surface water resources within the corresponding river sub-basins.

The functions of WRBM include:

1) the submission of the proposals to the State Water Agency, bodies of the executive power and local self-government regarding the development of state target and regional programs on management, use and reproduction of surface water resources;
2) the development of the management plans for the relevant river sub-basins;
3) under the coordination of the State Water Agency, the organization of the relevant basin council work;
4) the coordination of the work of water management organizations in terms of the integrated approach implementation in the water resource management according to the basin principle;
5) the summarizing of the results of water quality monitoring in control bodies in the areas of the main water intakes of complex purpose, inter-farm systems of inter-industry and agricultural water supply;
6) the conducting of an analysis of the surface water quality and notifying the executive authorities and local self-government bodies, the organization and development of the operational and long-term predictions of the changes in the water body ecological state, as well as the submission of the results to the State Water Agency;
7) the development of water management balance, the participation in solving issues related to inter-basin, inter-state distribution of river flow and border-line water usage;
8) the summary and analysis of the data about the water management and hydrological situation and the submission of the operational information to the State Water Agency;
9) the coordination of the measures as to the ecological sanitation of the surface water;
10) within the scope of competence, the ensuring of the control over the observance of the established modes of water reservoirs, water-economic systems and canals;
11) the generalization and analysis of the results of the public accounting of the water resource use and the submission of the proposals to the State Water Agency to fill/ add information to the state water cadastre in the chapters “Surface water objects” and “Water usage”;
12) the organization of the scientific, scientific-technical, investment, information, publishing activity, the promotion of the development and introduction of the updated information technologies and computer networks in the sphere of the development of water management and hydro-technical land amelioration, management, use and restoration of water resources (Figure 5);

13) the supply of the public with the information about the implementation of the public policy in the sphere of the development of water economy and hydro-technical land reclamation, management, use and restoration of the surface of water resources, as well as the implementation of the measures which popularize ecological knowledge, caring attitude to water resources, openness of the activity through the cooperation with mass media and the use of the means of information dissemination.

Figure 5. Structure of the administration of WRBM of the Prypiat river

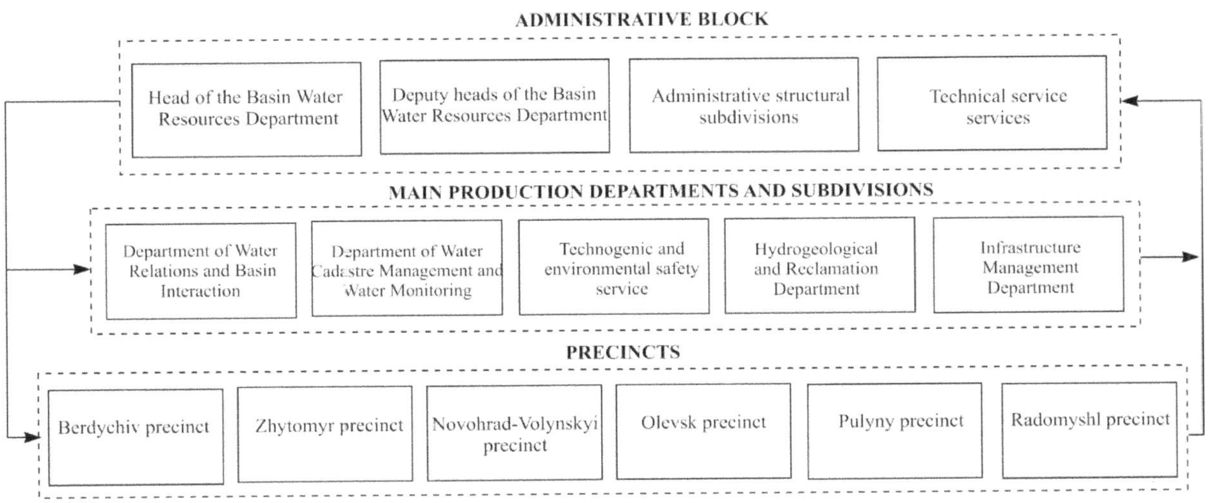

In 2021 the Basin administration of water resources of the Prypiat river focused its activity on satisfying the needs of the population and the branches of the economy of Zhytomyr region in water resources, their rational use, and the efficiency of the water-economic engineering infrastructure.

CONCLUSION

The strategic goal of the ecologically safe water resource usage is to ensure the priority of environmental protection functions over economic ones, the rational surface and underground water usage, the wide implementation of the water-saving technologies in all sectors of the economy, and therefore - sustainable water use with sustainable economic development. The main realization of such a strategy is the implementation of the Integrated water resources management principles.

The basin management principle is admitted as one of the effective methodological means of solving global environmental problems of water usage, water protection and water resource reproduction. The strategic goal of the water resource management according to the basin principle is to ensure basin balance in the water management development, water protection and water resource reproduction based on the legal principle consistency and water users' management actions according to the basin principle, aimed at sustainable water supply for the population and economic sectors, the implementation of promising technological standards for the water resource usage, prevention of the water harmful effects.

With help of an integrated approach, the water resource management is coordinated in different sectors of the economy or by the interested groups at different scales - from a local level to an inter-state level.

The integrated water resource management demands the inclusion in the processes of the national policy and legislative framework development, the creation of a more sophisticated management system and a more effective institutional and regulatory structure necessary for the development of more equitable and sustainable solutions. Within the framework of the Integrated water resources management of Ukraine, the following is envisaged: the creation of basin water resources management bodies; the creation of the structures responsible for the water resource reproduction; the development of regulatory and legal acts that ensure the development; the implementation of the economic regulation mechanisms; the development of target reconstruction and development programs; the industry financing program; the organization of environmental monitoring, etc.

The key issue of the prerequisite for the creation of the basin management system is the creation of such a management body for the entire river basin, which would include the representatives of all layers of the society affected by the water problems (both water consumers and those responsible for water supply). The purpose of creating such a representative body is to ensure the resolution of any conflicts of interests at the local level.

The Polesie region of Ukraine is characterized by a powerful and developed river hydrographic network, has trans-border significance, as it borders on the hydrographic network of the corresponding regions, which makes it extremely urgent to solve the issue of the integrated water resource management based on the basin principle.

The Basin Councils, including the Prypiat River Basin Council, are created with the aim of establishing an effective mechanism for the coordination of the efforts of border-line states for the ecological improvement of river basins, for the development and signing of the inter-governmental agreements and regulatory acts, the efficient information exchange and the support of a broad participation of all parties involved.

The Basin Councils carry out their activities in cooperation with the relevant Basin water resource departments, Regional water resource offices, the Ministry of Ecology and Natural Resources of Ukraine, the State Water Agency, the local bodies of the executive power, the territorial bodies of the central authorities of the executive power and local self-government bodies, exercising their powers within sub-basins, other interested organizations, institutions, enterprises, public associations.

REFERENCES

Charnyi D.V., Horuzhyi P.D. (1996). Accounting, quality improvement and water loss reduction in distribution systems of agricultural water supply. In *proceedings of the scientific-practical conference "Issues of the efficient use of water resources and land reclamation.* CG Space.

Deziron O.V. (2003). Water use in Ukraine: current state and development challenges. *Water management of Ukraine*, *3-4*, 4-8.

Khvesyk M., Smolenskyi E. (2000). Social-economic priorities of the environmentally safe development of water management of Ukraine. *Water management of Ukraine*, *3-4*, 11-15.

Voropay, G., Yatsyk, M., & Mozol, N. (2019). Current State And The Prospects Of Development Of Drainage Reclamation In A Changing Climate. *Land Reclamation and Water Management*, (2), 31 - 39. https://doi.org/10.31073/mivg201902-180.

Lazarchuk, N. A., Rokochinskiy, A. N., & Stashuk, V. A. (1991). Automated planning of water use on drainage-moistening systems. Land reclamation and water management, 11, 42-45.

Pavlov, V. I., & Stashuk, A. V. (2013). *Basin management of water resources: experience and priorities.* NUWMNU.

Shevchuk, V. Ya., Husev, M. V., Mazurkevych, O. O., Navrotskyi, V. M., Satalkin, Yu. M., Trehobchuk, V. M., & Yatsyk, A. V. (1997). *Economy and ecology of water resources of the Dnipro river.*

Single departmental administration in charge of the organization and monitoring of waters. Order of the Ministry ecological resources of Ukraine No 485 of 24.12.2001 – K.: 2001.

Kalinichenko, L. I., Karuk, B. P., & Tyshenko, O. I. (2000). *Management of water-economic complexes.* CSU.

Stashuk, V. A. (2006). Ecologic-economic principles of the basin management of water resources. SC Publisher Zoria.

Stashuk, V. A., Mokin, V. B., Hrebin, V. V., & Chunariov, O. V. (2014). Scientific principles of the rational use of water resources of Ukraine by a basin principle: Monograph. Hrin D.S.

Water code of Ukraine (with changes and amendments made by the Law of Ukraine of 21.09.2000 No1990-111). – 38 p. Honcharov S.M. & Dupliak V.D. (1992). *Principles of management and marketing.* – K.: NMKVO – 298 p.9.

Kozhushko, D., Stashuk, V., Khvesyk, A., & Rokochynski, A. (2018). *Water management in Ukraine: problems and development innovations: monograph.* Rivne.

Compilation of References

About approval of the Concept for the implementation of state policy in the field of climate change until 2030: Order of the Cabinet of Ministers N° 932-r of December 7, 2016 https://www.kmu.gov.ua/ua/npas/249573705 (access date 29.04.2020)

Adamuscin, A., Palicki, S., & Rącka, I. (2017). Real estate appraisal in central-eastern Europe: Comparative analysis of Poland and Slovakia. *Real Estate Management and Valuation*, *25*(4), 50–62. doi:10.1515/remav-2017-0029

Adger, W. (2006). Vulnerability. *Global Environmental Change*, *16*(3), 268–281. doi:10.1016/j.gloenvcha.2006.02.006

Afanasik, G. & Zhukov, L. (1983). Layered balance scheme for calculating the water regime of the soil for the purposes of its optimal management. *Issues of operation of drainage and humidification systems*, 15–22.

Afanasik, G., Armonik, O. & Skvortsov, N. (1976). Calculation of the water regime of reclaimed soils. *Constructions and calculations of drainage and humidification systems, 2*, 75–86.

Afanasik, G., Omelko, A., Zhukov, L. (1989). Water regime management of large reclamation systems. *Reclamation and environmental protection*, 17–26

Aidarov, I., Golovanov, A., & Nikolsky, Yu. (1990). *Optimization of reclamation regimes of irrigated and drained agricultural lands (recommendations)*. Agrometeoizdat.

Akinola O. V. & Adegoke J. (2019). Assessment of forest fire vulnerability zones in Missouri, United States of America. *International Journal of Sustainable Development & World Ecology, 26*(3), 251-257. doi:10.1080/13504509.2018.1551815

Alekseev, V. S., & Kommunar, G. M. (1974). Colmatage of filters and filter zones of water wells. *Water Resources*, *5*, 170–177.

Aleksyeyevskii, V., Skrypnyk, O., & Riabtseva, H. (1992). Assessment and control of changes in natural complexes under the influence of drainage. K: UkrNYYHyM.

Anderson, M., Bertha, T. (1988). *Hydrogeological forecasting: Per. with English*. Mir.

Armeanu, D., & Lache, L. (2009). The NPV criterion for valuing investments under uncertainty. economic computation and economic cybernetics studies and research. Academy of Economic Studies, 4(4), 133–143.

Armeanu, D., & Lache, L. (2009). The NPV criterion for valuing investments under uncertainty. *Economic computation and economic cybernetics studies and research. Academy of Economic Studies*, (4), 133–143.

Averyanov, S. F. (1982). *Outflow from canals and its influence on the mode of groundwater*. Kolos.

Aydarov, I., Golovanov, A., & Nikolsky, Yu. (1990). *Optimization of meliorative regimes of irrigated and drained agricultural land*. Agrometeoizdat.

Bagdonavicius, A., & Ramanauskas, R. (2004). Introducing a computerised market value-based mass appraisal system for real property taxation in Lithuania. *FIG Working Week*, Athens, Greece. https://www.fig.net/resources/proceedings/fig_proceedings/athens/papers/ts27/TS27_2_Bagdonavicius_Ramanauskas.pdf

Balligand, P., Le Gouellec, P., Dumont, M., & Grauby, A. (1978). Experience Gained in France on Heat Recovery from Nuclear Plants for Agriculture and Pisciculture. *Nuclear Technology*, *38*(1), 90–96. doi:10.13182/NT78-A16160

Baluk, S. A. (2015). Land reclamation. *Collective monograph.* Kherson: Grin D.S.

Bar-Mei, G. (2022). *Basics of Fluid Mechanic*. Potto Project Publication. doi:10.5281/zenodo.6462400

Barvika, S. (2020). The property tax in Latvia: system structure and current challenges [online]. *Working Paper WP20SB1*. Cambridge. Lincoln Institute of Land Policy. https://www.lincolninst.edu/sites/default/files/pubfiles/barvika_wp20sb1_0.pdf

Barvika, S., Rausis, A., & Berzina, I. (2013). Opportunities for the development of the Latvian property tax administration system through improvements in the property registration system and the implementation of European Union requirements for geospatial information [online]. *Proceedings REAL CORP 2013: Planning times,* Rome, Italy. https://programm.corp.at/cdrom2013/papers2013/CORP 2013_201.pdf

Barvinsky, A., & Tykhenko, R. (2015). *Otsinka i prohnoz yakosti zemel: Monografiya* [*Assessment and forecast of land quality: Monograph*]. Medinform.

Baryakhtar, V., Kuhar, V., Los, I., Poyarkov, V., Kholosha, V., & Shestopalov, V. (Eds.). (1998). *A comprehensive risk assessment of the accident at the Chernobyl NPP*. IAEA.

Bashkis, K. K. (1974). *On the issue of analysis of calculation methods of horizontal drainage*. Transactions of Lit. NIIGiM. Issues of land drainage using pottery drainage. -. *Jelgava*, *IX*, 5–24.

Bates, B. C., Kundzewicz, Z. W., & Palutikof, J. P. (2008). Climate Change and Water Resources. Intergovernmental Panel on Climate Change Technical Paper. IPCC Secretariat.

Bierman, H. J. R., & Smidt, S. (2006). The capital budgeting decision: Economic analysis of investment projects. 9th ed. Abingdonon-Thames. Routledge.

Bierman, H., Jr., & Smidt, S. (2006). The capital budgeting decision: economic analysis of investment projects. 9th ed.: New York. Routledge.

Bita, C., & Gerats, T. (2013, July). Plant tolerance to high temperature in a changing environment: Scientific fundamentals and production of heat stress-tolerant crops. *Frontiers in Plant Science*, *31*. Advance online publication. doi:10.3389/fpls.2013.00273 PMID:23914193

Blanc, E., & Schlenker, W. (2017, Summer). The use of panel models in assessments of climate impacts on agriculture. *Review of Environmental Economics and Policy*, *11*(2), 258–279. doi:10.1093/reep/rex016

Bohovin, A., Slyusar, I., & Tsarenko, M. (2005). Herbaceous biogeocenoses, their improvement and rational use. K.: Agrarian science.

Boichenko, S. (2005) *Hlobalni i rehionalni kolyvannia (zminy) klimatu ta mozhlyvi ekolohichni naslidky vid nykh na terytorii Ukrainy: napivempirychni modeli, stsenarii [Global and Regional Climate Fluctuations (Changes) and Their Possible Ecological Consequences on the Territory of Ukraine: Semiempirical Models, Scenarios]* [Doctoral dissertation, Odessa, Ukraine].

Boichenko, S., Voloshchuk, V., & Doroshenko, I. (2000). Global warming and its consequences on the territory of Ukraine. *Ukrainskyi heohrafichnyi zhurnal, 2*, 59–68.

Borowiec, J. (1990). *Torfowiska Regionu Lubelskiego. Prace Wydz. Nauk o Ziemi i Nauk Górniczych.* Wydaw. PWN.

Budagovsky, A. (1981). Scientific bases of an estimation of need for irrigation and calculation of irrigation systems. *Modern problems of hydrology of irrigated lands, 1*, 16–32.

Budz, O. (2000). *Rozrakhunok maksymalnykh vytrat vesnianoi poveni pravoberezhnykh prytok r. Prypiati: dys. kand. tekhn. n.*: hidravlika ta inzhenerna hidrolohiia [*Calculation of the maximum costs of the spring flood of the right-bank tributaries of the Pripyat River*] [Dissertation, Rivne, Ukraina].

Cabinet of Ministers of Ukraine. (2016). *On approval of the Concept for the implementation of state policy in the field of climate change until 2030.* Cabinet of Ministers of Ukraine. https://www.kmu.gov.ua/ua/npas/249573705

Cai, H.-y., Liu, Z.-l., Yang, G., & Di, X.-y. (2012). Study on forest fire direct loss assessment. *World Automation Congress 2012*, (pp. 1-4).

Celik, A., & Raper, R. L. (2012). Design and Evaluation of ground-driven rotary subsoilers. *Soil & Tillage Research, 124*, 203–210. doi:10.1016/j.still.2012.06.010

Chana Clara, S., Fakra Sirine, C., Edwards, D., & Emerson, D. (2009). Iron oxyhydroxide mineralization on microbial extracellular polysaccharides. *Geochimica et Cosmochimica Acta, 73*(13), 3807–3818. doi:10.1016/j.gca.2009.02.036

Charnyi D.V., Horuzhyi P.D. (1996). Accounting, quality improvement and water loss reduction in distribution systems of agricultural water supply. In *proceedings of the scientific-practical conference "Issues of the efficient use of water resources and land reclamation.* CG Space.

Chen, J. (2006). An analytical theory of project investment: A comparison with real option theory. *International Journal of Managerial Finance.*, *2*(4), 354–363. doi:10.1108/17439130610705535

Cherenkov, A., Cherenkova, L. & Lazarchuk, M. (1993). Layered-balance scheme of calculation of water regime in the conditions of close occurrence of groundwater level. *Hydromelioration and hydraulic engineering construction, 20*, 9–12.

Ciscar, J. C., Rising, J., Kopp, R. E., & Feyen, L. (2019). Assessing future climate change impacts in the EU and the USA: Insights and lessons from two continental-scale projects. *Environmental Research Letters, 14*(8), 084010. doi:10.1088/1748-9326/ab281e

Consensus on consensus: a synthesis of consensus estimates on human-caused global warming / Cook J. et al. // *Environmental Research Letters*. 2016. Vol. 11. 4. P. 1-7. https://doi.Org/10.1088/1748-9326/ll/4/048002

Czekalla, C., Mevius, W., & Hanert, H. (1985). Quantitative Removal of Iron and Manganese by Microorganisms in Rapid Sand Filters (In Situ Investigations). [Berlin.]. *Water Supply*, •••, 3.

Dalampakis, P., Gelegenis, J., Illias, A., Ladas, A., & Kolios, P. (2017). Technical and economic assessment of geothermal soil heating systems in row covered protected crops: A case study from Greece. *Applied Energy, 203*, 201–2018. doi:10.1016/j.apenergy.2017.06.029

Danylchenko, O. (2019). *Richkovi baseiny Sumskoi oblasti: heoekolohichnyi analiz*: monohrafiia [*River basins of the Sumy region: geoecological analysis*: monograph]. Sumy, Ukraina. 271

Datsko, L. V. (2016). Modern agricultural use of lands of the humid zone of Ukraine. *Reclamation and water management*. 2016. *Vip.*, *103*, 41–47.

Deziron O.V. (2003). Water use in Ukraine: current state and development challenges. *Water management of Ukraine, 3-4*, 4-8.

Dmitriev, A. F., Bezusyak, A. V., & Khlapuk, N. N. (1992). *Improvement of drainage and wetting systems*. Svit.

Dobrowolski, R., & Harasimiuk, M. (2002). Geologiczne uwarunkowania rozwoju rzeźby Polesia. [In:] S. Radwan et al. (eds.), Środowisko przyrodnicze Polesia – stan aktualny i zmiany.

Dobrowolski, R. (2006). *Glacjalna i peryglacjalna transformacja rzeźby krasowej północnego przedpola wyżyn lubelsko-wołyńskich (Polska SE, Ukraina NW)*. Wyd. UMCS.

Dong, Z., Pan, Z., Wang, S., An, P., Zhang, J., Zhang, J., Pan, Y., Huang, L., Zhao, H., Han, G., Wu, D., Wang, J., Fan, D., Gao, L., & Pan, X. (2016). Effective crop structure adjustment under climate change. *Ecological Indicators, 69*(October), 571–577. doi:10.1016/j.ecolind.2016.04.010

Doorenbos, J., & Kassam, A. H. (1979). Yield respons to water. FAO Irrigation and Drainage Paper. FAO.

Downey, L. A. (1972). Water-Yield relations fornonforage crops. *Journal of the Irrigation and Drainage Division, 98*(1, IR1), 107–115. doi:10.1061/JRCEA4.0000853

Duque, M., Zapico, I., Bugosh, N., Tejedor, M., Delgado, F., & Nicolau, M. (2021). A Somolinos quarry land stewardship history: From ancient and recent land degradation to sensitive geomorphic-ecological restoration and its monitoring. *Ecological Engineering, 170*(November), 106359. doi:10.1016/j.ecoleng.2021.106359

Dyatel, O. O., Sydorenko, O. O , Tsvetova, O. V., & Turaieva, O. V. (2019). Preservation of the natural geo-ecosystem of the cross-border territory of Volyn Polissya. *«Water for All»: dedicated to World Water Day: International. scientific-practical Conference,* (pp. 127-128).

Eckhardt, K., & Ulbrich, U. (2003). Potential Impacts of Climate Change on Groundwater Recharge and Streamflow in a Central European Low Mountain Range. *Journal of Hydrology (Amsterdam), 284*(1-4), 244–252. doi:10.1016/j.jhydrol.2003.08.005

Emerson, D., Field, E., Chertkov, O., Davenport, K. W., Goodwin, L., Munk, C., Nolan, M., & Woyke, T. (2013). Comparative genomics of freshwater Fe-oxidizing bacteria: implications for physiology, ecology, and systematics. Frontiers in Microbiology / Evolutionary and Genomic Microbiology, 4.

Eric E. Massey, 2012. The experience of the European Union in adapting to climate change and its application in Ukraine. Office of the Coordinator for Economic and Environmental Affairs of the OSCE P-40.

EU Water Framework Directive 2000/60/EC. (2006). Kyiv, Ukraina.

Farahnak-Ghazani, M., & Mirmohseni, M. Masoumeh Nasiri-Kenari. (2021). On Molecular Flow Velocity Meters. IEEE Transactions on Molecular, Biological and Multi-Scale Communications (pp. 224 – 238). IEEE. doi:10.1109/TMBMC.2020.3044772

Fijałkowski D. (1994). Flora roślin naczyniowych Lubelszczyzny. *Środowisko przyrodnicze Lubelszczyzny*. LTN.

Fijałkowski, D., & Izdebski, K. (2002). Flora Poleskiego Parku Narodowego, rośliny naczyniowe. In S. Radwan (ed.), Poleski Park Narodowy. Monografia przyrodnicza.

Formation of the regime of natural waters of the Shatsk lakes area in modern conditions. (2004). *Collective monograph (edited by M.I. Romashchenko, Y.Y. Bakhmachuk). Kyiv: Agricultural science.*

Frolenkova, N., & Rokochinskiy, A. (2015). The evaluation of environmental risks in the sphere of water and land reclamation [online]. *Oxford Journal of Scientific Research, 1*(9), 155–160. https://core.ac.uk/download/pdf/33693269.pdf#page=155

Frolenkova, N., Kozhushko, L., & Rokochinskiy, A. (2007). *Ekoloho-ekonomichne otsinyuvannya v upravlinni melioratyvnymy proektamy: Monografiya* [*Ecological and economic assessment in the management of reclamation projects: Monograph*]. NUVGP.

Frolenkova, N., & Rokochinskiy, A. (2007). *Ekoloho-ekonomichne otsinyuvannya v upravlinni melioratyvnymy proektamy*. NUVGP.

Frolenkova, N., Rokochinskiy, A., & Kozhushko, L. (2007). *Ecological and economic evaluation in the management of reclamation projects*. NUVGP.

Frolenkova, N., Rokochinskiy, A., Volk, P., Shatkovskyi, A., Prykhodko, N., Tykhenko, R., & Openko, I. (2020). Cost-effectiveness of investments in drip irrigation projects in Ukraine. [IJGE]. *International Journal of Green Economics, 14*(4), 315–326. doi:10.1504/IJGE.2020.112570

Fujikawa, Y., Sugahara, M., Hamasaki, T., Yoneda, D., Minami, A., Sugimoto, Y., & Iwasaki, H. (2010). Biological filtration using iron bacteria for simultaneous removal of arsenic, iron, manganese and ammonia: Application to waterworks facilities in Japan and developing countries. *J. Human Environ. Studies*, (9), 261–276.

Furdychko, O., Lavrov, O., & Konishchuk, V. (2010). Agroecological aspects of environmental protection on the basis of sustainable development. Agroecological journal, pp. 5-10.

Gadzala, S. (2017). Reclamation and development of Ukrainian Polesie: a collective monograph. NIH.

Gadzalo, Y., Stashuk, V., & Rokochinskiy, A. (2017). Melioratsiya ta oblashtuvannya Ukrayinskoho Polesie: Monografiya [Land reclamation and arrangement of the Ukrainian Polesie. Monograph]. Oldi-Plus.

Gaffarov Kh.R. (2016). Traction resistance of the central working body of the subsoiler. *Technical science. 2*(106), 130-132.

Gaffarov, Kh.R. (2016). What determines the quality and energy performance of machines and implements for deep tillage? *Technical science, 5* (109), 31-32.

Gaffarov, Kh.R. (2016). What determines the quality and energy performance of machines and implements for deep tillage? *Young Scientist. Technical science. 5*(109). 31-32.

Gai, C., Weng, W., & Yuan, H. (2011). GIS-Based Forest Fire Risk Assessment and Mapping. *Fourth International Joint Conference on Computational Sciences and Optimization,* (pp. 1240-1244). IEEE. 10.1109/CSO.2011.140

Geografica. (2023). Poliski landscape region. *Geographica.* http://geografica.net.ua/publ/galuzi_geografiji /fizichna_geografija_ukrajini/poliskij_landshaftnij_kraj/39-1-0-544

Giroud, J., Palmer, B., & Dove, J. E. (2000). Calculation of Flow Velocity in Pipes as a Function of Flow Rate. *Geosynthetics International*, *7*(4-6), 583–600. doi:10.1680/gein.7.0183

Global Forest Watch. (2018). Tree Cover Loss in [selected area name]. *Global Forest Watch.* www.globalforestwatch.org

Global temperature [Data set]. (n.d.). NASA. https://climate.nasa.gov/vital-signs/global-temperature/

Gohar, A., & Cashman, A. (2016). A methodology to assess the impact of climate variability and change on water resources, food security and economic welfare. Agricultural Systems, 147. doi:. 008. doi:10.1016/j.agsy.2016.05

Gooch, J. W. (2011). Kinematic Viscosity. In: Encyclopedic Dictionary of Polymers. Springer. doi:10.1007/978-1-4419-6247-8

Gopchak, I. (2018). Retrospective Analysis of the dynamics of changes in the surface water quality in the Western Bug river. *Land Reclamation and water Management.*

Gopchak, I. (2021). *Naukovi zasady zberezhennia ta vidrodzhennia malykh richok Zakhidnoho Polesie Ukrainy*: avtoref. dys. [Scientific principles of preservation and revival of small rivers of the Western Polesie of Ukraine. [Thesis, Kyiv, Ukraina].

Gopchak, I., Basiuk, T., Bialyk, I., Pinchuk, O., Gerasimov, I. (2019). Dynamics of changes in surface water quality indicators of the Western Bug River basin within Ukraine using GIS technologies. *Journal of Water and Land Development, 42* (VII–IX), 67–75. . doi:10.2478/jwld-2019-0046

Gorzhyk, I., Yemelyanova, Zh., Trofimova, I., & Shereshevskiy, A. (1998). Ukraine and Global Greenhouse Effect: Book. 2. Vulnerability and adaptation of environmental and economic systems to climate change. Kiev, Ukraine: Vydavnytstvo Ahentstva z ratsionalnoho vykorystannia enerhii ta ekolohii.

Grabovska, L. L., & Yefremova, O. O. (2010). *Environmental risk assessment in the field of drinking water supply in Ukraine*. Collection of scientific works of the National Academy of the State Border Guard Service of Ukraine. *Technical Sciences. Issue*, (54), 58–62.

Hadzalo, Ya., Stashuk, V., & Rokochynckyi, T. (2017). Land reclamation and development of the Ukrainian Polissia: a collective monograph. OLDI-PLUS.

Haka, S. F. (2006). A review of the literature on capital budgeting and investment appraisal: Past, present, and future musings. *Handbooks of Management Accounting Research*, *2*, 697–728. doi:10.1016/S1751-3243(06)02010-4

Hallbeck, L., & Pedersen, K. (1991). Autotrophic and mixotrophic growth of Gallionella ferruginea. *Journal of General Microbiology*, *137*(11), 2657–2661. doi:10.1099/00221287-137-11-2657

Hallberg, L., Ferris, R., & Grant, F. (2004). Biomineralization by Gallionella. *Geomicrobiology Journal*, *21*(5), 325–330. doi:10.1080/01490450490454001

Hall, J. V., Zibtsev, S. V., Giglio, L., Skakun, S., Myroniuk, V., Zhuravel, O., Goldammer, J. G., & Kussul, N. (2021). Environmentaland political implications of underestimated croplandburning in Ukraine. *Environmental Research Letters*, *16*(6), 064019. doi:10.1088/1748-9326/abfc04 PMID:34316296

Hamzaa M.A., Anderson W.K. (2005). Soil compaction in cropping systems: A review of the nature, causes and possible solutions. *Soil and Tillage Research, 82*(2), 121-145.

Hanks, R., & Ashcroft, J. (1985). *Applied soil physics. Soil moisture and temperature: Per. with English*. Gidrometeoizdat.

Hansen, M. C., Potapov, P. V., Moore, R., Hancher, M., Turubanova, S. A., Tyukavina, A., Thau, D., Stehman, S. V., Goetz, S. J., Loveland, T. R., Kommareddy, A., Egorov, A., Chini, L., Justice, C. O., & Townshend, J. R. G. (2013). High-Resolution Global Maps of 21st-Century Forest Cover Change. *Science*, *342*(6160), 850–853. doi:10.1126cience.1244693 PMID:24233722

Harasimiuk, M., & Dobrowolski, R. (2010). Geneza krajobrazów hydrogenicznych Polesia Zachodniego. [In:] T. Chmielewski, D. Piasecki (eds.), The Future of hydrogenic Landscapes In European Biosphere Reserves. Poleski PN, PAN Oddz. Lublin.

Havrysh, V. S. (2013). Scientific and methodical recommendations for applying deep loosening on drained mineral soils of the Western Polissia of Ukraine. Department of Defense.

Hebbink, A. J. (1999). Land drainage. In *Environmental Geology. Encyclopedia of Earth Science*. Springer. doi:10.1007/1-4020-4494-1_196

Himbarzhevsky, V. (1971). Cultural works on drained lands. *Visn. c.-g. science, 8*, 28-32.

Hud, V., Pinchuk, O., Martyniuk, P., & Volk, P. (2019). Mathematical modelling of heat transfer in a greenhouse with surface heating system. *Scientific Review Engineering and Environmental Sciences, 28*(4), 569–583. doi:10.22630/PNIKS.2019.28.4.52

IAAO. (2017). *Standard on mass appraisal of real property.* International Association of Assessing Officers. https://www.iaao.org/media/standards/StandardOnMassAppraisal.pdf

Iglesias, A., & Garrote, L. (2015). Adaptation strategies for agricultural water management under climate change in Europe. *Agricultural Water Management, 155*, 113–124. doi:10.1016/j.agwat.2015.03.014

IPCC. (2019). An IPCC special report on climate change, desertification, land degradation, sustainable land management, food security, and greenhouse gas fluxes in terrestrial ecosystems. In P.R. Shukla, J. Skea, E. Calvo Buendia, V. Masson-Delmotte, H.-O. Pörtner, D. C. Roberts, P. Zhai, R. Slade, S. Connors, R. van Diemen, M. Ferrat, E. Haughey, S. Luz, S. Neogi, M. Pathak, J. Petzold, J. Portugal Pereira, P. Vyas, E. Huntley, K. Kissick, M. Belkacemi, J. Malley, (eds.) *Climate Change and Land.* IPCC. https://www.ipcc.ch/srccl/

IPCC. (2020). *Climate Change and Land: An IPCC Special Report on climate change, desertification, land degradation, sustainable land management, food security, and greenhouse gas fluxes in terrestrial ecosystems.* IPCC. https://www.ipcc.ch/reports/

Irrigation and drainage strategy in Ukraine for the period up to 2030. (2019). FAO.

Isaeva, S., & Shevchenko, V. (2019). Principles of Ecological Monitoring of Agricultural Lands. *IOP Conference Series. Earth and Environmental Science, 272*(2), 022230. doi:10.1088/1755-1315/272/2/022230

Ivitsky, A. I. (1988). *Fundamentals of design and calculation of drainage and drainage-humidification systems.* Science and Technology.

Jatsiuk, M., Sydorenko, O., & Nechai, O. (2022b). Water accumulation capacity of forest zone (Polissia) of Ukraine under global warming conditions. 5th International Scientific and Technical conference «Innovative development of resource-saving technologies and sustainable use of natural resources». (pp. 31-33). Universitas Publishing Petroşani.

Jatsiuk, M., Tsvetova, O., & Turaieva, O. (2022a). Sustainability of natural waters within the area of potential impact of the «Khotyslavsky» quarry. 5th International Scientific and Technical conference «Innovative development of resource-saving technologies and sustainable use of natural resources», (pp. 76-78). Universitas Publishing Petroşani. .

Jatsyk, A., Pasheniuk, I., Gopchak, I., & Basiuk, T. (2019). Zoning of the territory of western polissya of Ukraine for anthrogenic loading and ecological status of small river basins. *Bichuk.*

Juszczyk, M., Výskalab, M., & Zimaa, K. (2015). Prospects for the use of BIM in Poland and the Czech Republic – Preliminary Research Result. *Procedia Engineering, 123*, 250–259. doi:10.1016/j.proeng.2015.10.086

Kalinichenko, L. I., Karuk, B. P., & Tyshenko, O. I. (2000). *Management of water-economic complexes.* CSU.

Kamra, S. K., Kumar, S., Kumar, N., & Dagar, J. C. (2019). Engineering and Biological Approaches for Drainage of Irrigated Lands. In J. Dagar, R. Yadav, & P. Sharma (Eds.), *Research Developments in Saline Agriculture.* Springer. doi:10.1007/978-981-13-5832-6_18

Karuk, B. (1992). System approach to assessing the impact on the environment of drainage reclamation facilities in Ukraine. *Theory and practice of ecological and land reclamation monitoring in the Ukrainian Polesie,* 22–32.

Kennedy, C. B., Scott, S. D., & Ferris, F. G. (2004). Hydrothermal phase stabilization of 2-line ferrihydrite by bacteria. *Chemical Geology*, *212*(3-4), 269–277. doi:10.1016/j.chemgeo.2004.08.017

Kharchenko, S. (1987). *Water regime management on reclaimed lands in the Non-Chernozem zone (Hydrological aspects).* Gidrometeoizdat.

Khimko, R., Merezhko, O., & Babko, R. (2003). *Mali richky – doslidzhennia, okhorona, vidnovlennia* [Small rivers - research, protection, restoration]. Ukraina.

Khomik, N. V. (2013). Water resources of Shatsk National Nature Park: current state, protection, management. Kyiv: Agricultural science.

Khomik, K. T. (1985). *Basics of calculating drainage systems.* Bureau of Scientific and Technical Information of the Ministry of Agriculture of the Estonian SSR.

Khvesyk M., Smolenskyi E. (2000). Social-economic priorities of the environmentally safe development of water management of Ukraine. *Water management of Ukraine*, *3-4*, 11-15.

Klymenko, M., Pryshchepa, A., & Vozniuk, N. (2006). Environmental monitoring. Vydavnychyi tsentr Akademia.

KMU. (2018). Postanova vid 7 lyutoho 2018 r. Pro provedennya zahalnonatsionalnoyi (vseukrayinskoyi) normatyvnoyi hroshovoyi otsinky zemel silskohospodarskoho pryznachennya ta vnesennya zmin do deyakykh postanov Kabinetu Ministriv Ukrayiny. No. 105 [Resolution of February 7, 2018 On conducting a nationwide (all-Ukrainian) normative monetary valuation of agricultural lands and making changes to some resolutions of the Cabinet of Ministers of Ukraine. No. 105]. Kabinet Ministriv Ukrayiny. [Access 30.04.2020]. Available at: https://zakon.rada.gov.ua/laws/show/105-2018-п#Text

Kolokytha, E., & Malamataris, D. (2020). Integrated Water Management Approach for Adaptation to Climate Change in Highly Water Stressed Basins. *Water Resources Management*, *34*(3), 1173–1197. doi:10.100711269-020-02492-w

Konishchuk, V. (2015). Concept and strategy of balanced development of wetland landscapes and peat ecosystems of Ukraine. K.

Konishchuk, V.V., Andrienko, T.L., Bondar, O.I., Grummo, D.G., Davidyuk, V.F., Yevstigneev, O.I., Konovalchuk, V.K., Landin, V.P., Lukash, O. .V., Serebryakov, V.V., Solovei, T.V., Tytar, V.M., Furdychko, O.I., Chobotko, H.M., & Shershun M.H. (2012). *Polesskaya ecological convention – an epistemological paradigm of formation, 9*, pp. 289-299.

Koptyuk R., Rokochinskiy A. (2008). Substantiation of necessity and modern approaches to optimization of structures of drainage systems taking into account the terrain. *Bulletin of the National University of Water Management and Environmental Management, 1* (41), 476-483.

Koptyuk, R., Budz, M., & Rokochinskiy, A. (2011). Influence of irregularity of relief on construction and reconstruction of drainage systems. *Agrarian Science*, (99), 98–102.

Koptyuk, R., & Rokochinskiy, A. (2014). Model forecasting of the water regime and technological water regulation of the drained lands with the terminal of the vestry. *Pour overlapping of the efficiency of the irritated earthlings.* Novocherkassk ROSNIPM.

Korobiichuk, I., Drevetsky, V., Kuzmych, L., & Kovela, I. (2020). The method of multy-criteria parametric optimization. *Advances in Intelligent Systems and Computing, 1140.* https://link.springer.com/chapter/10.1007/978-3-030-40971-5_9 doi:10.1007/978-3-030-40971-5

Korobiichuk, I., Kuzmych, L., Kvasnikov, V., & Nowak, P. (2017). The use of remote ground sensing data for assessment of environmental and crop conditions of the reclaimed land. Advances in Intelligent Systems and Computing (AISC), 550, 418–424. doi:10.1007/978-3-319-54042-9_39

Kotlyakova V. M. (2006). *Dictionary of modern geographical names*. Getty.

Koval S.I., & Zosimchuk O.A. (2014). Productivity of fodder crop rotations from rare fodder crops on drained peat soils of Western Polissia. *Bulletin of the National University of Water Management and Nature Management*, *65*(1), 64 - 72.

Kovalenko P., Rokochynskiy A., Jeznach J., Koptyuk R., Volk P., Prykhodko N., Tykhenko R. (2019). Evaluation of climate change in Ukrainian part of Polissia region and ways of adaptation to it. *Journal of Water and Land Development*, *41* (IV-VI), 77-82.

Kovalenko P.I. (2007). Conceptual approaches to land reclamation in the Ukrainian Polesie. *Amelioration of waterlogged lands, 2*.

Kovalenko P.I., Yatsyk M.V., Polyakov V.L. (1996). Management of moisture supply of agricultural crops on reclaimed land taking into account the dynamics of external environmental factors. *Land reclamation and water management, 82*.

Kovalenko, P. etc. (2019). Evaluation of climate change in Ukrainian part of Polissia region and ways of adaptation to it. *Journal of Water and Land Development, 41*(IV-VI), 77-82.

Kovalenko, P. etc. (2019). Evaluation of climate change in Ukrainian part of Polissia region and ways of adaptation to it. *Journal of Water and Land Development, 41*(IV-VI). 77-82.

Kovalenko, P., Rokochynskiy, A., Jeznach, J., Koptyuk, R., Volk, P., Prykhodko, N., Tykhenko, R. (2019). Evaluation of climate change in Ukrainian part of Polissia region and ways of adaptation to it. *Journal of Water and Land Development*, *41* (IV-VI), 77-82. . doi:10.2478/jwld-2019-0030

Kovalenko, P., Rokochinskiy, A., Mazhayskiy, Yu., Volk, P., Volk, L., & Chernikova, O. (2020). *Construction and agricultural drainage parameter optimization considering economic and environmental requirements*. Engineering for Rural Development. Jelgava. doi:10.22616/ERDev.2020.19.TF237

Kovalenko, P., Rokochinskiy, A., Volk, P., Turcheniuk, V., Frolenkova, N., & Tykhenko, R. (2021). Evaluation of ecological and economic efficiency of investment in water management and land reclamation projects. *Journal of Water and Land Development*, *48*(1–3), 81–87. doi:10.24425/jwld.2021.136149

Kozhushko, D., Stashuk, V., Khvesyk, A., & Rokochynski, A. (2018). *Water management in Ukraine: problems and development innovations: monograph*. Rivne.

Kozlovsky, B. I. (1995). *Scientific basis of monitoring drained lands*. Missioner.

Krainov, S. R., & Shvets, V. M. (1992). Hydrogeochemistry. Moscow: Nedra, 463.

Kravets, S. V., Kovan'ko, V. V., & Lukianchuk, O. P. (2015). Scientific bases of creation of earth-tiered cars and underground-moving devices: monograph. NUWEE.

Kryvoviaz, E., Openko, I., Tykhenko, R., Shevchenko, O., Tykhenko, O., Tsvyakh, O., & Chumachenko, O. (2020). Recovery of losses for inappropriate use of land. *Scientific Papers. Series E. Land Reclamation, Earth Observation & Surveying, Environmental Engineering, 9*, 175–182. http://landreclamationjournal.usamv.ro/pdf/2020/Art2 6.pdf

Kryvoviaz, E., Openko, I., Tykhenko, R., Shevchenko, O., Tykhenko, O., Tsvyakh, O., & Chumachenko, O. (2020). *Recovery of losses for inappropriate use of land*. Scientific Papers. Series E. Land Reclamation, Earth Observation & Surveying, Environmental Engineering. http://landreclamationjournal.usamv.ro/pdf/2020/Art26.pdf

Kurhak, V., Panasyuk, S., Asanishvili, N., Slyusar, I., & Slyusar, S. (2020). Influence of perennial legumes on the productivity of meadow phytocenoses. *Ukrainian Journal of Ecology*, *10*(6), 310–315. https://www.webofscience.com/wos/woscc/full-record/WOS:000608622800047. doi:10.15421/2020_298

Kurtener, D., & Chudnovsky, A. (1969). *Raschet i regulirovanie teplovogo rezhima v otkrytom i zashchishchennom grunte* [*Calculation and Control of the Thermal Regime in Open and Covered Soil*]. Hydrometeoizdat.

Kurtener, D., & Chudnovsky, A. (1979). *Agrometeorologicheskie osnovy teplovoi melioratsii* [*The Agrometeorological Foundation of Thermal Melioration of Soil*]. Hydrometeoizdat.

Kuzmych, L. V., Voropay, G. V., Shevchuk, S. A., Usatiy, S. V., Kozitsky, O. M., & Mozol, N. V. (2021c). Influence of climate change on the condition of the reclamation fund of the Ukrainian Polissya on the example of the Maryanivka drainage system. *Engineering: theory and practice: materials of the I International Correspondence Scientific and Practical Conference.* Polissya State University.

Kuzmych, L. V., Voropay, G. V., Shevchuk, S. A., Usatyi, S. V., Kozytsky, O. M., & Mozol, N. V. (2021b). The influence of climate change on the state of the reclaimed lands of the Ukrainian Polissya on the example of the drainage system "MARYANIVKA". *Proceedings of the IV Intern. scientific-practical. conf.* State Technical University.

Kuzmych, L., Furmanets, O., Usatyi, S., Kozytskyi, O., Mozol, N., Kuzmych, A., Polishchuk, V., & Voropai, H. (2022). Water Supply of the Ukrainian Polesie Ecoregion Drained Areas in Modern Anthropogenic Climate Changes. *Archives of Hydro-Engineering and Environmental Mechanics*, *69*(1), 79–96. doi:10.2478/heem-2022-0006

Kuzmych, L., Voropay, G., Moleshcha, N., & Babitska, O. (2021a). Improving water supply capacity of drainage systems at humid areas in the changing climate. *Archives of Hydro-Engineering and Environmental Mechanics.*, *68*(1), 29–40. doi:10.1515/heem-2021-0003

Kvartenko, A. N. (2016). Practical stabilization methods of groundwater in North-Western region of Ukraine. *Water Supply and Wastewater Removal: monograph*, 107–120. Lublin University of Technology.

Kvartenko, O. M. (2011). Characteristics of underground waters in the North-Western regions of Ukraine and technological schemes for their conditioning. Problems of water supply, sewerage, and hydraulics. *Scientific and Technical Journal*, (16), 32-40.

Kvartenko, O. M. (2018). System approach to substantiation of new and improvement of existing technologies of air conditioning of multicomponent groundwater. Municipal Economy of Cities. Series: Technical Sciences and Architecture, (140), 98 -103. Kharkiv National University.

Kvartenko, O. M. (2019). Development of scientific principles and improvement of technologies for treating multicomponent underground waters. Manuscript for academic degree of doctor of technical sciences in speciality 05.17.21 – technology of water treatment. KPI, Kyiv.

Kvartenko, O.M., & Safonov, R.V. (2015). Analysis of the degree of aggressiveness of groundwater in Rivne region. *Bulletin of National University of Water Environmental Engineering, 1*(69), 58-66.

Kwak, Y. H., & William, I. C. (2000). Calculating project management's return on investment. *Project Management Journal*, *31*(2), 38–47. doi:10.1177/875697280003100205

Kyoto protocol to the united nations framework convention on climate change. (1998). United Nations. https://unfccc.int/resource/docs/convkp/kpeng.pdf

Kyryliuk, O. (2007). Istoriia stanovlennia baseinovoho pidkhodu u heohrafii ta ekolohichnomu rusloznavstvi [The history of the formation of the basin approach in geography and ecological river studies]. Nauk. vypysky Vinnytsk. derzh. ped. un-tu im. Mykhaila Kotsiubynskoho. Seriia: Heohrafiia. Vyp., *14*, 40–47.

Łabędzki, L., Kaca, E., & Brandyk, A. (2021). Irrigation and Drainage in Polish Agriculture: State, Problems and Needs. In M. Zeleňáková, K. Kubiak-Wójcicka, & A. M. Negm (Eds.), *Quality of Water Resources in Poland. Springer Water.* Springer. doi:10.1007/978-3-030-64892-3_5

Lazarchuk, M. O., Cherenkov, A. V., & Rokochinskiy, A. M. (2009). Optimization of drainage systems calculation and management. Monograph, Rivne- NUWREM.

Lazarchuk, N. (1992). Optimization of the calculation of drainage systems and their management. Water, 10(1).

Lazarchuk, N. A., Rokochinskiy, A. N., & Stashuk, V. A. (1991). Automated planning of water use on drainage-moistening systems. Land reclamation and water management, 11, 42-45.

Lazarchuk, M. (1997). *Land drainage: A textbook.* IZMN.

Lazarchuk, N., Rokochinskiy, A., & Cherenkov, A. (1990). *Technical instructions for designing on a computer of calculated modes of moistening of agricultural crops on systems of bilateral action of the Ukrainian SSR: NTD 33.63-089-89.* Ukrgidrovodhoz.

Lazarchuk, N., Rokochinskiy, A., & Gus, V. (1991). *Technical instructions for the development of computer-aided system plans for water use for drainage and humidification systems of Ukraine: NTD 33-04-01-91.* Ukrgidrovodhoz.

Laziuta, M., Kabulova, B., & Gasanov, C. (2016). Research of soil tillage by advanced design slit ripper. *Agricultural Engineering. Research papers, 45*(2).

Laziuta, M., Kabulova, B., & Gasanov, C. (2016). Research of soil tillage by advanced design slit ripper. *Agricultural Engineering. Research papers, vol. 45, Nº 2. Policy, 59*, 152–164.

Li, X., Zhang, D., Zhang, R., & Osman, A. (2012). Performance of an oscillating subsoiler in reducing resistance. Presentation of the American Society of Agricultural and Biological Engineers, St. Joseph, Michigan.

Lihatsevich, A. (1991). On the assessment of the accuracy of the equations of the water balance of the irrigated field. *Melioration and water management, 9*, 25–26.

Likhatsevich, A. P. (2001). Land reclamation of Belarus. Minsk: BelNIIMiL

Li, X., Zhang, D., Zhang, R., & Osman, A. N. (2012). Performance of an oscillating subsoiler in reducing resistance. [St. Joseph, Michigan.]. *Presentation of the American Society of Agricultural and Biological Engineers., 12*, 1341191.

Lukyanchuk, O. P., Turcheniuk, V. O., & Prykhodko, N. V. etc. (2019). Necessity and possible approaches to applying deep loosening when cultivating rice. INMATEH - Agricultural Engineering, 57(1),. 199–207.

Lukyanchuk, O. P., Turcheniuk, V. O., & Prykhodko, N. V. etc. (2019). Necessity and possible approaches to applying deep loosening when cultivating rice. INMATEH - Agricultural Engineering, 57, 199–207.

Lukyanchuk, O. P., Ryzhyi, O. P., & Ihnatiuk, R. M. (2017). Design of a tiered operating unit for deep differentiated tillage. [Ukraine.]. *Scientific Bulletin of the National Mining University, Is., 4*, 43–48.

Luthy, R. G. (1964). New concept for iron bacteria control in water wells. *Water Well Journal., 24*, 29–30.

Lybbert, T., & Summer, D. (2015). Agricultural technologies for climate change in developing countries: Policy options for innovation and technology diffusion. *Food Policy, 37*(1), 114–123. doi:10.1016/j.foodpol.2011.11.001

Lyuty, G.G., & Sanina, I.V. (2011). Factors of increasing the density of underground waters in the process of exploitation of water intakes in Ukraine. *Collection of scientific works*, (2), 91-103. Ukrainian state Geological Survey Institute.

Malecha, Z. (2022). Turbulence and Fluid Mechanics. *Energies*, *15*(3), 1116. doi:10.3390/en15031116

Margen, P. (1978). Summary - Low-Temperature Nuclear Heat. *Nuclear Technology*, *38*(2), 321–330. doi:10.13182/NT78-A32030

Martyn, A., Openko, I., Ievsiuko, V., Shevchenko, O., & Ripenko, A. (2019). Accuracy of geodetic surveys in cadastral registration of real estate: value of land as determining factor. *Proceedings of the 18th International Scientific Conference on Engineering for Rural Development*, Jelgava, Latvia. https://www.tf.llu.lv/conference/proceedings2019/Papers/N236.pdf

Martyn, A., Openko, I., Ievsiukov, T., Shevchenko, O., & Ripenko, A. (2019). Accuracy of geodetic surveys in cadastral registration of real estate: value of land as determining factor. *18th International Scientific Conference. Engineering for Rural Development,* Jelgava, Latvia.

Martyn, A., Openko, I., Ievsiukov, T., Shevchenko, O., & Ripenko, A. (2019). Accuracy of geodetic surveys in cadastral registration of real estate: Value of land as determining factor. *18th International Scientific Conference. Engineering for Rural Development*, Jelgava, Latvia. 10.22616/ERDev2019.18.N236

Martyn, A., Shevchenko, O., Tykhenk, O. R., Openk, O. I., Zhuk, O., & Krasnolutsky, O. (2020). Indirect corporate agricultural land use in Ukraine: Distribution, causes, consequences. *International Journal of Business and Globalisation*, *25*(3), 378–395. doi:10.1504/IJBG.2020.109029

Martyn, A., & Tykhenko, O. (2015). *Formuvannya kadastrovo-reyestratsiynoyi systemy v Ukrayini: Monografiya [Formation of cadastral registration system in Ukraine: Monograph*]. Medinform.

Maslova, B. (1985). *Reclamation and water management. 3. Drainage: Handbook.* Agropromizdat.

Maslov, B. (1970). *Groundwater regime of wetlands and its regulation.* Kolos.

Maslov, B. S., Stankevich, V. S., & Chernenok, V. Ya. (1981). *Drainage and humidification systems.* Kolos.

Massey, E. (2012). *Experience of the European Union in adaptation to climate change and its application to Ukraine.* Office of the Coordinator of OSCE Economic and Environmental Activities. https://www.osce.org/ukraine/104019?download=true

Massey, E. E. (2012). *Experience of the European Union in adaptation to climate change and its application to Ukraine* [online]. Office of the Co-ordinator of OSCE Economic and Environmental Activities. https://www.osce.org/ukraine/104019?download=true

Massier, E. (2012). The experience of the European Union in adapting to climate change and its application in Ukraine. *OSCE Coordinator for Economic and Environmental Activities.* Biuro Koordynatora z ekonomichnoi ta dovkilnoi dialnosti OBSIE.

Matsniev, A., Protsenko, S., & Sablii, L. (2000). Monitoring and engineering methods of environmental protection. VAT Rivnenska drukarnia.

Mazhaiskiy, Y. A., Rokochinskiy, A. N., Volchek, A. A., Meshik, O. P., & Eznakh, E. (2017). Environmental management of Polesie. Ryazan: Meshcher. Branch of All-Russian Research Institute of Hydrotechnics and Melioration Named after A.N. Kostyakova.

Mazhayskiy, Y., & Rokochynskiy, A. etc. (2017). Environmental management of Polesie: monograf. Ryazan: Meshcher.f-VNIIIGiM them. A.N. Kostyakova.

Mazhayskiy, Y., & Rokochynskiy, A. etc. (2017). Environmental management of Polesie: monograph. Ryazan: Meshcher.f-VNIIIGiM them. A.N. Kostyakova,

Mazhayskiy, Y., Rokochinskiy, A., Volchek, A., Meshyk, O., & Jeznach, J. (2017). Environmental management of Polesie. VNIIGiM.

Mazhayskiy, Y., Rokochinskiy, A., Volchek, A., Meshyk, O., & Jeznach, J. (2017). Pryrodoobustroistvo Polesia. [Environmental management of Polesie]. VNIIGiM of A. Kostiakov.

Mazhayskiy, Y., Rokochinskiy, A., Volchek, A., Meshyk, O., & Jeznach, J. (Eds.). (2017). Environmental management of Polissya. VNIIGiM of A. Kostiakov.

Mazhayskiy, Y., Rokochinskiy, A., Volchek, A., Meshyk, O., & Jeznach, J. (Eds.). (2017). Pryrodoobustroistvo Polesia [Environmental management of Polissya]. Ryazan. VNIIGiM of A. Kostiakov.

Mazhayskiy, Yu., Rokochynskiy, A., Lukianchuk, O., Turcheniuk, V., Volk, P., & Chernikova, O. (2020). Deep loosening as effective adaptive agromeliorative practice on drained mineral soil of european polesie in variable climatic conditions. *Engineering for Rural Development.*, *19*, 28–35. doi:10.22616/ERDev.2020.19.TF009

Medvedev, V. V., Laktionova, T. M., & Pocheptsova, L. G. (2004). Influence of soil structure on filtration ability. [Ukraine.]. *Herald of Agrarian Sciences.*, *4*, 5–6.

Miernicki, E., Heald, A., Huff, K., Brooks, C., & Margenot, A. (2020). Nuclear waste heat use in agriculture: History and opportunities in the United States. *Journal of Cleaner Production*, *267*, 121918. doi:10.1016/j.jclepro.2020.121918

Minaev, I. V. (1977). *Design principles of drainage and drainage-humidification systems with optimal parameters.* Water Industry of Belarus.

Ministerstvo, A. P. P. U. (2017). *Nakaz vid 23 travnya 2017 r. Pro zatverdzhennya Poryadku normatyvnoyi hroshovoyi otsinky zemel silskohospodarskoho pryznachennya. No. 262 [Order dated May 23, 2017 On approval of the procedure for normative monetary valuation of agricultural lands. No. 262]* Ministerstvo ahrarnoyi polityky ta prodovolstva Ukrayiny. https://zakon.rada.gov.ua/ laws/show/z0679-17#Text

Mioduszewski, W., Kowalewski, Z., Zurawski, R., & Stankiewicz, J. (2010). Drainage system in the Kampinos National Park. *Journal of Water and Land Development*, *14*(1), 83–95. doi:10.2478/v10025-011-0007-3

Mitrofanov, E., & Mitrofanova, O. (2021).*Tile drainage monitoring in agricultural fields with an unmanned aerial system. IOP Conference Series: Earth and Environmental Science.*

Mohamed, S., & McCowan, A. K. (2001). Modelling project investment decisions under uncertainty using possibility theory. *International Journal of Project Management*, *19*(4), 231–241. doi:10.1016/S0263-7863(99)00077-0

Moore, F., & Lobell, D. (2014). Adaptation potential of European agriculture in response to climate change. *Nature Climate Change*, *4*(7), 610–614. doi:10.1038/nclimate2228

Moshynskyi V., Tsvyetova O. (2004). Methodology and prospects of ecological and reckamation monitoring in Ukrainian Polesie. *Reclamation and Water Management, 90*, 33-39.

Moshynskyi, V. (1995). Monitoring and assessment of the ecological and reclamation condition of drained lands of Rivne region. National University of Water and Environmental Engineering.

Moshynskyi, V. (2005). Methods of managing productivity and ecological sustainability of drained lands. Rivne: NUWM.

Moshynskyi, V., & Sasyuk, Z. (2010). Spatial extrapolation in monitoring data reproduction tasks. Rivne: NUWM.

Moszynski W. (2002). *Dynamika stezenia jonow w odwodmonych glebach norfowych terenow lesno-stepowych zachodniej Ukrainy*. Roczniki Gleboznawcze. – Warszawa.

Mouchet, P. (1995). Biological Filtration for Iron and Manganese Removal: Some Case Studies. *WQTC 95 (AWWA)*, New Orleans LA.

Mouchet, P. (1992). From Conventional to Biological Removal of Iron and Manganese in France. *Journal - American Water Works Association*, *84*(4), 158–167. doi:10.1002/j.1551-8833.1992.tb07342.x

Murashko, A. I. (1982). *Agricultural drainage in the humid zone. Methods of calculations of irrigation and drainage systems*. Kolos.

Nasedkin, I. (1981). Nutrition of waterlogged lands of Polesie with pressure waters and its study. *Reclamation and water management*. *53*, 25–29.

National report on drinking water quality and the state of drinking water supply in Ukraine in 2003. (2005). Rivne.

Natural Resources Management in Polesie. (2017). Monograph in 4 books (edited by Y.A. Mazhaiskiy, A.N. Rokochinskiy, A.A. Volchek, O.P. Meshik, E. Eznach). Ryazan: A.N. Kostyakov Meshchersky branch of VNIIGiM.

Nikoladze, G. I. (1987). Improving the quality of groundwater. Moscow: Stroyizdat.

NJD Science. (2020). Ecological state of the Oskil River in the Kharkiv Region. *Norwegian Journal of Development of the International Science*, *43*(3)

Novakovsky, L. (2015). Problemy metodychnoho zabezpechennya normatyvnoyi hroshovoyi otsinky zemel silskohospodarskoho pryznachennya [Problems of methodological support of normative monetary valuation of agricultural lands]. *Visnyk ahrarnoyi nauky*, *12*, 11–16.

Novokhatin, V., Osipova, N. (2020). *Monitoring of agrophysical properties of the active layer of drained long-frozen peat deposits in Western Siberia* / E3S Web of Conferences 164, 07016 Pages 6.

Nowak, M. (2005). Investment projects evaluation by simulation and multiple criteria decision aiding procedure. *Journal of Civil Engineering and Management*, *11*(3), 193–202. doi:10.1080/13923730.2005.9636350

NTD 33.63-074-87 (1987). *Guide to the design of drainage systems in the Ukrainian SSR*. Ukraine, Kiev: Ukrgidrovodhoz.

Odey Simon, O., Ovat Friday, A., & Okon Orok, O. Draughts, (2018). Power Requirements and Soil Disruption of Subsoilers. *International Journal of Emerging Engineering Research and Technology Volume* *6*(9), 24-38.

Odey Simon, O., Ovat Friday, A., & Okon Orok, O. (2018). Draughts, Power Requirements and Soil Disruption of Subsoilers. *International Journal of Emerging Engineering Research and Technology.*, *6*(90), 24–38.

Odey, S. O., & Manuwa, S. I. (2018). Subsoiler Development Trend in the Alleviation of Soil Compaction for Sustainable Agricultural Production. *International Journal of Engineering Inventions Volume* *7*(8).

Odey, S. O., & Manuwa, S. I. (2016). Design steps of narrow tillage tools for draught reduction and increased soil disruption - a review. *Agricultural Engineering International: CIGR Journal*, *18*(1), 91–102.

Ogbeche, O. S., Idowu, M. S., & Theophilus, E. (2018). Development and performance evaluation of instrumented subsoilers in breaking soil hard-pan. *Agricultural Engineering International: CIGR Journal*, *20*(3), 85–96.

Oleinik, A. (1978). Hydrodynamic calculations of vertical drainage. - Kjiv: Science Dumka, 202 p.

Oleinik, A. (1981). Geohydrodynamics of drainage. Science Dumka.

Oleynik, A. (1985). Regulation of water-air regime of soils on the background of drainage on the basis of mathematical modeling. *Melioration and water management, 54*, 72–75.

Oleynik, A., & Polyakov, V. (1987). *Drainage of wetlands*. Naukova dumka.

Oliynyk A. Ya. (1985). Regulation of the water-air regime of soils on the basis of mathematical modeling. *Improvement of the efficiency of drying-moistening systems*. K.: UkrNIIGyM.

Oliynyk, A. (1993). Methods of calculation and design of drainage and humidification systems of Ukraine: Reclamation systems and facilities. Magazine of Civil Engineering (96)4.

Oliynyk, A. Ya. (1981). *Hydrodynamics of drainage*. Naukova dumka.

Oliynyk, A. Ya., & Polyakov, V. P. (1987). *Drainage of waterlogged lands*. Naukova dumka.

Olszewski, M. (1978). *Waste heat utilization from electric generating plants*. OSTI.

Onyango, C. M., Imungi, J. K., Mose, L. O., & Harbinson, J. Olaf Van Koteen. (2009). Feasibility of commercial production of amaranth leaf vegetable by small scale farmers in Kenya. *African Crop Science Conference Proceedings,* 9, 767-772.

Openko, I., Stepchuk, Ya., Tykhenko, R., Shevchenko, O., & Tsvyakh, O. (2022). Modeling of indicators of economic efficiency of sectoral land use. *Proceedings of the 2022 International Conference. Economic science for Rural Development,* Jelgava, Latvia.

Openko, I., Tykhenko, R., Shevchenko, O., Tsvyakh, O., Stepchuk, Ya., Rokochinskiy, A., & Volk, P. (2022) Mathematical modeling of economic losses caused by forest fire in Ukraine. *21st International Scientific Conference Engineering for Rural Development,* Jelgava, Latvia. 10.22616/ERDev.2022.21.TF004

Openko, I., Kostyuchenko, Y. V., Tykhenko, R., Shevchenko, O., Tsvyakh, O., Ievsiukov, T., & Deineha, M. (2020). Mathematical modelling of postindustrial land use value in the big cities in Ukraine. *International Journal of Mathematical. Engineering and Management Sciences., 5*(2), 260–271. doi:10.33889/IJMEMS.2020.5.2.021

Openko, I., Shevchenko, O., Zhuk, O., Kryvoviaz, Y., & Tykhenko, R. (2017). Geoinformation modelling of forest shelterbelts effect on pecuniary valuation of adjacent farmlands. [IJGE]. *International Journal of Green Economics, 11*(2), 139–153. doi:10.1504/IJGE.2017.089015

Organization and management of ecological and reclamation monitoring on drained lands. (1997). *Drained lands*. K: Derzhvohosp Ukrainy.

Orlov, V. O. (2008). Deironing of groundwater by simplified aeration and filtration: monograph. Publishing center National University of Water Environmental Engineering.

Orlov, V. O., Martynov, S. Yu., Orlova, A. M., Zoshchuk, V. O., Minaeva, N. L., Kunitsky, S. O., Meddur, M. M., & Trokhimchuk, M. M. (2012). *Purification of natural water on expanded polystyrene filters: monograph/ under the general editorship of V.O. Orlov*. Publishing Center National University of Water Environmental Engineering.

Osadchyy, V. (2016). Processes determining surface water chemistry [Processes determining surface water chemistry]. *Volodymyr Osadchyy*. Springer.

Ostapchik, V. & Kostromin, V. (1976). Substantiation of the method of calculating the most probable seasonal dynamics of the water balance deficit of a given security. *Melioration and water management, 36*, 2–12.

Ostapchik, V. (1981). *Planning of irrigation regimes based on the biological method of calculating water consumption of agricultural crops*. Central Research Institute of the Ministry of Water Resources of the USSR.

Ostapchik, V., Kostromin, V., & Koval, A. (1989). *Information-advisory irrigation management system*. Harvest.

Palamarchuk, M., Zakorchevna, B. (2001). *Vodnyi fond Ukrainy*: dovidkovyi posibnyk [*Water fund of Ukraine*: refe rence guide]. Kyiv, Ukraina. 392 s.

Panas, R., Malanchuk, M. (2013). Osoblyvosti bonituvannya tekhnohennykh gruntiv [Features of assessment of technogenic soils]. *Heodeziya, kartohrafiya i aerofotoznimannya, 77*, 74–80.

Panchenko, A. N. (1999). The theory of soil pulverization by tillage implements): monograph. DGAU, 140 p. Dnepropetrovsk, Ukraine.

Panchenko, A. N. (1999). The theory of soil pulverization by tillage implements): monograph. DGAU.

Paris Agreement ratified by the Law N° 1469-VIII of 14.07.2016. https://ec.europa.eu/clima/policies/international/negotiations/paris_en

Pasaribu, K., Lambert, L., Lambert, D., English, B., Clark, C., Hellwinckel, C., Boyer, C., & Aaron Smith, S. (2021). Profitability of irrigating for corn, cotton, and soybeans under projected drought scenarios in the Southeastern United States. *Irrigation Science, 39*(3), 315–328. doi:10.100700271-020-00707-x

Pavlov, V., Fesina, Yu., Zaremba, V., & Mazuryk, S. (2006). Determinatsiya sil's'kohospodars'kykh zemel' na rynku nerukhomosti: Monografiya [Determination of agricultural land in the real estate market. Monograph]. Rivne: NUVGP.

Pavlov, V. I., & Stashuk, A. V. (2013). *Basin management of water resources: experience and priorities*. NUWMNU.

Peiqing, L. (2021). Experimental Fluid Mechanics. A General Theory of Fluid Mechanics. Springer. doi:10.1007/978-981-33-6660-2_5

Piao, S., Ciais, P., Huang, Y., Shen, Z., Peng, S., Li, J., Zhou, L., Liu, H., Ma, Y., Ding, Y., Friedlingstein, P., Liu, C., Tan, K., Yu, Y., Zhang, T., & Fang, J. (2010). The impacts of climate change on water resources and agriculture in China. *Nature, 467*(7311), 43–51. doi:10.1038/nature09364 PMID:20811450

Pierzgalski, E., Jeznach, J., Baryła, A., Brandyk, A., Stańczyk, T., Szejba, D., & Wiśniewski, S. (2012). *Weryfikacja systemów melioracyjnych pod kątem znaczenia dla bezpieczeństwa powodziowego w Regionie Wodnym Wisły Środkowej*. Maszynopis SGGW Warszawa.

Pinchuk, O. (2012) *Obhruntuvannia konstruktsii ta parametriv hidrotekhnichnoi systemy poverkhnevoho obihrivu hruntu obolonkamy-rukavamy pry vykorystanni skydnykh teplykh vod [Justification of Design and Parameters of Hydrotechnical Surface Soil Heating System Using Cover Sleeves by Means of Warm Waste Water].* [Dissertation, Rivne]

Pinchuk, O. (2015) Vozmozhnosti teplovoi melioratsii dlia obespecheniia prodovolstvennoi bezopasnosti strany [Possibilities of Heat Amelioration to Ensure the Country's Food Safety]. Prirodoobustroistvo: nauchno-prakticheskii zhurnal [Natural Habitat: Scientific and Practical Journal], 1, 6-11.

Pinck-Pfeiffer, S., Ragusa, S., Sztajnbok, P., & Vandevelde, I. (2000). Interrelationships between biological, chemical and physical processes as an analog to clogging in aquifer storage and recovery (ASP) wells. *Water Research, 34*(7), 2110–2118. doi:10.1016/S0043-1354(99)00356-5

Pivovar, N. G., & Bugai, V. L. Friedrichson [et al.]. (2000). Drainage with fiber filters to protect territories from flooding. Institute of Hydromechanics.

Pivovar, V. (2011). Deep rippers. *Information and analytical newspaper Agrobusiness today*. http://agro-business.com.ua/agro/mekhanizatsiia-apk/ item/847-hlybokorozpushuvachi.html

Platonov, V. (1982). *Optimization of conditions of moisture supply of agricultural crops*. Gidrometeoizdat.

Pledge, F. D. (1990). *Geography of Ukraine*. World.

Poliakov, V. L. (2014). Hydrodynamic deformations in drained soils: theory and applications. Kjiv: Agrar Media Group.

Poliakov, V.L. (2016). Calculation of the action of a single imperfect horizontal drain in an unlimited headless reservoir taking into account the aeration zone. *Hydrotechnical construction, 5*, 21–31.

Poliakov, V. L. (2012). Calculation of the steady-state action of the drain-humidifier system under conditions of intensive evaporation. *Problems of Water Supply, Water and Hydraulics. -. Kjiv: KNUBA*.

Poliakov, V. L. (2016). *Calculation of the established pressure-free groundwater flow for systematic drainage during water exchange with adjacent media*. Problems of Water Supply, Water and Hydraulics - Kjiv. *KNUBA*, *27*, 291–301.

Poliakov, V. L. (2016). On the modeling of controlled water regime in the soil column. *Reports of NAS of Ukraine*, *9*, 51–60.

Poliakov, V. L., & Kalugin, Yu. I. (2016). Mathematical modeling of water exchange between saturated and non-saturated zones of drained soil. *Reports of NAS of Ukraine*, *9*, 43–49.

Popescu, A., Dinu, T. A., & Stoian, E. (2019). Efficiency of the agricultural land use in the European Union. *Scientific Papers. Series Management, Economic, Engineering in Agriculture and Rural Development*, *19*(3), 475–486. http://managementjournal.usamv.ro/pdf/vol.19_3/Art62.pdf

Popescu, A., Dinu, T. A., Stoian, E., & Serban, V. (2020). Variation of the main agricultural crops yield due to drought in Romania and Dobrogea region in the period 2000-2019. *Scientific Papers. Series Management, Economic, Engineering in Agriculture and Rural Development*, *20*(4), 397–415. http://managementjournal.usamv.ro/pdf/vol.20_4/volume_20_4_2020.pdf

Postelnyak A. (2013). Evaluation of the accuracy of the heights of digital models of the relief SRTM and Aster GDEM. *Bulletin of geodesy and cartography*, *4*, 17-21.

Rabochev, I. S., & Pyagay, E. T. (1984). Optimization of drainage parameters on floodplain soil. Bulletin of agricultural science, 103-115

Raffel, M., Willert, C., Wereley, S., & Kompenhans, J. (2007). Examples of Application. In *Particle Image Velocimetry. Experimental Fluid Mechanics* (pp. 259–388). Springer. doi:10.1007/978-3-540-72308-0_9

Rahman, Md., Suwal, S., & Jäväjä, P. (2013). *Diverse approach of BIM in AEC industry: a study of current knowledge and practice.* The 30 th CIB W78 International Conference, Beijing, China.

Report on the state of the environment in Rivne region in 2014 . (2015). (RSER), Rivne.

Rex, L. (1995). *System research of meliorative processes and systems*. Aslan.

Rezaei Zaman, M., Morid, S., & Delavar, M. (2016). Evaluating climate adaptation strategies on agricultural production in the Siminehrud catchment and inflow into Lake Urmia, Iran using SWAT within an OECD framework. Agricultural Systems, 147, 98–110. doi:. 2016.06.001. doi:10.1016/j.agsy

Richling, A., & Solon, J. (1998). Ekologia krajobrazu. Warszawa: PWN.

Rodriguez S., Fathi N., Pourghasemi P. (2022). Theoretical Approach for the Fast Estimation of the Turbulent Kinematic Viscosity for Internal Flows. *Journal of Nuclear Engineering and Radiation Science.* . doi:10.1115/1.4054342

Rokochinskiy A. (2016). Systemna optymizatsiya vodorehulyuvannya yak neobkhidna umova stvorennya ta funktsionuvannya vodohospodars'ko-melioratyvnykh ob"yektiv na ekoloho-ekonomichnykh zasadakh [System optimization of water regulation as a prerequisite for the creation and operation of water management and reclamation facilities on ecological and economic grounds]. *Vodne hospodarstvo Ukrayiny, 104,* 67–71.

Rokochinskiy A., FrolenkovA N., & Koptiuk R. (2012). Investitsiyna otsinka proektiv optimizatsiї vodoregulyuvannya osushuvanih land of urahuvannyam mainly chinnikiv vplivu [Investment assessment project for optimizing water management of drained lands from the main bureaucrats*]. Tavriys'kyy naukovyy visnyk, 83,* 216–220.

Rokochinskiy A., Jeznach J., Volk P., Turcheniuk V., Frolenkova N., Koptyuk R. (2019). Reclamation projects development improvement technology considering optimization of drained lands water regulation based on BIM. *Scientific Review Engineering and Environmental Sciences, 3*(85), 432-443.

Rokochinskiy, A. & Zubyk, L. (2000). Model for forecasting the effectiveness of various technologies of water regulation of drained lands in projects of construction, reconstruction and operation of reclamation systems. *Bulletin of Rivne State Technical University: Coll. Science, 2*, 63–71.

Rokochinskiy, A., Bilokon, V., Frolenkova, N., Prykhodko, N., Volk, P., Tykhenko, R., & Openko, I. (2020). Implementation of modern approaches to evaluating the effectiveness of innovation for water treatment in irrigation. Journal of Water and Land Development, 45 (IV–VI), 119–125. DOI doi:10.24425/jwld.2020.133053

Rokochinskiy, A., Frolenkova, N., & Voloshchuk, V. (2008). Meteorological support of engineering and reclamation calculations in projects of construction and reconstruction of drainage systems. Manual to DBN B.2.4-1-99. Reclamation systems and structures, Rivne.

Rokochinskiy, A., Halik, O., Frolenkova, N., Voloshchuk, V., Shalai, S., Zubyk, Ya., Bezhuk, V., Zubyk, L., Pokladnov, Ye., Savchuk, T., Koptiuk, R., Nesteruk, L., Volk, P., & Kotiai, L. (2008). Guide to DBN V.2.4-1-99. Reclamation systems and structures. VAT.

Rokochinskiy, A., Jeznach, J., Volk, P., Turcheniuk, V., Frolenkova, N., & Koptiuk, R. (2018). Reclamation projects development improvement technology considering optimization of drained lands water regulation based on BIM. Scientific Review – Engineering and Environmental Sciences, 28(3), 432–443. doi:. 28.3.40. doi:10.22630/PNIKS.2019

Rokochinskiy, A., Stashuk, V., Duplyak, V., Frolenkova, N., & Koptyuk, R. (2011). Temporary recommendations for the forecast evaluation of the water regime and technologies of water regulation of drained land in projects of construction and reconstruction of reclamation systems. Rivne, Ukraine: NUVGP. .

Rokochinskiy, A., Volk, P., Frolenkova, N., Tykhenko, O., Shalai, S., Tykhenko, R., & Openko, I. (2021). Differentiation in the value of drained land in view of variable conditions of its use. Journal of Water and Land Development, 51 (X–XII), 172–178. doi:10.24425/jwld.2021

Rokochinskiy, A., Volk, P., Jeznach, J., Turcheniuk, V., Frolenkova, N. & Koptiuk, R. (2019). Reclamation projects development improvement technology considering optimization of drained lands water regulation based on BIM. *Scientific Review Engineering and Environmental Sciences, 3*(85), 432-443. doi:10.22630/PNIKS.2019.28.3.40

Rokochinskiy, A., Volk, P., Pinchuk, O., Turcheniu, K V., Frolenkova, N., Gerasimov, E. 2019. Forecasted estimation of the efficiency of agricultural drainage on drained lands. *Journal of Water and Land Development, 40*(I–III), 149–153. doi:. doi:10.2478/jwld2019-0016

Rokochinskiy, A. (2010). *Naukovi ta praktichni aspekti optimizacii vodoregulyuvannya osushuvanikh zemel' na ekologoekonomichnikh zasadakh: Monografiya* [*The scientific and practical aspects optimization of water regulation drained lands on environmental and economic grounds. Monograph*]. NUVGP.

Rokochinskiy, A. (2010). *Naukovi ta praktichni aspekti optimizacії vodoregulyuvannya osushuvanikh zemel' na ekologoekonomichnikh zasadakh: Monografiya [The scientific and practical aspects optimization of water regulation drained lands on environmental and economic grounds. Monograph*]. NUVGP.

Rokochinskiy, A. (2010). *The scientific and practical aspects optimization of water regulation drained lands on environmental and economic grounds*. NUVGP.

Rokochinskiy, A. M. (2010). *The scientific and practical aspects of optimization of water regulation of drained lands on ecological and economic bases: monograph*. NUWEE.

Rokochinskiy, A., Antonov, O., & Shalai, S. (2010). *Engineering surveys for water management and environmental construction*. NUVGP.

Rokochinskiy, A., Frolenkova, N., Turcheniuk, V., Volk, P., Prykhodko, N., Tykhenko, R., & Openko, I. (2021). The variability of natural and climatic conditions in investment projects in the field of nature management. *Journal of Water and Land Development*, *48*(1–3), 48–54. doi:10.24425/jwld.2021.136145

Rokochinskiy, A., Korobiichuk, I., Kuzmych, L., Volk, P., & Kuzmych, A. (2020). *The System Optimization of Technical*. Technological and Construction Parameters of Polder Systems., doi:10.1007/978-3-030-40971-5_8

Rokochinskiy, A., Stachuk, V., Frolenkova, N., Shalay, S., Koptyuk, R., & Volk, P. (2010). *Tymchasovi rekomendatsiyi z optymizatsiyi vodorehulyuvannya osushuvanykh zemel' u proektakh budivnytstva y rekonstruktsiyi vodohospodars'komelioratyvnykh ob˝yektiv [Temporary recommendations for optimization of water management of drained lands in projects of construction and reconstruction of water management and reclamation facilities*]. NUVGP.

Rokochinskiy, A., Volk, P., Frolenkova, N., Shalay, S., Koptyuk, R., Zayets, V., & Prykhodko, N. (2013). *Naukovo metodychni rekomendatsiyi do obgruntuvannya optymal'nykh parametriv sil's'kohospodars'koho drenazhu na osushuvanykh zemlyakh za ekonomichnymy ta ekolohichnymy vymohamy [Scientific and methodological recommendations for substantiation of optimal parameters of agricultural drainage on drained lands according to economic and environmental requirements*]. NUVGP.

Rokochinskiy, A., Volk, P., Pinchuk, O., Mendus, S., & Koptyuk, R. (2017). Comparative evaluation of various approaches to the foundation of parameters of agricultural drainage. *Journal of Water and Land Development*, *34*(1), 215–220. doi:10.1515/jwld-2017-0056

Rokochinsky, A. & Zubik, L. (1999). Principles of creation and implementation of models of optimization of water regulation of drained lands on a long-term basis. *Problems of reclamation and water management at the present stage: Sb. scientific of works Bel. s.-x. academy, 1*, 158–163.

Rokochynskiy, A. (2016). System optimization of water regulation on reclaimed lands. *Problems of rational use of natural resources and sustainable development of Polissya*. 111–114.

Rokochynskiy, A., Volk, P., Frolenkova, N., Prykhodko, N., Gerasimov, I., Pinchuk O. (2019). Evaluation of climate changes and their accounting for developing the reclamation measures in Western Ukraine. *Scientific Review Engineering and Environmental Sciences*, *1*(83), 3–13.

Rokochynskiy, A., Volk, P., Frolenkova, N., Prykhodko, N., Gerasimov, I., Pinchuk, O. (2019) Evaluation of climate changes and their accounting for developing the reclamation measures in western Ukraine. *Scientific Review Engineering and Environmental Sciences, 1*(83), 3-13. doi: . doi:10.22630/PNIKS.2019.28.1

Rokochynskiy, A., Volk, P., Pinchuk, O., Turcheniuk, V., Frolenkova, N., Gerasimov, I. (2019) Forecasted estimation of the efficiency of agricultural drainage on drained lands. *Journal of Water and Land Development*, *40* (I-III), 149-153. . doi:10.2478/jwld-2019-0016

Rokochynskiy, A., Turcheniuk, V., Prykhodko, N., Volk, P., Gerasimov, I., & Koç, C. (2020). Evaluation of climate change in the rice-growing zone of Ukraine and ways of adaptation to the predicted changes. *Agricultural Research*, *9*(4), 631–639. doi:10.100740003-020-00473-4

Rokochynskiy, A., Volk, P., & Kuzmych, L. (2022). *Formation of Water Consumption at Drained Lands in Climate Changes. Monograph.* LAMBERT Academic Publishing.

Rokochynskyi, A., Volk, P., Turcheniuk, V., Tokar, L., Volk, L., Mazhayskiy, Y., & Chernikova, O. (2020). Construction and agricultural drainage parameter optimization considering economic and environmental requirements. *Engineering for rural development.* Jelgava. http://www.tf.llu.lv /conference/proceedings 2020/Papers/TF010.pdf

Rokochynskyi, A., Volk, P., Turcheniuk, V., Tokar, L., Volk, L., Mazhayskiy, Y., & Chernikova, O. (2020). The drainage module is an important factor in the design of drainage systems reconstruction and construction projects in the polesia region. *Engineering for Rural Development.*, *19*, 36–47. doi:10.22616/ERDev.2020.19.TF010

Romashchenko M., Gusev Yu., Shatkovskiy A., Saidak R., Yatsyuk M., Shevchenko A. and Matiash T. (2020). Impact of climate change on water resources and agricultural production. *Reclamation and water resources management*, (1), 5 - 22. doi:10.31073/mivg202001-235

Romashchenko M.I., Husyev Yu. V., Shatkovskyi A. P., Saidak R. V., Yatsyuk M. V., Shevchenko A. M., Matiash T. V. (2020). The impact of modern climate change on water resources and agricultural production. *Land reclamation and water management*, *1*, 5–22.

Romashchenko, M. I. (2019). *Development of scenarios of climate change in Ukraine in the medium- and long-term using data from global and regional models.* UHMI. http: // uhmi.org.ua/project/rvndr/climate.pdf

Romashchenko, M. I., Saidak, R. V., Matiash, T. V., & Knysh, V. V. (2019). Influence of climatic changes on moisture supply of the territory of Ukraine and production of agricultural products. *"Water for All": dedicated to World Water Day: International. scientific-practical Conference: thesis add.* Kyiv, 2019. P. 179–180.

Romashchenko, M. I., Saydak, R. V., & Matyash, T. V. 2019. Development of irrigation and drainage as the basis of sustainable agriculture in Ukraine in climate change. *IX International scientific and technical conference*, 22-27. IX.

Romashchenko, M., Khvesik, M., & Mikhailov, Yu. (2015). Water strategy of Ukraine for the period till 2025 (scientific basis). Kiev, Ukraine

Romashchenko, M., Rokochinskiy, A., Halik, O., Savchuk, T. & Kolodich, O. (2007). Recent climate change and their manifestations from global to regional manifestations. *Hydromelioration and hydromelioration construction, 32,* 65–79.

Romashchenko, M., Rokochynsky, A., Halik, O., Savchuk, T. & Kolodych O. (2007). Modern climate change and their manifestations from global to regional manifestations. *Hydromelioration and hydraulic engineering construction: Coll. Science. work, 32*, 65–79.

Romashchenko, M., Sobko, O., Savchuk, D., Kulbida, M. (2003). *About some tasks of ag-ricultural science in relation to climate change.* The scientific report-information NAAS Kyiv.

Romashchenko, M. I., Yatsiuk, M. V., Sydorenko, O. O., Nechay, O. M., Voropay, G. V., Nasedkin, I. Y., Tsvetova, O. V., & Saydak, R. V. (2020). Reasons of Shatsk lakes shallowing and ways of water balance regulation. Kyiv. *Bulletin of Agrarian Science*, (8), 5–13.

Romashchenko, M., Sobko, O., Savchuk, D., & Kulbida, M. (2003). *About some problems of agrarian science in connection with climate change.* Institute of Hydrotechnics and Reclamation UAAN.

Rotjanakusol, T. & Laosuwan, T. (2018) Estimation of land surface temperature using Landsat satellite data: A case study of Mueang Maha Sarakham District. *Scientific Review – Engineering and Environmental Sciences, 27* (4), 401-409. doi:10.22630/PNIKS.2018.27.4.39

Rubenstein, D. A., Yin, W., & Frame, M. D. (2021). Fundamentals of fluid mechanics. In *Biofluid Mechanics: An Introduction to Fluid Mechanics* (pp. 17–70). Macrocirculation, and Microcirculation., doi:10.1016/B978-0-12-818034-1.00002-5

Rychagov, G. (2006). General geomorphology. Publishing House of Moscow University: Science.

Safonov, N. A., & Rusak, G. V. (1984). Self-cleaning plant for biological removal of iron from underground waters. In Preparation of water for drinking purposes Conference, Leningrad, LEBI, 162-167. .

Safonov, N. A., Kvartenko, A. N., & Safonov, A. N. (2000). *Self-cleaning water treatment plants (technology, design, calculation): monograph* (S. N. A. Rivne, Ed.). Publishing House Rivne State Technical University.

Savchyn, O. (2013). A little about Polesie and Volyn. *UA Chronicle.* http://hronosua.blogspot.com/2013/02/blog-post_7831.html

Schilling, K. E., & Dinsmore, S. (2018). Monitoring the Wildlife, Hydrology and Water Quality of Drained Wetlands of the Des Moines Lobe, Northern Iowa: Introduction to Special Feature. *Wetlands, 38*(2), 207–210. doi:10.100713157-017-0989-y

Schmidt-Thome, P. (2007). *The spatial effects and management of natural and technological hazards in Europe*. ESPON.

Seppänen, H. (1991). *Experiences of Biological iron and manganese removal in Finland.* Proc. IWEM ann. Sym. *15* (1), 9-11.

Seryy, A. (1986). *Bonitirovka pochv: Metodychni rekomendatsiyi* [*Assessment of soil: Methodical recommendations*]. USHA.

Seyedmohammadi, J., Esmaeelnejad, L., & Ramezanpour, H. (2016). Land suitability assessment for optimum management of water consumption in precise agriculture. *Modeling Earth Systems and Environment, 2*(3), 162. doi:10.100740808-016-0212-9

Shabanov, V. (1973). Bioclimatic substantiation by reclamation. Gidrometeoizdat.

Shalay, S., Rokochinskiy, A., Stashuk, V., & Bezhuk, V. (2004). *Tymchasovi rekomendatsiyi z obhruntuvannya efektyvnoyi proektnoyi vrozhaynosti na osushuvanykh zemlyakh pry budivnytstvi y rekonstruktsiyi melioratyvnykh system* [*Temporary recommendations for substantiation of effective project yield on drained lands during construction and reconstruction of reclamation systems*]. NUVGP.

Sharma, S.K., Petrusevski, B., & Schippers, J.C. (2005). Biological iron removal from groundwater. *Water Supply, 54*(4), 239-247.

Shawcroft, R., Lewon, E., Allen, L., Stewart, D., & Jensen, S. (1974). The soil-plant-atmosphere model and some of it's predictions. *Agricultural Meteorology, 14*(1/2), 287–307. doi:10.1016/0002-1571(74)90025-9

Shebeko, V. (1983). *Influence of drainage reclamation on the water regime of the territory*. Harvest.

Shebeko, V. (2000). *Water management calculations for reclamation of wetlands*. NIH.

Shebeko, V., Mozheem, I., & Kiselyova, A. (1980). *Methodical instructions and computer programs for designing the water regime of drained lands on the basis of water balance calculations*. NIH.

Shebeko, V., Zakrzhevsky, P., & Bragilevskaya, E. (1980). *Hydrological calculations in the design of drainage and dehumidification systems*. Gidrometeoizdat.

Shevchenko, A. (2002). Distribution and Migration of Radionuclides in Meliorated Areas. In V. Shestopalov (ed.) *Chernobyl disaster and groundwater*. Lisse/Abingdon/Exton(pa)/Tokyo.

Shevchenko, A.L. (2005). Activities on drying-moistening systems in case of radioactive soil contamination. *Reclamation and water management. 92*, 122-130.

Shevchenko, A.L., & Dolin, V.V. (2002). Natural and man-made prerequisites for the decontamination of the water basin of radiation-contaminated territories. In V. Sobotovych, G.N. Bondarenko (eds.) *Geochemistry of technogenic radionuclides*. Resp.

Shevchenko, O., Openko, I., Tykhenko, R., Tsvyakh, O., Zhuk, O., Kryvoviaz, E., Tykhenko, O., Bavrovska, N., Stepchuk, Ya., Rokochinskiy, A., & Volk, P. (2021). Assessment of economic losses caused by degradation processes of agricultural land use. *Scientific Papers. Series E. Land Reclamation, Earth Observation & Surveying, Environmental Engineering, 10,* 102–109. http://landreclamationjournal.usamv.ro/pdf/2021/Art15.pdf

Shevchenko, O., Openko, I., Zhuk, O., Kryvoviaz, Y., & Tykhenko, R. (2017). Economic assessment of land degradation and its impact on the value of land resources in Ukraine [online]. *International Journal of Economic Research (IJER), 14*(15), 93–100. https://serialsjournals.com/abstract/34405_ch_11_f_- _ivan_openko.pdf

Shevchenko, O., Openko, I., Zhuk, O., Kryvoviaz, Y., & Tykhenko, R. (2017). Economic assessment of land degradation and its impact on the value of land resources in Ukraine [online]. *International Journal of Economic Research (IJER), 14*(15), 93–100. https://serialsjournals.com/abstract/34405_ch_11_f_-_ivan_openko.pdf

Shevchenko, O., Openko, I., Zhuk, O., Kryvoviaz, Y., & Tykhenko, R. (2017). Economic assessment of land degradation and its impact on the value of land resources in Ukraine [online]. *International Journal of Economic Research (IJER), 14*, 93–100. https://serialsjournals.com/abstract/34405_ch_11_f_- _ivan_openko.pdf

Shevchenko, O., Shestopalov, V., Sakhatskiy, O., Nasedkin, I., Gudzenko, V., Akinfiev, G. (1999). Left bank flood plain: pathways to the decision of over moistening and increasing of ^{90}Sr outflow through the hole the dam N°7. *Bulletin of ecological state of Chornobyl Exclusion Zone,* (14), 51-57.

Shevchenko, O.L. (2018). Drainage systems in radioactively contaminated catchments. .M. Gadzala, V.A. Stashuka, A.M. Rokochinsky. (eds.) *Reclamation and development of the Ukrainian Polesie.* Y Kyiv-Rivne.

Shevchenko, O.L., Kozytskyi, O.M., Nasedkin, I.Yu., Akinfiev, G.O., Kireev, S.I., Sakhatskyi, O.I., Khodorovsky, A.Ya. (2001). Analysis of efficiency and operation options of the water protection complex of the Left Bank polder system. *Problems of the Chernobyl exclusion zone, 7,* 112–125.

Shevchenko, O.L., Kozytskyi, O.M., Nasedkin, I.Yu., Ryabtseva, G.P., Bublyas, V.M., Ivanushkina, N.I., Osadcha, N.M., & Syzonenko, V.P. (2011). *Patterns of migration of man-made radionuclides on reclamation systems of the Chernobyl Exclusion Zone (based on research results from 1986-2004).* Kherson.

Shevchenko, O.L., Sakhatsky, O.I., Khodorovsky, A.Ya., Azimov, O.T. (1999a). Justification of water protection measures on the left bank of the Pripyat River in the exclusion zone based on the use of multi-zone space survey materials. *New methods in space geology*. Kyiv: Ukraine.

Shevchenko,O.L., Kozytskyi,O.M. (2003). Overgrowth and siltation of riverbeds as an initial parameter in the prognostic model of radioactive contamination of reclamation channels. *Scientific works of the Ukrainian Research Hydrometeorological Institute*, *252*, 131–143.

Shevchenko, A. L., Charnyi, D. V., Akinfiev, G. A., & Kireev, S. I. (2016). FactorsGoverningStrontium-90 exportwith-SurfaceRunoffintheChernobylNPPRestrictedZone. *WaterResources*, *43*(3), 522–532. doi:10.1134/S0097807816010127

Shevchenko, A. L., Dolin, V. V., Nasedkin, I. Yu., Kireev, S. I., & Kozitskii, O. N. (2002). Hydrological Changes and their effect on the radiological characteristics in the Chernobyl abandoned Zone. *Water Resources*, *29*(4), 449–464. doi:10.1023/A:1019634609828

Shevchenko, O. L. (2013). Retrospective analysis of the priority of pollutant mobilization factors.*Bulletin of KNU named after T.* [In Ukraine]. *Shevchenko (Geology)*, *60*, 63–69.

Shevchenko, O. L., & Kireev, S. I. (2005). The method of integral assessment of the barrier capacity of catchments. [In Ukraine]. *Ecological Restoration*, *11*, 77–86.

Shevchenko, O., Openko, I., Tykhenko, R., Tsvyakh, O., Zhuk, O., Kryvoviaz, E., Tykhenko, O., Bavrovska, N., Stepchuk, Ya., Rokochinskiy, A., & Volk, P. (2021). Assessment of economic losses caused by degradation processes of agricultural land use. Scientific Papers. Series E. Land Reclamation, Earth Observation & Surveying. *Environmental Engineering.*, *X*, 102–109.

Shevchuk, V. Ya., Husev, M. V., Mazurkevych, O. O., Navrotskyi, V. M., Satalkin, Yu. M., Trehobchuk, V. M., & Yatsyk, A. V. (1997). *Economy and ecology of water resources of the Dnipro river.*

Shevchuk, V., Trofimova, I., & Trofimchuk, O. (2001). *Problems and strategy of Ukraine's implementation of the UN Framework Convention on Climate Change*. FAO.

Shih, S. F. (1983). Soil surface evaporation and water table depth. *Journal of Irrigation and Drainage Engineering*, *109*(4), 366–376. doi:10.1061/(ASCE)0733-9437(1983)109:4(366)

Shmatok, V. (1994). Qualitative changes in organic matter of dried peat soils under the influence of agricultural use. *Reclamation and water management, 80*, 39-40.

Shumakov, B. (1996). Melioration in the twenty-first century. *Reclamation and water management, 3*, 4–6.

Shumakov, B. B. (1996). *Melioration in the twenty-first century*. Reclamation and Water Management.

Single departmental administration in charge of the organization and monitoring of waters. Order of the Ministry ecological resources of Ukraine No 485 of 24.12.2001 – K.: 2001.

Sirin, A., Maslov, A., Makarov, D., Gulbe, Y., & Joosten, H. (2021). Assessing Wood and Soil Carbon Losses from a Forest-Peat Fire in the Boreo-Nemoral Zone. *Forests*, *12*(7), 880. doi:10.3390/f12070880

Skaggs, R. W. (1978). Effects of drain tube openings on water – table drawdown. *Journal of the Irrigation and Drainage Division*, *104*(1), 13–21. doi:10.1061/JRCEA4.0001183

Skripnik, O., Soroka, I., & Kubyshkin, V. (1992). *Technology of regulation of the water regime of drained lands*. Urozhay.

Slovoidilo.ua. (2019). Rynok zemli: vartist hektara v krayinakh Yevropy ta Pivnichnoyi Ameryky [Land market: the cost per hectare in Europe and North America]. *Slovoidilo.* https://www.slovoidilo.ua/2019/11/13/infografika/ekonomika/rynok-zemli-vartist-hektara-krayinax-yevropy-tapivnichnoyi-ameryky

Slyusar, I. (2013). Conceptual foundations of environmental use of land and water resources of the gumid zone of Polesie. In V. Kaminsky (ed.) Adaptive farming systems and modern technologies are the basis of rational land use of soil conservation and reproduction. VP Edelweiss.

Slyusar, I., Solyanik, O., Serbenyuk, V., Viryovka, V., & Tarasenko, O. (2020). Effect of the water regime, crop rotation and fertilizers in biogenic matters leaching into ground water and surface water. *Ukrainian Journal of Ecology*, *10*(3), 197–200. https://www.webofscience.com/wos/woscc/full-record/WOS:000576691500013. doi:10.15421/2020_154

Slyusar, I., Solyanik, O., Serbenyuk, V., Zadubinna, E., & Perets, S. (2021). Energy crops safe cultivation on drained organogenic soils. *Ukrainian Journal of Ecology*, *11*(1), 415–418. https://www.webofscience.com/wos/woscc/full-record/WOS:000631814800042. doi:0.15421/2021_60

Slyusar, I., & Solyanyk, O. (2005). *Features of the farming system on the drained lands of the gumid zone of Ukraine: problems, solutions. Zb. Ecology: problems of adaptive landscape farming*. State Ecological University.

Smith, K. A. (2000). *Soil and Environmental Analysis: Physical Methods, Revised, and Expanded* (2nd ed.). CRC Press., doi:10.1201/9780203908600

Snizhko, S. (2001). *Otsinka ta prohnozuvannia yakosti pryrodnykh vod [Evaluation and forecasting of the quality of natural waters]*. Kyiv.

Søgaard, G. E., Medenwaldt, R., & Abraham-Peskir, J. V. (2000). Conditions and rates of biotic and abiotic iron precipitation in selected Danish freshwater plants and microscopic analysis of precipitate morphology. *Water Research, 34* (10), 2675-2682.

Speight, J. G. (2017). Kinematic Viscosity. Rules of Thumb for Petroleum Engineers. Wiley. doi:10.1002/9781119403647.ch209

Stashuk, V. A. (2006). Ecologic-economic principles of the basin management of water resources. SC Publisher Zoria.

Stashuk, V. A., Mokin, V. B., Hrebin, V. V., & Chunariov, O. V. (2014). Scientific principles of the rational use of water resources of Ukraine by a basin principle: Monograph. Hrin D.S.

Stashuk, V., Mokin, V., Hrebin, V., Chunarov, O. (2014). *Naukovi zasady ratsionalnoho vykorystannia vodnykh resursiv Ukrainy za baseinovym pryntsypom*: *Monohrafiia* [*Scientific principles of rational use of water resources of Ukraine according to the basin principle*: *Monograph*]. Kherson, Ukraina.

Stashuk, V. (1993). *Assessment and forecast of conditions for the formation of the water regime of drained lands with different ways of its regulation: Dis. Cand. tech. sciences in the form of sciences. add.: 06.01.02*. UIIVG.

State Statistics Service of Ukraine. (2018). *Main indicators of forestry activity (1990 – 2017)*. UK Stats. http://www.ukrstat.gov.ua/operativ/operativ2014/sg/lis/lis_u/lgd2016_u.htm

Steinhauser, G., Brandl, A., & Johnson, T. E. (2014). Comparison of the Chernobyl and Fukushima Nuclear Accidents: A Review of the Environmental Impacts. *The Science of the Total Environment*, *470-471*, 470–471. doi:10.1016/j.scitotenv.2013.10.029 PMID:24189103

Stetsyuk M. A. (2015). *Promising payza* . The Ukrainian Farmer.

Strand, J. (2015). Modeling the Marginal Value of Rainforest Losses: A Dynamic Value Function Approach. *Policy Research Working Paper;No. 7445*. World Bank, Washington, DC.

Struts, V.A. (1984). Fire on peat deposits. Fire prevention measures. *Reclamation Encyclopedic reference book*. Minsk.: Belorus.

Sudong, Y. E., & Tiong, R. L. K. (2000). NPV-at-Risk method in infrastructure project investment evaluation. *Journal of Construction Engineering and Management*, *126*(3), 227–233. doi:10.1061/(ASCE)0733-9364(2000)126:3(227)

Sugier, P., & Różycki, A. (2010). Wybrane obszary hydrogeniczne Pojezierza Łęczyńsko-Włodawskiego. In A. Obidziński (Ed.), *Mazowsza na Polesie i Wileńszczyznę*. Zarząd Główny.

Supit, I., van Diepenb, C. A., de Wit, A. J. W., Wolf, J., Kabat, P., Baruthc, B., & Ludwig, F. (2012). Assessing climate change effects on European crop yields using the Crop Growth Monitoring System and a weather generator. *Agricultural and Forest Meteorology*, *164*, 96–111. doi:10.1016/j.agrformet.2012.05.005

Tamura, T., Tsunai, T., Ishimaru, Y., & Nakata, A. (1999). Iron and manganese removal by iron bacteria in groundwater. *Suido Kyokai Zasshi (J. Japan Water Works Assoc.)*, *68*, 1–13

Tedim, F., Xanthopoulos, G., & Leone, V. (2015). ForestFires in Europe: Facts and Challenges. Wildfire Hazards, Risks and Disasters, 77–99.

Teichroew, D., Robichek, A., & Montalbano, M. (1965). An analysis of criteria for investment and financing decisions under certaint. *Management Science*, *12*(3), 151–179. doi:10.1287/mnsc.12.3.151

The Ukrainian SSR. (1990). Series 3. Perennial data. Scientific and Applied Handbook on the Climate of the USSR. Russia, Lenynhrad: Hydrometeoyzdat.

Tishenko, O. (2014). Regulation of the water regime of soils on the two-way systems in the Polesie of Ukraine. *Water Statehood of Ukraine*, *6*(114), 35–39.

Tolstousov, A. (2007). *Ekolohichna entsyklopediia [Ecological encyclopedia]*. Kyiv, Ukraina.

Truskavetskyi R. (1992). Negative phenomena in soil formation on drained lands. *Soil fertility. Monitoring and management*. K.: Urozhai.

Truskavetskyi, R. (2010). *Peat soils and peatlands of Ukraine*. City Printing.

Tsipris, D., Seleznev, V., Plavnik, G., & Shtakovsky, A. (1986). Calculation of norms of water consumption for the set crop. *GiM*, *4*, 33–36.

Tsuman, N., Zinchenko, V. (2021). Agroecological problems of intensive agriculture on drained lands. *Plant Varieties Studying and Protection 17*(4).

Tsvetova, H., Riabtseva, G., & Turaeva, O. (2011). Monitoring of the chemical composition of several lakes of the Shatsk group in the Western Bug river basin exposed to recreational pressures. In K. Dygus (ed.) The Water environment. Problems of evaluation and protection: monograph. Warsawa.

Tsyupa, M., Bystrytsky, V., & Slusar, I. (1990). Agriculture on drained lands. Harvest..

Tugay, A. M., Oliynik, O. Ya., & Tugay, Ya. A. (2001). Productivity of water inlets in the minds of colmatage. Kharkiv: KHNAMG.

Turcheniuk, V., Rokochinskiy, A., Kuzmych, L., Volk, P., & Koptyuk, R. (2022). A Technological System for Using Waste Warm Water from Energy Facilities for Effective Agriculture. *Archives of Hydro-Engineering and Environmental Mechanics*, *69*(1), 13–25. doi:10.2478/heem-2022-0002

Turcheniuk, V., Rokochinskiy, A., Kuzmych, L., Volk, P., Koptyuk, R., Romanyuk, I., & Voropay, G. (2022). The efficiency of waste hot water utilization to improve the temperature conditions for growing plants. *Journal of Water and Land Development*, (54), 1–7. doi:10.24425/jwld.2022.141559

Tykhenko R., Tykhenko O., Openko I., Shevchenko O., Bavrovska N., Zhuk O., Tsvyakh O., Stepchuk Ya. (2021). The assessment of impact ecological stability of territory on the organization of rational land use of agricultural enterprises. *Scientific Papers Series "Management, Economic Engineering in Agriculture and Rural Development, 21*(2), 685–692.

Tykhenko, O. (2016). Ratsionalne vykorystannya silskohospodarskykh zemel: problemy ta perspektyvy yikh vyrishennya [Rational use of agricultural land: problems and prospects for their solution]. *Zemleustrij, kadastr i monitorynh zemel, 1–2,* 76–81. http://journals.nubip.edu.ua/index.php/Zemleustriy/article/viewFile/7204/6989

Tykhenko, R. (2010). Ekoloho-ekonomichna efektyvnist zemleustroyu v umovakh transformatsiyi zemelnykh vidnosyn v Ukrayini: Monografiya [Ecological and economic efficiency of land management in the conditions of transformation of land relations in Ukraine: Monograph]. Kyyiv. Anva-print.

Tykhenko, O. (2010). Porivnyalnyy analiz metodiv bonituvannya gruntiv Ukrayiny [Comparative analysis of soil grading methods in Ukraine]. *Visnyk Kharivskoho natsionalnoho ahrarnoho universytetu im. V. Dokuchayeva*, *6*, 226–230.

Tykhenko, R., Tykhenko, O., Openko, I., Shevchenko, O., Bavrovska, N., Zhuk, O., Tsvyakh, O., & Stepchuk, Ya. (2021). The assessment of impact ecological stability of territory on the organization of rational land use of agricultural enterprises. *Scientific Papers. Series Management, Economic, Engineering in Agriculture and Rural Development*, *21*(2), 685–692. http://managementjournal.usamv.ro/pdf/vol.21_2/volume_21_2_2021.pdf

Ubysz, B., & Szczygiel, R. (2002). Fire Situation in Poland. *International Forest Fire News*, *27*, 38–64.

Ukhurebor, K., Aigbe, U., Onyancha, R., & Ndunagu, J. (2022). An Overview of the Emergence and Challenges of Land Reclamation: Issues and Prospect. Hindawi. Applied and Environmental Soil Science. doi:10.1155/2022/5889823

UN. (1992). *United Nations framework convention on climate change.* United Nations. . https://unfccc.int/resource/docs/convkp/ conveng.pdf

UNFCC. (2020). United Nations framework convention on climate change. *Kyoto Protocol*, Kyoto. https://unfccc.int/resource / docs / convkp / kpeng.pdf (access date 30.04.2020).

United Nations Convention to Combat Desertification. (n.d.). *Documents.* UNCCD. https://www.unccd.int/official-documents

United Nations framework convention on climate change. (1992). United Nations. https://unfccc.int/resource/docs/convkp/conveng.pdf

United Nations. (2016). *Paris Agreement.* UN. https://treaties.un.org/doc/Treaties/2016/02/20160215%2006-03%20PM/Ch_XXVII-7-d.pdf

Untenecker, J., Tiemeyer, B., Freibauer, A., Laggner, A., & Luterbacher, J. (2017). Tracking changes in the land use, management and drainage status of organic soils as indicators of the effectiveness of mitigation strategies for climate change. *Ecological Indicators*, *72*, 459–472. doi:10.1016/j.ecolind.2016.08.004

Urban, D., Grzywaczewski, G., Jeznach, J., & Mażajski, J. (2020). Ochrona przyrody polesia lubelskiego, 854-854. .

Van der Pboeg, R. R., Kirkham, M. B., & Marquardt, M. (1999). The Golding equation for soil drainage: Its origin, evolution and use. *Soil Science Society of America Journal*, *63*(1), 33–39. doi:10.2136ssaj1999.03615995006300010006x

Van Genuchten, M. Th. (1980). A closed-form equation for predicting the hydraulic conductivity of unsaturated soils. *Soil Science Society of America Journal*, *44*(5), 892–898. doi:10.2136ssaj1980.03615995004400050002x

Várallyay, G. (2010). The impact of climate change on soils and on their water management. [Special Issue II]. *Agronomy Research (Tartu)*, 385–396.

Vasiliev, S. V., Verigin, N. V., & Glaser, B. A. (1970). *Methods of calculations of irrigation and drainage systems.* Kolos. [et al.]

Vedernikov, V. V. (1939). *The theory of groundwater flow and its application in the field of irrigation and drainage.* Gos-stroizdat.

Vetokhin, V.I. (2008). Design of deep rippers taking into account some aspects of soil deformation. *Machinery in agri-culture. production, branch mechanical engineering, automation, 20,* 104-109.

Volk, L., Bezusyak, O., & Volk, P. (2021). Improving the dimensioning of closed collecting and drainage network of drainage systems. *Land Reclamation and Water Management,* (1), 98–106. doi:10.31073/mivg202101-269

Voronin, I. V., Senkevich, A. A. & Bugaev, V. A. (1975). *Ekonomicheskaya effektivnost' v lesokhozyaystvennom i le-someliorativnom proizvodstvakh,* Moskow: Lesnaia.

Voropai H.V. (2021). Modern aspects of scientific support for the restoration of drainage systems of the humid zone of Ukraine. *Reclamation and water management, 1,* 39-51.

Voropai H.V., Yatsyk M.V., Mozol N.V. (2019). Current state and prospects of development of drainage reclamation in the conditions of climate change. *Reclamation and water management.*

Voropay, G., Yatsyk, M., & Mozol, N. (2019). Current State And The Prospects Of Development Of Drainage Reclamation In A Changing Climate. *Land Reclamation and Water Management,* (2), 31 - 39. https://doi.org/10.31073/mivg201902-180.

Vostrikov, V., Romaniuk, I., Pinchuk, O., & Vostrikova, N. (2008) Modeliuvannia teplovykh protsesiv v hrunti pry obihrivi vodonapovnenymy rukavamy [Modelling Thermal Processes in Soil While Heating with Water-Filled Sleeves]. Melioratsiia i vodne hospodarstvo: mizhvidomchyi tematychnyi naukovyi zbirnyk [Reclamation and Water Engineering: Interagency Thematic Collection of Research Papers], 96. Kyiv: Ahrarna nauka.

Vyshnevskyi, V. (2003). *Antropohennyi vplyv na richky Ukrainy: avtoref. dys. na zdobuttia nauk. stupenia doktora heohr. nauk: 11.00.11 [Anthropogenic influence on the rivers of Ukraine: author's abstract.* [Thesis, Geogr. Sciences].

Wang, L., Xu, N., Song, Y., Wang, Y., & Song, S. (2019). Research on investment decision of substation project based on life cycle cost. *IOP Conference Series. Earth and Environmental Science, 242*(2), 022016. doi:10.1088/1755-1315/242/2/022016

Ward, P., Flower, K., Cordingley, N., Weeks, C., & Micin, S. F. (2012). Soil water balance with cover crops and conser-vation agriculture in a Mediterranean climate. *Field Crops Research, 132,* 33–39. doi:10.1016/j.fcr.2011.10.017

Water code of Ukraine (with changes and amendments made by the Law of Ukraine of 21.09.2000 No1990-111). – 38 p. Honcharov S.M. & Dupliak V.D. (1992). *Principles of management and marketing.* – K.: NMKVO – 298 p.9.

Watkiss, P. (2015). The Cost of Climate Change in Europe. In K. Steininger, M. König, B. Bednar-Friedl, L. Kranzl, W. Loibl, & F. Prettenthaler (Eds.), *Economic Evaluation of Climate Change Impacts. Springer Climate.* Springer. doi:10.1007/978-3-319-12457-5_2

Wenyan, W., Bing, S., & Zhilu, L. (1994). Drain-spacing calculation considering influence of evaporation. *Journal of Irrigation and Drainage Engineering, 120*(3), 563–572. doi:10.1061/(ASCE)0733-9437(1994)120:3(563)

Willardson, L. S., & Walker, R. E. (1979). Synthetic drain envelope-soil interactions. *Journal of the Irrigation and Drainage Division, 105*(4), 367–373. doi:10.1061/JRCEA4.0001274

Yangol, A. (1970). *Bilateral humidity control during drying.* Kolos.

Yangol, A. (1970). *Two-sided regulation of the authorities for drainage. M.* Colossus.

Yangol, A. (1990). Degree of drainage of peat soils and its calculated elements. *Tr. UkrNIIGM, 70,* 18–26.

Yangol, A. M. (1965). *Humidification recommendations for drained lands.* Urogay.

Yangol, A. M. (1970). *Two-side regulation of moisture during drainage.* - Moscow: Kolos, 136p. Lennoz-Gratin, Ch. (1989). Effect of envelopes on flow pattern near drain pipe. *Journal of Irrigation and Drainage Engineering, 115*(4), 626–641.

Yatsiuk, M. V., Sydorenko, O. O., Nechay, O. M., Tsvetova, O. V., & Turaieva, O. V. (2022c). Impact of climate change on water availability in Ukrainian Polissya. *International. scientific-practical Conference.* Kyiv. https://doi.org/doi:10.31073/mivg2022

Yatsiuk, M. V., Sydorenko, O. O., Tsvetova, O. V., & Turaieva, O. V. (2021). Monitoring and features of hydrogeological and hydrological conditions of shatsky lakes in modern climate conditions. *Sciences of Europe*, (67), 51–59.

Yatsyk M.V., Voropai H.V., Moleshcha N.B. (2016). Improving water supply of reclamation systems of the humid zone. *Reclamation and water management, 103*, 63–68.

Yatsyk M.V., Voropai H.V., Topolnuk T.I., Shushkevich Yu.A. (2015). Methodological bases of increase of water supply of reclamation systems by creation of accumulating capacities. *Land reclamation and water management, 102,* 54–58.

Yatsyk, A. (1997). *Ecological bases of rational water use (on the example of rivers of Ukraine).* TSNIIKIVR.

Yatsyk, A. (2004). *Vodohospodarska ekolohiia [Water management ecology].* Kyiv.

Yatsyk, A., Hopchak, I., Basiuk, T. (2019). Normuvannia antropohennoho navantazhennia na baseiny malykh richok Ukrainy, yak osnova yikh zberezhennia ta vidrodzhennia [Normalization of the anthropogenic load on the basins of small rivers of Ukraine, as a basis for their preservation and revival]. *Hidrolohiia, hidrokhimiia i hidroekolohiia, 3*(54), 93-95.

Yatsyk, A., Hopchak, I., Pasheniuk, I., & Basiuk, T. (2015). Naukovi zasady normuvannia antropohennoho navantazhennia richkovykh baseiniv. *«ETEVK-2015»: zbirka dopovidei Mizhnarodnoho Konhresu.* [Scientific principles of normalization of anthropogenic load of river basins. *«ETEVK-2015»: a collection of reports of the International Congress,* Kyiv

Yatsyk, A., Yatsyk, I., Hopchak, I., & Basiuk, T. (2019). Assessment of state of aquatic ecosystems of volynsk region as to the level of anthropogenic load. *UDC.*

Yatsyk, M. V., Skrypnyk, O. V., Voroshnova, L. M., Voropay, G. V., & Mozol, N. V. (2009). Resource-saving technology of water regulation on reclaimed lands in the zone of excess moisture. Land reclamation and water management, 97, 68–76.

Yatsyk, A. (2001). *Environmental security in Ukraine.* Geneza.

Yatsyk, A. (2004). Water management ecology. Vol. 4. K. *Genesis (New York, N.Y.)*, 325–339.

Yatsyk, A., Hopchak, I., & Basiuk, T. (2018). Otsiniuvannia antropohennoho vplyvu na zemelni resursy v mezhakh baseiniv malykh richok Zakhidnoho Polesie Ukrainy [Assessment of anthropogenic impact on land resources within the basins of small rivers of the Western Polesie of Ukraine]. *Vyp., 108*, 66–71.

Yury, M. Rokochinskiy A., Aleksandr, A., Aleh, M., & Jerzy, J. (2020). POLESIE – Środowisko Melioracje. Monograph.

Zacon. (2020). *Irrigation and drainage strategy in Ukraine until 2030.* Cabinet of Ministers of Ukraine. https://zakon.rada.gov.ua/ laws / show /

Zakon Ukrayiny Pro otsinku zemel [Law of Ukraine on land valuation]. (2004). Vidomosti Verkhovnoyi Rady Ukrayiny. https://zakon.rada.gov.ua/laws/show/1378-15

Zelenin, A. N. (1969). Laboratory workshop on cutting soil: textbook. IOP.

Zhong, M., Fan, W., Liu, T., & Li, P. (2003). Statistical analysis on current status of China forest fire safety. *Fire Safety Journal*, *38*(3), 257–269. doi:10.1016/S0379-7112(02)00079-6

Zhukovsky, E. (1981). *Meteorologicheskaya informatsiya i ekonomicheskiye resheniya* [*Meteorological information and economic decisions*]. Gidrometeoizdat.

Zibtsev, S. V., & Borsuk, O. A. (2012). Protection of lisiv vid pozhezh u sviti that in ukraine – [wiki of the 21st century and prospect of development]. *Lisove i sadovo-parkove hospodarstvo, 1.* http://journals.nubip.edu.ua/index.php/ Lis/article/download/9630/8613

About the Contributors

Anatoliy Rokochinskiy is a Ph.D., Doctor of Sciences (Engineering), professor of the Department of Water Engineering and Water Technologies, National University of Water and Environmental Engineering,Ukraine. Field of activity: development of the scientific principles, methods and models for the justification of climatologically-optimal strategies for creating and managing complex natural and man-made objects in the field of water management, environmental protection, agriculture and the energy sector of the country through the use of modern computer and information technologies. Author of more than 390 scientific and scientific-methodological works. Teaching courses: Automation of the design of water management and melioration objects, Water Engineering and Water Technologies.

Lyudmyla Kuzmych is a Ph.D., Doctor of Science (Engineering), professor, the Chief Researcher of the Department of Drainage, the Institute of Water Problems and Land Reclamation of the National Academy of Agrarian Sciences of Ukraine. The specific areas of research concern studying the impact of various regime-forming factors on the formation of soil water regimes in modern economic conditions and climate change, to develop scientific and methodological principles of soil water regime management on reclaimed lands, water-energy-food nexus to ensure sustainable agricultural production taking into account climatic, economic, technical, technological and environmental factors. Lyudmyla Kuzmych is the author and co-author of about 200 scientific publications, including 4 manuals, 3 monographs, 10 patents, etc.

Pavlo Volk is a Ph.D., Doctor of Sciences (Engineering), professor of the Department of Water Engineering and Water Technologies, National University of Water and Environmental Engineering,Ukraine. Scientific activity is related to the development of scientific principles, methods, and models for substantiating optimal technical and technological solutions in projects of construction, reconstruction, and operation of water management reclamation and nature protection objects. These objects are complex natural-technical systems and ecological-economic systems. The development of new methods and models is carried out taking into account economic and environmental standards based on the use of modern computer and high-information technologies. 146 publications, which include: 6 collective monographs; 2 textbooks; 10 industry standards; 24 educational and methodical textbooks; 28 articles are included in the WoS and Scopus scientific and metric databases; 32 professional articles; 29 theses at conferences, 11 of which are included in WoS and Scopus; 10 Ukrainian utility model patents, 6 intellectual property certificates.

***

Tetiana Basiuk is a PhD Geographical Sciences, Associate Professor, Associate Professor of the Department of Geology and Hydrology, National University of Water and Environmental Engineering (Ukraine). Dissertation «Estimation of hydrological and gidroekologichnikh terms of development and reconstruction of the small hydroelectric power stations (on the example of the river Sonth Bug)» (2013). Research interests: ecological assessment of surface water quality, water management and ecological zoning river basins, assessment of anthropogenic load on river basins. Teaching courses: Constructive geography of water and mineral resources, Physical oceanology, Regional economic and social geography of the countries of the world.

Dmytro Charnyi was born on 2nd August 1966. Charnyi has been awarded qualification - Hydraulic Engineer,1990 Institute of Water Management Engineers, Rivne Ukraine. PhD in Engineering Sciences, "Improvement of water and energy saving methods in branched irrigation and water systems" (1994 -1999) Specialty 05.20.05 - Hydrotechnical reclamation Institute of Hydraulic Engineering and Land Reclamation of the National Academy of Agrarian Sciences of Ukraine. Academic title of Senior Researcher in Agricultural Reclamation (Engineering Sciences) Institute of Hydraulic Engineering and Reclamation of the National Academy of Agrarian Sciences of Ukraine December 15, (2005) Doctor of Engineering Sciences defended a thesis on specialty Agricultural Melioration (Engineering Sciences), "Development of theoretical principles and improvement of natural water treatment technology in agricultural water supply systems" (2015-2017) Institute of Water Problems and Land Reclamation of National Academy of Agrarian Sciences of Ukraine 11 of October 2017 Present post Leading researcher of the Department of Nuclear Physics since November 2020

Nadiia Frolenkova is a Ph.D., Associate Professor of the Department of Management, National University of Water and Environmental Engineering (Ukraine). Dissertation in the field of ecological and economic evaluation of investments in reclamation projects (2004). Research interests: project management, personnel management, educational management. Author of more than 50 scientific articles. Teaching courses: Management of environmental projects, Financial management, Personnel and information management, Management in the market of goods and services, Leadership.

Ievgenii Gerasimov is a Doctor of Engineering, Professor at the Department of Hydraulic construction and Hydraulic, National University of Water and Environmental Engineering (Ukraine). The main direction of the doctoral thesis is technical solutions for increasing the energy efficiency of close irrigation networks (2021). Research interests: energy efficiency of irrigation, transient processes in closed pipelines; water hammer in closed pipelines; reliability of pumping systems, integrated water resources management. Author more than 120 scientific articles. Teaching courses: Integrated Water Resources Management; Special flood protection constructions, Renewable sources of energy.

Grzegorz Grzywaczewski is a Ph.D., Professor of Department of Zoology and Animal Ecology, University of Life Sciences in Lublin (Poland). Thesis in the field of bird ecology and nature conservation. Research interests: biology and ecology of birds, importance, functions and restoration of wetlands, threats and nature conservation. Author of 150 publication, 73 scientific articles. Teaching courses: management of protected areas, vertebrates in Natura 2000 areas, zoology with ecology, forest zoology, environmental protection - nature conservation, ornithology, ecology and biology of birds and mammals.

Maksym Haponiuk is a graduate student of the department of water engineering and water technologies, National University of Water and Environmental Engineering (Ukraine).

Gopchak Igor is a doctor of Technical Science, Associate Professor, Professor of the Department of Geology and Hydrology, National University of Water and Environmental Engineering (Ukraine). Dissertation «Scientific principles of preservation and revival of small rivers of Western Polissya of Ukraine» (2021). Research interests: ecological assessment of surface water quality, water management and ecological zoning river basins, assessment of anthropogenic load on river basins. Teaching courses: Fundamentals of geophysics, Fundamentals of hydrology and meteorology, Management of spatial developmentю

Jerzy Jeznach is a doctor of Technical Sciences, Professor, Department Of Environmental Management, Warsaw University of Life Sciences, Member of the Presidium and scientific secretary of the Committee of Agronomic Sciences of the Polish Academy of Sciences. Author of more than 300 scientific works. The area of scientific interests is engineering, environmental protection and formation, land reclamation.

Roman Koptyuk is a Ph.D., Associate Professor of Department of water engineering and water technologies, National University of Water and Environmental Engineering (Ukraine). Thesis in the field of improving the methods of drainage systems calculation in the conditions of the developed topography of the area. Research interests: melioration, climate change, BIM and GIS technology. Author of more than 60 scientific works. Teaching courses: Design of water management and environmental systems, Basics of hydromelioration, Basics of automated design in water engineering, Automated design of constructions and systems, Fundamentals of rational nature management and nature management.

Petro Kovalenko is a doctor of technical sciences, professor, and academician of the National Academy of Agrarian Sciences of Ukraine. The author of more than 400 scientific works, 14 books, including 1 textbook, 24 copyright certificates and patents. Head of the scientific school, where about 50 doctoral theses and 380 candidate theses were defended. The author of fundamental and applied works on the management of technological processes on hydromelioration systems, the creation of perfect structures of water management complexes, and their automated management systems.

Leonid Kozhushko is a doctor of Technical Science, Professor of the Department of Management, National University of Water and Environmental Engineering (Ukraine). Honored Worker of Science and Technology of Ukraine. Research interests: engineering, production and environmental management, project management. Author of more than 180 scientific works. Teaching courses: International Management, Leadership.

Mariusz Kulik is a Ph.D., Assoc. Professor in the Department of Grassland and Landscape Planning, University of Life Sciences in Lublin (Poland). Thesis in the field of grazing management, meadow vegetation and preserving the post-bog habitats. Research interests: biodiversity of grasslands, use of native breeds of livestock for the protection of Natura 2000 habitats. Author of 80 scientific articles. Teaching courses: grassland science, silvopastoralism, agri-environmental schemes, phytosociology.

Anna Kuzmych is a Ph.D. Student at the Department of Water Engineering and Water Technologies, National University of Water and Environmental Engineering, Ukraine. Scientific activity is dedicated to the improvement of existing and the creation of new adaptive measures and means for moistening drained lands in the modern changing climatic conditions.

Oleksandr Kvartenko is a doctor of Technical Sciences, Associate Professor, Professor of Department of Water Supply, Sewerage and Drilling, National University of Water and Environmental Engineering, (Ukraine). Thesis in the field of technology of comprehensive water treatment of multicomponent underground waters from ammonia nitrogen, dissolved organics, Fe (II) compounds, phenols using combined biological and physico-chemical methods (2019). Research interests: development of modern water treatment technologies based on a combination of biological and physico-chemical methods. Author of 85 scientific articles. Teaching courses: Qualitative indicators of natural and wastewater, Rational use of water, Equipment and operation of water supply and sewerage systems, Water supply of industrial enterprises, Planning and management of water supply and water supply systems.

Olexander Lukyanchuk is a Ph.D., Associate Professor Department of building, road, reclamation, agricultural machinery and equipment, National university of water and environmental engineering (Ukraine). Thesis in the justification of the structural and technological parameters of the tiered deep soil loosener (2005). Research interests: technologies and technical means of energy-efficient moisture regulation in productive soil layers; deep differentiated development of soils and low-energy methods of loosening them. Author of 93 scientific works and 26 patents for inventions. Teaching courses: Methodology of scientific studies of working processes of machines, Fundamentals of automation of machine design, fundamentals of scientific research, CAD-CAM system, Complex mechanization in construction, reclamation machines and equipment for water management, Internet technologies

Zdzisław Michalczyk is a Ph.D., Professor Emeritus - M. Curie-Skłodowska University in Lublin. Scientific research focused on the conditions of occurrence and circulation of water, its quality and natural and anthropogenic transformation of the environment. Author of over 500 publications, including scientific articles, monographic studies, thematic maps and others. Didactic classes in hydrology and hydrogeology, focused on cognition and protection of the geographical environment.

Viktor Moshynskyi is a Ph.D., D.Sc. (agriculture), Professor, Honored Worker of Education of Ukraine, Rector of the National University of Water and Environmental Engineering. Main scientific interests: anthropogenic impact on natural systems; state of natural systems and their assessment; natural monitoring; landscape and ecological aspects of agricultural land use; mathematical modeling in agriculture and environmental protection activities; management in the field of agriculture and land reclamation; assessment of the condition, forecasting, and management of reclaimed territories. Author of 206 scientific articles, including 21 publications in international journals, which are indexed in the international scientometric database Scopus (Hirsch index – 9).

Ivan Openko is a doctor of Economic Sciences. Academic title: associate professor Membership in organizations: member of the Self-Regulating All-Ukrainian Public Organization “Association of Land Management Specialists of Ukraine”. Teaches disciplines: “Geodesy”, “Higher geodesy”, “Mathematical processing of geodetic measurements”, “GNSS - observation in applied tasks of geodesy”, “Fundamentals of engineering geodesy". Field of scientific interests: economics of nature use, GNSS technologies. The topic of the dissertation: "Ecological and economic principles of rational use and protection of lands under field protection forest plantations." The topic of the doctoral dissertation: "Ecological and economic principles of rational use and protection of forestry land in the conditions of decentralization of power" Total number of publications: more than 100. Work experience: working at the National Academy of Sciences of Ukraine at the Department of Geodesy and Cartography since 2013.

Mykola Panasiuk graduated in 1976 from the Faculty of Geology named after Taras Shevchenko National University, majoring in hydrogeology. Now I work as the head of the geoecological research sector at the Institute of Safety Problems of Nuclear Power Plants of the National Academy of Sciences of Ukraine. Since 1996, the geoecological research sector has been performing radiohydroecological monitoring around the destroyed 4th unit of the Chornobyl NPP.

Oleg Pinchuk is a Ph.D., Associate Professor at Hydroinformatics Department, National University of Water and Environmental Engineering (Ukraine). The main direction of the doctoral thesis is in the field of surface soil heating systems by waste warm water (2012). Research interests: technologies for using waste warm water from industrial and energy facilities; modeling of water phenomena and processes; integrated water resources management. Author more than 80 scientific articles. Teaching courses: Integrated Water Resources Management; Bases of Water Resources Management; Modeling of Water Phenomena And Processes; Accounting of Water Resources; Technology of Irrigation.

Vadym Poliakov is a doctor of technical sciences (1988), professor (2009), leading researcher at the Institute of Hydromechanics of the National Academy of Sciences of Ukraine and chief researcher at the Institute of Water Problems and Reclamation of the National Academy of Sciences of Ukraine. Author of two monographs, textbooks, and more than 300 scientific works in foreign and domestic publications. Conducts scientific research in the field of water purification problems using physicochemical and biological methods. Researches the dependence of the formation of water and nutrient regimes of agricultural lands in conditions of climate change using theoretical methods. Develops, using mainly analytical methods, engineering methods for forecasting the development of soil regimes against the background of reclamation drainage, substantiating the design of water treatment facilities (quick filters, biofilters), and technologies for the removal of various pollutants (organic, suspended, nitrogen).

Nataliia Prykhodko is a Ph.D., Associate Professor of the Department of Water Engineering and Water Technologies, National University of Water and Environmental Engineering (Ukraine). Thesis in the field of resource-saving parameters of water use technology in the rice irrigation system in the variable climate conditions (2016). Research interests: melioration, water resource saving technology, climate change. Author of 42 scientific articles. Teaching courses: Water resources use and protection, Agroengineering, Design of water management and environmental protection systems, Meliorative hydrogeology.

Ivan Romanyuk is a Ph.D., Associate Professor at the Hydroinformatics Department, National University of Water and Environmental Engineering (Ukraine). The main direction of the doctoral thesis is use a special hydrotechnical system with the discharge of warm water for soil heat amelioration by heat exchangers-sleeves (2007). Research interests: use of discharged warm water from industrial and energy facilities in agriculture; development and research of surface soil heating systems. Author more than 50 scientific articles. Teaching courses: Integrated Water Resources Management; Exploitation of Water Management Facilities; Exploitation of Hydromelioration Systems; Intellectual Property.

Larysa Semenko is a Doctor of Science, Associate Professor of the Department of Agrochemistry and Product Quality named after O.I. Dusychkina of the National University of Bioresources and Nature Management of Ukraine. Dissertation topic "Agrochemical evaluation of the use of fertilizers for the cultivation of early white cabbage on the dark gray podsolized soil of the Right Bank of the Forest Steppe of Ukraine" (2009). Author of 35 scientific articles. Training courses: Agrochemistry, Fertigation and irrigation, Agrochemical service in fruit and vegetable growing, logistics of cultivation and sale of plant products.

Hanna Serbeniuk is a Ph.D., Senior Lecturer of the Department of Agrosphere Ecology and Environmental Control of the National University of Life and Environmental Sciences of Ukraine. Dissertation in the field of growing grain crops according to different technological methods of cultivation. Scientific interests: agroecological assessment of technologies for growing crops. Author of more than 48 scientific papers. Training courses: Agroecology, Balanced nature management: Recreational resources of the biosphere, Soil and environmental monitoring and management of land resources in the agricultural sector.

Viktor Serbenyuk, Ph.D., Leading Researcher of the Department of Crop Rotations and Agriculture on Reclaimed Lands of the NSC "Institute of Agriculture of the National Academy of Sciences of Ukraine" (Ukraine). Dissertation in the field of agricultural reclamation on drained organogenic soils. Scientific interests: land reclamation agriculture, meadow growing, fodder production and environmental ecology. Author of more than 45 scientific papers.

Oleksandr Shevchenko is a PhD (Economics), Associate Professor of the Department of Geodesy and Cartography National University of Life and Environmental Sciences of Ukraine (Kyiv, Ukraine). Dissertation " Economical Efficiency of Soil Conservation Measures in Agricultural Land-Use" (2016). Research interests: land management, economics of nature use, land use, forest use, land protection, soil degradation, soil conservation measures. Author and co-author of more than 100 scientific and methodical papers. Teaching courses: «Geodesy», «Electronic geodetic devices», «Topographical and geodetic support of land and cadastral works».

Oleksii Shevchenko, Doctor of Geol. Sci., Professor, Chief researcher of Laboratory for the study of the impact of climate change on water resources of Ukrainian Hydrometeorological Institute National Academy of Sciences of Ukraine. The subject of the dissertation: "Radiohydrogeology of drainage landscapes of the Ukrainian Polissia (for example Chernobyl Exclusion zone)". He started the scientific-applied direction of "ameliorative radio-hydroecology". A leading specialist in the fields of radio-hydroecology, environmental audit, technical and economic substantiation of the feasibility of exploiting groundwater deposits, zoning and monitoring of drained lands, assessment of the groundwater balance based on regime observations, design of water intakes, etc. Training courses: Basics of moisture transfer in the aeration zone; Basics of groundwater migration; Exploratory and filtering observations; Methods of field studies of water features during hydrogeological surveying. Author of 190 scientific articles. Expert of the State Commission on Mineral Reserves of Ukraine. Member of the International Union of Radioecologists UIR.

Ivan Slyusar is a Doctor of Agricultural Sciences, Professor, Corresponding Member NAAS, Chief Researcher of the Department of Crop Rotations and Agriculture on Reclaimed Lands of the NSC "Institute of Agriculture of the National Academy of Sciences of Ukraine". Dissertation in the field of land reclamation farming on drained soils. Scientific interests: land reclamation soil science, agriculture, meadow growing, fodder production, environmental ecology. Author of 300 scientific works, including 29 monographs and manuals for universities.

Olena Solyanik is a Candidate of Agricultural Sciences, Senior Researcher of the Department of Crop Rotation and Agriculture on Reclaimed Lands of the NSC "Institute of Agriculture of the National Academy of Sciences of Ukraine". Dissertation in the field of fodder production. Scientific interests: land reclamation agriculture, fodder production, meadow growing and environmental ecology. Author of 70 scientific works, including 2 monographs in co-authorship.

Vasyl Stashuk is a Doctor of Technical Sciences, Professor, Academician of the National Academy of Agrarian Sciences of Ukraine. Author of 275 scientific articles. The main direction of scientific activity is the development of scientific principles of integrated management of water resources and the water management and reclamation complex of Ukraine according to sectoral and basin principles; the development of scientific and organizational principles for the restoration and sustainable development of hydrotechnical reclamations in the conditions of a market economy and global climate changes; development of scientific approaches and a set of measures for the protection of territories and settlements from the harmful effects of water and management of water resources in crisis situations.

Yanina Stepchuk received a junior specialist's degree in "Geodesy and Land Management" at the Kamianets-Podilskyi Industrial College in 2017. In 2021, she received a master's degree at the National University of Life and Environmental Sciences of Ukraine. From 2019 to 2021, she worked as a laboratory assistant at the Department of Geodesy and Cartography of the National University of Life and Environmental Sciences of Ukraine. Since 2022, I have been working as an assistant at the Department of Geodesy and Cartography of the National University of Life and Environmental Sciences of Ukraine

Piotr Sugier is a Ph.D., Associate Professor; Department of Botany, Mycology and Ecology, Institute of Biological Sciences, Maria Curie-Skłodowska University (Poland). Research interests: plant ecology, the role of the sediment seed-bank in the preservation and regeneration of macrophytes, factors modifying chemical properties of secondary metabolites of medicinal plants, vegetation of the abandoned waste deposits. Author of 164 publication.

Olena Sydorenko is a Ph.D. in agricultural sciences, Head of department of Ecology and nature management of the Institute of Water Problems and Land Reclamation NAAS, Ukraine. Olena Sydorenko have an experience and knowledge of modern methods and approaches for the ecological state of water objects and reclaimed lands in modern conditions. Scientific activity is dedicated to solving the current problems of substantializing directions for the methods and tools for monitoring of the environment, including monitoring of the quality of water objects, monitoring of the ecological and reclamation state of irrigated and drained lands, monitoring of soil conditions in the zones of influence of reclamation systems. Olena Sydorenko is interested in the development of integrated approaches to the consolidation of water and land resources. Have a practical experience in development of schemes for the integrated use and protection of water and land resources. She was engaged in work with the developing of operational and long-term forecasts, proposals on the main directions of land reclamation development, and the use and protection of reclaimed lands, including changes in the ecological status of water objects and reclaimed land.

Olena Tsvetova is a Ph.D. in Technical Sciences, leading researcher of the Department of Ecology and Nature Management of the Institute of Water Problems and Land Reclamation of the National Academy of Agrarian Sciences of Ukraine. Her scientific activity is aimed at the organization and monitoring of the natural environment, the development of ecological indicators, and the assessment of the ecological and reclamation state of reclaimed land. Olena Tsvetova's scientific interests are related to the development of measures to protect water bodies in transboundary territories. She is the author of 158 scientific articles.

Olga Turaieva is a Researcher of the Department of Ecology and Nature Management of the Institute of Water Problems and Land Reclamation of the National Academy of Agrarian Sciences of Ukraine. Her scientific activity is aimed at the topical issues of ensuring ecological stability and rational use of water bodies in the humid zone of Ukraine, development of scientific bases for adaptation of the ecosystems of Ukrainian Polissia to climate change, assessment of the impact of land use systems on the evolution of reclaimed soils, assessment of ecological sustainability of reclaimed landscapes in modern conditions. Olga Turaieva's scientific interests are related to the issues of scientific substantiation of the concept of the conservation program of the Shatsk lake area, conducting monitoring studies on the state of the natural environment of the territory of the Volyn Polissia, establishing the mechanisms of the impact of climate change on water resources, developing a methodology for the renaturalization of peatlands as well as organizational, ecological and economic principles of the consolidation of reclaimed land based on an assessment of their ecological and ameliorative state. She is the author of 92 scientific articles.

Vasyl Turcheniuk is a Doctor of Technical Sciences, Professor, Head of Department of Water Engineering and Water Technologies. National University of Water and Environmental Engineering (Ukraine). Thesis in the field of system optimization of water and energy use of rice irrigation systems on an ecological and economic (2018). Author of more than 150 scientific and scientific-methodical works. Research interests: melioration, water and energy resource saving technology, climate change. Teaching courses: Design of water management and environmental protection systems, Engineering protection of territories, Water engineering and water technologies.

Olha Tykhenko is a PhD (Agriculture), Associate Professor of the Department of Land Cadastre National University of Life and Environmental Sciences of Ukraine (Kyiv, Ukraine). Dissertation "The changes of phosphorus regime of meadow chernozem leached soil under conservation tillage influence in conditions Andruchevsky natural agricultural region" (2004). Research interests: land cadastre, protection and rational use of land, assessment of land. Author and co-author of more than 110 scientific and methodical papers. Teaching courses: Land Cadastre, Registration of Real Property Rights.

Ruslan Tykhenko is a PhD (Economics), Associate Professor of the Department of Land Resources Management. National University of Life and Environmental Sciences of Ukraine (Kyiv, Ukraine). Dissertation "Ecological and economic efficiency of land management in the conditions of transformation of land relations in Ukraine" (2009). Research interests: land management, economics of nature use, land use, water use, forest use, land protection. Author and co-author of more than 230 scientific and methodical papers. Teaching courses: Design engineering, Expertise of design and research documentation, Technologies for reproduction of land productivity, Geodetic Works and Land Management

Danuta Urban is a Ph.D., Professor of Institute of Soil Science, Engineering and Environmental Management, University of Life Sciences in Lublin (Poland). Thesis in the field: hydrogenic habitats and the genesis and evolution of valley peatlands. Research interests: transformations of marsh and post-marsh environments under the influence of human activity in the modern period and in the Holocene, biodiversity of peatlands, grasslands and forest ecosystems, management of protected areas and Natura 2000 areas, Author of 285 publications, 133 scientific articles. Teaching courses: phytosociology, forest typology, nature conservation.

Liubov Volk is a Ph.D., Associate Professor of Hydraulic Construction and Hydraulics Department, National University of Water and Environmental Engineering (Ukraine). Dissertation in the field of flood protection on mountain rivers (2015). Research interests: hydraulics of open and closed flows. Author of 72 scientific articles. Teaching courses: Hydraulic structures, Hydraulics, Special hydraulic structures, Automated design in hydraulic construction, Secondary resource storages.

Halyna Voropai is a Ph.D., the Head of the Department of Drainage, the Institute of Water Problems and Land Reclamation NAAS, Ukraine. Scientific activity is dedicated to solving the current problems of substantiating directions for the development of land reclamation of the humid zone in modern economic conditions, introducing into practice methods of managing technological processes on land reclamation systems, and restoring the effective use of the potential of drained lands. Halyna Voropai is interested in the development of constructive and technological approaches to ensure the effective functioning of drainage and humidification systems in the conditions of changes in the water resource potential of Polissia and the socio-economic conditions of agriculture there; justification and improvement of existing technologies for managing water regulation processes in canals and the collector network of drainage systems, technologies for managing the water regime of soils in accordance with the needs of crops; substantiation of the main approaches and the development of measures to increase the water supply of certain areas of reclaimed land.

Stepan Vozniuk is a Doctor of agricultural sciences, professor, honored worker of science and technology of Ukraine, active member of the Ukrainian Academy of Environmental Sciences, the International Academy of Environmental Sciences, the International Academy of Computer Sciences. Author of more than 350 scientific works. Dissertation in the field — hydromorphic soils of the Polesie and Forest Steppe of Ukraine

Artem Yatsyk is a Student National University of Water and Environmental Engineering (Ukraine). Specialty: 194 “Hydrotechnical construction, water engineering and water technologies”. Research interests: ecological assessment of surface water quality, water management and ecological zoning river basins, assessment of anthropogenic load on river basins.

Mykhailo Yatsyuk is a Ph.D. in Geographical Sciences in the field of Water Safety, Director of the Institute of Water Problems and Land Reclamation NAAS, Ukraine. Yatsiuk Mykhailo is the author of about 50 scientific papers, including monographs. The area of his scientific activity is improving water legislation, water safety, integrated water management, monitoring of water quality, minimizing the impact of harmful effects of water, assessment of the technical condition of engineering infrastructure of reclamation systems. He participated in the development of amendments to the Water Code of Ukraine, the Law of Ukraine “On Targeted National Program for the Development of Water Management and Environmental Improvement of the Dnieper River Basin for the Period up to 2021”, the Regulation on the State Water Resources Agency, other legislative and regulatory acts in the field of water and environmental legislation. He was an expert in the development of the management plans for the river basins of the Upper Dnieper Prut and Siret Rivers.

Index

E

F

G

H

I

K

L

M

N

O

P

R

S

T

V

W

Y

Recommended Reference Books

IGI Global's reference books are available in three unique pricing formats: Print Only, E-Book Only, or Print + E-Book.

Order direct through IGI Global's Online Bookstore at **www.igi-global.com** or through your preferred provider.

ISBN: 9781668445488
EISBN: 9781668445501
© 2022; 334 pp.
List Price: US$ **240**

ISBN: 9781668467503
EISBN: 9781668467527
© 2022; 296 pp.
List Price: US$ **240**

ISBN: 9781799889007
EISBN: 9781799889021
© 2022; 340 pp.
List Price: US$ **215**

ISBN: 9781799898511
EISBN: 9781799898535
© 2022; 276 pp.
List Price: US$ **250**

ISBN: 9781799896913
EISBN: 9781799896937
© 2022; 219 pp.
List Price: US$ **215**

ISBN: 9781668436868
EISBN: 9781668436875
© 2022; 2,064 pp.
List Price: US$ **2,085**

Do you want to stay current on the latest research trends, product announcements, news, and special offers?
Join IGI Global's mailing list to receive customized recommendations, exclusive discounts, and more.
Sign up at: **www.igi-global.com/newsletters.**

Publisher of Timely, Peer-Reviewed Inclusive Research Since 1988

IGI Global
PUBLISHER of TIMELY KNOWLEDGE

www.igi-global.com | Sign up at www.igi-global.com/newsletters | facebook.com/igiglobal | twitter.com/igiglobal | linkedin.com/igiglobal

www.ingramcontent.com/pod-product-compliance
Lightning Source LLC
LaVergne TN
LVHW062105300926
849161LV00010B/2379
* 9 7 8 1 6 6 8 4 8 2 4 8 3 *